ELECTRICITY

Principles and Applications

Fifth Edition

Richard J. Fowler

 **Glencoe
McGraw-Hill**

New York, New York   Columbus, Ohio   Woodland Hills, California   Peoria, Illinois

Cover photos: Jon Riley/Tony Stone Images; background: David McGlynn/FPG.

Text photos: Page x: (*top*) Lou Jones/The Image Bank; (*bottom*): © Cindy Lewis. **Page 3** (*upper left*): File photo; (*lower left*): COMSTOCK Inc.; (*lower right*): © 92 Al Francekevich Potential Energy, the Stock Market. **Page 4:** File photo. **Page 6** (*upper left*): Bettmann Archive; (*lower left*): Stock Imagery. **Page 8:** Pacific Northwest National Laboratory. **Page 10** (*left and right*): File photos. **Page 11** (*left*): COMSTOCK Inc.; (*top right*): Charles Thatcher/Tony Stone Worldwide; (*lower right*): Stock Imagery. **Page 17:** File photo. **Pages 20, 27, 32, and 41:** Bettmann Archive. **Page 68:** Electric Power Research Institute. **Page 101:** Bettmann Archive. **Page 143:** Courtesy of W. Atlee Burpee & Co. **Page 172:** Bettmann Archive. **Page 180:** Andy Sacks/Tony Stone Images. **Page 196:** B. Hathaway for Rod Millen Special Vehicles, http://www.rodmillen.com. **Page 210:** Don Mason/the Stock Market. **Page 240:** Bettmann Archive. **Page 269:** Electric Power Research Institute. **Page 293:** Elco, The Electric Launch Company Incorporated. **Page 326:** Bob Daemmrich, © The Stock Solution. **Page 354:** Courtesy of Fluke Corporation. **Page 358:** Wolf H. Hilbertz.

Library of Congress Cataloging-in-Publication Data

Fowler, Richard J.
    Electricity: principles and applications/Richard J. Fowler.—5th ed.
      p.   cm.
    Includes index.
    ISBN 0-02-804847-4
    1. Electricity.   I. Title.
    QC523.F75   1999
    621.3—dc21   98-35422
                                    CIP

Glencoe/McGraw-Hill

*A Division of The **McGraw·Hill** Companies*

Electricity: Principles and Applications, Fifth Edition

Send all inquiries to:
Glencoe/McGraw-Hill
8787 Orion Place
Columbus, OH 43240

ISBN 0-02-804847-4   (student edition)
ISBN 0-02-804849-0   (instructor's annotated edition)

    4 5 6 7 8 9   071   06 05 04 03 02 01 00

Contents

Editors' Foreword

The Glencoe *Basic Skills in Electricity and Electronics* series has been designed to provide entry-level competencies in a wide range of occupations in the electrical and electronic fields. The series consists of coordinated instructional materials designed especially for the career-oriented student. Each major subject area covered in the series is supported by a textbook, an experiments manual, and an instructor's productivity center. All the materials focus on the theory, practices, applications, and experiences necessary for those preparing to enter technical careers.

There are two fundamental considerations in the preparation of materials for such a series: the needs of the learner and needs of the employer. The materials in this series meet these needs in an expert fashion. The authors and editors have drawn upon their broad teaching and technical experiences to accurately interpret and meet the needs of the student. The needs of business and industry have been identified through personal interviews, industry publications, government occupational trend reports, and reports by industry associations.

The processes used to produce and refine the series have been ongoing. Technological change is rapid and the content has been revised to focus on current trends. Refinements in pedagogy have been defined and implemented based on classroom testing and feedback from students and instructors using the series. Every effort has been made to offer the best possible learning materials.

The widespread acceptance of the *Basic Skills in Electricity and Electronics* series and the positive responses from users confirm the basic soundness in content and design of these materials as well as their effectiveness as learning tools. Instructors will find the texts and manuals in each of the subject areas logically structured, well-paced, and developed around a framework of modern objectives. Students will find the materials to be readable, lucidly illustrated, and interesting. They will also find a generous amount of self-study and review materials and examples to help them determine their own progress.

Both the initial and on-going success of this series are due in large part to the wisdom and vision of Gordon Rockmaker who was a magical combination of editor, writer, teacher, electrical engineer and friend. Gordon has retired but he is still our friend. The publisher and editors welcome comments and suggestions from instructors and students using the materials in this series.

Charles A. Schuler,
Project Editor
and
Brian P. Mackin,
Editorial Director

Basic Skills in Electricity and Electronics

Charles A. Schuler, Project Editor

New Editions in This Series
Electricity: Principles and Applications, Fifth Edition, Richard J. Fowler
Electronics: Principles and Applications, Fifth Edition, Charles A. Schuler
Digital Electronics: Principles and Applications, Fifth Edition, Roger L. Tokheim
Other Series Titles Available:
Communication Electronics, Second Edition, Louis E. Frenzel
Microprocessors: Principles and Applications, Second Edition, Charles M. Gilmore
Industrial Electronics, Frank D. Petruzella
Mathematics for Electronics, Harry Forster, Jr.

Preface

Electricity: Principles and Applications, Fifth Edition, is written for students just beginning their study of electricity. No previous formal training in the subject is assumed. Only arithmetic and basic algebra are used in explaining and solving electrical problems. Although simultaneous equations are introduced in Chap. 6, they are defined and explained in some detail before actually being used. Similarly, the elements of trigonometry that relate to ac circuits are explained as needed.

This edition retains the logical, sequential order of topics found in the previous edition. The terms and concepts that are developed and discussed in Chap. 1 are needed to fully understand Chap. 2. The concepts of Chap. 2 are used in Chap. 3, and so on. The exception is Chap. 6, which can be studied out of sequence or omitted entirely without interrupting the continuity of the remaining chapters.

Chapters 1 through 6 of this text are devoted, in general, to the fundamentals of direct current, and Chaps. 8 through 13 focus on subjects usually associated with alternating current. This arrangement provides students with balanced coverage of basic concepts. The transition from direct current to alternating current through the study of magnetism and electromagnetism is distinct enough to allow use of the material in a traditional dc/ac sequence. However, all the material is structured to provide a unified introduction to the broad subject area called *electricity.*

Although no prior formal background in electricity is assumed, it is highly unlikely that a student will not have had extensive exposure to the world of electricity and electronics. For this reason, certain words associated with electricity are expected to have already entered the students' vocabulary, for example, *switch, plug, motor, computer, CPU,* and *television.* Nevertheless, strict technical definitions are used throughout to aid students to rid themselves of technical misconceptions.

Since concepts build from chapter to chapter, and within a chapter from section to section, students are provided with a number of stopping points for test questions and problems. These tests serve two purposes: They reveal areas of weakness that require restudy, and they serve as positive reinforcement for material students know. Answers to the tests are provided as the last item of each chapter.

Each chapter concludes with a summary of the key concepts covered in the chapter. The summaries also help identify areas of weakness and reinforce areas of strength. Students should not skim through this material lightly.

In addition to the tests that are strategically located within each chapter, a comprehensive list of review questions is provided at the end of each chapter. Direct answers, or exact procedures indicating how to obtain the answers, to the Critical Thinking Questions, which follow the sections on review questions, will not be found in the chapter. These questions require students to express their own ideas and beliefs or to develop procedures for solving problems that have not been specifically illustrated in the chapter.

For this edition, the text interior was redesigned to use color more effectively, while retaining the popular marginal color strip. Students can quickly find their chapter-ending assignments with the red strip. Highlighted key terms now appear in the margin of each page close to their point of text discussion. Color strips are also used as a quick thumb reference to find the numerous illustrative examples within a chapter. The text design allows for consistent color-coding of circuit components. This edition also contains informative features called "Did You Know?", "Job Tips," and "About Electronics," which should spark interesting in-class discussions.

But the changes are much more than cosmetic. For this edition, we sought the advice of instructors who have used the book for many years and of instructors who have used it for only a short period of time. They provided precise recommendations and consistent responses with their questionnaires. Their collective information and suggestions are included in this new edition.

Every chapter has undergone modifications, ranging from subtle changes in line art or problem additions to the addition of new sections. Chapter 4 contains an introduction to surface-mount devices (SMDs) and resettable fuses. Chapter 6 includes a brief introduction to Kirchhoff's voltage law. More information is included on magnetic quantities and units in Chap. 7. Chapter 10 provides an introduction to SMD capacitors. Off-center-tapped windings are now included in

Chap. 12. Numerous subtle changes in Chap. 14 result in improved coverage of dc motors. Throughout the book, there are numerous new questions and problems that reflect the changes within each chapter. New photographs and line drawings provide a timely updating of electrical concepts.

This truly is the age of electricity and electronics. The solid foundation in electricity provided by this text is likely to prove useful, whether a student intends to pursue a career in electricity and electronics or simply needs the material as background information for other pursuits. The content and organization of this book reflect the experience and feedback of years of classroom testing. The author welcomes comments and suggestions from students and teachers alike.

Acknowledgments

The author, project editor, and publisher would like to thank the instructors (listed at right) who provided general and detailed comments, and those who responded to the survey that was sent out while the book was being revised. Their comments and suggestions pro-vided the valuable input necessary to make a good book even better.

Jeff Anderson
Soldier Pond, ME

Tony Criswell
Southeastern Technical Institute
Vidalia, GA

Kenneth P. DeLucca
Millersville University
Millersville, PA

Gary G. Free
Mississippi Delta Community College
Moorhead, MS

Carl R. Goeckeler
Kansas City Power and Light Co. (KCPL)
Kansas City, MO

Clifton Ray Morgan
Northwest Kansas Technical School
Goodland, KS

Arlyn Smith
Alfred State College
Alfred, NY

Gary Wimer
Wayne County Schools Career Center
Smithville, OH

Richard J. Fowler

Electric and electronic circuits can be dangerous. Safe practices are necessary to prevent electrical shock, fires, explosions, mechanical damage, and injuries resulting from the improper use of tools.

Perhaps the greatest hazard is electrical shock. A current through the human body in excess of 10 milliamperes can paralyze the victim and make it impossible to let go of a "live" conductor or component. Ten milliamperes is a rather small amount of electrical flow: It is only *ten one-thousandths* of an ampere. An ordinary flashlight uses more than 40 times that amount of current!

Flashlight cells and batteries are safe to handle because the resistance of human skin is normally high enough to keep the current flow very small. For example, touching an ordinary 1.5-V cell produces a current flow in the microampere range (a microampere is one-millionth of an ampere). This amount of current is too small to be noticed.

High voltage, on the other hand, can force enough current through the skin to produce a shock. If the current approaches 100 milliamperes or more, the shock can be fatal. Thus, the danger of shock increases with voltage. Those who work with high voltage must be properly trained and equipped.

When human skin is moist or cut, its resistance to the flow of electricity can drop drastically. When this happens, even moderate voltages may cause a serious shock. Experienced technicians know this, and they also know that so-called low-voltage equipment may have a high-voltage section or two. In other words, they do not practice two methods of working with circuits: one for high voltage and one for low voltage. They follow safe procedures at all times. They do not assume protective devices are working. They do not assume a circuit is off even though the switch is in the OFF position. They know the switch could be defective.

As your knowledge and experience grow, you will learn many specific safe procedures for dealing with electricity and electronics. In the meantime:

1. Always follow procedures.
2. Use service manuals as often as possible. They often contain specific safety information. Read, and comply with, all appropriate material safety data sheets.
3. Investigate before you act.
4. When in doubt, *do not act*. Ask your instructor or supervisor.

General Safety Rules for Electricity and Electronics

Safe practices will protect you and your fellow workers. Study the following rules. Discuss them with others, and ask your instructor about any you do not understand.

1. Do not work when you are tired or taking medicine that makes you drowsy.
2. Do not work in poor light.
3. Do not work in damp areas or with wet shoes or clothing.
4. Use approved tools, equipment, and protective devices.
5. Avoid wearing rings, bracelets, and similar metal items when working around exposed electric circuits.
6. Never assume that a circuit is off. Double-check it with an instrument that you are sure is operational.
7. Some situations require a "buddy system" to guarantee that power will not be turned on while a technician is still working on a circuit.
8. Never tamper with or try to override safety devices such as an interlock (a type of switch that automatically removes power when a door is opened or a panel removed).
9. Keep tools and test equipment clean and in good working condition. Replace insulated probes and leads at the first sign of deterioration.
10. Some devices, such as capacitors, can store a *lethal* charge. They may store this charge for long periods of time. You must be certain these devices are discharged before working around them.
11. Do not remove grounds and do not use adaptors that defeat the equipment ground.
12. Use only an approved fire extinguisher for electrical and electronic equipment. Water can conduct electricity and may severely damage equipment. Carbon dioxide (CO_2) or halogenated-type extinguishers are usually preferred. Foam-type extin-

guishers may also be desired in *some* cases. Commercial fire extinguishers are rated for the type of fires for which they are effective. Use only those rated for the proper working conditions.

13. Follow directions when using solvents and other chemicals. They may be toxic, flammable, or may damage certain materials such as plastics. Always read and follow the appropriate material safety data sheets.

14. A few materials used in electronic equipment are toxic. Examples include tantalum capacitors and beryllium oxide transistor cases. These devices should not be crushed or abraded, and you should wash your hands thoroughly after handling them. Other materials (such as heat shrink tubing) may produce irritating fumes if overheated. Always read and follow the appropriate material safety data sheets.

15. Certain circuit components affect the safe performance of equipment and systems. Use only exact or approved replacement parts.

16. Use protective clothing and safety glasses when handling high-vacuum devices such as picture tubes and cathode-ray tubes.

17. Don't work on equipment before you know proper procedures and are aware of any potential safety hazards.

18. Many accidents have been caused by people rushing and cutting corners. Take the time required to protect yourself and others. Running, horseplay, and practical jokes are strictly forbidden in shops and laboratories.

Circuits and equipment must be treated with respect. Learn how they work and the proper way of working on them. Always practice safety: your health and life depend on it.

Electronics workers use specialized safety knowledge.

Chapter 1

Basic Concepts

Chapter Objectives

This chapter will help you to:

1. *Use* base units for specifying and calculating energy and work.
2. *Understand* energy conversion and conversion efficiency.
3. *List and explain* the characteristics of the major particles of an atom.
4. *Explain* the nature of electric charge.
5. *Discuss* several industrial applications of static electricity.

Electricity is a form of energy. The study of electricity is concerned primarily with learning how to control electric energy. When properly controlled, electricity can do much of the work required to keep our society going. However, uncontrolled electric energy, such as lightning, can be very destructive.

Electric energy is so much a part of our daily lives that we tend to take it for granted. Yet without it our lives would be quite different, and much harder. Electric energy lights our homes and industries, operates our computers, radios, and television sets, and turns the many motors used in clocks, washing machines, clothes dryers, vacuum cleaners, and so forth.

1-1 Work and Energy

Work consists of a force moving through a distance. *Energy* is the ability, or capacity, to do work. In other words, it takes energy to do work. For example, it requires energy to pull a boat out of the water onto the beach, and work is done in pulling the boat out of the water.

The energy required to pull the boat from the water comes from the human body. A force is required to overcome the friction of the boat on the sand. A force is also required to overcome the gravitational pull on the boat as it is raised out of the water. The work then consists of the force required to move the boat some distance as it is pulled onto the shore.

The symbol (or abbreviation) for either work or energy is W. The same symbol is used for work and energy because the two terms are so closely related to each other.

1-2 Unit of Energy

Base units are the terms used to indicate the amount of something. The *joule* is the base unit of energy and work. The symbol for the joule is J. Specifying energy in joules is the same as specifying butter in pounds or money in dollars. All are base units used to specify amount. Base units are important because nearly all relationships in electricity are expressed in base units.

A joule of energy (or work) is very small compared with the amount of energy you use each day. For example, an electric toaster uses approximately 100,000 joules of energy to make two slices of toast. It requires 360,000 joules to operate a small (100-watt) table lamp for 1 hour.

The work or energy involved in a mechanical system (such as pulling a boat) can be

Work (*W*)

Energy (*W*)

Base units

Joule (J)

determined by the following relationship:

$$\text{Work} = \text{force} \times \text{distance}$$

In the metric system, the base unit for force is the *newton* (approximately 0.2248 pounds). The base unit for distance is the *meter* (approximately 39.4 inches), and the base unit for work (energy) is the joule. The joule is equal to the newton-meter, which is a convenient unit for mechanical energy.

Let us return to the problem of pulling a boat out of the water and calculate how much work is done in performing the task.

Newton

Meter

Example 1-1

If it requires a steady force of 150 newtons to pull a boat, how much work is required to pull a boat 8 meters?

Given:	Force = 150 newtons
	Distance = 8 meters
Find:	Work
Known:	Work (*W*) = force × distance
	1 newton-meter = 1 joule (J)
Solution:	Work = 150 newtons × 8 meters
	= 1200 newton-meters
	= 1200 joules
Answer:	Work = 1200 joules
	W = 1200 J

Note the procedure used in solving the problem given in the above example. First, the information (values) *given* in the problem is listed. Next, the information you are required to *find* is recorded. Finally, the relationship (formula) between the two is written. For simple problems this formal procedure may seem unnecessary, but solving more complex problems will be easier if you establish the habit of using this formal procedure.

So far, the calculation of specific amounts of work or energy has been limited to mechanical examples. Once we learn some new terms, like *voltage, current,* and *power,* we will be able to solve problems involving electric energy.

In example 1-1, we found that the work done in moving the boat was 1200 joules (J). The amount of energy required to move the boat is

Converted energy

also 1200 J. Work and energy have the same base unit. They are basically the same thing. Work is the use of energy to perform some task. For example, a car battery has energy stored in it. When the car engine is started, energy from the battery is used to do the work of cranking the engine. The *work done* and the *energy used* are two ways of saying the same thing.

Example 1-2

It requires 500 joules of energy and 100 newtons of force to move an object from point A to point B. What is the distance between point A and point B?

Given:	Energy = 500 joules
	Force = 100 newtons
Find:	Distance
Known:	*W* = force × distance, and by rearranging
	$\text{Distance} = \dfrac{W}{\text{force}}$
	1 newton-meter = 1 joule
Solution:	$\text{Distance} = \dfrac{500 \text{ newton-meters}}{100 \text{ newtons}}$
	= 5 meters
Answer:	Distance = 5 meters

◼ TEST

Answer the following questions.

1. Define work and energy.
2. What is a base unit?
3. What is the base unit of energy?
4. How much energy does it take to push a car 130 meters with a steady force of 360 newtons?

1-3   Energy Conversion

One of the fundamental laws of classical physics states that, under ordinary conditions, energy can be neither created nor destroyed. The energy in the universe exists in various forms, such as heat energy, light energy, and electric energy. When we say we "use" electric energy, we do not mean that we have destroyed, or lost, the energy. We mean that we have *con-*

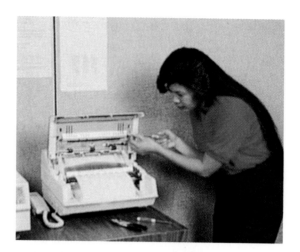

Technician examining a fax machine.

verted that electric *energy* into a more useful form of energy. For example, when we operate an electric lamp, we are converting electric energy into light energy and heat energy. We have used the electric energy in the sense that it no longer exists as electric energy, but we have not used up the energy. It still exists as heat energy and light energy.

The study of electricity deals with the study of converting energy from one form to another form. Electric energy itself is obtained by converting other forms of energy to electric energy. Batteries convert chemical energy to electric energy, solar cells convert light energy to electric energy, and generators convert mechanical (rotational) energy to electric energy.

We seldom use energy directly in the form of electric energy. Yet the electrical form of en-

ergy is very desirable because it can be easily moved from one location to another. Electric energy produced at an electric power plant many miles from your home can be easily transferred from the plant to your home. Once the electric energy is delivered to your home, it can be converted to a more useful form.

We have already noted that a light bulb converts electric energy into light energy and heat energy. Another familiar object that converts energy is the electric stove, which converts electric energy to heat energy. Changing electric energy to mechanical (rotational) energy with an electric motor is also a common conversion.

Although the process is very involved and complex, radio receivers convert electric energy into sound energy. A very small amount of the electric energy comes from electric signals sent through the air. The rest of the electric energy comes from a battery or an electric outlet. The sound energy radiates from the speaker. In a similar fashion, television receivers convert electric energy into sound energy and light energy.

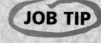
1-4 Efficiency

No conversion process is 100 percent efficient. That is, not all the energy put into a device or a system is converted into the form of energy we desire. When 1000 joules of electric energy

Lightning—an example of uncontrolled electric energy.

Electric energy is converted for use in the home.

is put into a light bulb, only about 200 joules of light energy is produced. The other 800 joules is converted into heat energy. We could say that the *efficiency* of the light bulb is low.

The efficiency of a system is usually expressed as a percentage. It is calculated by the formula

$$\text{Percent efficiency} = \frac{\text{useful energy out}}{\text{total energy in}} \times 100$$

By abbreviating percent efficiency to % eff. and using the symbol W for energy, we can write this formula as

$$\% \text{ eff.} = \frac{W_{out}}{W_{in}} \times 100$$

Let us determine the efficiency of the light bulb mentioned earlier.

Example 1-3

What is the efficiency of a light bulb that uses 1000 joules of electric energy to produce 200 joules of light energy?

Given: Energy in = 1000 joules
Energy out = 200 joules

Find: Percent efficiency

Known: $\% \text{ eff.} = \dfrac{W_{out}}{W_{in}} \times 100$

Solution: $\% \text{ eff.} = \dfrac{200 \text{ joules}}{1000 \text{ joules}} \times 100$
$= 0.2 \times 100$
$= 20$

Answer: Efficiency = 20 percent

Notice that in the efficiency formula both the denominator and the numerator have base units of joules. The base units therefore cancel, and the answer is a pure number (it has no units). We could reword the answer to read, "The efficiency of the light bulb is 20 percent" or, "The light bulb is 20 percent efficient."

Not all electric devices have such a low efficiency as the light bulb. Electric motors like those used in washing machines, clothes dryers, and refrigerators have efficiencies of 50 to 75 percent. This means that 50 to 75 percent of the electric energy put into the motor is converted to mechanical (rotational) energy. The other 25 to 50 percent is converted into heat energy.

Converting electric energy to light and heat energy.

Example 1-4

How much energy is required to produce 460 joules of light energy from a light bulb that is 25 percent efficient?

Given: $\% \text{ eff.} = 25$
Energy out = 460 joules

Find: Energy in

Known: $\% \text{ eff.} = \dfrac{W_{out}}{W_{in}} \times 100$

by rearranging:

$W_{in} = \dfrac{W_{out}}{\% \text{ eff.}} \times 100$

Solution: $W_{in} = \dfrac{460 \text{ joules}}{25} \times 100$
$= 18.4 \text{ joules} \times 100$
$= 1840 \text{ joules}$

Answer: Energy in = 1840 joules

So far, we have illustrated the efficiency of converting electric energy into other desired forms of energy. Of course, we are just as interested in the efficiency of converting other forms of energy into electric energy.

Example 1-5

What is the efficiency of an electric generator that produces 5000 joules of electric energy from the 7000 joules of mechanical energy used to rotate the generator?

Given: Energy in = 7000 joules
                 Energy out = 5000 joules

Find: Percent efficiency

Known: $\% \text{ eff.} = \dfrac{W_{out}}{W_{in}} \times 100$

Solution: $\% \text{ eff.} = \dfrac{5000 \text{ joules}}{7000 \text{ joules}} \times 100$

$$= 0.714 \times 100$$
$$= 71.4$$

Answer: Efficiency = 71.4 percent

◼ TEST

Answer the following questions.

5. List the forms of energy into which the electric energy from a car battery is converted.
6. What is the undesirable form of energy produced by both light bulbs and electric motors?
7. What happens to the temperature of an electric battery when the battery is discharging? Why?
8. An electric motor requires 1760 joules of electric energy to produce 1086 joules of mechanical energy. What is the efficiency of the motor?
9. A flashlight battery uses 110 joules of chemical energy to supply 100 joules of electric energy to the flashlight bulb. What is the efficiency of the battery?
10. How much mechanical energy will be provided by a motor that is 70 percent efficient and requires 1960 joules of electric energy?

1-5 Structure of Matter

All matter is composed of *atoms*. Atoms are the basic building blocks of nature. Regardless of their physical characteristics, glass, chalk, rock, and wood are all made from atoms. Rock is different from wood because of the type of atoms of which it is composed.

There are more than 100 different types of atoms. Matter composed of a single type of atom is called an *element*. Thus, there are as many elements as there are types of atoms. Some common elements are gold, silver, and copper.

There are thousands and thousands of different materials in the world. Obviously, most materials must be composed of more than one element. When different kinds of atoms combine chemically, they form materials called *compounds*. An example of a simple compound is water, which is composed of the elements oxygen and hydrogen. Many of the materials used in electronic circuits are composed of compounds.

To really understand electricity, we must "break the atom down" into still smaller particles. To understand electricity, we need to be familiar with the three major particles of the atom. These are the *electron, proton,* and *neutron*. A pictorial representation of a helium atom showing its three major particles is illustrated in Fig. 1-1. The center of the atom is called the *nucleus*. It contains the protons and neutrons. The electrons revolve around the nucleus in elliptical paths. The electron is much larger (nearly 2000 times larger) than either the proton or the neutron. Although larger, the electron is much lighter than either the proton or the neutron. (about $\frac{1}{2000}$ as heavy). Thus, the center (nucleus) of the atom contains most of the weight, and the electrons make up most of the volume. It should also be noted that the distance between the nucleus and the electron is very great compared

Element

Compounds

Electron

Proton

Neutron

Nucleus

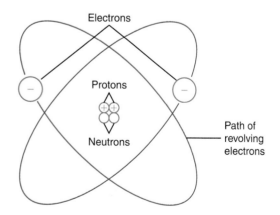

Fig. 1-1 Structure of an atom of helium.

Atoms

James Prescott Joule The SI (*Système Internationale*) unit of measure for electrical energy is the joule (J), named for James Prescott Joule (1 joule is equal to 1 volt-coulomb).

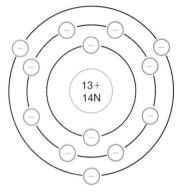

Fig. 1-2 Simplified presentation of an aluminum atom showing its 13 electrons (−), 13 protons (+), and 14 neutrons (N).

with the size of the electron. In fact, this distance is approximately 60,000 times greater than the diameter of the electron.

An analogy may help you to visualize the relative sizes of the atomic particles and the spaces between them. The simplest atom is the hydrogen atom, which contains one proton, one electron, and no neutrons. Let the nucleus of the hydrogen atom be represented by a common marble. The electron could then be represented by a 31-meter (100-foot) sphere located 1610 kilometers (1000 miles) from the marble. Although the distance between the nucleus and the electron is very great relative to the size of either, we must remember that these sizes and distances are submicroscopic. For example, the diameter of an electron is only 4×10^{-13} (0.0000000000004) centimeters.

Electrons rotate, or orbit, around the nucleus of the atom in much the same manner as the earth rotates around the sun (Fig. 1-1). In atoms that contain more than one electron (that is, in all atoms except hydrogen atoms), each electron has its own orbit. With proper coordination of the orbiting electrons, it is possible for two or more atoms to share common space. Indeed, in many materials, neighboring atoms share electrons as well as space.

Figure 1-2 represents the aluminum atom in a two-dimensional form. Remember that each electron is actually orbiting around the nucleus in its own elliptical path. The two electrons closest to the nucleus do not actually follow the same orbital path. Their orbital paths are merely the same average distance from the nucleus. The two electrons closest to the nucleus are said to occupy the first *shell,* or *orbit,* of the atom. This first shell of the atom can accommodate only two electrons. Atoms that have more than two electrons, such as the aluminum atom, must have a second shell, or orbit.

The second shell of the aluminum atom contains eight electrons. This is the maximum number of electrons that the second shell of any atom can contain. The third shell can contain a maximum of 18 electrons, and the fourth shell a maximum of 32 electrons. Since the aluminum atom (Fig. 1-2) has only 13 electrons, its third shell has 3 electrons.

Shell

Orbit

Technicians at work in their lab.

◼ TEST

Answer the following questions.

11. What are the major particles of an atom?
12. True or false. The diameter of the electron is small compared with the distance between the proton and the electron.

13. True or false. The diameter of the proton is greater than the diameter of the electron.
14. True or false. The electron weighs more than the proton.
15. True or false. All electrons in the second shell of an atom follow the same orbital path.

1-6   Electric Charge

Both electrons and protons possess electric charges, but these charges are of opposite *polarity*. Polarity refers to the type (negative or positive) of charge. The electron possesses a negative (−) charge, and the proton possesses a positive (+) charge. These electric charges create electric fields of force that behave much like magnetic fields of force. In Fig. 1-3, the lines with arrows on them represent the electric fields. Two positive charges or two negative

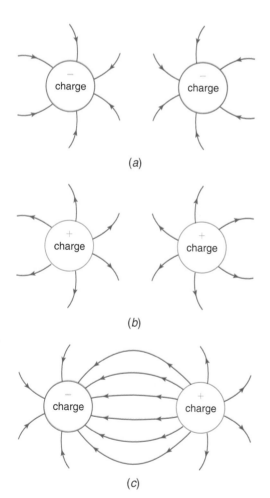

(a)

(b)

(c)

Fig. 1-3   Electric fields between charges. *(a)* and *(b)* Like charges repel each other. *(c)* Unlike charges attract each other.

charges repel each other. Two opposite electric charges attract each other (Fig. 1-3). The force of attraction between the positive proton and the negative electron aids in keeping the electron in orbit around the nucleus. The neutron in the nucleus of the atom has no electric charge. The neutron can be ignored when considering the electric charge of the atom.

An atom in its natural state always has a net electric charge of zero; that is, it always has as many electrons as it has protons. For example, look at the aluminum atom illustrated (in simplified form) in Fig. 1-2. It has 13 electrons orbiting around the nucleus, and the nucleus contains 13 protons in addition to the 14 neutrons. We can say that the aluminum atom is electrically neutral, even though individual electrons and protons are electrically charged.

1-7   Valence Electrons

The electrons in the outermost shell of an atom are called *valence electrons*. Valence electrons are atomic particles that are involved in chemical reactions and electric currents.

One of the forces that help to hold electrons in orbit is the force of attraction between unlike charges. The closer together two particles of opposite electric charges are, the greater the electrical attraction between them. Therefore, the attraction between the proton of the nucleus and the electron decreases as the electron gets farther from the nucleus. Thus, the valence electrons are held to the nucleus with less attraction than the electrons in the inner shells are. The valence electrons can be more easily removed from the parent atom than the electrons in the inner shells can.

All electrons possess energy. They possess energy because they have weight and they are moving. Thus, they are capable of doing work. Valence electrons possess more energy than electrons in the inner shells. In general, the farther the electron is from the nucleus of the atom, the more energy it possesses.

1-8   Free Electrons

Free electrons are valence electrons that have been temporarily separated from an atom. They are free to wander about in the space around the atom. They are unattached to any particular atom. Only the valence electrons are capa-

Polarity

Valence electrons

Free electrons

ble of becoming free electrons. Electrons in the inner shells are very tightly held to the nucleus. They cannot be separated from the parent atom. A valence electron is freed from its atom when energy is added to the atom. The additional energy allows the valence electron to escape the force of attraction between the electron and the nucleus. As a free electron, the electron possesses more energy than it did as a valence electron. One way to provide the additional energy needed to free an electron is to heat the atom. Another way is to subject the atom to an electric field.

1-9   Ions

When a valence electron leaves an atom to become a free electron, it takes with it one negative electric charge. This absence of one negative electric charge from the parent atom leaves that parent atom with a net positive charge. In the case of the aluminum atom, there would be 13 protons (positive charges) but only 12 electrons (negative charges). Atoms which have more than or less than their normal complement of electrons are called *ions*. When an atom loses electrons, it becomes a *positive ion*. Conversely, atoms with an excess of electrons contain a net negative charge and become *negative ions*. The amount of energy required to free a valence electron and create an ion varies from element to element.

The energy required to create a free electron is related to the number of valence electrons contained in the atom. In general, the fewer the electrons in the valence shell, the smaller the amount of additional energy needed to free an electron. The silver atom, illustrated in Fig. 1-4, with its single valence electron requires relatively little energy to free the valence electron. Carbon, with four valence electrons, requires much more energy to free an electron. Elements with five or more electrons in the outer shell do not readily release their valence electrons.

Negative ions are created when a atom accepts additional electrons. For example, in the

Ions

Positive ion

Negative ion

This single valence electron is loosely held to the atom

Fig. 1-4   Simplified silver atom.

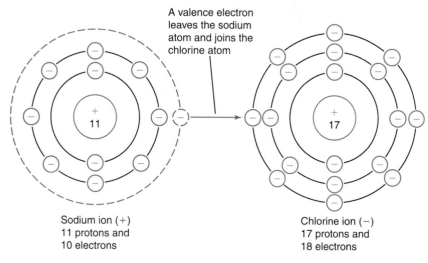

A valence electron leaves the sodium atom and joins the chlorine atom

Sodium ion (+)
11 protons and
10 electrons

Chlorine ion (−)
17 protons and
18 electrons

Fig. 1-5 Creation of positive (sodium) ions and negative (chlorine) ions.

compound sodium chloride (which is ordinary table salt), the sodium atoms share their lone valence electrons with the chlorine atoms to form salt crystals. When the sodium chloride is dissolved in water, the sodium and chlorine atoms separate from each other and the chlorine atom takes with it the sodium atom's valence electron. Thus the chlorine atom becomes a negative ion. At the same time, the sodium atom, which gave up an electron, becomes a positive ion (Fig. 1-5). The concept of ions is important in understanding electric circuits involving batteries and gas-filled devices.

■ TEST_____

Answer the following questions.

16. What polarity is the charge of an electron? A proton?
17. Is an atom electrically charged? Explain.
18. What is an atom called when it has lost a valence electron?
19. True or false. A free electron is at a higher energy level than a valence electron is.
20. True or false. An atom with seven valence electrons provides a free electron more readily than an atom with two valence electrons.
21. True or false. Ions can possess either a negative or a positive charge.

1-10 Static Charge and Static Electricity

Static electricity is a common phenomenon that all of us have observed. Probably the most dra-

matic example of static electricity is lightning. Static electricity is responsible for the shock you may receive when reaching for a metal doorknob after walking across a thick rug. It is also responsible for hair clinging to a comb and for the way some synthetic clothes cling to themselves.

All the above phenomena have one thing in common. They all involve the transfer of electrons from one object to another object or from one material to another material.

A *positive static charge* is created when a transfer of electrons leaves an object with a deficiency of electrons. A *negative static charge* results when an object is left with an excess of electrons. Static charges can be created by rubbing a glass rod with a piece of silk cloth. Some of the valence electrons from the glass rod become free electrons and are transferred to the cloth. The cloth takes on a negative charge, and the glass rod takes on a positive charge. The charges on the glass and the cloth tend to remain stationary, thus the name *static electricity*.

Positive static charge

Negative static charge

About ◀▬▶ Electronics

Niels Bohr In 1913 Niels Bohr theorized that the atoms of all substances contain negatively charged particles, called *electrons*, in orbit about positively charged particles, called *nuclei*. Bohr attributed the inward force that keeps the electrons from flying off into space to an electrical attraction between the nucleus and electron, caused by opposite polarity. (*Encyclopedia of Electronics*, Gibilisco and Sclater, McGraw-Hill, 1990)

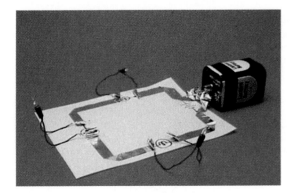

An example of a series circuit.

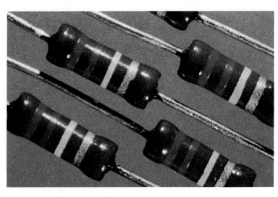

Typical resistors.

Induced charge

Static discharge

An object that possesses a static charge can attract objects that are not charged. This happens because the charged object, when placed near an uncharged object but not touching it, can induce a charge on the surface of the uncharged object (Fig. 1-6). The *induced charge,* because its polarity is the opposite of that of the charged object, is then attracted to the charged object. If the two objects—for example, the ball and the rod in Fig. 1-6—are allowed to touch, part of the positive charge of the rod is transferred to the ball. Then both objects have positive charges, and a force of repulsion results. The ball and the rod then move away from each other because their like charges repel each other.

Let us clarify how a charge is transferred from one object to the other, again using as an example the ball and the rod of Fig. 1-6. When the rod and the ball touch, protons *do not* travel from the rod to the ball. Remember, the electron is the only particle of the atom that can be easily detached and moved about. When the rod and the ball touch, electrons travel from the ball to the rod. This leaves the ball with a shortage of negative charge (electrons). The

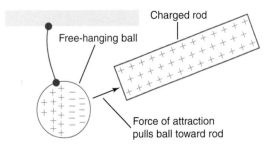

Fig. 1-6 Inducing a static charge. When the ball touches the rod, electrons are transferred to the rod. The ball is left with a positive charge.

ball still has its normal number of positive charges (protons). Since the ball now has more protons than electrons, it possesses a net positive charge.

1-11 Static Discharge

Static discharge occurs when the electric force field (Fig. 1-3) between a positive charge and a negative charge becomes too strong. Then electrons are pulled off the negatively charged object and travel through the air to the positively charged object. The observable spark is the result of the air path between the objects becoming ionized and deionized. When the air is ionized, electrons are raised to a higher energy level. When the air is deionized, the electrons return to a lower energy level. The difference in the two energy levels is given off as light energy when the electron returns to the lower energy level. Lightning is a column of ionized air caused by electrons traveling between a charged cloud and a portion of land with an opposite charge.

1-12 Uses of Static Electricity

Most of the useful applications of static electricity do not rely on discharges that ionize the air. Rather, they make use of the force of attraction between unlike charges or the force of repulsion between like charges. These forces are used to move charged particles to desired locations. For example, dust particles can be removed from air by this method. In Fig. 1-7, the air is forced between negatively charged rods and positively charged plates. A negative charge is transferred from the rods to the dust particles. When the airstream goes by the pos-

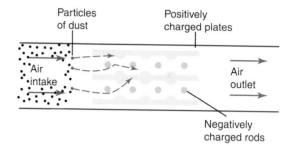

Fig. 1-7 Principle of a dust precipitator. Dust particles are given a negative charge so that they are attracted to positively charged plates.

itively charged plate, the negatively charged dust particles are pulled out. Electric devices of this type are often called *electrostatic precipitators*.

Static charges are also used in some spray-painting operations. The paint mist leaving the spray gun nozzle is negatively charged, and the object to be painted is positively charged. This process tends to give a uniform coat of paint

on an object, even one that has an irregular surface. As paint builds up on one part of the object, the charge on that part of the object is canceled. The force of attraction disappears. If excess paint builds up on part of the object, then that part becomes negatively charged and repels additional paint. The repelled paint is attracted to those parts of the object that are still positively charged.

Several properties of static electricity can be used to advantage in manufacturing abrasive paper (sandpaper). The backing paper is coated with an adhesive (glue) and given a static charge. The abrasive particles are given the opposite charge. As the paper is passed over the abrasive particles, the particles are attracted to the paper and adhere to it (Fig. 1-8). Once the abrasive particles are distributed on the adhesive backing paper, both the paper and the abrasive particles are given like charges. The like charges repel each other and try to push the abrasive particles from the paper. However, the adhesive is strong enough to hold the particles to the paper. The abrasive particles "stand up"

Electrostatic precipitators

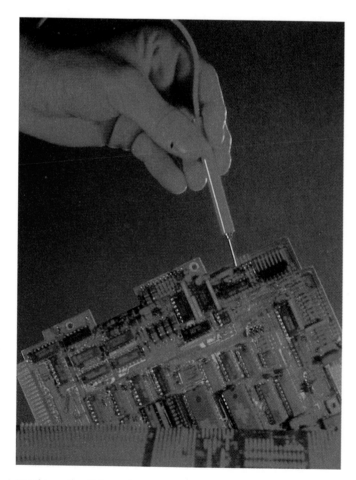

Modern circuit board.

Electrical power plant control room.

Technicians testing circuit boards.

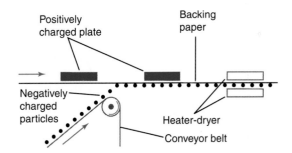

Fig. 1-8 Principles of making abrasive paper. Static charges position the abrasive particles so that their sharpest points are exposed.

Positively charged plate

Backing paper

Negatively charged particles

Heater-dryer

Conveyor belt

on the paper so that the sharpest point of each particle is exposed as the cutting surface. The adhesive material is then heated and hardened to hold the abrasive particles in place. The reason the abrasive particles stand up is that static charges concentrate on the sharpest point of an object. Since like charges repel each other, they push away from each other as far as possible. The sharpest point of the abrasive particle is the most highly charged part of the particle. Therefore, it is pushed farthest from the surface of the backing paper.

■ TEST

Answer the following questions.

22. What is a static charge?
23. List two industrial applications of static charges.
24. True or false. After a positively charged rod touches a neutral free-hanging ball, the ball will be repelled.
25. True or false. Static discharge causes electrons to transfer from one object to another object.
26. True or false. An object with a static charge can attract only other charged objects.

Summary

1. Energy is the ability to do work.
2. The symbol for energy is *W*.
3. Electricity is a form of energy.
4. Work equals force times distance.
5. The joule is the base unit of energy.
6. The joule is the base unit of work.
7. The symbol for the joule is J.
8. Under ordinary conditions, energy is neither created nor destroyed.
9. Percent efficient $= \dfrac{W_{out}}{W_{in}} \times 100$
10. All matter is made up of atoms.
11. Hydrogen has only one electron and one proton. All other atoms contain electrons, protons, and neutrons.
12. Protons and neutrons are found in the nucleus of the atom.
13. Electrons have a negative electric charge.
14. Protons have a positive electric charge.
15. Neutrons do not have an electric charge.
16. Atoms have an equal number of protons and electrons.
17. Valence electrons are found in the outermost shell of the atom.
18. Valence electrons can become free electrons when their energy level is raised.
19. Negative ions are atoms that have attracted and captured an extra (free) electron.
20. Positive ions are atoms that have given up an electron.
21. Static charges are the result of an object's possessing either more or fewer electrons than protons.
22. A negative charge means an excess of electrons.
23. Like charges repel each other.
24. Unlike charges attract each other.

Chapter Review Questions

For questions 1-1 to 1-7, determine whether each statement is true or false.

1-1. Opposite electric charges repel each other.
1-2. Free electrons are at a higher energy level than valence electrons are.
1-3. The same base unit is used for both work and energy.
1-4. Atoms have twice as many protons as electrons.
1-5. An object that has a static charge can possess either more or fewer electrons than protons.
1-6. The second shell, or orbit, of an atom can contain any number of electrons.
1-7. If an electric device is less than 100 percent efficient, part of the energy provided to it is destroyed.

For questions 1-8 to 1-11, choose the letter that best completes each sentence.

1-8. The symbol for energy is
   a. *D*
   b. *W*
   c. *N*
   d. *J*

1-9. The base unit of energy is the
   a. Ion
   b. Proton
   c. Joule
   d. Pound

1-10. The abbreviation for the base unit of energy is
   a. D
   b. W
   c. N
   d. J

1-11. A positive ion is an atom which has
   a. Captured one or more electrons
   b. Given up one or more electrons
   c. Captured one or more protons
   d. Given up one or more protons

Answer the following questions.

1-12. What is the efficiency of a motor which requires 914 base units of electric energy to produce 585 joules of mechanical energy?

1-13. How much work is done when an object is moved 3 meters by applying a force of 80 newtons?

1-14. How much energy is involved in question 1-13?

1-15. How much energy is required by an electric lamp that is 18 percent efficient and provides 5463 J of light energy?

Critical Thinking Questions

1-1. Discuss the probable changes in your lifestyle if the electric service to your home were disconnected for one week.

1-2. In addition to decreasing costs, why is it important to increase the efficiency of electric devices as much as is practical?

1-3. Why are large, continuously operated motors designed to be more efficient than small, intermittently operated motors?

1-4. Is the low efficiency of the lights in our homes a greater disadvantage during the winter months or the summer months? Why?

1-5. Many electric devices could be designed and constructed to operate more efficiently. Why aren't they?

Answers to Tests

1. Work is a force moving through a distance; energy is the ability to do work.

2. A base unit is a term that is used to specify the amount of something.

3. joule (J)

4. **Given:** Distance = 130 meters
   Force = 360 newtons
   Find: Work
   Known: Work = force × distance
   Solution: Work = 360 newtons × 130 meters
   Answer: Work = 46,800 joules

5. heat (due to inefficiencies of lights and motors), light, and mechanical energy.

6. heat energy

7. It increases, because inefficiency converts some of the battery's energy to heat.

8. **Given:** W_{out} = 1086 joules
   W_{in} = 1760 joules
   Find: Percent efficiency
   Known: $\% \text{ eff.} = \dfrac{W_{out}}{W_{in}} \times 100$

Solution: $\% \text{ eff.} = \dfrac{1086 \text{ joules}}{1760 \text{ joules}} \times 100$
$= 61.7$

Answer: Efficiency = 61.7 percent

9. **Given:** W_{out} = 100 joules
   W_{in} = 110 joules
   Find: Percent efficiency
   Known: $\% \text{ eff.} = \dfrac{W_{out}}{W_{in}} \times 100$

Solution: $\% \text{ eff.} = \dfrac{100 \text{ joules}}{110 \text{ joules}} \times 100$
$= 90.9$

Answer: Efficiency = 90.9 percent

10. **Given:** W_{in} = 1960 joules (J)
    $\% \text{ eff.}$ = 70
    Find: W_{out}
    Known: $\% \text{ eff.} = \dfrac{W_{out}}{W_{in}}$
    by rearranging:
    $W_{out} = \dfrac{W_{in} \times \% \text{ eff.}}{100}$

    Solution: $W_{out} = \dfrac{1960 \text{ J} \times 70}{100} = 1372 \text{ J}$

    Answer: Mechanical energy out = 1372 J

11. proton, electron, and neutron

12. T

13. F

14. F

15. F
16. Electron is negative, and proton is positive.
17. No, the negative charge of the electrons cancels the positive charge of the protons, leaving the atom neutral.
18. positive ion
19. T
20. F
21. T
22. A static charge is the excess or deficiency of electrons on an object.
23. removal of dust particles from air, spray painting, and manufacturing of abrasive paper
24. T
25. F
26. F

Chapter 2

Electrical Quantities and Units

Chapter Objectives

This chapter will help you to:

1. *Describe and correctly use* units of charge, current, voltage, resistance, and power.
2. *Describe* current in solids, liquids, and gases.
3. *Understand* the difference (and relationship) between power and energy.
4. *Convert* quantities from base units to submultiple or multiple units and vice versa.
5. *Express* the relationship between energy, charge, and voltage.
6. *List and explain* five ways of producing voltage.

If you were describing a shopping trip to the grocery store, you would probably use many quantities and units in your description. You might start out by telling the distance (a quantity) in blocks or miles (units) to the store. In describing what you purchased, you could use many quantities and units, such as pounds of sugar, heads of lettuce, cans of soup, quarts of milk, dozens of eggs, or bars of soap.

In describing electric circuits, you also need to use quantities and units. This chapter tells you about many of the basic electrical quantities and units.

2-1 Charge

Electric charge (Q)

 . . . that in Chap. 1 we defined an *electric charge* as the electrical property possessed by electrons and protons.

The proton has a positive charge and the electron a negative charge. However, we never did specify exact amounts of charge. That is, we never did define the base unit of charge so that we could specify exact amounts of charge.

2-2 Unit of Charge

Coulomb (C)

The base unit of charge is the *coulomb*. A coulomb of charge is the amount of charge possessed by 6.25×10^{18} (6,250,000,000,000,000,000) electrons. We do not use the charge on a single electron as the base unit because it is a very small charge—too small for most practical applications. The coulomb, the

Electric current (I)

base unit of charge, is also used in defining the base units of other electrical quantities, such as current and voltage. The coulomb is named after Charles Augustin Coulomb, a French physicist.

In electricity, we use many symbols (abbreviations) for electrical quantities and units. The symbol for charge is Q. The abbreviation for coulomb is C. Use of symbols in the study of electricity allows us to condense ideas and statements. For example, instead of writing, "the charge is 5 coulombs," we can just write, "$Q = 5$ C."

2-3 Current and Current Carriers

Electric current is the movement of charged particles in a specified direction. The charged particle may be an electron, a positive ion, or a negative ion. The charged particle is often referred to as a *current carrier*. The movement of the charged particle may be through a solid, a gas, a liquid, or a vacuum. In a solid, such as copper wire, the charged particle (current carrier) is the electron. The ions in a copper wire, and in other solids, are rigidly held in place by the atomic (crystalline) structure of the material. Thus, ions cannot be current carriers in solid materials. However, in both liquids and gases, the ions are free to move about and become current carriers.

The symbol for current is I. The symbol I was chosen because early scientists talked about the intensity of the electricity in a wire.

2-4   Current in Solids

When thinking about current, keep two points in mind. First, the effect of current is almost instantaneous. Current in a wire travels at nearly the *speed of light*: 186,000 miles per second [3×10^8 meters per second (m/s)]. Second, an individual electron moves much more slowly than the effect of the current. It may take minutes for an individual electron to travel a few feet in the wire.

The ideas of the "effect of the current being instantaneous" and the "individual electron moving much more slowly" are illustrated in Fig. 2-1. Suppose you had a very long cardboard tube with a diameter just large enough to pass a tennis ball. You lay the tube on the floor and fill it full of tennis balls. When you push an extra ball into one end of the tube, another ball immediately comes out the other end of the tube (Fig. 2-1). If you did not know that the tube was full of tennis balls, you might think that the tennis ball you pushed in one end of the tube traveled very quickly down the tube and out the other end. The effect (the one ball

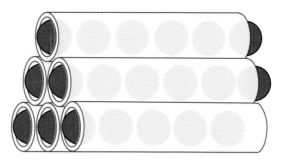

Fig. 2-2 Electron movement illustrated.

popping out the end) is very fast. Yet each tennis ball moves only a short distance.

Suppose you stacked up six tubes filled with balls (Fig. 2-2) and pushed balls first into one tube and then into another. Now you could have a steady stream of balls appearing at one end of the tubes, yet only the balls in one tube would be moving at any one time. Even within that tube each ball would move only a short distance. This is comparable to the way in which current carriers (electrons) move through a wire when current is flowing in the wire.

Assume you could look inside an aluminum wire and see the atoms and their particles (Fig. 2-3). The aluminum atom actually has 13 electrons and 13 protons, but, for simplicity, only 3 of each are shown. Now, suppose the ends of the wire are connected to a flashlight cell. The cell provides an electric field through the wire. The electric field frees some of the valence electrons of the aluminum atoms, as shown in Fig. 2-3, by giving them additional energy. At the moment that an individual electron is freed,

Speed of light

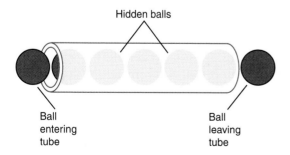

Fig. 2-1 Apparent speed illustrated. A ball exits the instant another ball enters the tube.

About ⬤ Electronics

Charles Augustin Coulomb
French natural philosopher Charles Augustin Coulomb developed a method for measuring the force of attraction and repulsion between two electrically charged spheres. Coulomb established the law of inverse squares and defined the basic unit of charge quantity, the coulomb.

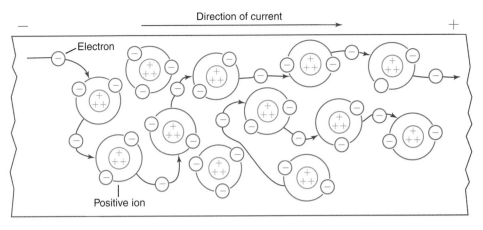

Fig. 2-3 Current in a solid. A free electron travels only a short distance before it combines with a positive ion.

it may be traveling in a direction opposite that of the main current. However, in the presence of the electric field, it soon changes its direction of movement. For every electron that is freed, a positive ion is created. This positive ion has an attraction for an electron. Eventually, one of the free electrons will migrate close to the positive ion. The electron will be captured by that positive ion, which then, of course, will become a neutral atom.

Notice that a free electron does not remain a free electron and travel the full length of the wire. Rather, it travels a short distance down the wire and is captured by one of the positive ions. At some later time this particular electron may again gain enough energy to free itself of its new parent atom. It then travels farther down the wire as a free electron. We can think of the electrons as hopping down the conductor from atom to atom to atom. As long as there is a new free electron created every time a free electron is captured, the net number of free electrons moving down the wire remains constant. Current continues to flow. It continues to flow in the same direction through the conducting wire. Current that travels in the same direction all the time is called *direct current,* which is abbreviated dc. It is the type of current you get from flashlight cells and batteries.

If you remember that the individual electron travels rather slowly from atom to atom, you will be able to understand alternating current. *Alternating current* (abbreviated ac) is the type of current you have in your home and your school. It is the type of current that periodically reverses the direction in which it is moving. The current in all the electric wires in your home reverses its direction every $\frac{1}{120}$ of a second. Some of the current in your television receiver reverses every $\frac{1}{67,000,000}$ of a second. Currents that reverse direction this often are easier to visualize if you think of the individual electrons as "swinging" back and forth between several atoms.

2-5 Current in Liquids and Gases

In gases, both positive ions and electrons are involved in current flow. When a gas is subjected to a strong electric field, the gas *ionizes.* Once ionized, the gas allows current to flow through it. Figure 2-4 illustrates current flow in ionized neon gas. The negative and positive signs indicate that the neon bulb is connected to a source of electric force, such as a battery. A neon atom has eight electrons in its outermost (valence) shell. When the atom is ionized, one electron is freed. The resulting positive neon ion travels toward the *negative plate* (Fig. 2-4). The resulting free electron travels toward the positive plate. Once the positive ion arrives at the negative plate, it receives an electron and becomes a neutral atom. It then drifts around in the glass enclosure until it is again ionized. The free electron is received by the *positive plate* and travels out the connecting wire. The system of Fig. 2-4 requires that electrons be supplied to the negative plate and removed from the positive plate. That job is done by the power supply (battery), which also supplies the electric field.

Ionizes

Negative plate

Direct current (dc)

Positive plate

Alternating current (ac)

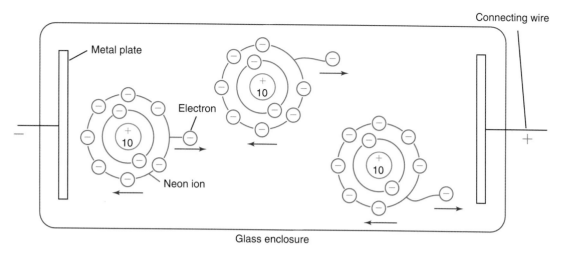

Fig. 2-4 Current in a gas. Both electrons and ions serve as current carriers.

Current flow in a liquid consists of both negative and positive ions moving through the liquid. A simplified diagram of current flow in a sodium chloride (table salt) solution is shown in Fig. 2-5. The abbreviation Na is the chemical symbol for sodium, and the plus sign (+) in the circle means a positive ion. Likewise, Cl is the chemical symbol for chlorine and the minus sign (−) in the circle means a negative ion. When an electric field is created between the plates, the positive sodium ions move to the negative plate and the negative chlorine ions move to the positive plate. Notice that the current in the liquid is composed entirely of ions. However, the current in the wires and plates connected to the liquid is composed of elec-

trons in motion. The changeover from electron-charge carriers to ion-charge carriers occurs at the surface where the plates and liquid meet. This change of carrier is actually more complex than implied by Fig. 2-5. The change also involves some ions created by the water in which the salt is dissolved. However, whether or not we account for the water ions, the end results are the same. That is, negative ions give up electrons at the positive plate, and positive ions pick up electrons from the negative plate. A liquid solution which is capable of carrying current is called an *electrolyte*. A solution of seawater is an electrolyte; it is a solution containing ionized substances.

One industrial use of current flow in electrolytes is electroplating. *Electroplating* is a process by which a thin layer of one type of metal can be plated (surface-covered) over another material. The other material may be another metal or a piece of plastic coated with a conductive material. Figure 2-6 illustrates the electroplating of copper onto iron. The electrolyte is copper sulfate, which ionizes into a copper ion (Cu^{++}) with two positive charges and a sulfate ion (SO_4^{--}) with two negative charges. The copper ions are attracted to the iron plate, where they pick up two electrons and adhere (stick to) to the iron plate as copper atoms. The sulfate ions move to the copper plate, where they chemically react with the copper to create more copper sulfate. The copper sulfate goes back into solution. The reaction that created the copper sulfate leaves two electrons on the copper plate. These two electrons move out through the wire connected to

Electrolyte

Electroplating

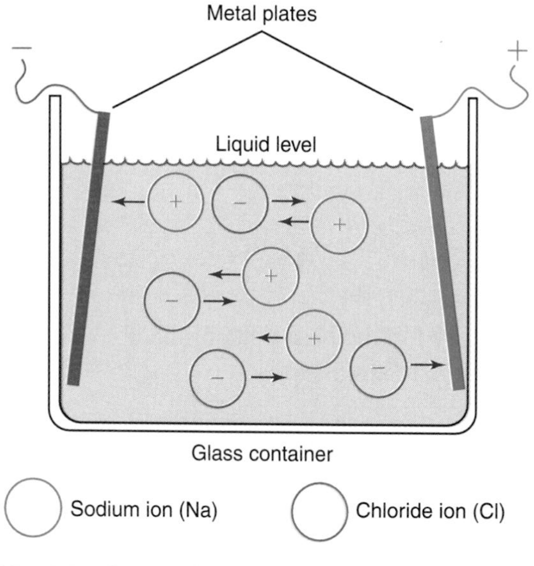

Fig. 2-5 Current in a liquid. Both positive and negative ions serve as current carriers.

Sodium ion (Na)   Chloride ion (Cl)

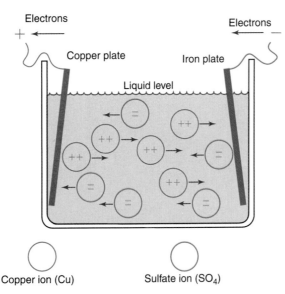

Electrons + ← Copper plate

Electrons ← − Iron plate

Liquid level

Copper ion (Cu)   Sulfate ion (SO_4)

Fig. 2-6 Electroplating copper onto an iron plate. Current in the liquid consists of ions in motion.

Coulombs per second

Ampere (A)

$$I = \frac{Q}{t}$$

Time (t)

Second (s)

$$A = \frac{C}{s}$$

the copper plate. Notice that this electroplating system is always in balance. That is, for every two electrons that enter the negative iron plate, two more electrons leave the positive copper plate. This type of balance is present in all electric devices which carry current.

2-6 Unit of Current—The Ampere

We have now developed a concept of electric current. Our next problem is to develop a logical unit for keeping track of the amount of current. The simplest method might be to keep track

of the number of electrons or of the number of coulombs that move down the conductor. However, this method would leave much to be desired. It would not take into account the amount of time required to move the charge. This would be somewhat comparable to tabulating automobile traffic without considering the time involved. For example, on a freeway, 1000 cars may pass a given point in 1 hour. On a two-lane country road it may take 20 hours for 1000 cars to pass a given point. Certainly the traffic is greater on the freeway. A more meaningful way to compare the traffic on these two roads would be to talk in terms of the number of cars per hour. Thus, the traffic on the freeway would be 1000 cars per hour while the traffic on the two-lane road would be 50 cars per hour.

In electricity, the amount of current is specified in terms of the charge and the time required to move the charge past a given point. The amount of electric current is therefore specified in *coulombs per second*. But, since coulombs per second is a rather long term, the base unit of current has been named the *ampere*. The ampere is the base unit of current. An ampere is equal to 1 coulomb per second. The ampere was chosen as the base unit of current in honor of André Marie Ampère, a French scientist who did some early work in the field of electricity.

The abbreviation for the ampere is A. To indicate that the current in a wire is 10 amperes, for example, we would write $I = 10$ A.

Notice that our definition of an ampere involves time. In electricity, *time* is represented by the symbol t. The base unit of time is the *second*, which we abbreviate as s. The relationship between time, charge, and current is

$$\text{Current } (I) = \frac{\text{charge } (Q)}{\text{time } (t)}$$

or

$$I = \frac{Q}{t}$$

Expressed in the base units of the above quantities, the relationship is:

$$\text{Ampere} = \frac{\text{coulomb}}{\text{second}}$$

or, in abbreviated form:

$$A = \frac{C}{s}$$

Example 2-1

How much time is required for 12.5×10^{18} electrons to leave the negative terminal of a battery if the current provided by the battery is 0.5 A?

Given: $Q = 12.5 \times 10^{18}$ electrons
$I = 0.5$ A

Find: Time (t)

Known: 1 coulomb (C) = 6.25×10^{18} electrons

$$I = \frac{Q}{t} \text{ and, rearranged,}$$

$$t = \frac{Q}{I}$$

Solution: Convert from electrons to coulombs

$$Q = \frac{12.5 \times 10^{18} \text{ electrons}}{6.25 \times 10^{18} \text{ electrons/C}}$$
$$= 2 \text{ C}$$

Then calculate time

$$t = \frac{Q}{I} = \frac{2C}{0.5A} = 4 \text{ s}$$

Answer: Time = 4 seconds

▪ TEST

Answer the following questions.

1. What is an electric charge?
2. The symbol for charge is _____.
3. A coulomb of charge is equal to the charge on _____ electrons.
4. The symbol for the base unit of charge is _____.
5. Either an _____ or an _____ can be a current carrier.
6. The current carrier in a copper wire is the_____.
7. True or false. An ionized gas has only one type of current carrier.
8. True or false. Random motion of a current carrier is considered to be an electric current.
9. True or false. The symbol for current is A.
10. Describe the way in which an electron travels through a copper wire.
11. How is alternating current different from direct current?
12. What is the base unit of current?
13. Define the ampere in terms of charge and time.
14. What is the abbreviation for the ampere?
15. Rewrite each of the statements below, using the correct symbols for the electrical quantities and units:
    a. The current is 8 amperes.
    b. The charge is 6 coulombs.
16. How much current is flowing when 16 coulombs pass a specified point in 4 seconds?

2-7 Voltage

Voltage is the electric pressure that causes current to flow. Voltage is also known as *electromotive force* (abbreviated emf), or *potential difference*. All these terms refer to the same thing, that is, the force that sets charges in motion. Potential difference is the most descriptive term because a voltage is actually a potential energy difference that exists between two points. The symbol for voltage is V. To really understand voltage, one must first understand what is meant by potential energy and potential energy difference. We must, therefore, extend the discussion of energy which we started in Chap. 1.

All energy is either potential energy or kinetic energy. *Kinetic energy* refers to energy in motion, energy doing work, or energy being converted into another form. When you swing a baseball bat, the baseball bat has kinetic energy. It does work when it hits the ball; that is, it exerts a force on the ball that moves the ball through a distance. Anything that has mass (weight) and is in motion possesses kinetic energy.

Potential energy is energy at rest. It is energy that can be stored for long periods of time in its present form. It is capable of doing work when we provide the right conditions to convert it from its stored form into another form. Of course, when it is being converted, it is changed from potential to kinetic energy. Water stored in a lake behind the dam of a hydroelectric plant possesses potential energy due to gravitational forces. The potential energy of the water can be stored for long periods of time. When the energy is needed, it is converted to kinetic energy by letting the water flow through the hydroelectric plant.

An electric charge possesses potential energy. When you scuff around on the carpet, you collect an electric charge on your body.

Voltage (*V*)

Electromotive force

Potential difference

Kinetic energy

Potential energy

That charge (static electricity) is potential energy. Then, when you touch some object in the room, a spark occurs. The potential energy becomes kinetic energy as the electric charge is converted to the light and heat energy of the spark.

An object, such as this book, resting on a table also possesses potential energy. The book is capable of doing work when it moves from the table to the floor. Thus, the book has potential energy with respect to the floor. When the book falls from the table to the floor, its potential energy is converted into mechanical energy and heat energy. Potential energy is dependent on mass. If we replace the book with an object that weighs more, then the potential energy increases. The *potential energy difference* is independent of the amount of mass. It is a function of the distance between the two surfaces and of the gravitational force. Knowing the potential energy difference that exists between two points, we can easily figure the potential energy possessed by an object of any given weight.

Electric fields

Potential energy difference

Example 2-2

What is the potential energy (with reference to the floor) of a 5.5-kilogram (kg) block resting on a table top if the potential energy difference between the tabletop and the floor is 8 joules per kilogram (J/kg)?

Given: Weight = 5.5 kilograms, potential energy difference = 8 J/kg
Find: Potential energy
Known: Potential energy = potential energy difference × weight
Solution: Potential energy = 8 J/kg × 5.5 kg
                            = 44 J
Answer: Potential energy = 44 joules

In the above discussions, we always started with the object on the tabletop and considered the energy it possessed with reference to the floor. We can just as well reverse the situation and consider the energy required to move the object from the floor to the tabletop. The potential energy difference is the same in either case. In the one case you remove energy from

the system; in the other you must put energy into the system. That is, you do work in lifting the object from the floor to the tabletop.

So far in our discussion of potential energy, we have been using mechanical examples. In these examples, the potential energy of the object and the potential energy difference between the floor and the tabletop are due to weight and gravitational forces. In electricity, the potential energy and the potential energy differences are due to *electric fields* and electric charges.

Voltage is a potential energy difference similar to the potential energy difference in the mechanical case discussed above. Instead of being moved by the force of gravity, electric charges are moved by the force of an electric field. In Fig. 2-7, the electron loses energy as it moves from a negatively charged point to a positively charged point. (This is the same as the book's losing energy as it moves from the tabletop to the floor.) The lost energy of the electron could be converted to heat and light as the electron moved through a lamp. There is a voltage (potential energy difference) between the negative and positive areas shown in Fig. 2-7. These negative and positive areas could represent the terminals of an electric battery. A lead-acid storage battery, such as the one shown in Fig. 2-8, is a common source of voltage. A potential energy difference (voltage) exists between the negative and the positive terminal of a battery. This voltage is the result of an excess of electrons at the negative terminal and a deficiency of electrons at the positive terminal. When electrons move from the negative terminal to the positive terminal of the battery, work is done. Energy is taken from the battery and converted to another form of energy.

Like the mechanical system, the electric system can also be reversed. That is, the electron can gain energy when it is moved from the

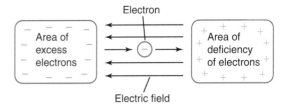

Fig. 2-7 Charge moving through an electric field. Part of the electron's energy is converted to another form of energy.

Fig. 2-8 Lead-acid battery. *(Mark Steinmetz)*

Example 2-3

Determine the potential energy (W) of a 6-V battery that has 3000 C of charge (Q) stored in it.

Given:	$V = 6$ V
	$Q = 3000$ C
Find:	W
Known:	$W = VQ$
Solution:	$W = 6$ V $\times$ 3000 C
	$= 18{,}000$ J
Answer:	Potential energy $= 18{,}000$ joules

positively charged point to the negatively charged point. This is what happens when a battery is charged. The battery charger forces electrons back through the battery in the reverse direction.

2-8 Unit of Voltage—The Volt

We need a unit to indicate the potential energy difference (voltage) between two points, such as the terminals of a battery. This unit must specify the energy available when a given charge is transported from a negative to a positive point. We already have the joule as the base unit of energy and the coulomb as the base unit of charge. Therefore, the logical unit of voltage is the *joule per coulomb*. The joule per coulomb is called the *volt*. The volt is the base unit of voltage. It is abbreviated V. A 12.6-V battery, like the ones used in automobiles, is shown in Fig. 2-8. In symbolic form we indicate the voltage of this battery as $V = 12.6$ V. A potential difference (voltage) of 12.6 V means that each coulomb of charge provides 12.6 J of energy. For example, 1 C flowing through a lamp converts 12.6 J of the battery's energy into heat and light energy.

The relationship between charge, energy, and voltage can be expressed as

$$\text{Voltage } (V) = \frac{\text{energy } (W)}{\text{charge } (Q)}$$

or, by rearranging,

$$W = V \times Q$$

This relationship can be used to determine electric energy in the same way as mechanical energy was found in example 2-2.

Notice in example 2-3 that multiplying volts by coulombs yields joules. This is because a volt is a joule per coulomb and the coulombs cancel:

$$\frac{\text{Joule}}{\text{Coulomb}} \times \frac{\text{coulomb}}{1} = \text{joule}$$

2-9 Polarity

Polarity is a term that is used in several ways. We can say that the polarity of a charge is negative, or we can say that the polarity of a terminal is positive. We can also use the term to indicate how to connect the negative and positive terminals of electric devices. For example, when putting new cells in a transistor radio, we must install the cells correctly (Fig. 2-9). The positive terminal of one cell connects to the positive terminal of the radio, and the negative terminal of the other cell connects to the negative terminal of the radio.

Polarity

Joule per coulomb

Volt (V)

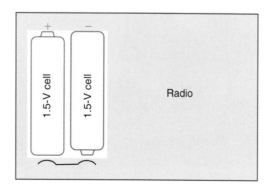

Fig. 2-9 When cells in a radio are installed, the correct polarity must be observed.

$W = V \times Q$

Polarized

Thermocouples

Crystals

Solar cells

Straight polarity

Reverse polarity

Seebeck effect

Piezoelectric effect

Electric generator

Photovoltaic

Electric cell

Electric devices which have negative and positive identifications on their terminals are said to be *polarized*. When connecting such devices to a source of voltage (such as a battery), we must observe the polarity markings. Again, the negative terminal of the device is connected to the negative terminal of the source, and the positive to the positive. If polarity is not observed (that is, if a positive is connected to a negative), the device will not function and may be damaged or ruined.

In electric-arc welding, the welder can weld with either *straight polarity* or *reverse polarity*. The *straight* and *reverse* refer to whether the negative or the positive terminal of the voltage source is connected to the welding rod.

2-10 Sources of Voltage

Voltage can be created by a number of techniques. All involve the conversion of some other form of energy into electric energy. All of them create a voltage by producing an excess of electrons at one terminal and a deficiency of electrons at another terminal.

The most common way of producing a voltage is by an *electric generator*. Generators convert mechanical energy into electric energy. Large generators like the one shown in Fig. 2-10 produce the voltage (and energy) that is provided to our homes, schools, and industries. Most of these generators are turned by mechanical devices, such as steam turbines.

The *electric cell* is the next most common source of voltage. Cells convert chemical energy into electric energy. Several cells can be connected together to form a battery. A wide variety of batteries and cells are manufactured. They range from dry cells which weigh a few grams to industrial batteries which weigh hundreds of kilograms.

Other devices that produce voltage are *thermocouples, crystals,* and *solar cells.*

YOU MAY RECALL . . . that static electricity, which also produces voltage, was discussed in Chap. 1.

The thermocouple converts heat energy into electric energy. This process is known as the *Seebeck effect.* Thermocouples are used extensively for measuring temperatures, especially high temperatures.

Crystals, like those in Fig. 2-11, produce voltage by the *piezoelectric effect.* A voltage is produced when a varying pressure is applied to the surface of the crystal. Crystals are used in such devices as phonograph pickup cartridges and microphones. In a microphone, the sound energy of the voice is first converted to mechanical energy by a diaphragm which applies pressure to the crystal. Thus, crystals convert mechanical energy into electric energy.

Solar cells are semiconductor devices. They convert light energy into electric energy. *Photovoltaic* is the term used to describe this conversion process. Solar cells can be used to provide the voltage needed to operate exposure meters in photography or a satellite communications system.

Fig. 2-10   Electric generator shown being shipped on a railroad flatcar. *(Courtesy of General Electric)*

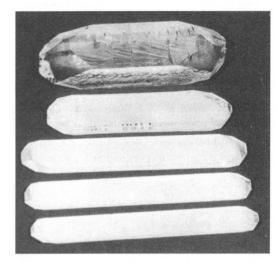

Fig. 2-11   Uncut quartz crystals. *(Courtesy of CTS Knights, Inc.)* When stressed, quartz produces voltage.

Answer the following questions.

17. A potential energy difference between two points in a circuit is called _____ .
18. The base unit of voltage is the _____ .
19. The symbol or abbreviation for voltage is _____ .
20. _____ is abbreviated emf.
21. The symbol or abbreviation for the base unit of voltage is _____ .
22. The electrolyte used to electroplate copper onto iron is _____ .
23. Explain the difference between kinetic energy and potential energy.
24. Why is it incomplete to say, "The potential energy difference of the desk top is 9 joules per kilogram"?
25. Why is it incomplete to say, "The voltage at point *A* is 18 volts"?
26. Define the base unit of voltage in terms of energy and charge.
27. Define *polarized*.
28. List five devices that can produce a voltage. Also specify what type of energy they convert into electric energy.
29. What is the potential energy difference between two points if 100 J of energy is required to move 5 C of charge from one point to the other point?

2-11 Resistance

The opposition a material offers to current is called *resistance*. The symbol for resistance is *R*. All materials offer some resistance to current. However, there is extreme variation in the amount of resistance offered by various materials. It is harder to obtain free electrons (current carriers) from some materials than others. It requires more energy to free an electron in high-resistance materials than in low-resistance materials. Resistance converts electric energy into heat energy when current is forced through a material.

2-12 Conductors

Materials that offer very little resistance (opposition) to current are called *conductors*. Copper, aluminum, and silver are good conductors. They have very low resistance. In general, those elements which have three or fewer electrons in the valence shell can be classified as conductors. However, even within those elements classified as conductors, there is wide variation in the ability to conduct current. For example, iron has nearly six times the resistance of copper, though they are both considered conductors. Although silver is a slightly better conductor than copper, it is too expensive for common use. Aluminum is not as good a conductor as copper, but it is cheaper and lighter. Large aluminum conductors are used to bring electric energy into homes.

Superconductivity is the condition in which a material has no resistance. For many years, superconductivity could be demonstrated only at temperatures close to absolute zero, which is about $-273°C$ (degrees Celsius) or $-460°F$ (degrees Fahrenheit). Further research led to the development of new materials that exhibit superconductivity at temperatures well above absolute zero. The aim has always been to find materials that will superconduct at room temperature. Such materials would, obviously, greatly improve the efficiency of any electric system in which they could be used.

Superconductivity

2-13 Insulators

Materials that offer a high resistance to current are called *insulators*. Even the best insulators do release an occasional free electron to serve as a current carrier. However, for most practical purposes we can consider an insulator to be a material that allows no current to flow through it. Common insulator materials used in electric devices are paper, wood, plastics, rubber, glass, and mica. Notice that common insulators are not pure elements. They are materials in which two or more elements are joined together to form a new substance. In the process of joining together, elements share their valence electrons. This sharing of valence electrons is called *covalent bonding*. It takes a lot of added energy to break an electron free of a covalent bond.

Insulators

Resistance (R)

Covalent bonding

2-14 Semiconductors

Between the extremes of conductors and insulators are a group of elements known as *semiconductors*. Semiconductor elements have four valence electrons. Two of the best-known semiconductors are silicon and germanium.

Conductors

Semiconductors

Semiconductors are neither good conductors nor good insulators. They allow some current to flow, yet they have a considerable amount of resistance. Semiconductors are extremely important industrial materials. They are the materials from which electronic devices such as transistors, integrated circuits (ICs), and solar cells are manufactured.

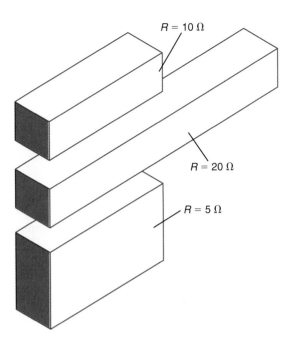

Fig. 2-12 Effects of length and cross-sectional area on resistance.

2-15 Unit of Resistance—The Ohm

Ohm (Ω)

So far we have discussed the amount of resistance in terms of *low resistance* and *high resistance*. In order to work with electric circuits, we must be able to state the amount of resistance more specifically. The unit used to specify the amount of resistance is the *ohm*. The ohm is the base unit of resistance. The symbol used as an abbreviation for ohm is Ω (the capital Greek letter *omega*). The ohm is named in honor of Georg Ohm, who worked out the relationship between current, voltage, and resistance. The ohm can be defined in several ways. First, it is the amount of resistance that allows 1 A of current to flow when the voltage is 1 V. Second, it is the amount of resistance of a column of mercury 106.3 centimeters (cm) in length, 1 millimeter square (mm²), and at a temperature of 0°C. From this second definition of the ohm, you can see that the amount of resistance of an object is determined by four factors: (1) the type of material from which the object is made, (2) the length of the object, (3) the cross-sectional area (height × width) of the object, and (4) the temperature of the object. The amount of resistance of an object is directly proportional to its length and inversely proportional to its cross-sectional area. For example, if the length of a piece of copper wire is doubled, then its resistance is also doubled (Fig. 2-12). If the cross-sectional area of the copper wire is made twice as great, then its resistance is one-half its former value. The shaded ends of the conductors in Fig. 2-12 are the cross-sectional areas of the conductors.

No simple relationships exist between resistance and temperature. The resistance of most materials increases as the temperature increases. However, with some materials, such as carbon, the resistance decreases as the temperature increases.

Temperature coefficient of resistance

2-16 Temperature Coefficient

The change in resistance corresponding to a change in temperature is known as the *temperature coefficient of resistance*. Each material has its own temperature coefficient. Carbon has a negative temperature coefficient (the resistance decreases as the temperature increases), while most metals have a positive temperature coefficient (the resistance increases as the temperature increases). A temperature coefficient is expressed as the number of ohms of change per million ohms per degree Celsius, which is abbreviated ppm/°C. For example, carbon has a negative temperature coefficient of 500 ppm°C at 20°C. That is, a piece of carbon that has 1,000,000 Ω of resistance at 20°C has 1,000,500 Ω at 19°C. At 18°C it has 1,001,000 Ω. In many electric and electronic devices, changes in resistance due to changes in temperature are so small that they can be ignored. In those devices in which small changes in resistance are important, such as electric meters, special low-temperature-coefficient materials are used. One of these materials is constantan (a mixture of copper and nickel). Constantan has a positive temperature coefficient at 18 ppm/°C at 20°C. Notice that the temperature coefficient is defined at a specific temperature. This means that the temperature coefficient itself changes with temperature.

However, these changes are extremely small over the range of temperature in which most electric devices operate. A table of temperature coefficients of resistance for some common materials is in Appendix E.

2-17 Resistivity

The characteristic resistance of a material is given by its *resistivity* or its *specific resistance*. These terms mean the same thing. The resistivity of a material is just the resistance (in ohms) of a specified-size cube of the material, usually a 1-cm, 1-m, or 1-ft cube. Resistivity ratings allow us to compare the abilities of various materials to conduct current. Figure 2-13 illustrates the way in which resistivity is determined in the base unit of ohm-centimeters ($\Omega \cdot$ cm). Annealed copper at 20°C has a resistivity of 1.72×10^{-6} (0.00000172) $\Omega \cdot$ cm. This means that a cube of copper 1 cm on each side has 1.72×10^{-6} Ω of resistance between any two opposite faces of the cube. A table of resistivity (in ohm-centimeters) is shown in Appendix D. The lower the resistivity of a material, the better a conductor it is.

The relationship of resistance to length, cross-sectional area, and resistivity is given by the formula.

$$\text{Resistance } (R) = \frac{\text{resistivity } (K) \times \text{length } (L)}{\text{area } (A)}$$

or, using only symbols,

$$R = \frac{KL}{A}$$

Resistance is in ohms if all other quantities are also in base units, that is, if resistivity is in ohm-centimeters, length in centimeters, and area in square centimeters.

A common unit used in electric wiring is the circular mil. Many wire tables give the diameter of wire in circular mils. The resistance formula can be used with this unit if K is given in ohm-circular mils per foot, L in feet, and A in circular mils.

Resistivity

Specific resistance

Example 2-4

What is the resistance, at 20°C, of an electric motor winding that uses 200 m of copper conductor which is 0.26 centimeter square (cm²) (approximately ³⁄₁₆ in. by ³⁄₁₆ in.)? The resistivity of copper at 20°C is 0.00000172 $\Omega \cdot$ cm.

Given: K = 0.00000172 ohm-centimeter
L = 20,000 centimeters
A = 0.26 square centimeter

Find: R

Known: $R = \dfrac{KL}{A}$

Solution: $R =$
$$\frac{0.00000172 \, \Omega \cdot \text{cm} \times 20,000 \text{ cm}}{0.26 \text{ cm}^2}$$
$$= 0.132 \, \Omega$$

Answer: Resistance = 0.132 ohm

$R = \dfrac{KL}{A}$

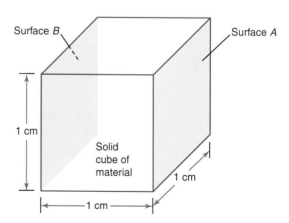

Surface B
Surface A
1 cm
Solid cube of material
1 cm
1 cm

Fig. 2-13 Resistivity equals the resistance between surface *A* and surface *B*.

About ⎓ Electronics

Alessandro Volta In 1796 Italian physicist Alessandro Volta developed the first chemical battery, which provided the first practical source of electricity. Named for Volta, the volt is the base unit of voltage.

2-18 Resistors

Many electronic devices, such as radio and television receivers, use large amounts of resistance to control the current. When we need resistance to control current, we use resistors. *Resistors* are physical devices that are manufactured in a wide variety of shapes and sizes. They are available in resistance values that range from less than 1 Ω to many millions of ohms.

In most electric and electronic devices, the resistance of the conductors is so small compared with the resistance in the other parts of the device that the resistance of the conductors can be ignored. In most cases the resistance of the conductors is not calculated or figured into the design of the device.

■ TEST_____

Answer the following questions.

30. _____ can be defined as opposition to electric current.
31. _____ energy is converted into heat energy when current flows through resistance.
32. Materials which have no free electrons are called _____ .
33. Semiconductors have _____ valence electrons.
34. The _____ is the base unit of resistance.
35. _____ is the symbol for the base unit of resistance.
36. An ohm is equal to a _____ per ampere.
37. List the four factors that determine the resistance of an object.
38. What is meant by the statement, "This material has a negative temperature coefficient of 250 ppm/°C at 20°C"?
39. Classify the following materials as either insulators or conductors:
    a. Iron   e. Aluminum
    b. Rubber   f. Glass
    c. Paper   g. Copper
    d. Silver   h. Mica
40. Give an example of a material that has a negative temperature coefficient.
41. Define *resistivity.*
42. What is a common unit used to specify resistivity?
43. What is a resistor?
44. Aluminum has a resistivity of 17.0 ohm-circular mil per foot at 20°C. What is the resistance, at 20°C, of 200 ft of aluminum wire with a cross-sectional area of 6530 circular mils?

2-19 Power and Energy

Power refers to how rapidly energy is used or converted to another form of energy. Since energy is the ability to do work, we may also say power is concerned with how fast work is done. We will combine these two ideas in our definition of power. *Power* is the *rate of using energy* or doing work. The symbol for power is *P.*

The power needed to do a specified amount of work depends on how much time it takes to do the work. Suppose two workers are going to load 100 bricks onto a truck. They agree to each load one-half of the bricks. The first worker carries five bricks at a time. He finishes his half of the work in 10 trips to the truck. He is finished in 40 minutes (4 min per trip). The second worker carries two bricks at a time, and she finishes in 25 trips. She requires 100 min (4 min per trip). Both workers have done the same amount of work (loaded 50 bricks). However, carrying five bricks requires more power than carrying two bricks; so the first worker used more power than the second worker. He, the first worker, has done more work per unit of time.

The companies from which we buy our electricity are very often named *power companies.* However, when we pay our electrical bill, we are paying for electric energy, not electric power. The electric power company is less concerned with the power we use (how rapidly we consume electric energy) than with the amount of energy we use.

2-20 Unit of Power

YOU MAY **RECALL** . . . that we use the joule as our base unit of energy and the second as our base unit of time.

Therefore, the logical unit for power is the *joule per second* (J/s). In honor of a Scottish scientist and inventor, James Watt, the joule per second has been named the *watt.* The base unit of power, therefore, is the watt. One watt

---

Resistors

Power (*P*)

Rate of using energy

Joule per second

Watt (W)

is equal to 1 J/s. The abbreviation for the watt is W. The relationship between power, energy, and time is

$$\text{Power } (P) = \frac{\text{energy } (W)}{\text{time } (t)}$$

By rearranging terms, we can state the relationship as $W = Pt$.

Let us use these relationships to solve some electrical problems. As with other symbols in electricity, the letter "W" is used to denote two entirely different things. Don't confuse W, meaning energy, for W, meaning watts.

Example 2-5

What is the power rating of an electric device that converts 940 J of energy in 10 s?

Given: $W = 940$ J and $t = 10$ s
Find: P

Known: $P = \dfrac{W}{t}$

Solution: $P = \dfrac{940 \text{ J}}{10 \text{ s}} = 94$ J/s

$\qquad\qquad = 94$ W

Answer: Power rating = 94 watts

Example 2-6

How much energy would be required to operate a 60-W lamp for 30 min?

Given: $P = 60$ W and $t = 30$ min
Find: W
Known: $W = Pt$
Solution: Since there are 60 sec in 1 min, 30 min equals
$\qquad\qquad 30 \times 60 = 1800$ s
$\qquad\qquad W = 60 \text{ W} \times 1800 \text{ s}$
$\qquad\qquad\quad = 108{,}000$ wattseconds
$\qquad\qquad\quad = 108{,}000$ J
Answer: Energy required = 108,000 joules

Notice that in example 2-6 we first converted time from minutes to the base unit of seconds. By doing this, the answer came out in joules—the base unit of energy. When you are first learning electrical relationships, it is best if you work with all quantities in base units,

whenever possible. Also notice in example 2-6 that watt × second = wattsecond. One watt-second is equal to 1 J. This can be shown by substituting joule per second for watt:

$$\text{Wattsecond} = \frac{\text{joule}}{\text{second}} \times \frac{\text{second}}{1} = \text{joule}$$

The term *wattsecond* is often used to express the amount of energy.

Since electric power companies charge their customers for energy use, electrical bills must be based on a unit of energy. The joule is much too small to be practical. Imagine getting a bill for several hundred million joules each month! Instead, power companies use the *kilowatthour*, which is equal to 3.6 million J.

W = Pt

Wattsecond

Kilowatthour

2-21 Efficiency

YOU MAY RECALL ... that, in Sec. 1-4, we dealt with the concept of efficiency. In that section we viewed efficiency in terms of useful energy out and total energy in.

Efficiency can also be viewed in terms of power. The formula is the same except power is substituted for energy:

$$\% \text{ eff.} = \frac{P_{\text{out}}}{P_{\text{in}}} \times 100$$

Example 2-7

What is the efficiency of a radio receiver that requires 4 W of electric power input to deliver 0.5 W of power output?

Given: $P_{\text{in}} = 4$ W
$\qquad\qquad P_{\text{out}} = 0.5$ W
Find: % eff.

Known: $\% \text{ eff.} = \dfrac{P_{\text{out}}}{P_{\text{in}}} \times 100$

Solution: $\% \text{ eff.} = \dfrac{0.5 \text{ W}}{4 \text{ W}} \times 100$

$\qquad\qquad\quad = 0.125 \times 100 = 12.5$

Answer: Efficiency = 12.5 percent

The efficiency formula can also be arranged to allow us to determine the amount of input power.

$$P_{\text{in}} = \frac{P_{\text{out}}}{\% \text{ eff.}} \times 100$$

Example 2-8

A stereo amplifier is producing 50 W of output power. How much power input is required if the amplifier is 30 percent efficient?

Given: $P_{out} = 50$ W

% eff. = 30

Find: P_{in}

Known: $P_{in} = \dfrac{P_{out}}{\% \text{ eff.}} \times 100$

Solution: $P_{in} = \dfrac{50 \text{ W}}{30} \times 100$

$= 1.7 \text{ W} \times 100$

$= 170$ W

Answer: Power input = 170 watts

■ TEST

Answer the following questions.

45. True or false. Power and work mean the same thing.
46. True or false. The base unit of power is the joule-second.
47. True or false. A joule per second equals one watt.
48. True or false. A wattsecond equals a joule.
49. True or false. Energy is equal to power divided by time.
50. True or false. A kilowatthour is equal to 3.6 million J.
51. An 850-W toaster requires 4 min to make a slice of toast. How much energy is required to make a slice of toast? Give your answer in joules.
52. A transistor receiver requires 6020 J to operate for 50 min. What is the power rating of the receiver?

About ◁▭▷ Electronics

Thomas Seebeck In 1921 Thomas Seebeck discovered a thermoelectric phenomenon that became known as the Seebeck effect. He found that near a closed circuit composed of two linear conductors of different metals, a magnetic needle would be deflected if the two junctions were at different temperatures. Further, if the cooler junction were to become warmer, the direction of deflection would be reversed. (*McGraw-Hill Concise Encyclopedia of Science and Technology, Second Edition*, Parker, McGraw-Hill. 1989.)

53. What is the efficiency of an electric motor that requires 1200 W of electric power to deliver 800 W of mechanical power?

2-22 Powers of 10

By now you may have noticed that some very large numbers are used in electricity. These numbers are easier to manage if they are expressed in powers of 10. *Powers of 10* refer to numbers that are exponents of the base 10. *Base 10* refers to the number system we use in our everyday life. The *exponent* (or power) refers to the number of times that the digit 1 is multiplied by 10. For example, 10^2 means that 1 is multiplied by 10 twice:

$$1 \times 10 \times 10 = 100$$

Ten to the first power (10^1) is $1 \times 10 = 10$; $10^0 = 1$. Powers of 10 and their base-10 equivalents that are commonly used in electricity are listed in Table 2-1. Notice that the power (exponent) of 10 tells you how many places to move the decimal place. For example, 10^4 can be changed to a number without an exponent by writing a 1 and moving the decimal point four places to the right. Thus, 10^4 is equal to 10,000. Also, a number can be converted to a power of 10 by making the exponent (power) equal to the number of places you move the decimal point. Thus, 2100 can be expressed as 2.1×10^3, or 105,000 can be written 1.05×10^5. Notice that it is conventional to move the decimal so that there is only one digit to the left of it. The number multiplied by the power of 10 is called the *coefficient*. In the above examples, 1.05 and 2.1 are the coefficients.

Numbers smaller than 1 are expressed in powers of 10 with negative exponents. The negative exponent tells you how many times 1 is to be divided by 10. Thus 10^{-2} is the same as $1 \div 10 \div 10 = 0.01$. Notice again that the easy way to handle exponents is by moving the decimal point as many places as the exponent. When the exponent (power) is negative, move the decimal to the left. To convert 10^{-3}, just write a 1 and move the decimal three places to the left. This procedure yields 0.001, which is equal to 10^{-3}. To convert to a negative power of 10, just move the decimal to the right of the first digit larger than zero. The negative exponent equals the number of places you moved the decimal place. For example, 0.000000054 is equal to 5.4×10^{-8}, and 0.03816 is equal to 3.816×10^{-2}.

Power of 10	Number (Base-10) Equivalent
10^{12}	1,000,000,000,000
10^{11}	100,000,000,000
10^{10}	10,000,000,000
10^{9}	1,000,000,000
10^{8}	100,000,000
10^{7}	10,000,000
10^{6}	1,000,000
10^{5}	100,000
10^{4}	10,000
10^{3}	1,000
10^{2}	100
10^{1}	10
10^{0}	1
10^{-1}	0.1
10^{-2}	0.01
10^{-3}	0.001
10^{-4}	0.0001
10^{-5}	0.00001
10^{-6}	0.000001
10^{-7}	0.0000001
10^{-8}	0.00000001
10^{-9}	0.000000001
10^{-10}	0.0000000001
10^{-11}	0.00000000001
10^{-12}	0.000000000001

TABLE 2-1 Powers of 10 and Base-10 Equivalents

The big advantage to expressing numbers in powers of 10 is that it simplifies arithmetic involving large numbers. To multiply two numbers expressed in powers of 10, just multiply the coefficients and then *add* the powers. The product is the new coefficient multiplied by the new power of 10. Some examples are

$$10^4 \times 10^2 = 10^6$$
$$10^{-2} \times 10^4 = 10^2$$
$$10^{-5} \times 10^3 = 10^{-2}$$
$$1.4 \times 10^2 \times 1.2 \times 10^6 = 1.4 \times 1.2 \times 10^2 \times 10^6$$
$$= 1.68 \times 10^8$$
$$6.3 \times 10^3 \times 8.4 \times 10^4 = 6.3 \times 8.4 \times 10^3 \times 10^4$$
$$= 52.92 \times 10^7$$
$$= 5.292 \times 10^8$$

To divide with powers of 10, first divide the coefficients. Then subtract the exponent in the divisor from the exponent in the dividend. This procedure is illustrated below:

$$(1 \times 10^4) \div (1 \times 10^2) = 1 \times 10^2$$
$$(4 \times 10^3) \div (2 \times 10^{-2}) = (4 \div 2)(10^3 \div 10^{-2})$$
$$= 2 \times 10^{3-(-2)}$$
$$= 2 \times 10^5$$
$$(6 \times 10^{-10}) \div (4 \times 10^{-8}) =$$
$$(6 \div 4)(10^{-10} \div 10^{-8})$$
$$= 1.5 \times 10^{-2}$$

To either add or subtract numbers expressed as powers of 10, both numbers must have the same exponent. The exponent then remains the same. For example:

$$(2.4 \times 10^6) + (3.5 \times 10^6) = 5.9 \times 10^6$$
$$(2.4 \times 10^4) + (3.5 \times 10^5) = (0.24 \times 10^5)$$
$$+ (3.5 \times 10^5)$$
$$= 3.74 \times 10^5$$
$$(3.8 \times 10^3) - (1.6 \times 10^3) = 2.2 \times 10^3$$

◢ TEST

Answer the following questions.

54. Express the following numbers in powers of 10:
    a. 180
    b. 42,000
    c. 2,000,000

55. Convert the following powers of 10 to ordinary numbers:
    a. 3.1×10^3
    b. 10^4
    c. 2.46×10^3

56. Convert the following powers of 10 to ordinary numbers:
    a. 10^{-4}
    b. 2.81×10^{-3}
    c. 6.3×10^{-4}
    d. 6.3×10^2

57. Convert the following numbers to powers of 10:
    a. 0.0000001
    b. 0.028
    c. 0.0072
    d. 1000

58. Solve the following problems:
    a. $2 \times 10^8 \times 4 \times 10^3$
    b. $1.4 \times 10^2 \times 2.8 \times 10^{-3}$
    c. $\dfrac{6.6 \times 10^4}{3 \times 10^{-2}}$
    d. $\dfrac{4 \times 10^{-3}}{2 \times 10^2}$
    e. $(4 \times 10^3) + (6 \times 10^4)$

Visit the Website for the National Institute of Standards and Technology (NIST) for additional information.

2-23 Multiple and Submultiple Units

Multiple and submultiple units

Prefixes

Multiples of three

For some applications of electricity, the base unit of a quantity may seem to be very large. For other applications, the same base unit may seem rather small. For example, in solid-state devices, we work with currents of less than 0.0000001 A. In an aluminum reduction plant, currents are greater than 110,000 A. Although these numbers could be shortened by expressing them in powers of 10, they would still be long expressions. Also, they would be long when spoken. For example, 1.1×10^5 A would be spoken as "one point one times 10 to the fifth amperes." To avoid such long expressions, scientists use *prefixes* to indicate units that are smaller and larger than the base unit.

The prefixes and their symbols commonly used in electricity are shown in Table 2-2.

TABLE 2-2 Prefixes and Symbols			
Prefix	Symbol	Number (Base 10)	Power of 10
Mega	M	1,000,000	10^6
Kilo	k	1000	10^3
Base unit		1	10^0
Milli	m	0.001	10^{-3}
Micro	μ	0.000001	10^{-6}
Nano	n	0.000000001	10^{-9}
Pico	p	0.000000000001	10^{-12}

Also shown in Table 2-2 are the relationships (both powers of 10 and base 10) of the prefixes to the base unit. Notice that adjacent prefixes are related by factors of 1000. Adjacent prefixes are either 1000 times larger or 1/1000 as large as their neighbor.

Multiple and submultiple units are designated by adding the appropriate prefix to the base unit. Now we can specify the 110,000 A used in an aluminum plant as 0.11 megampere (MA) or 110 kiloamperes (kA). The 0.0000001 A used in a solid-state device can be written as 0.1 microampere (μA). Some other examples of conversions between units are

$$2200 \text{ ohms } (\Omega) = 2.2 \text{ kilohms (k}\Omega)$$
$$0.083 \text{ watt (W)} = 83 \text{ milliwatts (mW)}$$
$$450,000 \text{ volts (V)} = 450 \text{ kilovolts (kV)}$$
$$2.7 \times 10^6 \text{ ohms } (\Omega) = 2.7 \text{ megohms (M}\Omega)$$
$$3700 \text{ microamperes } (\mu\text{A}) =$$
$$3.7 \text{ milliamperes (mA)}$$
$$6,800,000 \text{ ohms } (\Omega) = 6.8 \times 10^6 \text{ ohms } (\Omega)$$
$$= 6.8 \text{ megohms (M}\Omega)$$

Notice that in all the above examples the conversion is made by moving the decimal point either three places of *multiples of three* places. If you are converting from a smaller unit to a larger unit, the decimal point is moved to the left. Remember that going from *micro* to *milli* is going from a smaller to a larger unit. When converting from a larger unit to a smaller unit, you shift the decimal point to the right.

It is important that you become very familiar with converting from one unit to another. A repair manual for an electric device may list a resistor as 2.2 kΩ. When the technician goes to order a replacement resistor, the parts manufacturer may list the resistor as 2200 Ω. The technician must make the conversion.

■ TEST _____

Answer the following questions.

59. Complete the conversions listed below:
    a. 120 millivolts = _____ volt
    b. 3800 ohms = _____ kilohms
    c. 490 microamperes = _____ ampere
    d. 5.6×10^5 ohms = _____ megohm
    e. 6000 millicoulombs = _____ coulombs

James Watt The unit of electrical power, the watt, is named for James Watt. One watt equals 1 joule of energy transferred in 1 second.

About ⟨▭⟩ Electronics

60. Complete the following conversions:
    a. 53 mA = _____ A
    b. 4.7 kΩ = _____ Ω
    c. 0.4 V = _____ mV

2-24 Special Units and Conversions

In this book, standard metric units are used whenever practical. However, there are some areas of electrical work in which nonmetric units are so common that we must consider them.

As noted earlier, electric power companies sell their electric energy by the kilowatthour. A *kilowatthour* (kWh) = 1000 watthours (Wh). An hour = 3600 s. Therefore, a watthour = 3600 wattseconds. And finally, a kilowatthour = 3,600,000 wattseconds or joules. As you can see, the joule would be a small unit of energy for a power company to use. Consider, for instance, that many homes use more than 1000 kWh of energy every month. Remember that the wattsecond and watthour are perfectly usable units of energy, even though the base unit is joule.

Example 2-9

How much energy is used by a 1200-W heater in 4 hours of continuous operation?

Given:	$P = 1200$ W
	$t = 4$ hours
Find:	Energy
Known:	$W = Pt$
Solution:	1200 W × 4 hours = 4800 Wh
Answer:	Energy used = 4800 watthours or 4.8 kilowatthours

The output power of an electric motor (and many mechanical devices) is specified in *horsepower* (hp) rather than in watts. One horsepower is equal to 746 W. When figuring the efficiency of a motor, you must first convert the output power to watts.

Horsepower (hp)

hp = 746 W

Example 2-10

What is the efficiency of a ¾-hp motor that requires an input of 1000 W of electric power?

Given:	Power input = 1000 W
	Power output = ¾ hp
Find:	Efficiency
Known:	% eff. = $\dfrac{P_{out}}{P_{in}} \times 100$
	1 hp = 746 W
	¾ hp = 0.75 hp
Solution:	$P_{out} = 0.75 \text{ hp} \times 746 \dfrac{\text{W}}{\text{hp}}$
	= 559.5 W
	% eff. = $\dfrac{559.5 \text{ W}}{1000 \text{ W}} \times 100$
	= 55.95
Answer:	Efficiency = about 56 percent

Kilowatthour (kWh)

■ TEST_____

Answer the following questions.

61. Name two ways to condense a large value of an electrical quantity, such as 3,600,000 ohms.
62. What is horsepower?
63. How much power is required by a ½-hp motor that is 62 percent efficient?

Summary

1. Charge is the electrical property of electrons and protons.
2. One coulomb (C) is the charge possessed by 6.25×10^{18} electrons.
3. Current is the movement of charge in a specified direction.
4. Current can flow in solids, gases, liquids, and a vacuum.
5. In solids the current carriers are electrons.
6. In gases the current carriers are both electrons and ions.
7. In liquids and current carriers are ions—both positive and negative.
8. Current travels at approximately the speed of light.
9. Individual current carriers (electrons) travel much slower than the speed of light.
10. Direct current (dc) never reverses direction.
11. Alternating current (ac) periodically reverses direction.
12. A liquid containing ions is an electrolyte.
13. An ampere (A) is 1 coulomb per second (C/s).
14. Voltage is a potential energy difference between two points.
15. Energy = voltage × charge
16. Polarity indicates whether a point is negative or positive.
17. Sources of voltage include generators, batteries, thermocouples, solar cells, and crystals.
18. Resistance is opposition to current.
19. Resistance converts electric energy to heat energy.
20. Conductors are materials that have low resistance.
21. Silver, copper, and aluminum (in that order) are the best conductors.
22. Insulators do not allow any current to flow (for practical purposes).
23. An ohm (Ω) is 1 volt per ampere (V/A). It is the resistance of a specified column of mercury at a specified temperature.
24. The resistance of an object is determined by the resistivity, length, cross-sectional area, and temperature of the material.
25. Resistance is directly proportional to length.
26. Resistance is inversely proportional to cross-sectional area.
27. Most conductors have a positive temperature coefficient.
28. A temperature coefficient of resistance specifies the number of ohms of change per million ohms per degree Celsius (abbreviated ppm/°C).
29. Resistivity (specific resistance) of a material is the resistance of a specified-size cube of the material.
30. Resistors are devices used to control current.
31. Power is the rate of doing work or converting energy.
32. A watt (W) is 1 joule per second (J/s).
33. Energy = power × time
34. % eff. $= \dfrac{\text{power output}}{\text{power input}} \times 100$
35. A kilowatthour (kWh) is a unit of energy.
36. Horsepower is a unit of power.
37. 1 horsepower (hp) = 746 watts (W)
38. Table 2-3 is a summary of the units and symbols used throughout this book.

TABLE 2-3	Symbols and Abbreviations		
Quantity	**Symbol**	**Base Unit**	**Abbreviation**
Charge	Q	Coulomb	C
Current	I	Ampere	A
Time	t	Second	s
Voltage	V	Volt	V
Energy	W	Joule	J
Resistance	R	Ohm	Ω

Chapter Review Questions

For questions 2-1 to 2-11, determine whether each statement is true or false.

2-1. Neutrons can be current carriers in an electric circuit.

2-2. One wattsecond is equal to 1 joule.

2-3. 5 microamperes = 0.005 ampere

2-4. Semiconductors generally have five valence electrons.

2-5. The liquid in a battery is called an electrolyte.

2-6. A potential energy difference between two points is called power.

2-7. Copper is a better conductor than aluminum.

2-8. $3.2 \times 10^3 = 3200$

2-9. $6 \times 10^8 \times 3 \times 10^{-6} = 1.8 \times 10^3$

2-10. 0.420 watts = 42 milliwatts

2-11. Most conductors have a negative temperature coefficient.

For questions 2-12 to 2-18, choose the letter that best completes each sentence.

2-12. A joule per second defines a
   a. Volt
   b. Ampere
   c. Watt
   d. Ohm

2-13. A coulomb per second defines a
   a. Volt
   b. Ampere
   c. Watt
   d. Ohm

2-14. Electric energy is converted into heat energy by
   a. Voltage
   b. Charge
   c. Power
   d. Resistance

2-15. Charge has base units of
   a. Coulombs
   b. Watts
   c. Protons
   d. Amperes

2-16. Voltage has base units of
   a. Ohms
   b. Amperes
   c. Watts
   d. Volts

2-17. Resistance has base units of
   a. Ohms
   b. Amperes
   c. Watts
   d. Volts

2-18. Current has base units of
   a. Ohms
   b. Amperes
   c. Watts
   d. Volts

Answer the following questions.

2-19. What are the four factors that determine the resistance of an object?

2-20. What is the symbol for each of the following terms?
   a. Volt   e. Watt
   b. Ampere   f. Charge
   c. Resistance   g. Milliampere
   d. Ohm   h. Horsepower

2-21. How much energy does a motor use in 6 hours if it requires 800 watts to operate the motor?

2-22. What is the percent efficiency of a 1-horsepower motor that requires 1.1 kilowatts to operate?

2-23. What is the power rating of a television receiver that uses 270,000 joules to operate for 1.5 hours?

2-24. An electronic amplifier produces 60 W of output power. If it is 53 percent efficient, how much power input power does it require?

2-25. A 12.6-V battery forces 3 coulombs through a load. How much energy does the load convert?

2-26. A lamp requires 0.167 kilowatthour to operate for 40 minutes. What is the power rating of the lamp?

2-27. How many seconds is required by a 120-W device to convert 2 Wh of energy?

Critical Thinking Questions

2-1. Explain why efficiency can be calculated using either P_{out} and P_{in} or W_{out} and W_{in}.

2-2. Why does the temperature of a solid conductor increase when the conductor is carrying current?

2-3. Why does a long conductor have more resistance than a short conductor of the same material, cross-sectional area, and temperature?

2-4. Which base unit is represented by the following:

$$\frac{\text{Newton-meter} \times \text{ampere}}{\text{Coulomb}}$$

Explain how you arrived at your answer

2-5. Prove that a joule per coulomb is equal to a watt per ampere.

2-6. Do you think horsepower is as meaningful a unit of power as the watt? Why?

2-7. What is the power rating of a light that is 20 percent efficient and produces 27 Wh of light energy in 18 minutes? Show the conversions and formulas you used.

Answers to Tests

1. An electric charge is the electrical property exhibited by protons and electrons.
2. Q
3. 6.25×10^{18}
4. C
5. electron, ion
6. electron
7. F
8. F
9. F
10. The electron travels from atom to atom. The individual electron moves down the wire slowly.
11. Alternating current periodically reverses its direction, while direct current flows continuously in the same direction.
12. ampere
13. An ampere is equal to 1 coulomb per second.
14. A
15. a. $I = 8A$
    b. $Q = 6$ C
16. 4 A
17. voltage
18. volt
19. V
20. Electromotive force
21. V
22. copper sulfate
23. Kinetic energy is energy being

converted or energy in use. Potential energy is energy at rest or stored energy.
24. Because potential energy differences can exist only between two points. The reference point (floor) must be specified.
25. Because voltage is a potential energy difference. This means that voltage always exists between two points.
26. The volt is equal to 1 joule per coulomb.
27. Polarized means that an electric device has a positive and a negative terminal.
28. battery—chemical energy crystal (microphone, cartridge)—mechanical energy generator—mechanical energy thermocouple—heat energy solar cell—light energy
29. 20 V
30. Resistance
31. Electric
32. insulators
33. four
34. ohm
35. Ω
36. volt
37. resistivity of the material from which the object is

made, length of the object, cross-sectional area of the object, and temperature of the object.
38. The statement means that the resistance of the material decreases 250 ohms for every million ohms for each degree Celsius that the temperature increases above 20°C.
39. conductors: iron, silver, aluminum, and copper; insulators: rubber, paper, glass, and mica
40. carbon
41. Resistivity is the characteristic resistance of a material. It is the resistance of a specified-size cube of the material. The cube is usually defined as being either 1 cm, 1 m, or 1 ft on each side.
42. ohm-centimeter
43. A resistor is a physical device designed to provide a controlled amount of resistance.
44. 0.52 Ω
45. F
46. F
47. T
48. T
49. F
50. T

51. **Given:** $P = 850$ W, $t = 4$ min
    Find: W
    Known: $W = Pt$
    60 seconds = 1 min

    Solution: $t = 4$ min $\times 60 \dfrac{\text{s}}{\text{min}} = 240$ s

    $W = 850 \times 240 = 204{,}000$ J

    Answer: Energy = 204,000 joules

52. **Given:** $W = 6020$ J, $t = 50$ min
    Find: P

    Known: $P = \dfrac{W}{t}$

    Solution: $t = 50$ min $\times 60 \dfrac{\text{s}}{\text{min}} = 3000$ s

    $P = \dfrac{6020}{3000} = 2$ W

    Answer: Power = 2 watts

53. **Given:** $P_{in} = 1200$ W, $P_{out} = 800$ W
    Find: % eff.

    Known: % eff. $= \dfrac{P_{out}}{P_{in}} \times 100$

    Solution: % eff. $= \dfrac{800}{1200} \times 100 = 66.7$

    Answer: % eff. = 66.7%

54. a. 1.8×10^2
    b. 4.2×10^4
    c. 2×10^6

55. a. 3100

    b. 10,000
    c. 2460

56. a. 0.0001
    b. 0.00281
    c. 0.00063
    d. 630

57. a. 1×10^{-7} or 10^{-7}
    b. 2.8×10^{-2}
    c. 7.2×10^{-3}
    d. 1×10^3 or 10^3

58. a. 8×10^{11}
    b. 3.92×10^{-1}
    c. 2.2×10^6
    d. 2×10^{-5}
    e. 6.4×10^4

59. a. 0.12 V
    b. 3.8 kΩ
    c. 0.00049 A
    d. 0.56 MΩ
    e. 6 C

60. a. 0.053 A
    b. 4700 Ω
    c. 400 mV

61. By using powers of 10 (3.6×10^6 ohms) or by using multiple units (3.6 megohms)

62. Horsepower is a nonmetric unit of power. It is equal to 746 watts.

63. 601.6 W

Chapter 3

Basic Circuits, Laws, and Measurements

Chapter Objectives

This chapter will help you to:

1. *Understand* the relationship between schematic diagrams and physical circuits.
2. *Use* Ohm's law to calculate the current, voltage, and resistance in simple electric circuits.
3. *Calculate* the power of a circuit when any two of the three quantities voltage, current, and resistance are known or can be determined.
4. *Calculate* the cost of operating an electric device for a specified length of time.
5. *Measure* the current, voltage, and resistance in electric circuits without damaging the meter or circuit.
6. *Understand* the relationship between scales and ranges on multi-scale, multi-range meters.

You are now familiar with electrical quantities and units. Now you are ready to explore the circuits, laws, and devices used to control and measure these quantities and units.

3-1 Circuit Essentials

Most complete electric circuits contain six parts:

1. An *energy source* to provide the voltage needed to force current (electrons) through the circuit
2. *Conductors* through which the current can travel
3. *Insulators* to confine the current to the desired paths (conductors, resistors, etc.)
4. A *load* to control the amount of current and convert the electric energy taken from the energy source
5. A *control device,* often a switch, to start and stop the flow of current
6. A *protection device* to interrupt the circuit in case of a circuit malfunction

Simple circuit

The first four of the above six parts are essential parts. All complete circuits use them. The control device (item 5) is occasionally omitted. Protection devices (item 6) are often omitted

from circuits. A complete electric circuit has an uninterrupted path for current (electrons) to flow from the negative terminal of the energy source through the load and control device to the positive terminal of the energy source.

The simplest electric circuit contains only one load, one voltage course, and one control device. It is sometimes referred to as a *simple circuit* to distinguish it from more complex circuits. A single-cell flashlight is an example of a simple circuit. Figure 3-1(*a*) shows (in cross section) the construction details of such a flashlight. Electron flow (current) in the flashlight circuit can be traced by referring to Fig. 3-1. Electrons leave the negative end of the cell, travel through the spring, the metal case, the leaf spring of the switch, the metal reflector, and the light bulb, and back to the positive end of the cell. Note that the spring, case, and reflector are conductors for the flashlight circuit. It is quite common in electric devices for structural parts of the device to serve also as con-

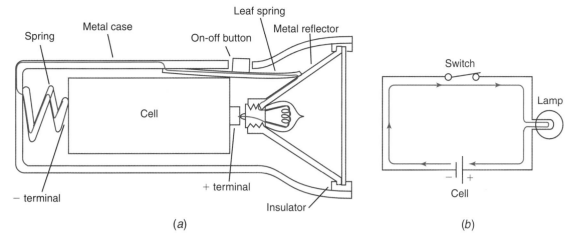

Fig. 3-1 Flashlight. (*a*) Cross-sectional drawing. (*b*) Schematic diagram.

ductors. For example, both an automobile frame and the chassis of many radio receivers serve as circuit conductors.

3-2 Circuit Symbols and Diagrams

In describing an electric circuit, you will find it more convenient to use symbols to represent electric components than to draw pictures of the components. A resistor and the symbol used to represent it are shown in Fig. 3-2. This same symbol is used for all fixed resistors regardless of the material from which they are made. Other common electric components and their symbols are shown in Fig. 3-3. It should be noticed that there is no symbol to distinguish insulated from noninsulated conductors. Insulation is assumed to be present wherever it is needed to keep components and conductors from making electric contact. In constructing an electric circuit that is described by electrical symbols, the circuit builder must determine where insulation is needed.

A drawing that uses only symbols to show how components are connected together is called a *schematic diagram*. A schematic diagram shows only how the parts are electrically interconnected. The physical size and the mechanical arrangements of the parts are in no way indicated. Also, accessories, such as battery or lamp holders, are not indicated. Unless the schematic diagram is accompanied by some form of pictorial drawing, the physical arrangement of the electric components is left to the discretion of the circuit builder.

A schematic diagram of the flashlight illustrated in Fig. 3-1(*a*) is shown in Fig. 3-1(*b*). Current can also be traced in the schematic diagram. The line with the arrowheads indicates the direction of current flow. The electrons (current) flow from the negative end of the cell through the closed switch and the lamp and back to the positive end of the cell. The conductor between the negative terminal of the cell and the switch in Fig. 3-1(*b*) represents the spring and the metal case in Fig. 3-1(*a*). The contact between the positive end of the cell and the light bulb in Fig. 3-1(*a*) is represented by the long line in the schematic diagram. Notice that lines in a schematic do not neces-

Schematic diagram

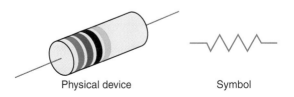

Fig. 3-2 Resistor and its symbol.

Electrical values

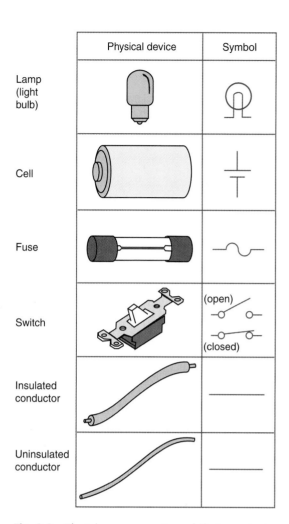

Physical device		Symbol
Lamp (light bulb)		
Cell		
Fuse		
Switch		(open) (closed)
Insulated conductor		
Uninsulated conductor		

Fig. 3-3 Electric components and their symbols.

Common ground

sarily indicate a wire, but they do indicate a path for current to flow through.

The *common ground symbol* shown in Fig. 3-4 is often used in schematic diagrams—especially in schematic diagrams for complex circuits and systems. The symbol does not represent any specific electric component. Rather, it represents a common electric point in an electric circuit (or an electric system) which is a common connecting point for many components. For example, the metal frame or chassis of an automobile is the common ground for the many electric circuits used in the automobile. Usually, the negative terminal of the car battery is connected only to the frame or chassis. Then, any circuit that needs an electric connection to the negative terminal of the battery can be physically (and electrically) connected to any convenient spot on the frame or chassis. This idea is illustrated in Fig. 3-4(*a*), where the two ground symbols tell us that there is a conductive path between them. Figure 3-4(*b*)

shows another way to indicate that a circuit is connected to a common ground.

The electrical values of the components used in the circuit can also be included on the schematic diagram. This is done in one of two ways. In the first method, shown in Fig. 3-5(*a*), the values of the components are printed beside the symbols for the components. In the second method, shown in Fig. 3-5(*b*), an identifying letter or symbol is printed beside each component and the values of the components are given in an accompanying parts list.

■ TEST_____

Answer the following questions.

1. What are the six parts of a complete circuit?
2. True or false. Component values are always given on a schematic diagram.
3. True or false. The conductors in an electric circuit are always insulated wire conductors.
4. Draw the symbol for a lamp, a resistor, and a conductor.
5. What is a schematic diagram?

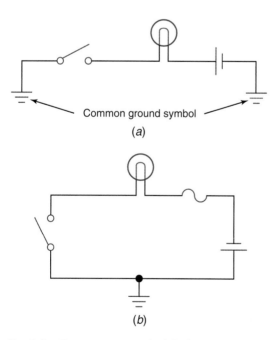

Common ground symbol

(*a*)

(*b*)

Fig. 3-4 Common grounds. (*a*) There is a conductive path between the common grounds. (*b*) The switch and the positive terminal are connected to the common ground.

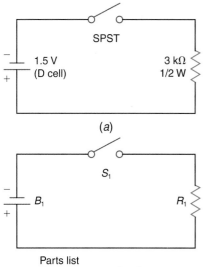

SPST

1.5 V
(D cell)

3 kΩ
1/2 W

(a)

S_1

B_1

R_1

Parts list
R_1 Resistor, 3 kΩ, 1/2 W
S_1 Switch, SPST
B_1 Cell, 1.5 V, size D

(b)

Fig. 3-5 Specifying component values. (a) Values given on the diagram. (b) Values given in a parts list.

6. True or false. The chassis or frame of a device can serve as a conductor for more than one circuit.

7. True or false. A schematic diagram is used to show both the electrical and the mechanical layout of a circuit.

3-3 Calculating Electrical Quantities

... that in Chaps. 1 and 2 you learned how to use relationships between electrical quantities to calculate other electrical quantities.

In this section we work with more relationships of electrical quantities. These new relationships emphasize quantities which can be easily measured or are commonly specified by manufacturers of electrical products.

Ohm's Law

The relationship between current (*I*), voltage (*V*), and resistance (*R*) was discovered by a German scientist named Georg Ohm. This relationship is named *Ohm's law* in his honor. Ohm found that the current in a circuit varies directly with the voltage when the resistance is

kept constant. While keeping the resistance constant, Ohm varied the voltage across the resistance and measured the current through it. In each case, when he divided the voltage by the current, the result was the same. In short, this is Ohm's law, which can be stated as, "The current is directly proportional to the voltage and inversely proportional to the resistance."

Written as a mathematical expression, Ohm's law is

$$\text{Resistance } (R) = \frac{\text{voltage } (V)}{\text{current } (I)} \quad \text{or} \quad R = \frac{V}{I}$$

$R = \dfrac{V}{I}$

The above equation allows you to determine the value of the resistance when the voltage and the current are known.

Of course, Ohm's law can be rearranged to solve for either current or voltage. The rearranged relationships are

$$\text{Current } (I) = \frac{\text{voltage } (V)}{\text{resistance } (R)} \quad \text{or} \quad I = \frac{V}{R}$$

$I = \dfrac{V}{R}$

and

$$\text{voltage } (V) = \text{current } (I) \times \text{resistance } (R)$$

or

$$V = IR$$

$V = IR$

An aid to remembering the Ohm's law relationships is shown in the divided circle of Fig. 3-6. To use the aid, just cover the quantity you want to find and perform the multiplication or

Ohm's law

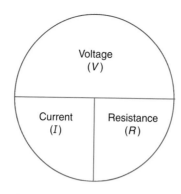

Voltage
(V)

Current
(I)

Resistance
(R)

Fig. 3-6   Ohm's law circle.

division indicated. Cover the *V* of Fig. 3-6, and the remainder of the circle indicates *I* times *R*. Thus, voltage (*V*) equals current (*I*) times resistance (*R*). Cover the *R*, and the remainder of the circle shows voltage (*V*) divided by the current (*I*). Finally, if the (*I*) is covered, the indicated operation is to divide the voltage (*V*) by resistance (*R*).

Example 3-1

How much current (*I*) flows in the circuit shown in Fig. 3-7?

Given: Voltage (*V*) = 2.8 volts (V)
Resistance (*R*) = 1.4 kilohm
(1.4 kΩ)

Find: Current (*I*)

Known: $I = \dfrac{V}{R}$, 1.4 kΩ = 1400 Ω

Solution: $I = \dfrac{2.8\ V}{1400\ \Omega} = 0.002$ ampere (A)

Answer: The current is 0.002 A.

Example 3-2

A lamp has a resistance of 96 ohms. How much current flows through the lamp when it is connected to 120 volts?

Given: $R = 96\ \Omega$
$V = 120\ V$

Find: *I*

Known: $I = \dfrac{V}{R}$

Solution: $I = \dfrac{120\ V}{96\ \Omega} = 1.25\ A$

Answer: The current through the lamp equals 1.25 A.

Notice in the preceding examples that the answers for the current are in their base units. This is because both voltage and resistance are also in base units. Remember that an ohm is defined as 1 volt per ampere. Therefore, Ohm's law expressed in base units is

$$1\ \text{ampere} = \frac{1\ \text{volt}}{1\ \text{volt/ampere}}$$

This expression reduces to 1 ampere = 1 ampere, which shows that proper units were used.

Example 3-3

The manufacturer specifies that a certain lamp will allow 0.8 ampere of current when 120 volts is applied to it. What is the resistance of the lamp?

Given: Current (*I*) = 0.8 ampere (A)
Voltage (*V*) = 120 volts (V)

Find: Resistance (*R*)

Known: $R = \dfrac{V}{I}$

Solution: $R = \dfrac{120\ V}{0.8\ A} = 150$ ohms (Ω)

Answer: The resistance of the lamp is 150 Ω.

Example 3-4

How much voltage is required to cause 1.6 amperes in a device that has 30 ohms of resistance?

Given: $R = 30\ \Omega$
$I = 1.6\ A$

Find: *V*

Known: $V = IR$

Solution: $V = 1.6\ A \times 30\ \Omega = 48\ V$

Answer: The voltage applied to the device must be 48 V.

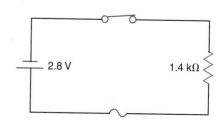

2.8 V   1.4 kΩ

Fig. 3-7   Circuit diagram for example 3-1.

Example 3-5

The current flowing through a 10-kΩ resistor is 35 mA. What is the potential energy difference (voltage) across the resistor?

Given: $R = 10\text{ k}\Omega = 10,000\ \Omega$
$I = 35\text{ mA} = 0.035\text{ A}$
Find: V
Known: $V = IR$
Solution: $V = 0.035\text{ A} \times 10,000\ \Omega$
$= 350\text{ V}$
Answer: The voltage across the resistor is 350 V.

Calculating Power

 YOU MAY **RECALL** . . . that in Chap. 2 we learned how to find the amount of power when energy and time were known.

Now we are going to work with the relationship between current, voltage, and power. Since current and voltage are easily measured quantities, you will be using this relationship quite often in electrical work.

Power is equal to current times voltage. Expressed as a formula, we have

Power (P) = current (I) × voltage (V)

or

$$P = IV$$

Power is in its base unit of watts when voltage is in volts and current is in amperes.

Example 3-6

What is the power input to an electric heater that draws 3 amperes from a 120-volt outlet?

Given: Current (I) = 3 amperes (A)
Voltage (V) = 120 volts (V)
Find: Power (P)
Known: $P = IV$
Solution: $P = 3\text{ A} \times 120\text{ V}$
$= 360\text{ watts (W)}$
Answer: The power input to the electric heater is 360 W.

By rearranging the power formula, we can solve for current when power and voltage are known. The formula is

Current $(I) = \dfrac{\text{power }(P)}{\text{voltage }(V)}$ or $I = \dfrac{P}{V}$ $I = \dfrac{P}{V}$

This formula is used to find the current in a conductor feeding a load of specified power.

The voltage needed to provide a specified current and power can be found using:

Voltage $(V) = \dfrac{\text{power }(P)}{\text{current }(I)}$ $V = \dfrac{P}{I}$

or

$$V = \frac{P}{I}$$

Example 3-7

How much current flows through a 120-volt, 500-watt lamp?

Given: Voltage (V) = 120 volts (V)
Power (P) = 500 watts (W)
Find: Current (I)
Known: $I = \dfrac{P}{V}$
Solution: $I = \dfrac{500\text{ W}}{120\text{ V}} = 4.17\text{ A}$
Answer: The current flowing through the lamp is 4.17 A.

Example 3-8

The heating element in a clothes dryer is rated at 4 kilowatts (kW) and 240 volts (V). How much current does it draw? $P = IV$

Given: $P = 4\text{ kW} = 4000\text{ W}$
$V = 240\text{ V}$
Find: I
Known: $I = \dfrac{P}{V}$
Solution: $I = \dfrac{4000\text{ W}}{240\text{ V}} = 16.7\text{ A}$
Answer: The current drawn by the clothes dryer is 16.7 A.

Using Ohm's law and the power formula, we can find the power when resistance and current or voltage are known.

Example 3-9

Find the power used by the resistor in the circuit shown in Fig. 3-8.

Given: $V = 1.5$ V
$R = 10\ \Omega$

Find: P

Known: $P = IV,\ I = \dfrac{V}{R}$

Solution: $I = \dfrac{1.5\text{ V}}{10\ \Omega} = 0.15$ A

$P = 0.15\text{ A} \times 1.5\text{ V} = 0.225$ W

Answer: The power used (dissipated) by the resistor is 0.225 W.

The procedure used in example 3-9 is a two-step process. If Ohm's law and the power formula are combined, the procedure can be shortened to one step. The combining of Ohm's law and the power formula yields two formulas. From Ohm's law we know

$$I = \frac{V}{R}$$

Substituting for I in the basic power formula ($P = IV$), we find that

$$P = \frac{V}{R} \times V$$

$$P = \frac{V^2}{R}$$

Thus, we can solve for the power if we know the voltage and the resistance.

From Ohm's law we also know

$$V = I \times R$$

Again, substituting for V in the basic power formula, we see that

$$P = I \times I \times R$$
$$P = I^2R$$

$P = \dfrac{V^2}{R}$

$P = I^2R$

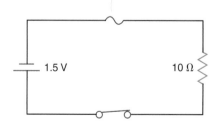

Fig. 3-8 Circuit diagram for examples 3-9, 3-10, and 3-13.

Thus, we can solve for the power, given the current and the resistance.

Example 3-10 is solved using one of the above formulas.

Example 3-10

Find the power dissipated (used) by the resistor in Fig. 3-8 using the appropriate power formula.

Given: $V = 1.5$ V
$R = 10\ \Omega$

Find: P

Known: $P = \dfrac{V^2}{R}$

Solution: $P = \dfrac{(1.5)^2}{10} = \dfrac{1.5 \times 1.5}{10}$
$= 0.225$ W

Answer: The power dissipated by the resistor is 0.225 W.

Example 3-11

How much power is dissipated when 0.2 ampere of current flows through a 100-ohm resistor?

Given: $R = 100\ \Omega$
$I = 0.2$ A

Find: P

Known: $P = I^2R$

Solution: $P = 0.2^2 \times 100$
$= 0.2 \times 0.2 \times 100$
$= 0.04 \times 100$
$= 4$ W

Answer: The resistor uses 4 W of power.

Calculating Energy

YOU MAY RECALL . . . from Chap. 2 that energy is equal to power times time.

You have just learned that power is equal to current times voltage. Thus, energy can also be determined by knowing current, voltage, and time.

Energy can also be determined if resistance and voltage or current are known. Of course, it is always necessary to know the amount of time the power is being used.

Example 3-12

How much energy is converted by a device that draws 1.5 amperes from a 12-volt battery for 2 hours?

Given: $I = 1.5$ A
$V = 12$ V
$t = 2$ hours (h)
Find: Energy (W)
Known: $W = Pt$, $P = IV$
Solution: $P = 1.5$ A $\times$ 12 V $= 18$ W
$W = 18$ W $\times$ 2 h $= 36$ Wh
Answer: The energy is 36 Wh.

Example 3-13

In Fig. 3-8, how much energy is taken from the battery by the resistor if the switch remains closed for 30 minutes?

Given: $V = 1.5$ V, $R = 10$ Ω
$t = 30$ min
Find: W
Known: $W = Pt$
$P = \dfrac{V^2}{R}$
1 joule = 1 wattsecond
1 min = 60 s
Solution: $P = \dfrac{(1.5)^2}{10} = \dfrac{2.25}{10}$
$= 0.225$ W
$W = 0.225$ W $\times$ 1800 s
$= 405$ Ws
$= 405$ joules (J)
Answer: The energy drawn from the cell is 405 J.

Calculating Cost

The cost of electric energy can be determined from the amount of energy used and the cost rate. The cost rate is usually specified in *cents per kilowatthour.* Specifying the cost rate of energy in cents per kilowatthour is like specifying the cost rate of gasoline in cents per gallon. The cost of anything is equal to the total quantity times the cost rate. If potatoes cost 10¢ per pound (cost rate) and you buy 10 pounds, the cost is $1. That is, the cost is equal to the rate times the quantity. The cost of electric energy is

Cost = rate $\times$ energy
= cents per kilowatthour
$\times$ kilowatthours

Cost = rate $\times$ energy

Example 3-14

What is the cost of 120 kWh of energy if the rate is 6¢ per kWh?

Given: $W = 120$ kWh
Rate = 6¢ per kWh
Find: Cost
Known: Cost = rate $\times$ energy
Solution: Cost = 6¢ per kWh $\times$ 120 kWh
$= 720$¢ $= \$7.20$
Answer: The cost is $7.20.

Example 3-15

What is the cost of operating a 100-watt lamp for 3 hours if the rate is 6¢ per kWh?

Given: $P = 100$ W
$t = 3$ h
Rate = 6¢ per kWh
Find: Cost
Known: Cost = rate $\times$ energy, $W = Pt$
Solution: $W = 100$ W $\times$ 3 h
$= 300$ Wh $= 0.3$ kWh
Cost = 6¢ per kWh $\times$ 0.3 kWh
$= 1.8$¢
Answer: It costs 1.8¢ to operate the lamp for 3 hours.

Example 3-16

An electric iron operates from a 120-volt outlet and draws 8 amperes of current. At 9¢ per kWh, how much does it cost to operate the iron for 2 hours?

Given: $V = 120$ V
$I = 8$ A
$t = 2$ h
Rate = 9¢ per kWh
Find: Cost
Known: Cost = rate $\times$ energy
$W = Pt$, $P = IV$
Solution: $P = 8$ A $\times$ 120 V $= 960$ W
$W = 960$ W $\times$ 2 h $= 1920$ Wh
$= 1.92$ kWh
Cost = 9¢ per kWh $\times$ 1.92 kWh
$= 17.3$¢
Answer: The cost is 17.3¢.

Cents per kilowatthour

Notice the order in which the "Known" in example 3-16 is listed. The procedure outlined below was used in developing the order in which the "Known" formulas are listed.

1. Write the formula needed to solve for the quantity to be found. Look at the quantities to the right of the equals sign in this formula. If one of these quantities is not listed in the "Given" row, write the formula needed to find it ($W = Pt$).
2. Look at the right-hand half of the formula just written ($W = Pt$). If there is a quantity there that is not listed in the "Given," then write the formula needed to find it ($P = IV$).
3. Look at the right-hand half of ($P = IV$). The "Given" lists both (I and V). The problem can now be solved.

The above procedure should always be used to reduce complex problems into a series of simple steps.

■ TEST_____

Answer the following questions.

8. Express the answer to example 3-6 in kilowatts.
9. Express the answer to example 3-10 in milliwatts.
10. Refer to Fig. 3-5(*a*). How much current flows when the switch is closed?
11. What is the resistance of a semiconductor device that allows 150 milliamperes of current when 600 millivolts of voltage is connected to it?
12. What is the power of a hair dryer that operates from 120 volts and draws 4.5 amperes of current?
13. How much current is required by a 1000-watt toaster that operates from a 120-volt outlet?
14. What is the power rating of an automobile tape deck that requires 2.2 amperes from a 12.6-volt battery?
15. A miniature lamp has 99 ohms of resistance when lit. It operates from a 6.3-volt battery. What is the power rating of the lamp?
16. An electric heater draws 8 amperes from a 240-volt source. How much energy does it convert in 9 hours?
17. The current through a 100-ohm resistor is 200 milliamperes. How much energy does the resistor convert to heat in 10 minutes?
18. An electric water heater has a 3000-watt heating element. The element is on for 3 hours. What is the cost if the rate is 4¢ per kilowatthour?
19. A string of Christmas tree lights draws 0.5 amperes from a 120-volt outlet. At 8¢ per kilowatthour, how much does it cost to operate the lights for 40 hours?

3-4 Measuring Electrical Quantities

Most electrical quantities are measured with a device called a *meter*. Voltage is measured with a *voltmeter*, current with an *ammeter*, resistance with an *ohmmeter*, and power with a *wattmeter*.

Panel Meters

A meter that measures only one of the above quantities is called a *panel meter*. Panel meters are often permanently connected (wired) into a circuit to provide continuous monitoring of an electrical quantity. The meters may be either analog or digital.

Analog Panel Meter An analog panel meter is shown in Fig. 3-9. With this meter, the needle can point anywhere on the scale. When an analog meter is read, the reading is generally taken to the nearest minor-division mark. If the pointer is halfway between marks, it is read as

Fig. 3-9 Panel meter. *(Courtesy of Triplett Corporation.)*

Voltmeter

Ammeter

Ohmmeter

Wattmeter

Panel meter

a half-division. Before reading a meter scale, you must figure out the value of each division of the scale. Look at the scale on the meter in Fig. 3-9. Notice there is a heavy line halfway between 0 and 10. This heavy line represents five units. Now count the number of divisions (called *minor divisions*) between 0 and 5. Since there are five minor divisions between 0 and 5, each minor division has a value of one unit. Therefore, each minor division on the scale represents 1 microampere. Suppose the needle (pointer) on the meter in Fig. 3-9 is pointing to the second mark to the right of the 30 mark. The meter would be indicating 32 microamperes.

Digital Panel Meter The digital panel meter (DPM) shown in Fig. 3-10 eliminates the need to decide which mark is closest to the pointer. There is no guesswork in trying to decide if the meter reading is 1.999 or 1.998.

Digital meters are usually specified by the number of digits in their readout. When the most significant (left-most) digit can be only a 0 or a 1, it is counted as a *half-digit*. The 2-volt DPM in Fig. 3-10 is a 3½-digit meter. Even though it is called a 2-volt meter, it can measure a maximum of 1.999 volts.

Multimeters

Often a single meter serves as a *voltmeter,* an *ammeter,* and an *ohmmeter.* Meters which are capable of measuring two or more electrical quantities are known as *multimeters.* Multimeters use the same basic mechanism to indicate the amount of a quantity as panel meters. However, the multimeter also includes some circuitry (switches, resistors, etc.) inside its housing or case.

Fig. 3-10 Digital panel meter. *(Courtesy of Data Technology Corporation.)*

Analog Multimeter Like most analog multimeters, the meter in Fig. 3-11 has multiple scales printed on its face and can measure current, voltage, and resistance. An analog meter that can measure these three quantities is often called a *volt-ohm-milliammeter (VOM).*

Although the various analog multimeters may look different, all of them have *functions, ranges,* and *scales. Function* refers to the *quantity being measured. Range* refers to the *amount* of the quantity that can be measured. Which *scale* of the meter is used depends on both the function and the range to which the meter is set. Proper use of the multimeter involves selection of the correct function, range, and scale. Once you understand the relationships between (1) function and scale and (2) range and scale, using any multimeter is possible.

The VOM (multimeter) in Fig. 3-11 has five functions (ac voltage, dc voltage, dc current,

Volt-ohm-milliammeter (VOM)

Functions

Ranges

Scales

Half-digit

Multimeters

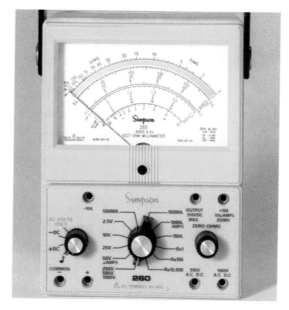

Fig. 3-11 Multimeter (VOM). *(Courtesy of Simpson Electric Company.)*

resistance, and continuity). The function switch (on the left side of the meter) has four positions. One position (AC VOLTS ONLY) is used for measuring ac voltage. Two positions (− DC and + DC) are used for measuring dc voltage, dc current, and resistance. Which quantity is being measured in either of these two positions is determined by the position of the range switch (located in the center of the meter). The fourth position (note symbol) is used for checking continuity. In this position, a tone is emitted whenever there is a low-resistance path for current between the test-probe tips.

Except for the resistance function, the range indicates the *maximum* amount of a quantity that can be measured on a given range setting. For the ohm function, the range indicates the amount by which the ohm scale is to be multiplied.

The ohm scale on the meter in Fig. 3-11 is different from most voltage and current scales in three ways. First, it is reverse-reading. Second, it is nonlinear. Third, the number of minor divisions between the heavy lines (and the numbered lines) is not the same throughout the scale. Therefore, the value of a minor division varies across the scale.

Example 3-17

The meter in Fig 3-11 is set to the R × 100 range. The needle points to the second mark to the left of the 20 mark. How much resistance is the meter indicating?

Solution: Between 20 and 30 there are five divisions; each division is worth 2 units (30 − 20 = 10; 10 ÷ 5 = 2). The scale indicates 24 units (20 + 2 + 2). Since the meter is on the R × 100 range

$$R = 24 \times 100 \ \Omega$$
$$= 2400 \ \Omega$$

Answer: The meter indicates 2400 Ω.

Digital multimeters (DMMs)

When more than one scale is available for a given function and range, select the scale that ends in a number equal to the range or a power of 10 of the range.

Example 3-18

Refer to Fig. 3-11. Assume that the function switch is on + DC, the range switch is on 25 V, and the needle points one division left of the 200 mark on the 0 to 250 scale.

Solution: The heavy mark between 150 and 200 represents 175. Between 175 and 200 there are five divisions; so, each division is worth 5 units. The scale reading is, therefore, 195. To make the scale fit the range, the scale must be divided by 10. Thus, the measured voltage is 19.5 V (195 ÷ 10 = 19.5).

Answer: The meter indicates 19.5 V dc.

Example 3-19

Assume the same conditions as in example 3-18 except that the range is 2.5 V. How much voltage does the meter indicate?

Solution: Read the same scale as in example 3-18. Now the scale reading of 195 must be divided by 100 to make the 250 scale match the 2.5-V range. Therefore, the measured voltage is 1.95 V (195 ÷ 100 = 1.95).

Answer: The voltage is 1.95 V dc.

Notice in Fig. 3-11 that there is a separate (slightly nonlinear) scale for the 2.5 V ac range. Having a separate scale for the lowest ac voltage range is very common for analog multimeters.

Another common feature of multimeters is having two or more ranges for one setting of the range switch (see Fig 3-11). These additional ranges are selected by inserting the test leads in the appropriate (labeled) jacks.

Digital Multimeter   With *digital multimeters (DMMs)*, like those shown in Figs. 3-12 and 3-13, there are no scales to worry about. The decimal point in the display changes when the range is changed so the display reading never has to be multiplied or divided to obtain the

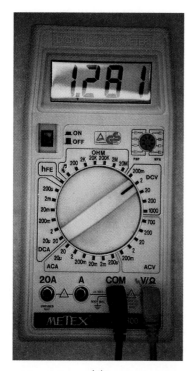

(a)

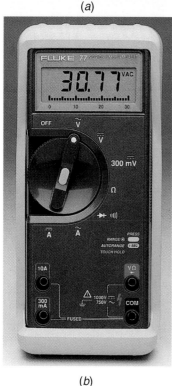

(b)

Fig. 3-12 Digital multimeters. These are hand-held (portable) units. (a) Note the combined function and range switch. (b) This meter is autoranging. (Courtesy of Fluke Corporation.)

correct reading. Thus, the resistance ranges on the DMM tell the maximum resistance a range can measure rather than indicating a multiplier.

Fig. 3-13 Digital multimeter with separate push-button switches for function and range.

The operator must select the correct function and ranges for the meters in Figs. 3-12(a) and 3-13. The meter in Fig. 3-13 uses separate push buttons to select the functions and the ranges, while the meter in Fig. 3-12(a) uses a common rotary switch. More sophisticated DMMs automatically select the correct range, so the operator must only ensure that the meter is on the correct function. Meters of this type [see Fig. 3-12(b)] are said to be *autoranging*.

Autoranging

Using Multimeters

All electrical quantities to be measured are applied to the meter through test leads. A typical test lead is illustrated in Fig. 3-14. For most measurements, the test leads plug into the two jacks in the lower right-hand corner of the meter shown in Fig. 3-11. A black lead goes in the COMMON − jack, and a red lead in the + jack. The black lead is negative and the red lead positive when the function switch is in the + DC position. When the function switch is in the − DC position, the black lead (COMMON −) is positive. Of course, the red lead also reverses polarity. The + DC position

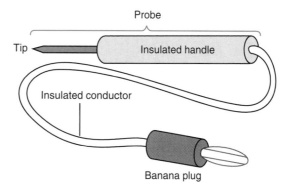

Fig. 3-14 Test lead.

should be used except when special measurements are being made. In this way, the red lead is positive. Red is the color often used to indicate positive in electric circuits.

The DMM has no + DC and − DC switch positions. If reverse polarity is applied to a DMM, a minus (−) sign appears ahead of the digits. The magnitude of the indicated value is still correct.

Measuring Resistance

When resistance is measured, the VOM must first be ohms-adjusted. The *ohms-adjust control* is labeled "zero OHMS" in Fig. 3-11. This control is rotated until the meter indicates zero on the ohms scale *when the test-lead tips are touching each other.* The ohms-adjust control must be adjusted for each range of the ohms function. Thus, every time the range is changed, the leads must be touched together and the meter adjusted to zero. No such adjustment is needed with the DMM.

The ohmmeter function of any multimeter uses a cell, battery, or power supply inside the meter housing. That is, it has its own source of energy. Therefore, any other energy source must be disconnected from any circuit in which resistance is to be measured. Never measure the resistance of a load when power (the energy source) is connected to the circuit. Doing this damages the ohmmeter. Figure 3-15(*a*) illustrates the correct technique for measuring the resistance of a lamp. Notice in Fig. 3-15(*b*) the symbols used for the ohmmeter and the test-lead connection.

The procedure used in measuring resistance is as follows:

1. Remove power from the circuit.
2. Select an appropriate range in the ohms function. The appropriate range is the one that gives the best resolution.
3. When using the VOM, short (touch) the test leads. Turn the ohms-adjust control until the pointer reads 0 ohms.
4. Connect, or touch, the test leads to the terminals of the device whose resistance is to be measured. Except for some electronic components, the polarity of the ohmmeter leads is unimportant.

When measuring resistance, do not touch the metal parts of the test leads with your hands. If you do, you will be measuring your body's resistance as well as the circuit's resistance. This will not harm you, but you will not obtain the correct resistance reading for the circuit.

Measuring Voltage

Voltage measurements are the easiest electrical measurements to make. They are also the most common. Voltage measurements are made with power connected to the circuit. The following procedure is used to make voltage measurements:

1. Select the correct voltage function (ac or dc) for the type of voltage used in the circuit.
2. Select a range that is greater than the expected voltage.

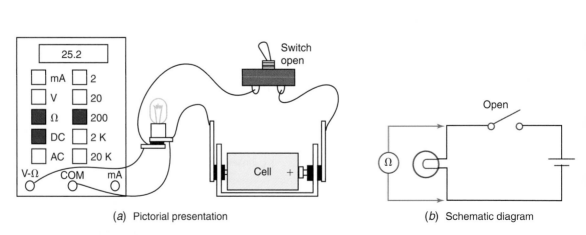

(a) Pictorial presentation   (b) Schematic diagram

Fig. 3-15 Measuring resistance. Notice that the power source is disconnected from the load by the open switch.

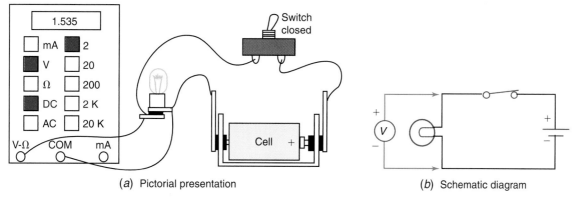

(a) Pictorial presentation

(b) Schematic diagram

Fig. 3-16 Lamp voltage being measured. The switch must be closed.

3. Determine the polarity of the voltage to be measured by looking at the schematic diagram or at the battery terminals. This step is omitted when measuring ac because the polarity reverses every fraction of a second.

4. Connect the negative (black) lead of the multimeter to the negative end of the voltage to be measured. Touch (or connect) the positive (red) lead of the meter to the positive end of the voltage. In other words, observe polarity when measuring voltage with a multimeter or voltmeter. If you do not, the meter pointer of the VOM may bend when it tries to rotate counterclockwise.

Figure 3-16 shows the correct connections for measuring dc voltage. In this figure, the meter is measuring the voltage across the lamp. Notice that the switch is in the closed position. If the switch were open, there would be no voltage across the lamp and the meter would indicate 0 volts. A load has a voltage across it only when current is flowing through it. If the meter were connected across the cell, as in Fig. 3-17, the meter would indicate the cell's voltage regardless of whether the switch were open or closed. As long as the switch is closed, the

meter reading is the same in Figs. 3-16 and 3-17. That is, the voltage output of the cell appears across the lamp when the switch is closed.

Measuring Current

Current measurements are made much less frequently than either resistance or voltage measurements. This is because the circuit usually has to be physically interrupted to insert the meter. In Fig. 3-18, the circuit has been physically interrupted by disconnecting one end of the lead between the cell and the lamp. The meter is then connected between the end of the wire and the lamp. As shown in Fig. 3-19, the meter can just as well be connected on either side of the switch. All three meter locations in Fig. 3-19 are correct, and all three locations yield the same results. Remember, the meter must be inserted into the circuit so that the circuit current flows through the meter as well as the load.

In using the current function of a DMM or a VOM, follow these steps:

1. Select the current function.
2. Select a range that is greater than the expected current.
3. Physically interrupt the circuit.
4. Observing polarity, connect the DMM or VOM between the points created by the interruption. Correct polarity can be determined by tracing current (electron flow). As indicated in Fig. 3-19, current enters the negative terminal and leaves the positive terminal of the meter.

Current measurements

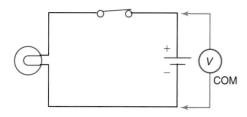

Fig. 3-17 Cell voltage being measured. The switch may be either open or closed.

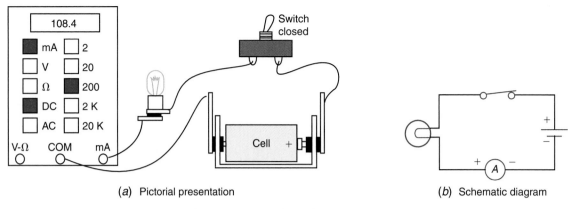

<table>
</table>

(a) Pictorial presentation   (b) Schematic diagram

Fig. 3-18 Current being measured. Current must flow through both the meter and the load.

A meter can be easily damaged, and so it should be used with care. Reverse polarity on a VOM may bend the pointer. However, most damage to multimeters and ammeters occurs when they are incorrectly connected to the circuit. If you connect an ammeter, or a multimeter on the current function, into a circuit in the same way you connect a voltmeter, you may destroy the meter. Even if the meter is not completely destroyed, its accuracy will be greatly decreased.

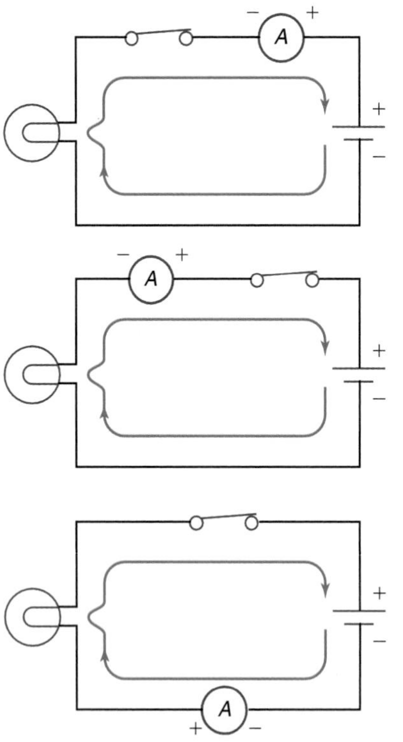

■ TEST_____

Answer the following questions.

20. In Fig. 3-9 the needle rests three divisions to the right of the 30 mark. How much current is the meter indicating?

21. Refer to Fig. 3-11. The needle points one mark to the right of 20. The range is set on R × 100. What is the resistance?

22. Refer to Fig. 3-11. Assume the meter is on the DC function and the 100-mA range. The needle is pointing to 20 on the third scale from the top. How much current is indicated by the meter?

23. True or false. The resistance function of the DMM should be adjusted whenever a new range is selected.

24. True or false. A 10-volt DPM that can indicate a maximum voltage of 9.99 volts is a 2½-digit meter.

25. True or false. Most DMMs are autofunctioning.

26. Summarize the procedure for measuring resistance with the VOM.

27. Summarize the procedure for measuring voltage.

28. Summarize the procedure for measuring current.

29. What does the polarity switch on a VOM do?

30. True or false. Of the possible misuses of an ammeter, reverse polarity is usually the most damaging.

Fig. 3-19 Ammeter location. The location of the ammeter with respect to the lamp or the switch does not change the amount of current.

Summary

1. Electric components are represented by symbols in schematic diagrams; sometimes their electrical values are also included on the diagrams.
2. Sometimes the physical structure of a device also serves as a conductor in a circuit.
3. Current is directly proportional to voltage and inversely proportional to resistance.
4. The Ohm's law formulas are

$$V = IR$$
$$I = \frac{V}{R}$$
$$R = \frac{V}{I}$$

5. The power formulas are

$$P = IV$$
$$P = I^2R$$
$$P = \frac{V^2}{R}$$

6. Energy equals power multiplied by time.

7. The cost of energy equals energy times rate.
8. A panel meter measures only one quantity.
9. Multimeters measure several quantities.
10. VOMs have functions, ranges, and scales.
11. The ohmmeter function of the VOM must be adjusted every time the range is changed.
12. No ohms adjustment is required with a DMM.
13. Polarity must be observed when measuring dc voltage.
14. Polarity must be observed when measuring direct current.
15. Ammeters are easily damaged by incorrect connections to a circuit.
16. Multimeters may have separate switches for selecting function and range.
17. Many digital meters have automatic range and polarity switching.
18. A readout digit that will always be either a 0 or a 1 is specified as a half-digit.

Chapter Review Questions

For questions 3-1 to 3-12, determine whether each statement is true or false.

3-1. The metal chassis of an electric device is often used as a conductor.
3-2. Current is directly proportional to voltage and inversely proportional to resistance.
3-3. A watt is equal to 1 volt divided by 1 ampere.
3-4. A 60-watt, 120-volt lamp requires more current than a 40-watt, 120-volt lamp.
3-5. A 60-watt, 120 volt lamp has more resistance than a 100-watt, 120-volt lamp.
3-6. An electric circuit must be physically interrupted to measure voltage in the circuit.
3-7. An ohmmeter contains its own energy source.
3-8. Always observe polarity when measuring current.
3-9. Correct polarity for an ammeter requires that current (electrons) enter the positive terminal of the meter.

3-10. All DMMs are autoranging.
3-11. A VOM has only one ohmmeter scale.
3-12. The polarity switch on the VOM is set in the negative position to make the common negative jack the negative terminal of the VOM.

Answer the following questions.

3-13. List the four essential parts of an electric circuit.
3-14. Draw the schematic symbols for the following components:
   a. Lamp
   b. Conductor
   c. Resistor
   d. Cell
   e. Open switch

3-15. What is the resistance of a lamp which draws 240 milliamperes when connected to a 12.6-volt battery?

3-16. How much current does a 500-W lamp draw from a 120-V source?

3-17. How much power does a heater require if it draws 11 amperes from a 240-volt circuit?

3-18. A toaster draws 5 amperes from a 120-volt outlet. How much energy does it use in 2 hours?

3-19. How much would it cost to operate the toaster in question 3-18 if energy cost 10¢ per kWh?

3-20. How much voltage is required to force 30 milliamperes of current through a 1-kilohm resistor?

3-21. Measuring which quantity (voltage, current, or resistance) requires physically interrupting the circuit?

3-22. Measuring which quantity (voltage, current, or resistance) requires removing power from the circuit or component?

Critical Thinking Questions

3-1. Using base unit equivalents, prove that current in amperes multiplied by voltage in volts yields power in watts.

3-2. Some of the example circuits in this chapter include fuses. Yet, the fuses were ignored in calculating resistance, voltage, and current. Does this mean that fuses have no resistance? Explain.

3-3. From the way an ammeter is used to measure current in a circuit, would you conclude that it has a very high or a very low internal resistance? Why?

3-4. Assume that you are purchasing a multimeter for personal use. Would you buy a DMM or VOM? Why?

3-5. It costs $1.20 to operate a heater for 6 hours on a 240-V source. What is the resistance of the heater if energy costs 8¢/kWh?

3-6. Prove that energy will be in joules when current is in amperes, time is in seconds, and the formula is $W = I^2Rt$.

3-7. Explain how you could measure the current in Fig. 3-15 without disconnecting any circuit conductors.

3-8. When you measure the resistance of a 60-W, 120-V lamp it is approximately 17 Ω. Using its rated wattage and voltage, calculate its resistance. Explain the large difference between the measured and calculated values.

3-9. Explain what will happen to the power of a circuit if the source voltage is doubled and the resistance is not changed.

Answers to Tests

1. source of energy, conductors, insulators, load, control device, and protection device

2. F

3. F

4.

   Lamp   Resistor   Conductor

5. A schematic diagram is a drawing which uses symbols for electric components. It indicates the connections between the components.

6. T

7. F

8. 0.36 kW

9. 225 mW

10. **Given:** $V = 1.5$ V, $R = 3$ kΩ $= 3000$ Ω
    Find: I
    Known: $I = \dfrac{V}{R}$
    Solution: $I = \dfrac{1.5 \text{ V}}{3000 \text{ Ω}} = 0.0005$ A
    Answer: The current is 0.0005 A, or 0.5 mA.

11. **Given:** $I = 150$ mA $= 0.15$ A
    $V = 600$ mV $= 0.6$ V
    Find: R
    Known: $R = \dfrac{V}{I}$
    Solution: $R = \dfrac{0.6 \text{ V}}{0.15 \text{ A}} = 4$ Ω
    Answer: The resistance is 4 Ω.

12. **Given:** $V = 120$ V, $I = 4.5$ A
    Find: P
    Known: $P = IV$
    Solution: $P = 4.5$ A $\times$ 120 V $= 540$ W
    Answer: The dryer is a 540-W dryer.
13. **Given:** $P = 1000$ W, $V = 120$ V
    Find: I
    Known: $I = \dfrac{P}{V}$
    Solution: $I = \dfrac{1000 \text{ W}}{120 \text{ V}} = 8.3$ A
    Answer: The current for the 1000-W toaster is 8.3 A.
14. **Given:** $I = 2.2$ A, $V = 12.6$ V
    Find: P
    Known: $P = IV$
    Solution: $P = 2.2$ A $\times$ 12.6 V $= 27.72$ W
    Answer: The tape deck requires 27.72 W.
15. **Given:** $R = 99$ Ω, $V = 6.3$ V
    Find: P
    Known: $P = \dfrac{V^2}{R}$
    Solution: $P = \dfrac{(6.3)^2}{99} = 0.4$ W
    Answer: The power rating of the lamp is 0.4 W.
16. **Given:** $I = 8$ A, $V = 240$ V, $t = 9$ h
    Find: W
    Known: $W = Pt$, $P = IV$
    Solution: $P = 8$ A $\times$ 240 V $= 1920$ W
    $W = 1920$ W $\times$ 9 h
    $= 17{,}280$ Wh $= 17.28$ kWh
    Answer: The energy converted is 17.28 kWh.
17. **Given:** $R = 100$ Ω, $I = 200$ mA $= 0.2$ A, $t = 10$ min $= 600$ s
    Find: W
    Known: $W = Pt$, $P = I^2R$, Ws $=$ J
    Solution: $P = (0.2)^2 \times 100 = 4$ W
    $W = 4$ W $\times$ 600 s
    $= 2400$ Ws $= 2400$ J
    Answer: The resistor converts 2400 J of electric energy to heat energy.
18. **Given:** $P = 3000$ W, $t = 3$ h, rate $= 4$¢ per kWh
    Find: Cost
    Known: Cost $=$ rate $\times W$, $W = Pt$

Solution: $W = 3000$ W $\times$ 3 h $= 9000$ Wh
$= 9$ kWh
Cost $= 4$¢ per kWh $\times$ 9 kWh
$= 36$¢
Answer: The cost is 36¢.

19. **Given:** $I = 0.5$ A, $V = 120$ V, rate $= 8$¢ per kWh, $t = 40$ h
    Find: Cost
    Known: Cost $=$ rate $\times W$, $W = Pt$, $P = IV$
    Solution: $P = 0.5$ A $\times$ 120 V $= 60$ W
    $W = 60$ W $\times$ 40 h $= 2400$ Wh
    $= 2.4$ kWh
    Cost $= 8$¢ per kWh $\times$ 2.4 kWh
    $= 19.2$¢
    Answer: The cost is 19.2¢.

20. 33 microamperes
21. 1900 ohms (19 $\times$ 100)
22. 40 milliamperes
23. F
24. F
25. F
26. Remove the power from the circuit, select the ohms function, select the correct range, and connect the test leads to the device or component being measured. If a VOM is used, the meter must also be ohms-adjusted.
27. Select the correct voltage function, select the correct range, determine the polarity of the voltage, and—observing polarity—connect the meter to the circuit.
28. Select the current function, select the correct range, interrupt the circuit, and—observing polarity—connect the meter between the interrupted points.
29. The polarity switch reverses the polarity of the meter jacks to which the test leads are connected.
30. F

Chapter 4

Circuit Components

Chapter Objectives

This chapter will help you to:

1. *Identify* common electric components and their schematic symbols.
2. *Measure and specify* wire size for electric conductors.
3. *Understand* the operating principles of electric components.
4. *Interpret and specify* the ratings of components.
5. *Understand* the terminology used to describe circuit components and faults.
6. *Use* the resistor color code to determine resistance and tolerance.

 . . . that in the previous chapter you learned about the requirements for a complete electric circuit.

In this chapter you will learn about some of the components (parts) used to construct electric circuits.

4-1 Batteries and Cells

Batteries and cells are chemical devices that provide dc voltage. Since voltage is a potential energy difference, batteries and cells are also known as sources of electric energy. And, since power is the rate of using energy, these devices are also called power sources. They are used to power electric devices.

Terminology and Symbols

A *cell* is an electrochemical device consisting of two electrodes made of different materials and an electrolyte. The chemical reaction between the *electrodes* and the *electrolyte* produces a voltage.

A *battery* consists of two or more cells electrically connected together and packaged as a single unit. Although technically a battery has two or more cells, the term *battery* is often used to indicate either a single cell or a group of cells.

The schematic symbols for a cell and a battery are shown in Fig. 4-1. The voltage rating of the cell or battery is specified next to the

symbol. The short line in the symbol is always the negative terminal. Polarity indicators (negative and positive signs) are not always shown with the symbol.

Cells and batteries are classified as either primary or secondary. *Primary cells* are those cells that are not rechargeable. That is, the chemical reaction that occurs during discharge is not easily reversed. When the chemicals used in the reactions are all converted, the cell is fully discharged. It must then be replaced by a new cell. Included in the primary cell category are the following types: carbon-zinc, zinc chloride, most alkaline, silver oxide, mercury, and lithium cells.

Secondary cells may be discharged and recharged many times. The number of discharge/charge cycles a cell can withstand depends on the type and size of the cell and on the operating conditions. The number of cycles will vary from fewer than 100 to many thousands. Secondary cells include the following types: rechargeable alkaline, lead-acid, nickel-cadmium, and nickel-iron cells.

Cells and batteries are also classified as ei-

Primary cells

Cell

Electrodes

Electrolytes

Battery

Secondary cells

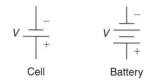

Fig. 4-1 Cell and battery symbols.

ther dry or wet. Historically, a *dry cell* was one that had a paste or gel electrolyte. It was semi-sealed and could be used in any position. With newer designs and manufacturing techniques, it is possible to completely (hermetically) seal a cell. With complete seals and chemical control of gas buildup, it is possible to use liquid electrolyte in dry cells. Today the term *dry cell* refers to a cell that can be operated in any position without electrolyte leakage.

Wet cells are cells that must be operated in an upright position. These cells have vents to allow gases generated during charge or discharge to escape. The most common wet cell is the lead-acid cell.

Ratings of Cells and Batteries

Cell (or battery) capacity refers to the amount of energy the cell can provide under specified conditions. The conditions specified are temperature, current drain, discharge schedule (minutes per hour), and final output voltage when discharge is complete. The amount of energy available varies widely as these specified conditions change. For example, a typical flashlight cell provides approximately twice as much energy when the discharge rate is changed from 300 to 50 mA. For the same cell, the capacity decreases about 30 percent when the temperature is decreased from 21 to 5°C. In general, a cell (or battery) provides the most energy when the temperature is high and the discharge schedule, current drain, and final voltage are low.

The capacity of a cell (or battery) is expressed in *ampere-hours* (Ah). When the voltage of the cell is specified, ampere-hour becomes a unit of energy. The energy of a cell is

$$W = Pt$$

Since $P = IV = VI$ we have

$$W = VIt$$

If t is given in hours and I is in amperes, then the energy of a cell is expressed in watthours.

Internal Resistance

The output voltage from a cell varies as the load on the cell changes. *Load on a cell* refers to the amount of current drawn from the cell. As the load increases, the voltage output drops, and vice versa. The change in output voltage is caused by the *internal resistance* of the cell. Since materials from which the cell is made are not perfect conductors, they have resistance. Current flowing through the external circuit also flows through the internal resistance of the cell. According to Ohm's law, a current flowing through a resistance (either external or internal) results in a voltage drop ($V = IR$). Any voltage developed across the internal resistance is not available at the terminals of the cell. The voltage at the terminals is the voltage due to the chemical reactions *minus the voltage dropped across the internal resistance.* The terminal voltage of a cell, therefore, depends on both the internal resistance of the cell and the amount of load current.

In some applications, the changes in cell terminal voltage are so small that they make no practical difference. In other applications, the changes are very noticeable. For example, while an automobile engine is starting up, the battery output voltage changes from about 12.6 to 8 V.

As a cell is discharged, its internal resistance increases. Therefore, its output voltage decreases for a given value of load current.

Example 4-1

What is the terminal voltage of a 6-V battery with an internal resistance of 0.15 Ω when the load draws a current of 3.4 A?

Given:	No-load terminal voltage $(V_{NLT}) = 6$ V
	Internal resistance $(R_1) = 0.15\ \Omega$
	Load current $(I_L) = 3.4$ A
Find:	Loaded terminal voltage (V_{LT})
Known:	$V_{LT} = V_{NLT} - (I_L \times R_1)$
Solution:	$V_{LT} = 6$ V $- (3.4$ A $\times 0.15\ \Omega)$
	$\quad\quad = 5.49$ V
Answer:	The terminal voltage of the loaded battery is 5.49 V.

Polarization

When a primary cell discharges, gas ions (such as hydrogen) are created around the positive electrode. The gas ions polarize the cell and reduce the cell's terminal voltage. If the gas ions are allowed to build up, the polarization can become so great that the cell is useless. Therefore, cells include a *depolarizing agent* in their chemical composition. The depolarizing agent is a chemical compound that reacts with the polarizing gas to remove it.

Depolarizing agent

■ TEST_____

Answer the following questions.

1. What is a cell?
2. What is a battery?
3. Describe the basic structure of a cell.
4. Define the following terms:
    a. Primary cell
    b. Secondary cell
    c. Dry cell
    d. Wet cell
    e. Ampere-hour
5. List four types of secondary cells.
6. List six types of primary cells.
7. Internal resistance causes a battery's terminal voltage to _____ .
8. The development of hydrogen ions on the positive electrode of a cell is called _____ .

Sulfuric acid (H_2SO_4)

Lead sulfate ($PbSO_4$)

4-2 Lead-Acid Cells

The lead-acid cell is a very common secondary cell. It is the power source for the electric system of most cars, trucks, boats, and tractors. It can provide the large currents (hundreds of amperes) needed to crank internal-combustion engines.

A lead-acid cell produces about 2.1 V. Higher voltages are obtained by connecting cells together to form batteries. A 12-V automobile battery actually has a nominal voltage of 12.6 V because it contains six cells.

The structure of a lead-acid cell is shown in Fig. 4-2. Notice that the cell terminals connect to a *group of plates*. The groups are tied to-

Group of plates

gether (both electrically and mechanically) at the top. The meshing together of the positive and negative groups effectively provides one large negative and one large positive plate.

Chemical Action

When a load is connected to the cell of Fig. 4-3(a), current flows. Electrons leave the negative terminal (plate), flow through the load, and return to the positive terminal of the cell. At the surface of the positive plate shown in Fig. 4-3(a), a molecule of lead peroxide (PbO_2) becomes three ions. Notice that each ion of oxygen (O) has two excess electrons while the ion of lead (Pb) has a deficiency of only two electrons. Thus, the plate is left with a deficiency of two electrons; that is, it has a positive charge. In the electrolyte, two molecules of *sulfuric acid* (H_2SO_4) provide four hydrogen (H) ions and two sulfate (SO_4) ions. The four hydrogen ions combine with the two oxygen ions to form two molecules of water (H_2O). One of the sulfate ions combines with the lead ion at the positive plate to form a molecule of *lead sulfate* ($PbSO_4$).

On the surface of the negative plate in Fig. 4-3(a), an atom of lead becomes a positive ion. In becoming a positive ion, the lead atom leaves two electrons on the plate. Thus, the plate has a negative charge. The lead ion combines with a sulfate ion from the electrolyte and forms a molecule of lead sulfate. The result of these reactions, shown in Fig. 4-3(b), is that the plates change to lead sulfate and the electrolyte changes to water.

The above-described chemical reaction continues until one of two things happens:

1. The load is removed; the chemical reaction stops. At this time, the charges on the plates and the charges of the electrolyte ions are in a state of equilibrium (balance).
2. All the sulfuric acid solution has been converted to water and/or all the lead and lead peroxide have been converted to lead sulfate. In this case, the battery is fully discharged.

Recharging

A lead-acid cell is recharged by forcing a reverse current through the cell. That is, elec-

---

About ◄▮▮► Electronics

Exercise Caution with Batteries Although low-voltage, high-current sources such as lead-acid batteries cannot deliver an electric shock, they can cause severe burns when shorted by jewelry such as rings and watchbands.

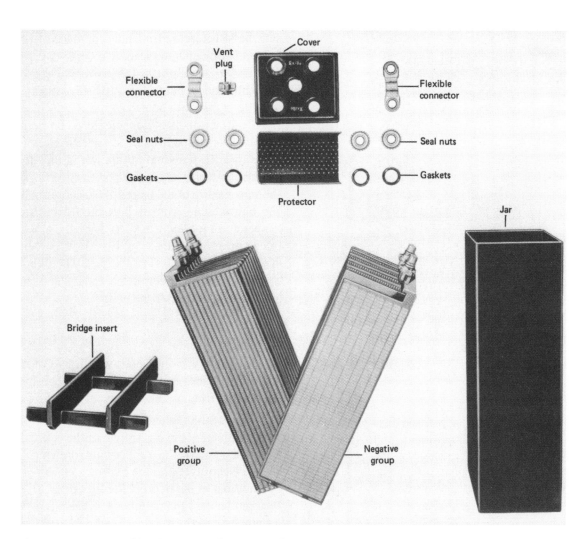

Fig. 4-2 Structure of lead-acid cell. *(Courtesy of Exide Power Systems.)*

trons enter the negative plate and leave the positive plate. When a cell is being charged, all the chemical reactions described above are reversed. The water and lead sulfate are converted back into sulfuric acid, lead, and lead peroxide. Cells are charged by connecting them to a voltage source that is greater than the cell's voltage. In other words, the cell becomes the load rather than the energy source. As the load, the cell is converting electric energy (from the other voltage source) into chemical energy.

The six-cell (12-V) battery in Fig. 4-4 is being recharged. Notice that the negative terminal of the charger is connected to the negative terminal of the battery. Since the charger's voltage is greater than the battery's voltage, current flows in the direction indicated in Fig. 4-4. The exact voltage of the charger depends on the condition of the battery and on the *rate of charge* desired. The charger voltage is adjusted

to provide the desired charge current. The manufacturer's recommendation should be followed in determining the rate of charge for a specific type of battery or cell. Too fast a charge rate must be avoided because it will overheat the battery.

A battery should not be overcharged. Overcharging can weaken the plate structure of the battery. When a battery is overcharged, water from the electrolyte is converted into hydrogen gas and oxygen gas. Having to add more than normal amounts of water to a battery is an indication that it is being overcharged.

Some of the newer lead-acid batteries are sealed. There is no provision for adding water. Such batteries require no maintenance other than keeping them clean—especially their terminals.

Most jobs require teamwork. You should work on your interpersonal skills and learn to accept criticism.

Rate of charge

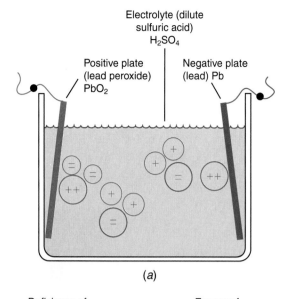

Electrolyte (dilute sulfuric acid) H_2SO_4

Positive plate (lead peroxide) PbO_2

Negative plate (lead) Pb

(a)

Hydrometer

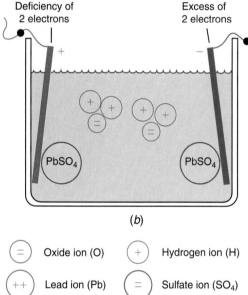

Deficiency of 2 electrons

Excess of 2 electrons

$PbSO_4$ $PbSO_4$

(b)

(=) Oxide ion (O)	(+) Hydrogen ion (H)
(++) Lead ion (Pb)	(=) Sulfate ion (SO_4)

Fig. 4-3 Lead-acid cell. (a) Chemical reaction. (b) Results.

Specific Gravity

The charge of a battery can be determined by measuring the specific gravity of the electrolyte. The specific gravity of a substance is the ratio of its weight to the weight of water. If a substance has a specific gravity of 1.251, it is 1.251 times as heavy as water. Sulfuric acid is heavier than water. Therefore, the more sulfuric acid in the electrolyte, the higher the specific gravity of the electrolyte. Since the amount of sulfuric acid increases as the battery is charged, specific gravity indicates the state of charge of the battery. The specific gravity of a fully charged cell is adjusted at the factory to

suit the structure and intended purpose of the cell. For fully charged lead-acid cells, the specific gravity ranges from 1.21 to 1.28. The typical automotive battery is fully charged at a specific gravity of about 1.26.

A lead-acid cell is considered to be completely discharged when the specific gravity drops to 1.12. It should not be left in this discharged state for extended periods. To do so shortens the life of the battery. Discharged cells and batteries need to be protected from low temperature. A completely discharged cell will freeze at about $-9°C$ ($16°F$). At 50 percent discharge it freezes at $-24°C$ ($-11°F$). When a cell freezes, the electrolyte expands and can break the jar (case) of the cell.

Hydrometer

Specific gravity is measured with a hydrometer like the one shown in Fig. 4-5. The hydrometer is used as follows:

1. Squeeze the rubber bulb. Insert the flexible tube into the electrolyte of the cell. Slowly release the rubber bulb, drawing the electrolyte into the hydrometer. When the float in the hydrometer lifts free, remove the flexible tube from the electrolyte and finish releasing the rubber bulb.
2. Read the specific gravity from the float at the top surface of the electrolyte.
3. Reinsert the flexible tube into the cell and slowly squeeze the bulb to return the electrolyte.
4. When finished testing, flush the hydrometer (both inside and out) with clean water.

When the specific gravity of a cell cannot be raised by charging to within 0.05 of the manufacturer's specification, it is a questionable cell. It will probably completely fail in the near future.

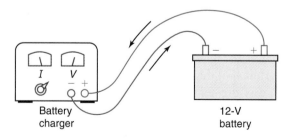

Battery charger

12-V battery

Fig. 4-4 Charging a battery. The charger voltage must exceed the battery voltage.

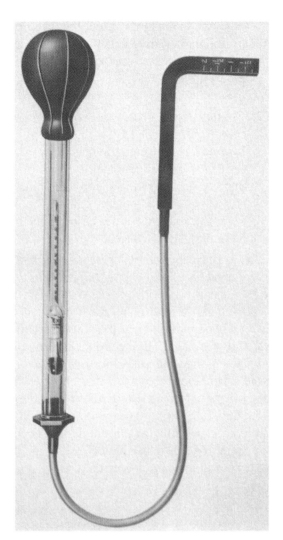

Fig. 4-5 Hydrometer. *(Courtesy of Exide Power Systems.)*

Battery Safety

If not properly handled, lead-acid cells and *batteries can be dangerous*. The acid used in the electrolyte can cause skin burns and burn holes in clothing. It is extremely harmful to the eyes. Always wear safety glasses when working with lead-acid cells and batteries. If acid does come in contact with the skin or clothing, immediately flush with clean water. Then wash with soap and water, except for the eyes. If the acid is in the eyes, get medical attention immediately after repeated flushing with water. Wash hands in soap and water after handling batteries.

The gases released from charging batteries are explosive. Charge batteries in a ventilated area where there are no sparks or open flames.

Answer the following questions.

9. Describe the chemical reaction that occurs in a lead-acid cell as it is discharged.
10. Describe how a lead-acid battery is recharged.
11. List two precautions to follow to prevent battery damage when charging a battery.
12. Define *specific gravity.*
13. A _____ is used to measure specific gravity.
14. In cold weather a _____ lead-acid battery can freeze.
15. The acid in a lead-acid cell is _____ .
16. List the safety rules that apply to handling and charging lead-acid batteries.

4-3 Nickel-Cadmium Cells

Nickel-cadmium cells are dependable, rugged secondary cells. As vented wet cells, they have been in limited use for many years in many types of service. When technological advances made it possible to produce a nickel-cadmium dry cell, the popularity of the cell greatly increased. The dry cell is hermetically sealed and can be operated in any position. Figure 4-6 illustrates the structure of a typical sealed nickel-cadmium cell. These cells are most commonly available in button or cylindrical shape and in capacities ranging from a few milliampere-hours to 5 or 6 Ah.

Sealed nickel-cadmium cells have a life ranging from several hundred to several thousand *charge/discharge cycles*. They have an extremely long *shelf life* (storage life) and can be stored either charged or discharged. They can be stored at temperatures ranging from -40 to $60°C$ (-40 to $140°F$).

Other notable characteristics of the nickel-cadmium cell are

1. Good low-temperature operating characteristics.
2. High initial cost but low operating cost.
3. Very low internal resistance. Thus, they can provide high currents with little drop in output voltage.
4. Nearly constant output voltage (approximately 1.2 V) as the cell is discharged by a moderate load. See Fig. 4-7 for a general comparison with other types of cells.

Nickel-cadmium cells

Charge/discharge cycles

Shelf life

Batteries can be dangerous

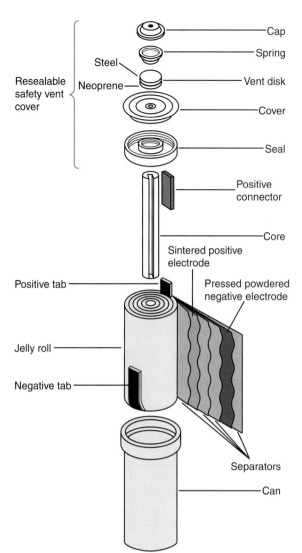

Zinc chloride cells

Alkaline cells

Fig. 4-6 Nickel-cadmium cell. *(Courtesy of Union Carbide Corporation.)*

4-4 Carbon-Zinc and Zinc Chloride Cells

Carbon-zinc cells

Primary alkaline cells

The *carbon-zinc cell* is the most common of the dry cells. It is also the least expensive of the primary cells.

The composition of a carbon-zinc cell is shown in Fig. 4-8. The zinc can is the negative electrode, and the manganese dioxide is the positive electrode. The carbon rod makes electric contact with the manganese dioxide and conducts current to the positive terminal. However, the carbon rod is not involved in the chemical reaction that produces the voltage. The electrolyte for the chemical system is a solution of ammonium chloride and zinc chloride.

Secondary alkaline cells

Although extensively used because it is cheap, the carbon-zinc cell has many weaknesses. Some of these weaknesses are

1. Poor low-temperature operating characteristics.
2. Gradual decrease in output voltage as the cell discharges (Fig. 4-7).
3. Low energy-to-weight ratio and low energy-to-volume ratio. Other types of primary cells have ratios two to three times as great as the carbon-zinc cell.
4. Poor efficiency under heavy loads due to high internal resistance.

Carbon-zinc cells and batteries are available in a wide range of sizes and shapes. Batteries with capacities of 30 Ah are available.

Zinc chloride cells are constructed much like carbon-zinc cells, but they use a modified electrolyte system. The electrolyte is a solution of zinc chloride. Because only zinc chloride is used in the electrolyte, the zinc chloride cell has several advantages over the standard carbon-zinc cell. First, the zinc chloride cell is more efficient at higher current drain. Second, the output voltage does not decrease as rapidly as it does in the carbon-zinc cell.

4-5 Alkaline—Manganese Dioxide Cells

Alkaline cells use the same electrodes as carbon-zinc cells (zinc and manganese dioxide). However, a solution of potassium hydroxide is used as the electrolyte. There are two types of alkaline cells—primary and secondary (rechargeable).

Alkaline cells produce about 1.5 V. The cell voltage decreases gradually as the cell is discharged. However, the change is not as great as it is with carbon-zinc cells (Fig. 4-7).

The *primary alkaline cell* is more expensive than the carbon-zinc cell. However, it has several advantages. It can be discharged at high current drains and still maintain reasonable efficiency and cell voltage. It can operate effectively at lower temperatures ($-30°C$ versus $-7°C$ for carbon-zinc cells). It has a lower internal resistance and a higher energy-to-weight ratio. An alkaline cell stores at least 50 percent more energy than a carbon-zinc cell of the same size.

Secondary alkaline cells are much cheaper than nickel-cadmium cells. They have better

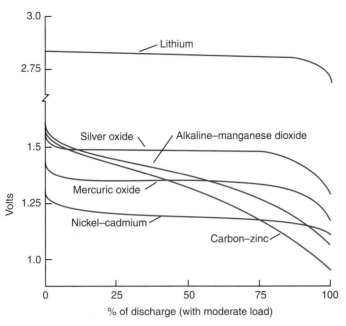

Fig. 4-7   Comparison of various cell voltages as the cells are discharged.

charge retention (ability to remain charged when stored) and a wider operating temperature range than nickel-cadmium cells do. The internal resistance of the two types of cells is comparable. However, the alkaline cell has several shortcomings not found in the nickel-cadmium cell:

1. The cycle life (number of discharge/recharge cycles) is fewer than 75.
2. The cell voltage is not as constant (Fig. 4-7).

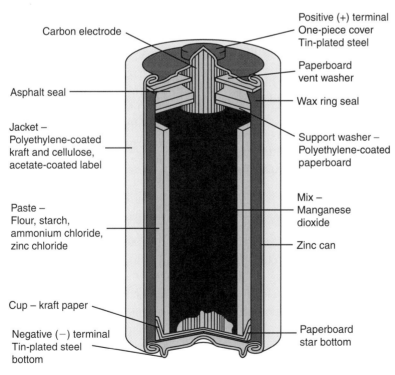

Fig. 4-8   Cutaway view of carbon-zinc cell. *(Courtesy of Union Carbide Corporation.)*

3. The cycle life is more dependent on proper discharge and recharge of the cell. Discharging beyond rated capacity and extended overcharging both shorten the cycle life.

4. Charging circuits for optimum cycle life are more complex.

4-6   Mercuric Oxide Cells

Mercuric oxide cells

Mercuric oxide cells (commonly called mercury cells or batteries) have some distinct advantages over the primary cells discussed above:

1. The cell voltage is more uniform as the cell discharges (Fig. 4-7).
2. The capacity is less dependent on the discharge rate.
3. The energy-to-volume ratio is two to four times as great as that of either the alkaline or the carbon-zinc cell.
4. The energy-to-weight ratio is higher.

The mercury cell is mechanically rugged, has low internal resistance, and can operate at high temperatures (54°C). Since the output voltage (1.35 V) is so constant over most of its discharge, the mercury cell is used as a *voltage standard*. (A standard is used to check the accuracy of other voltage sources and measuring devices.)

Voltage standard

Lithium cells

Some of the disadvantages of the mercury cell are its poor low-temperature (0°C) performance and its relatively high cost. Both the initial cost and the operating cost are high compared with costs of carbon-zinc and alkaline cells.

Figure 4-9 shows the internal structure of a mercury cell. Mercury cells and batteries are available in a wide range of capacities (up to 28 Ah) and shapes.

4-7   Silver Oxide Cells

Silver oxide cells

Light loads

Silver oxide cells are much like mercury cells. However, they provide a higher voltage (1.5 V) and they are made for *light loads*. The loads can be continuous, such as those encountered in hearing aids and electronic watches. Like the mercury cell, the silver oxide cell has good energy-to-weight ratios, poor low-temperature response, and flat output voltage characteristics. The structures of the mercuric and silver oxide

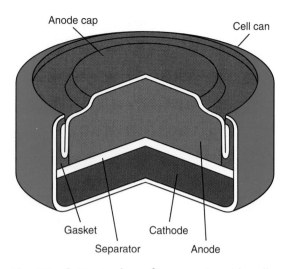

Fig. 4-9   Cutaway view of a mercuric oxide cell. *(Courtesy of Union Carbide Corporation.)*

cells are very similar. The main difference is that the positive electrode of the silver cell is silver oxide instead of mercuric oxide. Figure 4-7 compares the voltage characteristics of the silver oxide cell with those of other cells.

4-8   Lithium Cells

The *lithium cell* is another type of primary cell. It is available in a variety of sizes and configurations. Compared with other cells, it is quite expensive. Depending on the chemicals used with lithium, the cell voltage is between 2.1 and 3.8 V. Note that this voltage is considerably higher than that of other primary cells. Lithium cells operate at temperatures ranging from −50 to 75°C. They have a very constant output voltage during discharge. Two of the advantages of lithium cells over other primary cells are

1. Longer shelf life—up to 10 years
2. Higher energy-to-weight ratio—up to 350 Wh/kg

■ TEST

Answer the following questions.

17. Can nickel-cadmium cells be damaged by leaving them discharged?
18. Compare the nickel-cadmium and rechargeable alkaline cell in terms of
    a. Voltage during discharge
    b. Cycle life
    c. Cost

19. Compare alkaline and carbon-zinc cells in terms of
    a. Efficiency at high current loads
    b. Voltage at high current loads
    c. Energy storage
20. The _____ cell is the least expensive of the various types of cells.
21. The _____ cell is expected to provide several hundred charge/discharge cycles.
22. The _____ cell has the highest output voltage of the various types of cells.

4-9 Miniature Lamps

Miniature lamps are used extensively in electric and electronic circuits. They are used for indicator lights and instrument illumination in automobiles, aircraft, home appliances, coin-operated machines, and all kinds of electronic instruments. Miniature lamps are also used in flashlights, lanterns, and signal lights. Both *neon glow lamps* and *incandescent lamps* are used as indicators.

Incandescent Lamps

The heart of the incandescent lamp is the *tungsten filament* (wire), which emits light when heated by an electric current. When the tungsten wire is long, thin, and coiled, as in Fig. 4-10, it is often supported in the center.

The tungsten filament is enclosed in a glass bulb from which the air is evacuated. Often the air is replaced by an inert gas. As the last of the air is removed from the bulb, the bulb is sealed. It is sealed by melting the glass around the evacuation hole. The seal on the lamp in Fig. 4-10 was made on the small tip of glass at the bottom of the bulb.

The heavy wire leads connecting to the filament carry current to the filament. For this lamp, these leads are also the terminals which make electrical contact with the lamp holder. This configuration of terminal connections is called a *wedge base*. It provides for a simple, compact lamp holder.

The two principal bases used in miniature incandescent lamps are the screw-type-base and the *bayonet base*. The lamp on the left in Fig. 4-11 has a screw-type base. The screw is stamped on the portion of the base called the shell. The next lamp in Fig. 4-11 has a bayonet base. The bayonet-base lamp usually has

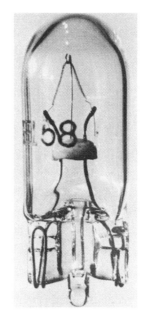

Fig. 4-10 Wedge-base lamp. *(Courtesy of General Electric.)*

one or two small projections on the shell base. These projections slide into grooves in the bayonet lamp holder. When the lamp is in place, it is given a slight twist to lock it in the holder. Lamp bases are usually made of brass for good conductivity and long wear.

Figure 4-11 shows a variety of other base and bulb styles used with miniature lamps. All the bases provide two contacts which make connections to the filament. Each base is designed to best fulfill a specific need. For example, the base may provide for exact location of the lamp. This is important when the light is to be focused. Some lamps, like the one on the right in Fig. 4-11, are designed to be soldered into the circuit. Some bases are designed to withstand high temperature. Others are made to withstand severe vibration.

Miniature lamps are available in a wide range of voltage and *current ratings*. At rated voltage, a lamp draws its rated current, provides its specified light, and operates for a specified time, usually called the *life* of the lamp.

Operating above *rated voltage* increases the current draw and light output (specified in units called *lumens*). However, operating with overvoltage decreases the life expectancy so greatly

JOB TIP

One of the common characteristics of all valuable employees is pride in their work.

Neon glow lamps

Incandescent lamps

Tungsten filament

Wedge base

Current ratings

Life

Bayonet base

Rated voltage

Lumens

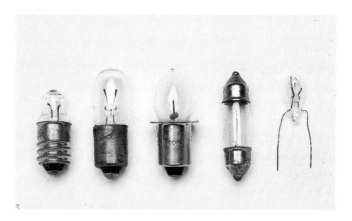

Fig. 4-11   Miniature lamps. These lamps are about 2.5 cm long. *(Mark Steinmetz)*

that it is not recommended. Operation below rated voltage is permissible. Under this condition, current decreases, light output decreases, and expected life greatly increases. The efficiency in terms of light output per watt input is reduced when operating below rated voltage.

Sometimes the manufacturer specifies the voltage and power rating of a lamp. In this case, the current can be determined by using the power relationship $I = P/V$.

Example 4-2

How much current is drawn by a type-1992 lamp which is rated at 14 V and 35 W?

Given: Voltage (V) = 14 V
　　　　　Power (P) = 35 W

Find: Current (I)

Known: $I = \dfrac{P}{V}$

Solution: $I = \dfrac{35\ \text{W}}{14\ \text{V}} = 2.5\ \text{A}$

Answer: The lamp draws 2.5 A.

Flasher lamps

Bimetallic strip

A few miniature lamps are made with built-in flasher units. *Flasher lamps* are used as warning indicators, such as seat belt and emergency brake warnings in automobiles. The flasher action is caused by a *bimetallic strip* located next to the filament. This bimetallic strip is one of the conductors carrying current to the filament, as shown in Fig. 4-12(*a*). When the filament heats the bimetallic strip, the strip

bends out as shown in Fig. 4-12(*b*) and opens the circuit, turning off the lamp. When the lamp is off, the strip cools and straightens out. When the strip returns to its normal shape, it makes contact with the filament again and turns on the lamp. This cycle repeats as long as the lamp is connected to a power source.

Figure 4-13 illustrates the structure of a bimetallic strip, which is composed of two different metals. The two metals are rigidly joined where they meet. The metals have different temperature coefficients of expansion. That is, one of the metals increases its length more rapidly than the other as the temperature increases. This unequal expansion causes the strip to curve when heated, as shown in Fig. 4-13(*b*).

The resistance of a lamp filament when cool is only a fraction of the resistance of the filament when it is hot. Therefore, a lamp draws a

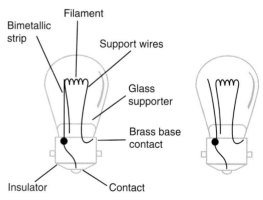

Fig. 4-12   Flasher lamp. The bimetallic strip turns the lamp off and on as it warms up and cools down. (*a*) Lamp on. (*b*) Lamp off.

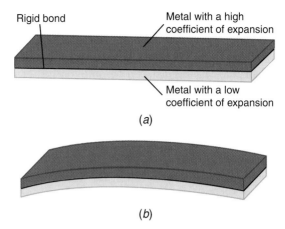

Fig. 4-13 Bimetallic strip. (*a*) At room temperature. (*b*) Above room temperature.

large current when it is first turned on. However, the filament comes up to operating temperature in a few milliseconds, so this large current does not last long. The reason for the change in resistance of the lamp is the temperature coefficient of resistance of the tungsten filament. Although the coefficient of tungsten is not much higher than that of copper, the temperature change is tremendous.

Measuring the resistance of a lamp with an ohmmeter indicates only whether or not the lamp is burned out. The indicated resistance will be far less than the actual *hot resistance* of the lamp. This is because the current from the meter is too small to bring the lamp up to operating temperature.

Neon Glow Lamps

The lamp in Fig. 4-14 is a neon glow lamp. The glass bulb is filled with neon gas. When neon gas is not ionized, it is an insulator. Once it ionizes, it is a fair conductor.

Ionization of the neon in the lamp in Fig. 4-14 occurs between the two *metal electrodes* inside the bulb. If the ionization voltage is dc, only one electrode glows. If it is ac, both electrodes glow. It requires between 70 and 95 V to ionize a neon bulb. After ionization occurs, it requires 10 to 15 percent less voltage, or from 60 to 80 V, to maintain ionization.

Because the internal resistance of an ionized lamp drops to a very low value, a resistor must be connected in series with the lamp. The resistor (Fig. 4-15) limits the current through the lamp to a safe value. In Fig. 4-15 the neon bulb

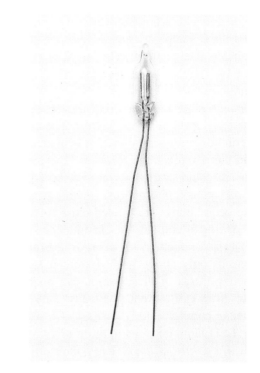

Fig. 4-14 Neon lamp. Note the absence of a filament in the neon lamp. (*Mark Steinmetz*)

and the resistor are represented by their schematic symbols. Some neon lamps are manufactured with a resistor in the base. In this case no *external resistor* is needed.

Neon lamps are made in a number of different bulb and base styles. Most of the styles are the same as for the incandescent lamps. Neon lamps require very little power to operate. Most are less than ½ W and require no more than 2 mA.

Light-Emitting Diodes

The *light-emitting diode* (LED) is often used as an indicator lamp. It is also used for nu-

External resistor

Hot resistance

Light-emitting diode

Ionization

Metal electrodes

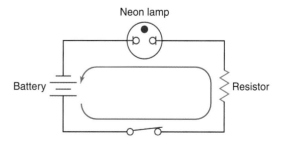

Fig. 4-15 Neon lamp and current-limiting resistor. Without the resistor, the neon lamp would immediately burn out.

meric readouts on calculators and other digital devices. LED lamps are small, low-powered, and reliable. They can operate for years without failure. They are available in either a wire-terminal or a socket-type base.

Although most LEDs are red, they are also available in green, blue, and yellow. The color is determined by the combination of materials used in the manufacturing process. The materials used include various compounds of gallium, arsenic, and phosphorus.

Current-limiting resistors

Like neon lamps, LEDs need *current-limiting resistors* to control the current once the lamp starts conducting. However, the typical LED needs only about 1.5 V before it starts conducting. Most of these lamps can operate on 5 to 40 mA of current. Typically, they are operated at about 20 mA.

Polarized lamp

The LED lamp operates only on direct current. It is a *polarized lamp*. The polarity of its leads need to be determined before it is connected to a voltage source.

Replacement lamp

Working with Miniature Lamps

Some miniature lamps operate at very high temperatures. A quartz lamp can operate with a bulb temperature of 350°C. Wedge-based quartz lamps can operate at temperatures as high as 450°C. At these temperatures, dirt on the surface of the glass can be quite harmful to the lamp. Needless to say, severe burns can result from touching lamps operating at such high temperatures.

When connecting wire-terminal lamps, like the neon lamp in Fig. 4-14, you must be careful in bending the wire leads. Do not bend the leads within 3 mm (⅛ in.) of the glass bulb. Also, do not solder within 3 mm of the glass. Either bending or soldering too close to the bulb can damage the metal-to-glass seal.

Lamp Safety

Following the procedures listed below will reduce the probability of accidents when working with lamps:

1. Turn off the power to a lamp before changing the lamp.
2. Allow a lamp to cool before changing it.
3. Be certain the *replacement lamp* has the correct voltage. If the voltage of the power source greatly exceeds the voltage rating of the lamp, the lamp may explode.

? Did You Know?

No More Tangles Fluctuations in electric power service can cause havoc with sensitive machinery in industries such as the rug industry. Prior to the development of the Westinghouse dynamic voltage restorer (DVR), rug workers had to clean up a mess of yarn on the way to the winder every time there was a brief power outage. The DVR (pictured) works by injecting energy into a power line to fix deviations from sags, swells, and transients that occur because of storms or dust.

4. When replacing lamps that fit tightly in their sockets, wear a glove (or wrap the bulb in a heavy cloth) in case the bulb breaks.

■ TEST

Answer the following questions.

23. What material is used in the filament of a miniature lamp?
24. Name two types of miniature lamps used as indicators.
25. What makes a flasher lamp blink off and on?
26. How much current does a 12-V, 3-W lamp draw?
27. The _____ is a polarized indicator lamp.
28. Tungsten has a _____ temperature coefficient.
29. With _____ voltage, only one electrode of neon lamps glows.
30. What is the purpose of the resistor in a neon lamp circuit?
31. What precaution should be observed when using wire-terminal lamps?
32. Why is it dangerous to put a 6-V lamp in a 28-V circuit?

4-10 Resistors

One of the most common and most reliable electric-electronic components is the resistor. The resistor is used as a load, or part of the load, in most electronic circuits. Its major purposes are to control current and divide voltage. Some types of resistors can operate at temperatures as high as 300°C (572°F). Resistors are manufactured in a wide range of resistance values. Resistances of less than 1 Ω to more than 100 MΩ are available. In fact, devices that have a resistance equivalent to a conductor but have the physical size and shape of a resistor are also manufactured. These devices are known as *zero-ohm resistors*. They are used in place of a short piece of wire (jumper wire) on a circuit board because they are easier to insert by automated machines.

Classification and Symbols

Resistors can be classified into four broad categories: fixed, variable, adjustable, and tapped. Sketches and symbols for these four categories are shown in Fig. 4-16. Notice that the variable and the adjustable resistor use the same symbol.

As shown in Fig. 4-16, there are two types of variable resistors. The *potentiometer* has three terminals. Rotation of the shaft changes the resistance between the middle terminal and the two end terminals.

Most potentiometers are *linear.* That is, one degree of shaft rotation results in the same change of resistance regardless of the shaft location. Other potentiometers have *nonlinear tapers.* This means that the rate of change of

Potentiometer

Nonlinear tapers

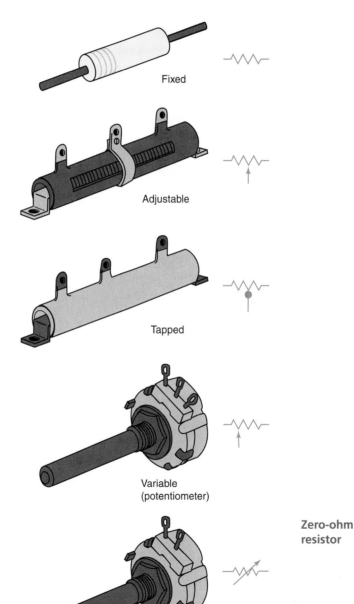

Fixed

Adjustable

Tapped

Variable
(potentiometer)

Variable
(rheostat)

Zero-ohm resistor

Fig. 4-16 Classification and symbols of resistors.

Tapped resistors

resistance varies as the shaft is rotated. A variety of tapers are available. Tapered potentiometers are sometimes used as volume and tone controls in stereo amplifiers.

Potentiometers are often ganged together so that the resistance in several circuits can be changed simultaneously. Figure 4-17 shows some *ganged potentiometers*. The left potentiometer in Fig. 4-17 is a dual-shaft unit. One shaft controls the ganged potentiometers, and the other shaft controls the switch. A triple-shaft potentiometer is also shown in Fig. 4-17. One shaft operates the switch, and the other shafts control the potentiometers. Often the same shaft operates both the switch and the potentiometer when the two are ganged together.

Ganged potentiometers

Power rating

Rheostats

Rheostats (Fig. 4-16) have only two terminals. Turning the shaft changes the resistance between these two terminals. Rheostats are used to adjust the current in a circuit to a specified value. Very often potentiometers are used as rheostats by not using one of the end terminals.

Adjustable resistors

Adjustable resistors serve about the same functions as potentiometers and rheostats. However, adjustable resistors are used in high-power circuits. They are used only when infrequent changes in resistance are required. Unlike rheostats and potentiometers, adjustable resistors are not usually adjusted while the circuit is in operation.

A single-tap resistor is illustrated in Fig. 4-16. Multiple-tap resistors are also available.

Tapped resistors, like adjustable resistors, are mostly used in high-power circuits (greater than 2 W).

Power Ratings

A resistor has a power rating as well as a resistance rating. The *power rating* of a resistor indicates the amount of power the resistor can safely dissipate without being destroyed. As current passes through the resistor, heat is produced. The resistor is converting electric energy into heat energy. If the current were allowed to increase, the heat would burn up the resistor. Thus, some safe level of heat must be specified, and this is what the power rating does.

There is no relationship between the resistance rating and the power rating of a resistor. The same resistance value can be obtained in resistors ranging in power ratings from less than 1 W to many watts. The power rating of a resistor is primarily determined by the physical size of the resistor and the type of materials used. Figure 4-18 shows, from top to bottom, resistors in the ¼-W, ½-W, 1-W, and 2-W size. The three resistors with the squared-off ends are carbon-composition resistors. Those with the rounded ends are metal-film resistors, which, for a given power rating, are slightly smaller than the carbon-composition resistors.

The power rating of a resistor is assigned by the manufacturer under specified conditions.

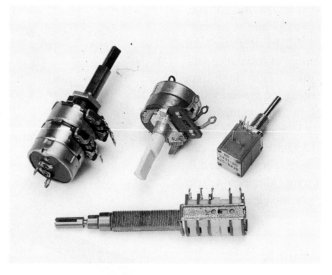

Fig. 4-17 Potentiometers. The potentiometer on the bottom row has three concentric shafts. *(Mark Steinmetz)*

These conditions include free air circulation around the resistor and resistor leads soldered to sizable terminals. Often these conditions are not fully satisfied in the devices that use resistors. Therefore, a resistor with a power rating higher than the value calculated for the circuit is used. As a rule of thumb, many circuit designers specify a power rating twice as great as the calculated value. In most applications this is a reasonable *safety factor*.

Example 4-3

How much power is used (dissipated) by a 1000-Ω resistor when 100 V is connected to it? What power rating would be specified for the resistor?

Given: Resistance (R) = 1000 Ω
Voltage (V) = 100 V

Find: Power (P)

Known: $P = \dfrac{V^2}{R}$

Solution: $P = \dfrac{(100\ V)(100\ V)}{1000\ \Omega} = \dfrac{10{,}000}{1000}$
= 10 W

Answer: The calculated power is 10 W. Using the rule of thumb mentioned above, the specified power rating should be twice the calculated value, or
Resistor rating = 10 W $\times$ 2
= 20 W
Therefore, use a 20-W resistor.

Resistor Tolerance

It is very difficult to mass-produce resistors that have exactly the same resistance value. Therefore, manufacturers specify tolerances for their resistors. *Tolerance* is specified as a percentage of the nominal (stated) resistance. For example, a 10 percent 1000-Ω resistor can have a resistance anywhere between 900 (1000 − 10 percent of 1000) and 1100 Ω (1000 + 10 percent of 1000).

Common tolerances for resistors are 10, 5, 2, and 1 percent. However, resistors are available with tolerances of less than 0.01 percent. Of course, low-tolerance resistors are more expensive than high-tolerance resistors.

Example 4-4

What is the maximum power a 250-Ω ± 10 percent resistor will dissipate when it is drawing 0.16 A of current?

Given: Resistor tolerance (T_R) = 10 percent
Nominal resistance (R_N) = 250 Ω
Current (I) = 0.16 A

Find: Power (P)

Known: $P = I^2 R$, $R_{max} = R_N + (T_R \times R_N)$,
and 10 percent = 0.1

Solution: R_{max} = 250 Ω + (0.1 $\times$ 250 Ω)
= 275 Ω
P = 0.16 A $\times$ 0.16 A $\times$ 275 Ω
= 7.04 W

Answer: The maximum power dissipated is 7.04 W.

Safety factor

Types of Resistors

Resistors are grouped according to the type of material or process used to make the resistive element. The major types and their characteristics are listed below.

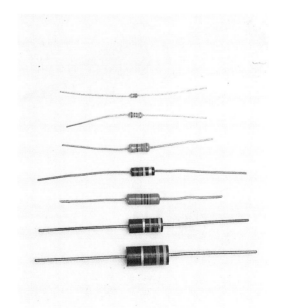

Fig. 4-18 Relative size and power rating of resistors. The physical size of a resistor determines its power rating rather than its resistance. *(Mark Steinmetz)*

Tolerances

Resistors Can Become Very Hot! A power resistor can operate at a temperature high enough to cause severe burns if touched before it has had time to cool.

Cermet

Power resistors

Deposited-film resistors

Carbon-film resistors

Metal-film resistors

Conductive plastics

Carbon-Composition The resistance element is made of finely powdered carbon held together with an inert binding material. The resistance of the element is determined by the ratio of carbon to binding material. The size determines the power rating, not the resistance. Carbon composition is used for both fixed and variable resistors. A fixed carbon-composition resistor is shown in Fig. 4-19(*a*). These resistors are reliable and relatively inexpensive. They are available in a wide range of resistances and in power ratings up to 2 W.

Wire-Wound Wire-wound resistors are used for all classes of resistors—variable, fixed, adjustable, and tapped. In fact, nearly all tapped and adjustable resistors are wire-wound and are classified as *power resistors* (above 2 W). The wire for wire-wound resistors is made of alloys such as nickel-chromium or copper-

nickel. The wire is wound (evenly spaced) on an insulator form, and the ends are connected to solder-coated copper leads. For fixed resistors, the entire assembly except for the leads is coated with insulating material.

Fixed wire-wound resistors come in many sizes and shapes. Some look just like carbon-composition resistors. Wire-wound resistors are made with power ratings that are in excess of 1000 W.

Wire-wound resistors have low temperature coefficients and good stability and can be made to close tolerances.

Cermet Cermet is a mixture of fine particles of glass (or ceramic) and finely powdered metal (or oxides), such as silver, platinum, or gold. This mixture, in a paste form, is applied to a base material made of ceramic, glass, or alumina. Leads are attached to the form and to the cermet mixture, and the whole device is fired (baked) in a kiln. The firing fuses everything together. The cermet is put on the substrate in a spiral ribbon. This effectively makes a long, thin resistive element for the resistor. The structure of a cermet (or cermet film, as it is sometimes called) resistor is shown in Fig. 4-19(*b*).

Cermet is used in variable as well as fixed resistors. It has a low temperature coefficient, and it is very stable and rugged. As a potentiometer, cermet offers fine adjustment and long life. As a fixed resistor, it is compact and durable.

Deposited-Film Deposited-film resistors are made like and look like the resistor in Fig. 4-19(*b*). The film is applied to the substrate by vaporizing (in a vacuum) the film material. The material forms a very thin, uniform film on the substrate. The film is often deposited in the form of a spiral ribbon.

Either carbon or metal may be deposited as a film. One of the nickel-chromium alloys is often used as the metal to be deposited. Deposited-film resistors are often referred to as either *carbon-film* or *metal-film resistors*. These film resistors have characteristics much like the cermet resistors. They operate well at high frequencies.

Conductive-Plastic Conductive plastics are used as the resistive element in some potentiometers. They are made by combining carbon powder with a plastic resin, such as polyester

Fig. 4-19 Resistors. The internal structure of various resistors is quite different. (*a*) Carbon-composition resistor. *(Courtesy of Allen-Bradley.)* (*b*) Cermet or film resistor.

Fig. 4-20   Surface-mount resistors. Notice the tinned end terminals.

or epoxy. The carbon-resin mixture is applied to a substrate, such as ceramic. Conductive-plastic potentiometers are relatively cheap. They wear very well in applications that require much adjusting of the potentiometer.

Surface-Mount   Many circuit components, especially resistors, capacitors, and solid-state devices, are made without leads for connecting them to tie points or holes in a circuit board trace (conductor). Such components are referred to as *surface-mount devices* (SMDs). They are also called *chip components.* Instead of having leads, these components have conductive terminal pads that make contact with (and are soldered to) the traces on the printed circuit board. In general, an SMD is smaller than its counterpart with leads, and it is also easier to attach to the printed circuit board.

Surface-mount resistors (*chip resistors*) are shown in Fig. 4-20. Because they are very small, their power rating is usually no greater than $\frac{1}{4}$ W. However, a wide range of resistance values is available.

The resistance value of a chip is indicated by a three- or four-digit number printed on the chip. With a three-digit number, the first two digits are the digits of the value, and the third (last) digit gives the power of 10 by which the first two digits are to be multiplied. The 391 printed on the chips in Fig. 4-20 indicates a resistance of 390 Ω ($39 \times 10^1 = 390$). With a four-digit number, the first three digits are the digits of the value, and the fourth digit is the power of 10 multiplier. For example, a resistor

chip labeled 5363 would have a value of 536,000 Ω (536 kΩ).

Color Codes

The resistance and tolerance of many fixed resistors are specified by color bands around the body of the resistors. Both resistance and tolerance are indicated by the color, number, and position of the bands. See Fig. 4-21 for the values assigned to the various colors. Fig. 4-21 also details the meaning of the location and number of color bands.

As shown in Fig. 4-21, the first band is always closer to one end of the resistor. Since 5 and 10 percent resistors are made only in values with two significant figures, only four color bands are needed to mark them. Three significant figures are common with 1 and 2 percent resistors; therefore, five color bands are needed to identify their resistance and tolerance. Examples of the use of the color code are given in Table 4-1.

To distinguish wire-wound resistors from other types of resistors, many manufacturers make the first color band on wire-wound resistors about twice as wide as the other color bands.

Some resistors have only three color bands. These are resistors with tolerances of ± 20 percent. They are not very common anymore.

Also, some 5 and 10 percent resistors manufactured to military specifications have five color bands. In this case, the first four bands are read the same as with the four-band system (the fourth band is either gold or silver). The fifth band indicates reliability. The *reliability* of a resistor tells what percentage of the resistors fail within 1000 h.

Careful examination of Fig. 4-21 reveals that the color codes can specify resistances from 0.10 to 999×10^9 Ω. However, as shown in Fig. 4-22, only certain values are commonly available for each tolerance. Figure 4-22 shows only the significant figures. Thus, for resistors with 10 and 5 percent tolerances, resistances of 0.22, 2.2, 22, 220, 2200, and so on, are available. In 2 and 1 percent tolerances, 2.26, 22.6, 226, and so on, are available. Note from Fig. 4-22 that only the 10/100 series is available in all tolerances.

Color code

Surface-mount device

JOB TIP

Learn how to disagree with fellow workers without offending them.

Chip resistor

Reliability

TABLE 4-1 Color Code Examples

Color of Band					Resistance	Tolerance
1st	**2d**	**3d**	**4th**	**5th**	**(Ω)**	**(%)**
Yellow	Violet	Orange	Silver	None	47,000	10
Red	Red	Red	Gold	None	2200	5
Brown	Black	Gold	Gold	None	1	5
Orange	White	Silver	Silver	None	0.39	10
Green	Blue	Red	Red	Brown	56,200	1
Violet	Gray	Violet	Silver	Red	7.87	2

Resistor Networks

Resistor networks are composed of many film resistors, such as cermet, on a single substrate. The resistors and substrate are housed in either a dual in-line package (DIP) or a single in-line package (SIP) like those in Fig. 4-23. These resistor networks are often used in digital electronic circuits on printed-circuit boards.

Special Resistors

A number of special resistors are manufactured. Three of them and their typical applications are listed below.

Fuse Resistor Fuse resistors have low resistance (less than 200 Ω) and limited current-carrying capacity. They are wire-wound resistors. These resistors are used in circuits that require high initial currents but low operating currents. If the current remains high, the resistor burns out (opens) and protects the circuit.

Thermal Resistor These resistors are commonly called *thermistors*. They are resistors which have a very high temperature coefficient. The large changes in resistance for small changes in temperature make the thermistor useful for measuring temperature. The thermistor is also used to compensate for other temperature-induced changes in electronic circuits.

Voltage-Sensitive or Voltage-Dependent Resistors (VDR) Also called *varistors*, these devices are used to protect circuits from very

Margin notes: Resistor networks · DIP · SIP

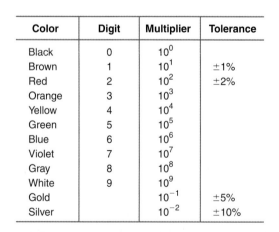

Color	Digit	Multiplier	Tolerance
Black	0	10^0	
Brown	1	10^1	$\pm 1\%$
Red	2	10^2	$\pm 2\%$
Orange	3	10^3	
Yellow	4	10^4	
Green	5	10^5	
Blue	6	10^6	
Violet	7	10^7	
Gray	8	10^8	
White	9	10^9	
Gold		10^{-1}	$\pm 5\%$
Silver		10^{-2}	$\pm 10\%$

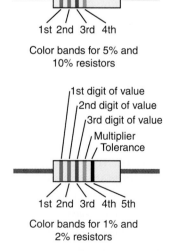

Color bands for 5% and 10% resistors

Color bands for 1% and 2% resistors

Fig. 4-21 Resistor color code. The 1 and 2 percent resistors have five color bands.

±1%	±2%	±5%	±10%	±1%	±2%	±5%	±10%
100	100	10	10	316	316		
102				324			
105	105			332	332	33	33
107				340			
110	110	11		348	348		
113				357			
115	115			365	365	36	
118				374			
121	121	12	12	383	383		
124				392		39	39
127	127			407	407		
130		13		412			
133	133			422	422		
137				432		43	
140	140			442	442		
143				453			
147	147			464	464		
150		15	15	475		47	47
154	154			487	487		
158				499			
162	162	16		511	511	51	
165				523			
169	169			536	536		
174				549			
178	178			562	562	56	56
182		18	18	576			
187	187			590	590		
191				604			
196	196			619	619	62	
200		20		634			
205	205			649	649		
210				665			
215	215			681	681	68	68
221		22	22	698			
226	226			715	715		
232				732			
237	237			750	750	75	
243		24		765			
249	249			787	787		
255				806			
261	261			825	825	82	82
267				845			
274	274	27	27	866	866		
280				887			
287	287			909	909	91	
294				931			
301	301	30		953	953		
309				976			

Fig. 4-22 Commonly available resistances in various tolerances. Numbers in tolerance columns represent only the significant digits of the resistance values available.

sudden increases in source voltage. If the sudden increase lasts only for a fraction of a second, the varistor limits the voltage to a safe value. These devices are often connected directly across the line input to voltage-sensitive devices such as computers.

■ TEST_____

Answer the following questions.

33. List the four categories of resistors.
34. A two-terminal variable resistor is called a
   _____ .

Fig. 4-23 Resistor networks. Two views of a DIP are shown on the left. The SIP is on the right. *(Mark Steinmetz)*

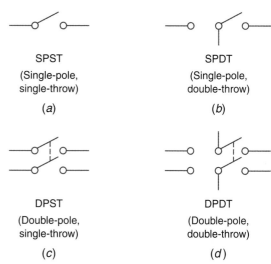

SPST (Single-pole, single-throw)	SPDT (Single-pole, double-throw)
(a)	(b)
DPST (Double-pole, single-throw)	DPDT (Double-pole, double-throw)
(c)	(d)

Fig. 4-25 Switch symbols and names.

35. A three-terminal variable resistor is called a _____.
36. The physical size of a given type of resistor determines the resistor's _____.
37. What are the resistance and tolerance of a resistor with the following color-band combinations:
    a. Brown, black, orange, silver
    b. Red, red, silver, gold
    c. Blue, gray, green
    d. Yellow, violet, black, silver
    e. Gray, red, green, orange, brown
38. What are the minimum and maximum permissible resistances of a 470-Ω, 5 percent resistor?
39. List four resistive materials used in variable resistors.
40. What is the power dissipated by a 270-Ω resistor with 6 V across it?
41. What type of resistor is generally used for high-power applications?

4-11 Switches

SPDT

There are many types of switches available for electric circuits (Fig. 4-24). They all perform the same basic function of opening or closing circuits. The type used in a given application is often a matter of style and/or convenience of operation. When the switching requirements are complex, the choice narrows to the rotary switch.

Types, Symbols, and Uses

Some of the more common types of small switches are shown in Fig. 4-24. These types are fairly simple switches which usually control only one or two circuit paths. Figure 4-25 shows the schematic symbols of these switches. The dashed line in Fig. 4-25(c) and (d) means that the two poles are mechanically connected but electrically insulated. Figure 4-26 shows the internal structure of a toggle switch similar to those used as wall switches in the home.

Controlling a lamp from two different locations is illustrated in Fig. 4-27. The circuit uses two single-pole, double-throw (*SPDT*) switches. As shown, the lamp would be off. Changing the position of either switch turns the lamp on. The lamp can be turned on or off by either

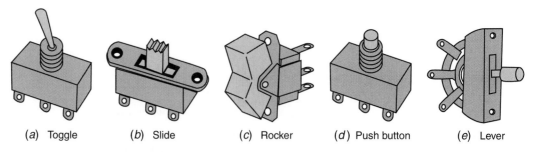

(a) Toggle   (b) Slide   (c) Rocker   (d) Push button   (e) Lever

Fig. 4-24 Some common types of switches.

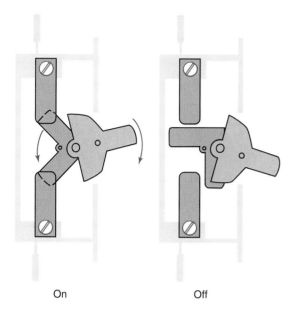

On   Off

Fig. 4-26 Internal structure of a switch.

switch regardless of the position of the other switch.

Some switches are constructed so that they always return to the same position when released by the operator. These are called *momentary-contact* switches. They can be either *normally open* (*NO*) or *normally closed* (*NC*). Schematic symbols for such switches are shown in Fig. 4-28. This type of switch is sometimes called *spring-loaded* because a spring returns it to its normal position.

Rotary switches are illustrated in Fig. 4-29. They range in complexity from a simple on-off switch to a multideck switch. Figure 4-30 shows the symbols and terminology used for rotary switches.

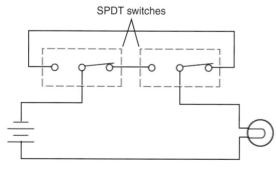

SPDT switches

Fig. 4-27 Controlling a lamp from two different locations using SPDT switches. The lamp can be turned off or on with either switch.

(*a*) Normally closed   (*b*) Normally open

Fig. 4-28 Symbols for momentary-contact switches.

Rotary switches come in either a *shorting* or a *nonshorting* configuration (Fig. 4-31). The shorting configuration is sometimes referred to as a *make-before-break* switch. This means the pole makes contact with the new position before it breaks contact with the old position. The wiper in Fig. 4-31(*a*) comes in contact with position 2 before it loses contact with position 1.

An example of the use of a rotary switch is shown in Fig. 4-32. This switch is a nonshorting switch. In the position shown, the amplifier receives its input from the CD player.

Ratings

Whenever an electric circuit is broken, an arc (ionization of the air) occurs. The greater the current or the voltage, the larger the arc. *Arcing* causes switch contacts to erode away. Therefore, switches have both a voltage and a current rating. Exceeding the manufacturer's ratings shortens the life of the switch. It can also be dangerous to the operator if the ratings are greatly exceeded. An arc can char and break down the insulation in the switch and thus put the operator in contact with the circuit.

Switches are usually given multiple current and voltage ratings. For example, a switch may be rated for 3 A at 250 V ac, 6 A at 125 V ac, and 1 A at 120 V dc. In general, a switch can handle higher alternating currents than direct currents.

Shorting

Nonshorting

Make-before-break

Normally open

Normally closed

Arcing

Spring-loaded

Rotary switches

▪ TEST

Answer the following questions.

42. What do the dashed (broken) lines in a switch symbol indicate?
43. Name at least four common types of switches.
44. Switches are rated for both _____ and _____.
45. When used to describe a switch, NO means _____.

Fig. 4-29 Rotary switches. These switches can provide many switching functions. *(Mark Steinmetz)*

46. A make-before-break switch is also called a _____ switch.

4-12 Wires and Cables

A *cable* is a multiple conductor device. The conductors of the cable are insulated from each other and are held together as a single unit. A *wire* is a single conductor. It may or may not be insulated. Typical cables are shown in Fig. 4-33.

A wire may be solid or stranded. A *solid conductor* is made from a single piece of low-resistance material, such as copper, aluminum, or silver. A *stranded conductor* is made from a

Solid conductor

Stranded conductor

Cable

Wire

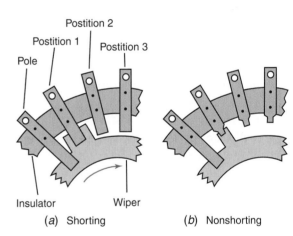

(a) Shorting   (b) Nonshorting

Fig. 4-31 Shorting and nonshorting rotary switches.

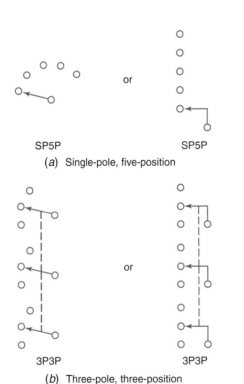

SP5P   or   SP5P

(a) Single-pole, five-position

3P3P   or   3P3P

(b) Three-pole, three-position

Fig. 4-30 Rotary switch symbols and names.

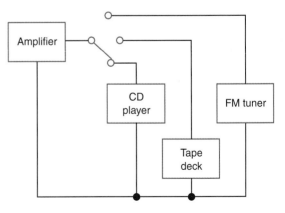

Fig. 4-32 Using a rotary switch to select various input devices.

number of wires twisted or braided together to form a single conductor. Stranded wire is more flexible than solid wire.

Smaller conductors used in electric wire and cable are often coated with a thin layer of solder. Bare copper conductors oxidize quite easily. The oxide must be removed before conductors can be joined by soldering. Thus, tinned wires make soldering easier.

Electric Cables

Cables may be either shielded or unshielded (Fig. 4-33). The *shielding* helps to isolate the conductors from electromagnetic fields in the vicinity of the cable. Either individual conductors or the whole cable may be shielded.

A wide variety of cables are manufactured for general-purpose and audio use. Shielded cables are used for such things as connecting a CD player or tape deck to an amplifier. Unshielded cables are used for such things as extension cords, telephone cords, and doorbell circuits.

Coaxial cable (Fig. 4-34) is a special-purpose, shielded cable used to connect antennas to receivers or transmitters. These cables are used for amateur radio, citizens band radio, and television antenna systems.

Much of the cable used with digital electronic systems is produced in a *flat ribbon* form. Figure 4-35 shows a typical example and

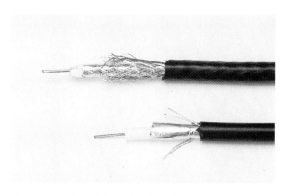

Fig. 4-34 Coaxial cable. *(Mark Steinmetz)*

also illustrates how this cable can be connected to flat cable plugs.

Power cables carrying very high currents often include a hollow tube through which oil is pumped to cool the cables.

Conductor (Wire) Specifications

The material used in making conductors for electric cables and wires is most often copper. However, aluminum is also used. The size of a round conductor (wire) is specified by a *gage number*. Each gage number represents a different diameter of wire. The diameter of wire is specified in mils. A mil is equal to 0.001 in. Thus, each gage number represents a certain diameter in mils. Electric wire is measured with the *American wire gage* (AWG), which is the same as the Brown and Sharpe gage used by machinists.

The smaller the gage number, the larger the diameter of the wire. A typical wire gage, such as that shown in Fig. 4-36(*a*), gives both the gage number and the diameter. The diameter is often printed on the reverse side of the gage. Figure 4-36(*b*) illustrates how to use the gage. The width of the slot leading to the hole in the gage indicates the gage size. The hole is larger

Shielding

Gage number

Coaxial cable

American wire gage

Flat ribbon

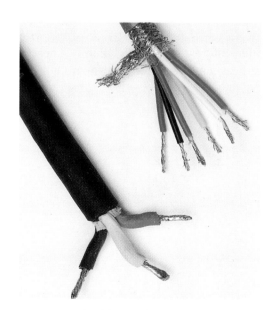

Fig. 4-33 Unshielded and shielded electric cables. *(Mark Steinmetz)*

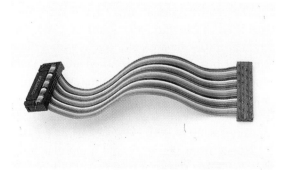

Fig. 4-35 Flat cable. *(Mark Steinmetz)*

(a)

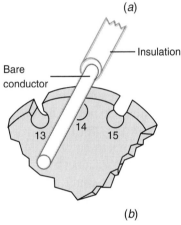

Insulation

Bare conductor

13   14   15

(b)

Fig. 4-36   American wire gage. (a) Gage.
(*Courtesy of L. S. Starrett Company.*)
(b) Using a gage.

Circular mils

than the wire and merely allows the wire to move freely through the slot. The slot through which the solid, bare copper wire passes smoothly but snugly represents the wire size. Stranded wire cannot be measured with the wire gage.

The cross-sectional area of a conductor is specified in *circular mils* (cmil or CM). One circular mil is the area of a circle with a diameter of 1 mil. The cross-sectional area of a conductor is equal to the diameter (in mils) squared. A conductor that is 30 mils in diameter would have a cross-sectional area of 900 cmil ($30 \times 30 = 900$).

Notice from Fig. 4-37 that a circular mil is not equal to a square mil. This could make the calculation of the resistance of a wire a diffi-

cult job. However, wire tables which list the resistance per 1000 ft of wire for various gage numbers are readily available. Such a table is provided in Appendix C. This table also gives the diameter and area of wires of different gages. Notice from the wire table that an increase of three gage numbers results in the area being reduced by about half. Of course, the resistance doubles when the cross-sectional area is halved. This leads to a simple method for remembering wire areas and resistances. Number 10 wire is about 10,000 cmil and 1 Ω/1000 ft. Thus, no. 13 wire (an uncommon size) is about 5000 cmil and 2 Ω/1000 ft. Three sizes smaller, no. 16 wire (a common size for extension cords) is about 2500 cmil and 4 Ω/1000 ft. Check these values against those in the table in Appendix C.

The amount of current a copper wire can safely carry depends on the cross-sectional area of the wire and how much heat the wire can dissipate before the insulation is damaged and the metal is affected. This in turn depends on where and how the wire is run. When air is freely circulating around a conductor, it can carry more current than when air circulation is restricted. For example, a no. 12 copper conductor used to wire a house can safely carry 20 A under certain conditions. The same size wire used in a transformer may carry only 4 A. In the house, the wire has about 325 cmil/A, while in the transformer it has more than 1500 cmil/A. In the house, 100 ft of the wire is strung over a large area. There is a lot of air for heat (created by the current in the wire) to transfer into. In the transformer, 100 ft of the wire is coiled into a very small volume. Therefore, less heat can be transferred away from the wire

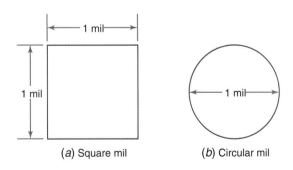

(a) Square mil   (b) Circular mil

Fig. 4-37   Square mil and circular mil. Notice that a circular mil has less cross-sectional area than a square mil.

in the transformer, and so the wire must carry less current.

Insulation Specifications

The insulation used on electric wires and cables has a number of important ratings. Each insulating material has a *maximum temperature rating*. This is the maximum continuous temperature to which the insulation should be exposed. Typical maximum temperatures range from 60°C for some thermoplastic compounds to 250°C for extruded polytetrafluoroethylene.

Voltage ratings for cables and wires depend on both the type of insulation and the thickness of the insulation. The voltage rating of an insulated wire is the maximum voltage which the insulation can continuously withstand. Voltage ratings vary from several hundred volts for appliance cords to several hundred thousand volts for high-voltage cables.

Insulating materials are also rated as to their relative resistance to damage by various materials and environments. They are rated on such things as resistance to water, acid, abrasion, flame, and weather (sun and extreme cold, for example).

Insulating materials used on cables and wires include neoprene, rubber, nylon, vinyl, polyethylene, polypropylene, polyurethane, varnished cambric, paper, silicone rubber, and many more.

Electric Wires

Besides building wires, which come in sizes from 18 AWG to as large as 2,000,000 cmil, electric wire is manufactured in an almost endless variety of sizes, styles, and ratings. Some of the common names used in identifying broad groups of electric wires are listed and defined below.

1. *Busbar wire.* An uninsulated, usually tinned, solid copper wire. Available in a wide range of sizes.
2. *Test-prod wire.* Primarily used to make test leads. High-voltage insulation (from 5000 to 10,000 V). Very fine strands for good flexibility.
3. *Hookup or lead wire.* Insulated wire used to connect electric components and devices together. May be either solid or stranded. Usually tinned. Available in a

wide variety of sizes and types of insulation.
4. *Magnet wire.* Used in magnetic devices, such as motors, transformers, speakers, and electromagnets. Always insulated, but the insulation is very thin and often has a copper color. Available in a wide variety of sizes and insulating materials.
5. *Litz wire.* A small, finely stranded, low-resistance, insulated wire. Used for making certain coils and transformers for electronic circuits.

■ TEST

Answer the following questions.

47. The cross-sectional area of round conductors is specified in _____ .
48. The _____ is used to measure the size of solid round conductors.
49. List at least three common insulating materials used on wires and cables.
50. Where is magnet wire used?
51. Which will carry more current, a no. 20 or no. 18 wire?

4-13 Fuses and Circuit Breakers

Protection of a circuit from excessive current is provided by a fuse or a circuit breaker. Protection is needed because an excessively high current can damage either the load or the source (or both). Higher than normal currents result from a defective circuit which has lower than normal resistance.

Shorts and Opens

In electricity and electronics, the term *short* or *short circuit* means that an undesired conductive path exists in a circuit. A partially shorted circuit [Fig. 4-38(*a*)] has only part of the load shorted out (bypassed). Partially shorted circuits have lower than normal resistance. When all the load is shorted out [Fig. 4-38(*b*)], the circuit is said to have a *dead short* or *direct short*. Such circuits have almost no resistance.

Figure 4-38(*c*) shows another way the term *short* is used. When a conductive path develops between the electric part of a device and the frame or housing that encloses the device, the device is said to be shorted or *grounded* to the frame.

Maximum temperature rating

*inter***NET** **CONNECTION**
You may be interested in visiting the Website for the American National Standards Institute.

Short circuit

Dead short

Grounded

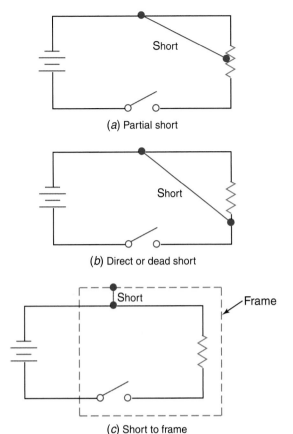

(a) Partial short

(b) Direct or dead short

(c) Short to frame

Fig. 4-38 Short circuits.

The resistors in Fig. 4-38 could just as well be heating elements in a toaster or coffee pot. The battery could be a 120-V outlet. If not properly wired and protected, a shorted circuit can cause a severe (even lethal) shock or start a fire.

The circuits in Fig. 4-38 could be protected by a fuse or a circuit breaker, as shown in Fig. 4-39. Note the symbols for fuses and circuit breakers. If a short occurs in one of the circuits in Fig. 4-39, the fuse or circuit breaker opens. In electricity, *open* means that the path for current has been interrupted or broken. When a fuse opens, we say, "The fuse has blown."

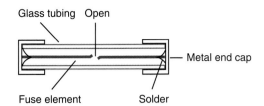

Fig. 4-40 Open fuse.

When a circuit breaker opens, we say, "The circuit breaker has tripped."

Traditional Fuses

A *fuse* opens when the current through it exceeds its current rating for a period of time. The fuse opens when its conducting element becomes hot enough to melt (Fig. 4-40). The element heats up because it has resistance. A 1-A fuse has approximately 0.13 Ω of resistance. Since $P = I^2R$, more current means more power. More power means more joules (of heat energy) per second converted by the fuse element.

A variety of fuses and fuse holders are shown in Fig. 4-41. The fuse in Fig. 4-41(*a*) is typical of those used in automobiles, airplanes, and electronic equipment. Figure 4-41(*b*) shows a subminiature fuse and holder of the type used on printed-circuit boards. The two prongs of the holder are to be inserted and soldered in the holes directly below them. Other styles of subminiature fuses and holders are illustrated in Fig. 4-41(*c*). The holder on the right is an indicating holder. When the fuse blows (opens), the lamp glows to indicate the fuse has blown. The fuse pictured in Fig. 4-41(*d*) is a pigtail type. Pigtail fuses are soldered directly into the circuit. They do not need a fuse holder.

Fuses have three important ratings, or characteristics, that need to be considered when buying or specifying fuses. These ratings apply

Fuse

Open

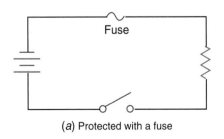

(a) Protected with a fuse

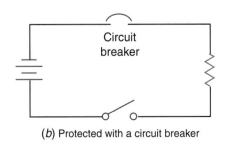

(b) Protected with a circuit breaker

Fig. 4-39 Circuits protected against excessive current.

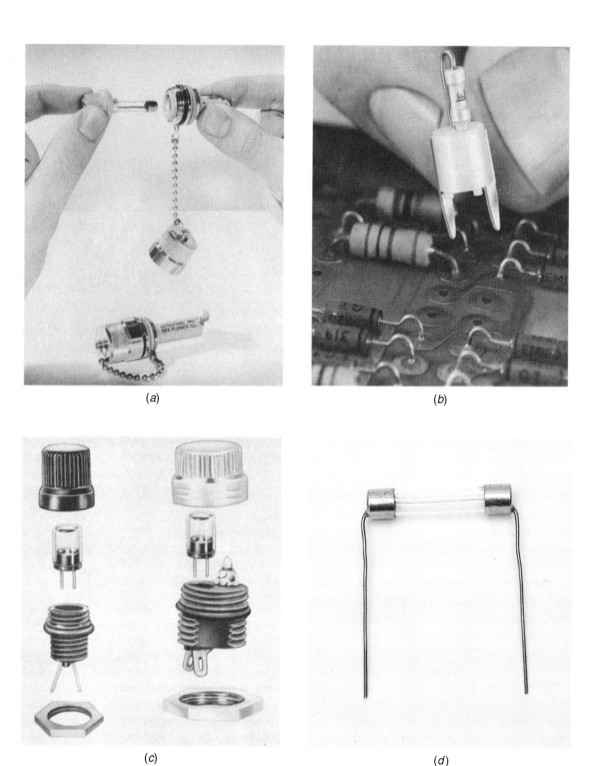

(a)

(b)

(c)

(d)

Fig. 4-41 Fuses and fuse holders. (a) Typical small glass fuse and holder. (b) Subminiature fuse and circuit board holder. (c) Subminiature fuses and holders. (d) Pigtail fuse. (Parts *a, b, and c courtesy of Littlefuse, Inc.; part d Mark Steinmetz*)

to all fuses regardless of size, shape, or style. These three characteristics are (1) a current rating, (2) a voltage rating, and (3) a blowing (fusing) characteristic.

Current ratings of fuses have already been mentioned. The current rating of a fuse is given for specific conditions of air circulation and temperature. These conditions are rarely obtained in typical applications of fuses. Therefore, fuses are usually operated well be-

Current rating

low their rated current. This practice avoids unnecessary blowing of fuses while still providing short-circuit protection.

Voltage rating

Voltage ratings for fuses are necessary for the same reasons that switches have voltage ratings. When a fuse opens, arcing and burning of the element occurs. If the current and voltage are high enough, the arcing/burning can continue until the fuse caps and holder are melted. Thus, fuses need voltage ratings. Voltage ratings of fuses are based on a power source capable of providing 10,000 A of current when dead-shorted. A 250-V fuse protects any 250-V power source that can provide no more than 10,000 A. Of course, the current rating of the fuse must be less than the current capabilities of the power source. Many power sources (small batteries and transformers, for example) can provide only a few amperes of current when short-circuited. With limited-current power sources, the voltage ratings of fuses are often exceeded. It is fairly common for electronic circuits with more than 400 V to be fused with 250-V fuses.

Slow-blow fuses

Blowing characteristic

The *blowing characteristic* of a fuse indicates how rapidly the fuse blows (opens) when subjected to specified overloads. There are three general categories of blowing characteristics of fuses. These are *fast-blow, medium-blow,* and *slow-blow.* These categories are sometimes referred to as *short-time lag, medium-time lag,* and *long-time lag,* respectively. All three categories respond rapidly (approximately 1 ms) to extreme overloads (more than 10 times the rated current). All three categories respond about the same to very small overloads. At 1.35 times (135 percent) the rated current, they all take a minute or two to open. In between these two extremes, the three categories differ dramatically. For example, at 5 times rated current, a slow-blow may take more than 1 s to blow while a fast-blow will usually open in less than 1 ms. Under the same conditions, a medium-blow would take about 10 ms.

Fast-blow fuses

Instrument fuses

Fast-blow fuses, sometimes called *instrument fuses,* are used to protect very sensitive devices, such as electric meters. The physical appearance of a fast fuse is the same as that of a medium fuse.

Medium-blow fuses

Medium-blow fuses (Fig. 4-42) are general-purpose fuses. They are used when the initial (turn on) current of a device is about the same as the normal operating current.

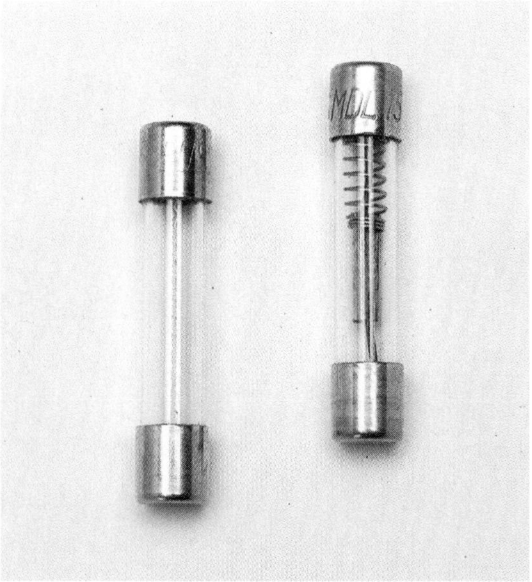

Fig. 4-42   Types of fuses. Medium-blow fuse on the left. Slow-blow fuse on the right. *(Mark Steinmetz)*

Slow-blow fuses are used whenever short-duration surge currents are expected. An electric motor is a good example of a device that requires much higher current to start than to continue operating. Its starting current may be five or six times the normal operating current. Therefore, a motor is often fused with a slow-blow fuse. A typical slow-blow fuse is shown in Fig. 4-42. When subjected to an extreme overload (short circuit), the thin part of the element burns open. When the overload is less severe and of long duration, the solder (holding the spring and element together) melts.

Always replace fuses with the style and ratings specified by the manufacturer of the equipment. Do not ignore the voltage ratings of fuses. If a medium or fast fuse is replaced by a slow-blow fuse, the equipment may be damaged before the fuse can respond. When ordering fuses, specify either current, voltage, blowing characteristic, physical dimensions, and style, or current and manufacturer's type number.

Fuses can be tested with an ohmmeter. If the fuse is blown, the ohmmeter will indicate an infinite resistance. The meter will show a very low resistance if the fuse is good. Fuses rated at less than ¼ A should not be checked on the $R \times 1$ range. The ohmmeter may provide enough current to blow the fuse. A higher range of the ohmmeter will provide less current and still indicate if the fuse has continuity, that is, if the fuse element still provides a continuous path. When checking a fuse with an ohmmeter, always remove the fuse from the circuit.

Resettable Fuses

A *resettable fuse* (in a surface-mount package) and its schematic symbol are shown in Fig. 4-43. Resettable fuses differ from traditional fuses in a number of ways:

1. The resettable fuse does not open (produce an open circuit); instead it trips. When it trips, its internal resistance greatly increases, and this reduces the circuit current to a value that is safe for the load and/or the source.

2. Once the circuit fault is corrected, this type of fuse resets itself and its internal resistance returns to a low value.
3. After a circuit fault is corrected, the resettable fuse does not need to be replaced the way a traditional fuse does.

Resettable fuses are available with either fast or medium time lag. Like traditional fuses, resettable fuses have both a current and a voltage rating. They are available with either axial or radial leads, as well as in the surface-mount package shown in Fig. 4-43. Notice how small this surface-mount device is.

Resettable fuses

Circuit Breakers

Circuit breakers have one big advantage over traditional fuses. When they open (trip), they can be reset. Nothing needs to be replaced. Once the overload that caused the breaker to trip has been corrected, the breaker can be reset and used again. There are two types of trip mechanisms: thermal and magnetic.

Thermal circuit breakers are the more common type for such things as small motors, household circuits, and battery chargers. Thermal circuit breakers can have either manual or automatic reset. The automatic reset breaker resets after the breaker has cooled. Cooling takes a few minutes. Automatic reset is used where overloads are self-correcting. Thermal breakers use bimetallic strips or disks as their current-sensing element. Therefore, they have long-time lag characteristics.

Thermal circuit breakers

YOU MAY **RECALL** . . . that the operating principle of a bimetallic strip is illustrated in Fig. 4-13 and described in Sec. 4-9.

For manual reset breakers, a mechanical mechanism locks the bimetallic strip in the open position once the breaker is tripped.

Two styles of manual reset thermal breakers are shown in Fig. 4-44. The one in Fig. 4-44(*b*) is a combination switch/breaker. The breaker is reset by moving the handle all the way to the *off* position and then to the *on* position. The breaker in Fig. 4-44(*a*) is a typical push-button breaker. When the breaker is tripped, the button extends out so that the white band shows. Pushing the button resets the breaker.

The external appearance of a magnetic cir-

(*a*)

(*b*)

Fig. 4-43 Resettable fuse. (*a*) SMD resettable fuse. *(Courtesy of Raychem Corp.)* (*b*) Symbol for a resettable fuse.

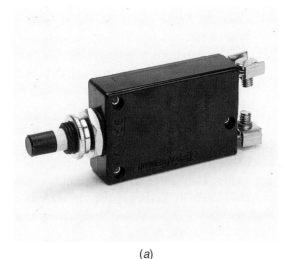

(a)

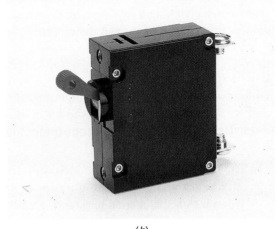

(b)

Fig. 4-44   Thermal circuit breakers. (*a*) Push-button reset. (*b*) Switch/breaker. *(Mark Steinmetz)*

Magnetic circuit breakers

cuit breaker is like that of the thermal breaker. However, its internal structure, principle of operation, and characteristics are quite different. Magnetic breakers can be made with time lags ranging from a few milliseconds to many seconds.

Figure 4-45 shows a cutaway view of a *magnetic circuit breaker* in the tripped position. The four shields between the breaker contacts help to reduce the arc that may occur when the contacts are opened. The three small terminals located between the two breaker terminals connect to a SPDT switch. Although the switch is operated by the breaker mechanism, it is electrically isolated from the circuit protected by the breaker. Thus, this switch can be used to control a separate circuit which remotely indi-

cates whether or not the breaker is tripped. The large circuit breakers that protect high-current loads or conductors often have both thermal and magnetic mechanisms. The thermal trip opens when the overload is small and a quick break in the circuit is not needed. A large rush of current, however, causes the breaker to open quickly via the magnetic trip mechanism.

4-14   Other Components

Capacitors, inductors, and transformers are other commonly used circuit components. However, these devices cannot be adequately appreciated until one develops an understanding of alternating current and magnetism. Capacitors, inductors, and transformers are discussed in detail in later chapters.

Fig. 4-45   Magnetic circuit breaker. *(Courtesy of Heinemann Electric Company.)*

■ TEST _____

Answer the following questions.

52. Define the following terms:
    a. Open circuit
    b. Short circuit
    c. Slow-blow
53. What is an indicating holder?
54. Three major ratings of a fuse are
    _____ , _____ , and _____ .
55. Fuses protect a circuit from _____ .
56. Two types of circuit breakers are the
    _____ type and the _____ type.
57. What is an instrument fuse?

Summary

1. Cells are energy or power sources.
2. Cells provide dc voltage.
3. A cell contains an electrolyte and two electrodes.
4. A battery is made from two or more cells.
5. Primary cells are not rechargeable.
6. Secondary cells are rechargeable.
7. Dry cells may use either a paste or a liquid electrolyte.
8. Gases are produced when cells are charged or discharged.
9. The energy storage capacity of a cell or battery is expressed in ampere-hours.
10. The energy available from a battery is dependent on temperature, rate of discharge, and final voltage.
11. Internal resistance causes a cell's voltage to decrease as the current increases.
12. Polarization refers to the buildup of gas ions around an electrode.
13. Common secondary cells are lead-acid, nickel-cadmium, and rechargeable alkaline.
14. Common primary cells are carbon-zinc, alkaline, mercury, and silver oxide.
15. When a lead-acid cell is discharging, sulfuric acid is being converted to water.
16. Specific gravity is measured with a hydrometer.
17. Safety glasses should be worn when working with lead-acid cells and batteries.
18. Electrolyte on the skin or clothing should be immediately removed by flushing with lots of water.
19. Carbon-zinc cells are relatively inexpensive.
20. Alkaline cells store more energy than carbon-zinc cells. The former are more efficient at high current drains.
21. Both mercury and silver oxide cells have nearly constant output voltage.
22. Incandescent miniature lamps have tungsten filaments.
23. Miniature lamps have both a current and a voltage rating.
24. Bimetallic strips bend because the two metals have different coefficients of expansion.
25. The resistance of a lamp when hot is greater than its resistance when cold.
26. Only one electrode glows when a neon lamp is operated on direct current.
27. A neon lamp circuit must have a resistor to limit the current through the lamp.
28. A potentiometer is a variable resistor.
29. A potentiometer can also be used as a rheostat.
30. The power rating of a resistor is independent of its resistance.
31. Common types of resistors are carbon-composition, cermet, wire-wound, deposited-film, and conductive-plastic.
32. Resistance values and tolerances are indicated with a color code on the body of a resistor.
33. Thermistors have high temperature coefficients.
34. Common types of small switches are rotary, toggle, slide, rocker, and push-button.
35. Spring-loaded, momentary-contact switches can be either normally open or normally closed.
36. Rotary switches can handle complex switching involving many circuits.
37. Switches have both a current and a voltage rating.
38. Cables are multiple conductors.
39. Wires are single conductors.
40. Conductors may be stranded or solid.
41. Shielded wires and cables are used to help isolate a conductor's electromagnetic fields.
42. Coaxial cables are used to connect antennas to receivers or transmitters.
43. The American wire gage is the standard used to specify the size of a conductor.
44. The cross-sectional area of a conductor is specified in circular mils.
45. When the gage of a conductor is decreased by three gage numbers, its cross-sectional area doubles. Its resistance is half as much.
46. Insulation used on conductors has both a temperature and a voltage rating.
47. Circuit-breaker mechanisms work on either a magnetic or a thermal principle.
48. Fuses are thermally operated devices.
49. A blown fuse results in an open circuit.

50. Instrument (fast-blow) fuses are used to protect electric meters.
51. Resettable fuses do not produce an open circuit when tripped.
52. SMDs are smaller than traditional components.
53. Fuses and breakers have both current and voltage ratings.
54. The characteristics of various dry cells are given in Table 4-2.

TABLE 4-2 Dry Cell Characteristics						
Cell Type	Rated Voltage	Voltage during Discharge	Cycle Life	Initial Cost	Internal Resistance	Energy-to-Weight Ratio
Carbon-zinc	1.5	Decreases		Low	Med	Med
Alkaline	1.5	Decreases		Med	Low	Med
Mercury	1.35	Almost constant		High	Low	High
Silver oxide	1.5	Almost constant		High	Low	High
Lithium	2.5–3.6	Almost constant		High	Med high	Very high
Rechargeable alkaline	1.5	Decreases	Low	Med	Very low	Low
Nickel-cadmium	1.25	Decreases slightly	High	High	Very low	Low

Chapter Review Questions

For questions 4-1 to 4-21, determine whether each statement is true or false.

4-1. The amount of energy stored in a cell is specified in ampere-hours.

4-2. A device which contains two or more cells is called a battery.

4-3. All secondary cells are wet cells.

4-4. The negative and positive terminals of a cell are produced by polarization of the cell.

4-5. Specific gravity is a number which represents the ratio of the weight of a substance to the weight of air.

4-6. Lead sulfate is being converted to lead peroxide when a cell is being charged.

4-7. The specific gravity of a lead-acid cell decreases when the cell is being charged.

4-8. The resistance of a lamp is greatest when the lamp is hot.

4-9. Copper is commonly used for the filaments of miniature lamps.

4-10. Both electrodes glow when direct current is applied to a neon lamp.

4-11. A neon lamp circuit uses a resistor only if the voltage in the circuit is more than 20 percent above the lamp's rated voltage.

4-12. The control element in a flashing miniature lamp is a bimetallic strip.

4-13. Light-emitting diodes operate as well on alternating as on direct current.

4-14. The abbreviation NC denotes *no connection* when it is used in switch specifications.

4-15. The deposited-film method is usually used to make power resistors.

4-16. The greater the resistance of a resistor, the larger its physical size.

4-17. The dashed lines between the poles of a switch symbol indicate that the poles are electrically connected to each other.

4-18. When a shorting-type switch is rotated, all positions of the switch are momentarily shorted together.

4-19. Switches have both a current and a voltage rating.

4-20. A circular mil is equal to a square mil.

4-21. The larger the diameter of a wire, the larger the gage number of the wire.

For questions 4-22 to 4-26, choose the letter that best completes each sentence.

4-22. Cells that cannot be recharged are classified as
   a. Primary cells
   b. Secondary cells
   c. Dry cells
   d. Wet cells

4-23. Which one of the following cells is rechargeable?
   a. Nickel-cadmium
   b. Mercury
   c. Lithium
   d. Carbon-zinc
4-24. Which one of the following cells provides the
least constant output voltage?
   a. Silver oxide
   b. Lithium
   c. Carbon-zinc
   d. Nickel-cadmium
4-25. A variable resistor with three terminals is a
   a. Varistor
   b. Thermostat
   c. Rotary resistor
   d. Potentiometer
4-26. A switch with five poles and four positions
would most likely be a
   a. Toggle switch

   b. Slide switch
   c. Rotary switch
   d. On-off switch

Answer the following questions:

4-27. What is the nominal resistance of a resistor
that is color-banded yellow, violet, gold, gold?
4-28. What is the nominal resistance of a resistor
that is color-banded red, red, blue, black,
brown?
4-29. Name two types of circuit breakers.
4-30. What are the three electrical ratings used in
specifying fuses?
4-31. Describe a short circuit.
4-32. What type of fuse would you use to protect an
electric motor?
4-33. What is the minimum resistance of a 4700-Ω,
10 percent resistor?
4-34. Which protection device does not produce an
open circuit when tripped?

Critical Thinking Questions

4-1. What are the resistance and power of the load
in example 4-1?
4-2. Would you expect the internal resistance of a
carbon-zinc cell to increase or decrease with
decreasing temperature? Why?
4-3. What assumption must be made about the
power sources used in the examples in Chap.
4? Why?
4-4. Discuss other factors besides those dealt with
in this chapter that might be considered in selecting cells and batteries.
4-5. Why do switches have a higher ac than dc rating?

4-6. Why does only one electrode in a neon
lamp glow when the power source is direct
current?
4-7. What is the maximum power dissipated by a
100-Ω, 5 percent resistor connected to a 50-V
source with 5 Ω of internal resistance?
4-8. Determine the minimum voltage across a 50-Ω,
10 percent resistor connected to a 60-V source
with 3 Ω of internal resistance.
4-9. A 24-V battery stores 1.728×10^7 J of energy.
Determine its Ah rating.

Answers to Tests

1. A cell is a device
that converts chemi-
cal energy to electric
energy. A cell is also
known as an energy
or power source.
2. A battery is an elec-
tric device composed
of two or more cells.

3. A cell is composed
of two electrodes
and an electrolyte.
4. a. Primary cell is a
cell that is not
rechargeable.
   b. Secondary cell is
a cell that is
rechargeable.

   c. Dry cell is a cell
that can be oper-
ated in any posi-
tion. It is sealed.
   d. Wet cell is a
vented cell that
must be operated
in an upright
position.

   e. Ampere-hour is a
rating of a cell
which indicates
the amount of
energy it stores.
5. rechargeable alka-
line, lead-acid,
nickel-cadmium, and
nickel-iron.

6. alkaline, carbon-zinc, mercury, silver oxide, zinc chloride, and lithium.

7. decrease

8. polarization

9. Lead and lead peroxide are changed to lead sulfate while sulfuric acid is changed to water. This process leaves one plate deficient in electrons and the other plate with an excess of electrons.

10. The battery is recharged by using the battery as a load rather than as a source. Electrons are forced into the negative terminal and out of the positive terminal. The chemical reaction is also reversed.

11. Charge at, or below, the rate recommended by the manufacturer, and do not overcharge the battery.

12. Specific gravity is the ratio of the weight of a substance to the weight of water.

13. hydrometer

14. discharged

15. sulfuric acid

16. Wear safety glasses, avoid spilling the electrolyte, and charge in a spark-free, ventilated area.

17. no

18. a. The nickel-cadmium has a more constant output voltage.
    b. The nickel-cadmium provides 3 to 30 times as many cycles.
    c. The alkaline is much cheaper.

19. a. The alkaline cell is much better for providing high current loads.
    b. Alkaline cells maintain higher voltage under very heavy current loads.
    c. Alkaline cells store more energy for a given weight.

20. carbon-zinc

21. nickel-cadmium

22. lithium

23. tungsten

24. incandescent and neon glow

25. A bimetallic strip which acts like a switch as it heats and cools

26. 0.25 A

27. LED

28. positive

29. dc

30. The resistor limits current through the lamp.

31. Do not bend or solder leads within 3 mm of the bulb.

32. The lamp may explode.

33. variable, adjustable, tapped, and fixed

34. rheostat

35. potentiometer

36. power rating

37. a. $10,000 \, \Omega \pm 10$ percent
    b. $0.22 \, \Omega \pm 5$ percent
    c. $6,800,000 \, \Omega \pm 20$ percent
    d. $47 \, \Omega \pm 10$ percent
    e. $825 \, k\Omega \pm 1$ percent

38. Minimum resistance is 446.5 Ω, and maximum resistance is 493.5 Ω.

39. carbon-composition, conductive-plastic, cermet, deposited-film, and wire

40. 0.133 W

41. wire-wound

42. The dashed line indicates that two poles are mechanically connected but electrically isolated.

43. rotary, slide, rocker, toggle, and push-button

44. voltage, current

45. normally open

46. shorting

47. circular mils

48. American wire gage

49. vinyl, rubber, neoprene, nylon, asbestos, and other plastic materials

50. electromagnets, motors, generators, speakers, transformers

51. no. 18 wire

52. a. Open circuit means the conductive path for current has been broken.
    b. Short circuit means the circuit has an abnormally low resistance.
    c. Slow-blow means a fuse can stand a short-term overload without blowing.

53. An indicating holder is a fuse holder which lights up when the fuse has blown.

54. current, voltage, blowing characteristic

55. excessive current

56. thermal, magnetic

57. a fast-blow fuse

Chapter 5

Multiple-Load Circuits

Chapter Objectives

This chapter will help you to:

1. *Identify and classify* multiple-load circuits.
2. *List and understand* the characteristics of series, parallel, and series-parallel circuits.
3. *Measure* correctly the current, voltage, and/or resistance in any part of a multiple-load circuit.
4. *Calculate* power, current, voltage, and/or resistance for the total circuit or any load in a multiple-load circuit.
5. *Understand* Kirchhoff's laws and use them in conjunction with Ohm's law to solve circuit problems.
6. *Convert* from resistance to conductance and vice versa.
7. *Understand* the relationship between maximum power transfer and efficiency.

A great majority of electric circuits operate more than one load. Circuits which contain two or more loads are called *multiple-load circuits.* A multiple-load circuit can be a series circuit, a parallel circuit, or a series-parallel circuit.

5-1 Subscripts

Notice the symbols R_1, R_2, and R_3 in Fig. 5-1. The *subscripts* "1," "2," and "3" are used to identify the different-load resistors in the circuit. Two-level subscripts are also used in these types of circuits. For example, I_{R_2} is used to indicate the current flowing through resistor R_2. The symbol V_{R_1} indicates the voltage across resistor R_1. Similarly, P_{R_3} is used to specify the power used by R_3.

Electrical quantities for a total circuit are identified by the subscript T. Thus, the voltage of the battery of Fig. 5-1 is indicated by the symbol V_T. The symbol I_T indicates the source (battery) current, and R_T represents the total resistance of all the loads combined. The power from the battery or source is identified as P_T.

5-2 Power in Multiple-Load Circuits

The total power taken from a source, such as a battery, is equal to the sum of the powers used by the individual loads. As a formula, this statement is written

$$P_T = P_{R_1} + P_{R_2} + P_{R_3} + \text{etc.}$$

The "+ etc." means the formula can be expanded to include any number of loads. Regardless of how complex the circuit, the above formula is appropriate.

Subscripts

5-3 Series Circuits

A *series circuit* contains two or more loads but only one path for current to flow from the source voltage through the loads and back to the source. Figure 5-1 is an example of a series circuit.

Series circuit

Current in Series Circuits

In Fig. 5-1, the battery current I_T flows through the first load R_1, the second load R_2, and the third load R_3. If 1 A flows through R_1, then 1 A also flows through R_2 and R_3. And, of course, the battery provides 1 A of current. In

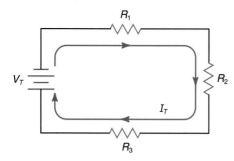

Fig. 5-1 Series circuit. There is only one path for current.

symbolic form, the current relationship in a series circuit is

$$I_T = I_{R_1} = I_{R_2} = I_{R_3} = \text{etc.}$$

One path for current

Current in a series circuit can be measured by inserting a meter in series. Since there is only *one path for current*, any part of the circuit can be interrupted to insert the meter. All the meters in Fig. 5-2 give the same current reading provided the voltage and the total resistance are the same in each case.

Resistance in Series Circuits

Total resistance

The *total resistance* in a series circuit is equal to the sum of the individual resistances around the series circuit. This statement can be written as

$$R_T = R_1 + R_2 + R_3 + \text{etc.}$$

This relationship is very logical if you remember two things: (1) resistance is opposition to current and (2) current has to be forced through all the resistances in a series circuit.

The total resistance R_T can also be determined by Ohm's law if the total voltage V_T and total current I_T are known. Both methods of determining R_T can be seen in the circuit of Fig. 5-3. In this figure, the "2 A" near the ammeter symbol means that the meter is indicating a current of 2 A. The resistance of each resistor is given next to the resistor symbol. Using Ohm's law, the total resistance is

$$R_T = \frac{V_T}{I_T} = \frac{90\,\text{V}}{2\,\text{A}} = 45\,\Omega$$

Using the relationship for series resistance yields

$$\begin{aligned}
R_T &= R_1 + R_2 + R_3 \\
&= 5\,\Omega + 10\,\Omega + 30\,\Omega \\
&= 45\,\Omega
\end{aligned}$$

The total resistance of a series circuit can be measured by connecting an ohmmeter across the loads as in Fig. 5-4(*a*). The power source must be disconnected from the loads. Individual resistances can be measured as shown in Fig. 5-4(*b*) and (*c*).

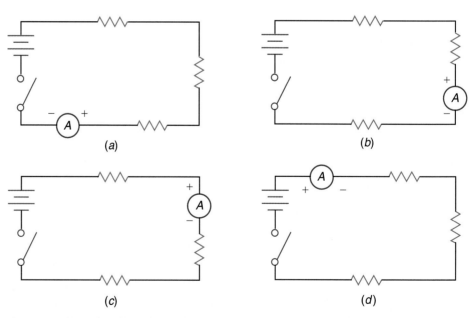

(a)

(b)

(c)

(d)

Fig. 5-2 Measuring current in a series circuit. The ammeter will indicate the same value of current in any of the positions shown in diagrams (a) to (d).

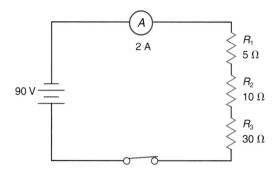

Fig. 5-3 Total resistance can be found by using Ohm's law or by adding individual resistances.

Voltage in Series Circuits

The battery voltage in Fig. 5-3 divides up across the three load resistors. It always di-

vides up so that the sum of the individual load voltages equals the source of voltage. That is,

$$V_T = V_{R_1} + V_{R_2} + V_{R_3} + \text{etc.}$$

This relationship is often referred to as *Kirchhoff's voltage law.* Kirchhoff's law states that "the sum of the voltage drops around a circuit equals the applied voltage."

Kirchhoff's voltage law

For Fig. 5-3, this relationship can be verified by using Ohm's law to determine the individual voltages:

$$V_{R_1} = I_{R_1} \times R_1 = 2\text{ A} \times 5\ \Omega = 10\text{ V}$$
$$V_{R_2} = I_{R_2} \times R_2 = 2\text{ A} \times 10\ \Omega = 20\text{ V}$$
$$V_{R_3} = I_{R_3} \times R_3 = 2\text{ A} \times 30\ \Omega = 60\text{ V}$$
$$\begin{aligned} V_T &= V_{R_1} + V_{R_2} + V_{R_3} \\ &= 10\text{ V} + 20\text{ V} + 60\text{ V} \\ &= 90\text{ V} \end{aligned}$$

Notice from the above calculations that individual voltages are directly proportional to individual resistances. The voltage across R_2 is twice the voltage across R_1 because the value of R_2 is twice the value of R_1.

Voltage Drop and Polarity

The voltage, or potential energy difference, across a resistor is referred to as a *voltage drop.* We can say a voltage develops across the resistor. That is, part of the potential energy difference of the source develops, or appears, across each resistor as a smaller potential energy difference. There is a distinction between source voltage and the voltage across the loads. The source voltage provides the electric energy, and the load voltage converts the electric energy into another form.

Voltage drop

The voltage drop across a resistor has *polarity.* However, in this case the polarity does not necessarily indicate a deficiency or excess of electrons. Instead, it indicates the direction of current flow and the conversion of electric energy to another form of energy. Current moves through a load resistor from the negative polarity to the positive polarity. This means that electric energy is being converted to another form. The current *through* a battery moves from the positive to the negative polarity. Thus, the battery is providing the electric energy.

Polarity

Polarity signs are shown on the resistors in Fig. 5-5. The current through R_1 and R_2 flows from the negative end of the resistors to the

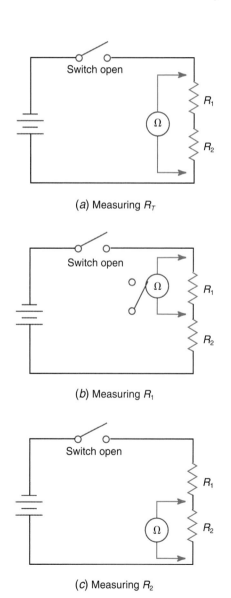

(a) Measuring R_T

(b) Measuring R_1

(c) Measuring R_2

Fig. 5-4 Measuring resistance in series.

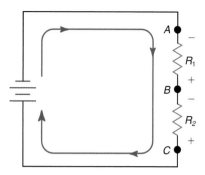

Fig. 5-5 Voltage polarity in a series circuit.

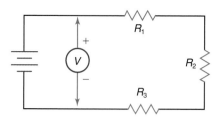

Fig. 5-6 Measuring the total voltage of a series circuit.

positive end. Current inside the battery flows from the positive to the negative. However, the external current still flows from the negative terminal to the positive terminal of the battery.

Notice in Fig. 5-5 that point *B* is labeled both negative (−) and positive (+). This may seem contradictory, but it is not. Point *B* is *positive with respect to point A,* but *negative with respect to point C.* It is very important to understand the phrase "with respect to." Remember that voltage is defined as "a potential energy difference between two points." Therefore, it is meaningless to speak of a voltage or a polarity at point *B*. Point *B* by itself has neither voltage nor polarity. But, with respect to either point *A* or point *C*, point *B* has both polarity and voltage. Point *B* is positive

with respect to point *A*; this means that point *B* is at a lower potential energy than point *A*. It also means that electric energy is being converted to heat energy as electrons move from point *A* to point *B*.

Measuring Series Voltages

The total voltage of a series circuit must be measured across the voltage source, as shown in Fig. 5-6. Voltages across the series resistors are also easily measured. The correct connections for measuring series voltages are illustrated in Fig. 5-7. The procedure is to first determine the correct function, range, and polarity on the meter. Then touch the meter leads to the two points where voltage is to be measured. Voltmeters have a very high internal resistance, so high that connecting them to a circuit has no noticeable effect on many circuits. For now,

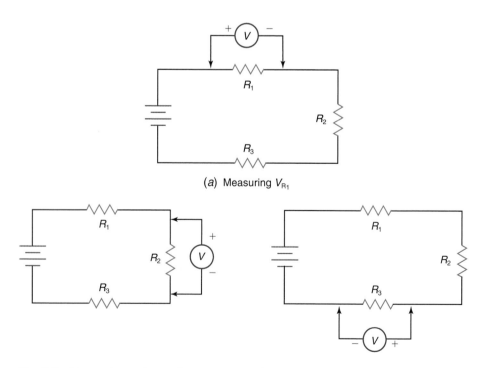

Fig. 5-7 Measuring voltage drops across each resistor in a series circuit.

assume that meters (both ammeters and volt-meters) do not change the circuits in which they are used. Also, assume that the voltage source has insignificant internal resistance.

Open in Series Circuits

In a series circuit, if any part of the circuit is *open,* current stops flowing and voltage and power are removed from all loads. This is one of the weaknesses of a series circuit. For example, when one lamp in a series (such as Christmas tree lights) burns out (opens), all lamps go out.

An easy way to determine which load in a series is open is to measure the individual voltages. The load that is open will have a voltage drop equal to the entire source voltage. In Fig. 5-8(a) a meter connected across either of the good resistors reads 0 V. Since R_2 is open, no current flows in the series circuit. A meter across the open resistor R_2, however, as in Fig. 5-8(b), reads approximately 50 V. When a voltmeter is connected across R_2 as in Fig. 5-8(b), a small current flows through R_1, the voltmeter, and R_3. In most circuits, the internal resistance of the voltmeter is very, very high compared with the other series loads. Therefore, nearly all the battery voltage is developed (and read) across the voltmeter. It should be noted that 50 V is also across R_2, in Fig. 5-8(a). It has to be to satisfy Kirchhoff's voltage law. That is, $V_{R_1} + V_{R_2} + V_{R_3}$ has to equal the 50 V of the source.

The diagrams of Fig. 5-8 show the open resistor. However, in a real physical circuit, R_2 may look just like R_1 and R_3. There may be no physical evidence to indicate that it is open. A technician repairing the circuit would have to interpret the voltmeter readings to conclude that R_2 is open.

Shorts in Series Circuits

When one load in a series circuit is *shorted out,* the other loads *may continue to operate.* Or, one of the other loads *may open* because of increased voltage, current, and power. In Fig. 5-9(a), each lamp is rated at 10 V and 1 A. Each lamp has a power rating of 10 W ($P = IV = 1$ A $\times$ 10 V). When lamp L_2 shorts out, as shown in Fig. 5-9(b), the 30-V source must divide evenly between the two remaining lamps. This means each remaining lamp must drop 15 V. Since the lamp voltages increase by 50 percent, the lamp currents also increase by 50 percent to 1.5 A (this assumes the individual lamp resistance does not change). With 15 V and 1.5 A, each lamp has to dissipate 22.5 W. One or both of the lamps will soon burn out (open).

The effects of one shorted load in a series can be summarized as follows:

1. The total resistance decreases.
2. The total current increases.
3. The voltage across the remaining loads increases.
4. The power dissipation of the remaining load increases.
5. The total power increases.

Shorted

Open

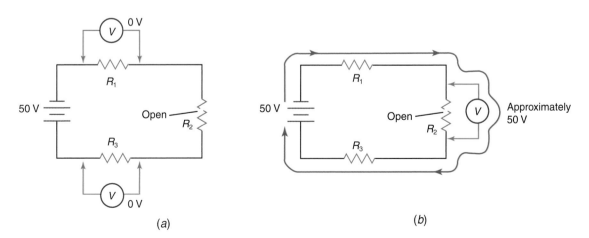

Fig. 5-8 Voltage drops across (*a*) normal loads and (*b*) open loads.

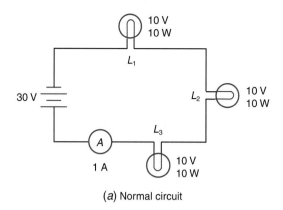

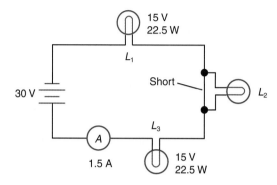

Fig. 5-9 Effects of a shorted series load.

6. The resistance, voltage, and power of the shorted load decrease. If the short is a dead short, these quantities decrease to zero.

Solving Series-Circuit Problems

Proper use of Ohm's law and the series relationships of current, voltage, resistance, and power solves most series problems. In using Ohm's law, you should develop the habit of *using subscripts* for voltage, current, and resistance. Without subscripts, it is easy to forget which voltage, current, or resistance to substitute into the formula. If you are using Ohm's law to find the voltage across R_1, write

$$V_{R_1} = I_{R_1}R_1$$

If you are calculating the total voltage, you should write

$$V_T = I_T R_T$$

Example 5-1

As shown in Fig. 5-10, an 8-V, 0.5-A lamp is to be operated from a 12.6-V battery. What resistance and wattage rating are needed for R_1?

Given: $V_T = 12.6$ V

$V_{L_1} = 8$ V

$I_{L_1} = 0.5$ A

Find: R_1, P_{R_1}

Known: $R_1 = \dfrac{V_{R_1}}{I_{R_1}}$

$V_T = V_{L_1} + V_{R_1}$

$I_T = I_{R_1} = I_{L_1}$

$P_{R_1} = V_{R_1}I_{R_1}$

Solution: $I_{R_1} = I_{L_1} = 0.5$ A

$V_T = V_{L_1} + V_{R_1}$

Therefore,

$V_{R_1} = V_T - V_{L_1}$

$V_{R_1} = 12.6$ V $- 8$ V $= 4.6$ V

$R_1 = \dfrac{4.6\text{ V}}{0.5\text{ A}} = 9.2 \ \Omega$

$P_{R_1} = V_{R_1}I_{R_1}$

$= 4.6$ V $\times 0.5$ A

$= 2.3$ W

Answer: The calculated values for R_1 are 9.2 Ω and 2.3 W. Use a 5-W resistor to provide a safety factor.

Example 5-2

Find the total current and total resistance of the circuit in Fig. 5-11. Also determine the voltage across each resistor.

Given: $V_T = 90$ V
$R_1 = 35\ \Omega$
$R_2 = 70\ \Omega$
$R_3 = 45\ \Omega$

Find: $I_T, R_T, V_{R_1}, V_{R_2}, V_{R_3}$

Known: $I_T = \dfrac{V_T}{R_T}$

$R_T = R_1 + R_2 + R_3$
$V_{R_1} = I_{R_1} R_1$
$I_T = I_{R_1} = I_{R_2} = I_{R_3}$

Solution: $R_T = 35\ \Omega + 70\ \Omega + 45\ \Omega$
$= 150\ \Omega$
$I_T = \dfrac{90\text{ V}}{150\ \Omega} = 0.6$ A
$V_{R_1} = 0.6\text{ A} \times 35\ \Omega = 21$ V
$V_{R_2} = 0.6\text{ A} \times 70\ \Omega = 42$ V
$V_{R_3} = 0.6\text{ A} \times 45\ \Omega = 27$ V

Answer: $I_T = 0.6$ A, $R_T = 150\ \Omega$
$V_{R_1} = 21$ V
$V_{R_2} = 42$ V
$V_{R_3} = 27$ V

After you solve a complex problem, it is a good idea to *cross-check* the problem for mathematical errors. This can usually be done by checking some relationship *not used* in originally solving the problem. The problem of example 5-2 can be cross-checked by using Kirchhoff's voltage law:

$$V_T = V_{R_1} + V_{R_2} + V_{R_3}$$
$$= 21\text{ V} + 42\text{ V} + 27\text{ V}$$
$$= 90\text{ V}$$

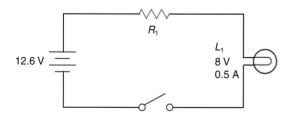

Fig. 5-10 Circuit diagram for example 5-1.

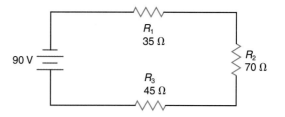

Fig. 5-11 Circuit diagram for example 5-2.

Since 90 V was specified for V_T, the cross-check verifies that at least the sum of the individual voltages is as it should be.

Voltage-Divider Equation

When you want to find the voltage across only one of the resistors in a series circuit, you can use the *voltage-divider equation*. This equation in its general form is

$$V_{R_n} = \frac{V_T R_n}{R_T}$$

where R_n is any one of the resistors in the series circuit. The logic of this equation is obvious if it is written in the form

$$V_{R_n} = \frac{V_T}{R_T} \times R_n = I_T \times R_n = I_{R_n} \times R_n$$

To illustrate the usefulness of the voltage-divider equation, let us use it to solve for V_{R_2} in Fig. 5-11. If we remember that $R_T = R_1 + R_2 + R_3$ for Fig. 5-11, we can write the voltage divider equation as

$$V_{R_2} = \frac{V_T R_2}{R_1 + R_2 + R_3}$$
$$= \frac{90\text{ V} \times 70\ \Omega}{35\ \Omega + 70\ \Omega + 45\ \Omega} = 42\text{ V}$$

Voltage-divider equation

Cross-check

Estimations, Approximations, and Tolerances

In a series circuit, the resistor with the most resistance dominates the circuit. That is, the highest resistance drops the most voltage, uses the most power, and has the most effect on the total current. Sometimes one resistor is so much larger than the other resistors that its value almost determines the total current. For example, resistor R_1 in Fig. 5-12 dominates the circuit. Shorting out R_2 in the circuit would increase the current from about 18 mA to only

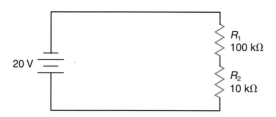

Fig. 5-12 Dominant resistance. Circuit current and power are largely determined by R_1.

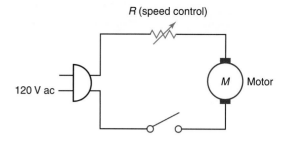

Fig. 5-13 Motor-speed control.

20 mA. Assume R_1 and R_2 are ± 10 percent resistors. Then R_1 could be as low as 90 kΩ and R_2 as low as 9 kΩ. If this were the case, the current in Fig. 5-12 would still be about 20 mA. The presence or absence of R_2 has no more effect on the current than the tolerance of R_1 does. Therefore, a good *estimate* of the current in Fig. 5-12 could be obtained by ignoring R_2. This estimate would be close enough for such things as

1. Determining the power rating needed for the resistors by using $P = I^2 R$
2. Determining which range of an ammeter should be used to measure the current
3. Estimating the power required from the battery

When estimating current in a series circuit, ignore the lowest resistance if it is less than the tolerance of the highest resistance.

Applications of Series Circuits

One application of series circuits has already been mentioned—Christmas tree lights. Other applications include (1) motor-speed controls, (2) lamp-intensity controls, and (3) numerous electronic circuits.

A simple *motor-speed control* circuit is shown in Fig. 5-13. This type of control is used on very small motors, such as the motor on a sewing machine. With a sewing machine, the variable resistance is contained in the foot control. The circuit provides continuous, smooth control of motor speed. The lower the resistance, the faster the motor rotates. A big disadvantage of this type of motor control is that it is inefficient. Sometimes the resistance converts more electric energy to heat than the motor converts to mechanical energy.

The intensity of an indicator lamp is often controlled by a *variable* series *resistor*. Such

circuits are used to illuminate dials and meters on radio and navigation equipment in airplanes. A typical circuit is illustrated in Fig. 5-14. In this circuit, increasing the resistance decreases the total current and the lamp intensity. Notice that the symbol for the variable resistor in Fig. 5-14 is different from the one used in Fig. 5-13. Either symbol is correct. They both are symbols for a *rheostat*.

A portion of a *transistor* circuit is shown in Fig. 5-15. Resistor R_1 is in series with the collector and emitter of the transistor. The transistor acts like a variable resistor. Its resistance is controlled by the current supplied to the base. Thus, controlling the current to the base controls the current through the transistor. This makes it possible for a transistor to *amplify*, or increase, the small current presented to the base.

▪ TEST

Answer the following questions.

1. Write the symbol for
   a. Voltage across resistor R_4
   b. Source current
   c. Current through resistor R_2
2. True or false. In any multiple-load circuit the total power is the sum of the individual powers.
3. True or false. When resistors are in series, they share a common current.

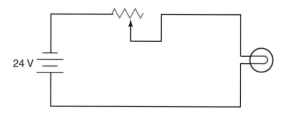

Fig. 5-14 Lamp-intensity control.

Estimate

Rheostat

Transistor

Amplify

Motor-speed control

Variable resistor

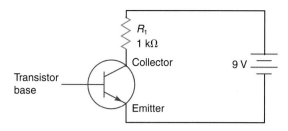

Fig. 5-15 Series resistor in a transistor circuit.

4. True or false. In a series circuit, the source current is equal to the current through any load.
5. Give two formulas for determining the total resistance of a series circuit.
6. Can the resistance of a resistor in a series circuit be measured without completely removing it from the circuit?
7. Write the formula that shows the relationship between total and individual voltages in a series circuit.
8. Which drops more voltage in a series circuit, a 100-Ω resistor or a 56-Ω resistor?
9. Does a negative-polarity marking on a voltage drop indicate an excess of electrons at that point?
10. A series circuit contains two resistors. One resistor is good, and the other is open. Across which resistor will a voltmeter indicate more voltage?
11. A series circuit contains three resistors: R_1, R_2, and R_3. If R_2 shorts out, what happens to each of the following:
    a. Current through R_1
    b. Total power
    c. Voltage across R_3
12. A 125-Ω resistor (R_1) and a 375-Ω resistor (R_2) are connected in a series to a 100-V source. Determine the following:
    a. Total resistance
    b. Total current
    c. Voltage across R_1
    d. Power dissipated by R_2
13. Refer to Fig. 5-10. Change L_1 to a 7-V, 150-mA lamp. What resistance is now needed for R_1? How much power does the battery have to furnish?
14. If the resistance in Fig. 5-14 decreased, what would happen to the intensity of the lamp?
15. If the transistor in Fig. 5-15 opened, how much voltage would be measured across R_1?

16. If the transistor in Fig. 5-15 shorted, how much current would flow through R_1?

5-4 Maximum Power Transfer

Maximum power transfer refers to getting the maximum possible amount of power from a source to its load. The source may be a battery, and the load a lamp, or the source could be a guitar amplifier and the load a speaker.

Maximum power transfer occurs when the source's internal opposition to the current equals the load's opposition to the current. Resistance is one form of opposition to current; *impedance* is another form of opposition. You will learn more about impedance when you study alternating current. In dc circuits, maximum power transfer occurs when resistances are matched.

Referring to Fig. 5-16 will aid you in understanding resistance matching and power transfer. In this figure, the internal resistance of the battery is represented by R_B. The battery B_1 represents what is called a *constant voltage source* or an *ideal voltage source*. That is, it represents a voltage source that has no internal resistance. The dotted line around R_B and B_1 means that R_B and B_1 together behave like a real battery. Resistance R_B and battery B_1 do not actually exist as separate components. They cannot be separated. Thus, they are enclosed in dotted lines. Together they form the source. The load is R_1. When R_1 equals R_B, the maximum amount of power is transferred from the source to the load. This statement can be proved by an example. First assign fixed values to R_B and B_1. Then calculate the power dissipated by R_1

Maximum power transfer

Impedance

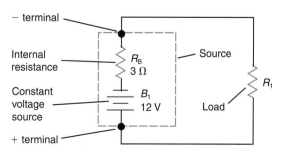

Fig. 5-16 Maximum power transfer occurs when R_B and R_1 are equal.

for various values of R_1. For example, when R_1 is 9 Ω, the series calculations are

$$R_T = R_1 + R_B = 9 \text{ Ω} + 3 \text{ Ω} = 12 \text{ Ω}$$
$$I_T = \frac{V_T}{R_T} = \frac{12 \text{ V}}{12 \text{ Ω}} = 1 \text{ A}$$
$$P_{R_1} = I_{R_1}^2 R_1 = (1 \text{ A})^2 \times 9 \text{ Ω} = 9 \text{ W}$$
$$P_{R_B} = I_{R_B}^2 R_B = (1 \text{ A})^2 \times 3 \text{ Ω} = 3 \text{ W}$$
$$P_T = P_{R_1} + P_{R_B}$$
$$= 9 \text{ W} + 3 \text{ W} = 12 \text{ W}$$

Calculations for other values of R_1 have been made and recorded in Table 5-1. Notice that the maximum power dissipation occurs when R_1 is 3 Ω.

Also shown in Table 5-1 are the power dissipated within the source and the total power taken from the source. Notice that when maximum power transfer occurs, the efficiency is only 50 percent. Of the 24 W furnished by the source, only 12 W is used by the load. As the load gets larger, the efficiency improves and the power transferred decreases.

■ TEST

Answer the following questions.

17. When is maximum power transferred from a source to a load?
18. For high efficiency, should the load resistance be equal to, less than, or greater than the internal resistance of the source?

5-5 Parallel Circuits

Parallel circuits

More than one path

Branch

Parallel circuits are multiple-load circuits which have *more than one path* for current. Each different current path is called a *branch*. The circuit in Fig. 5-17 has three branches. Current from the battery splits up among the three branches. Each branch has its own load,

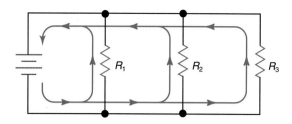

Fig. 5-17 Parallel circuit. There is more than one path for current to take.

and each branch is independent of all other branches. The current and power in one branch are not dependent on the current, resistance, or power in any other branch.

In Fig. 5-18, switch S_2 in the second branch controls the lamp in that branch. However, turning switch S_2 on and off does not affect the lamps in the other branches. Figure 5-18 illustrates the way in which the lamps in your home are connected. In a house, the electric circuits are parallel circuits. The fuse in Fig. 5-18 protects all three branches. It carries the current for all branches. If any one branch draws too much current, the fuse opens. Also, if the three branches together draw too much current, the fuse opens. Of course, when the fuse opens, all branches become inoperative.

Voltage in Parallel Circuits

All voltages in a parallel circuit are the same. In other words, the source voltage appears across each branch of a parallel circuit. In Fig. 5-19(*a*), each voltmeter reads the same voltage. Rearranging the circuit in Fig. 5-19(*a*) yields the circuit shown in Fig. 5-19(*b*). It is easier to see in Fig. 5-19(*b*) that each branch of the circuit receives the total battery voltage. For a parallel circuit, the relationship of source voltage to load voltage is expressed as

TABLE 5-1	Calculated Values for Fig. 5-16				
R_1 (Ω)	R_T (Ω)	I_T (A)	P_{R_1} (W)	P_{R_B} (W)	P_T (W)
1	4	3.00	9.00	27.00	36.00
2	5	2.40	11.52	17.28	28.80
3	6	2.00	12.00	12.00	24.00
4	7	1.71	11.76	8.82	20.57
5	8	1.50	11.25	6.75	18.00
6	9	1.33	10.67	5.33	16.00

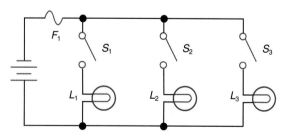

Fig. 5-18 Independence of parallel branches. Opening or closing S_2 has no effect on L_1 or L_3.

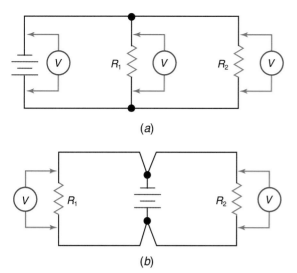

(a)

(b)

Fig. 5-19 Voltage measurement in parallel circuits. All voltmeters will indicate the same value of voltage.

$$V_T = V_{R_1} = V_{R_2} = V_{R_3} = \text{etc.}$$

In a parallel circuit, the voltage measured across the load does not change if the load opens. If R_2 in Fig. 5-19(a) were open, the voltmeter across R_2 would still measure the voltage of the battery.

Current in Parallel Circuits

The relationship of the currents in a parallel circuit is as follows:

$$I_T = I_{R_1} + I_{R_2} + I_{R_3} + \text{etc.}$$

In other words, the total current is equal to the sum of the individual *branch currents*. Figure 5-20 shows all the places where an ammeter could be inserted to measure current in a parallel circuit. The current being measured at each location is indicated inside the meter symbol.

The total current I_T in Fig. 5-20 splits into two smaller currents at junction 1 (I_{R_1} and $I_{R_2} + I_{R_3}$). A similar division of current occurs in the other three junctions in Fig. 5-20. The various currents entering and leaving a junction are related by *Kirchhoff's current law*. This law states that "the sum of the currents entering a junction equals the sum of the currents leaving a *junction*." No matter how many wires are connected together at a junction, Kirchhoff's current law still applies. Thus, in Fig. 5-20, the following relationships exist:

$$I_T = I_{R_1} + I_A$$
$$I_A = I_{R_2} + I_{R_3}$$
$$I_{R_2} + I_{R_3} = I_B$$
$$I_{R_1} + I_B = I_T$$
$$I_B = I_A$$

In Fig. 5-21, five wires are joined at a junction. If three of the wires carry a total of 8 A into the junction, the other two must carry a total of 8 A out of the junction. Therefore, the unmarked wire in Fig. 5-21 must be carrying 3 A out of the junction.

Branch currents

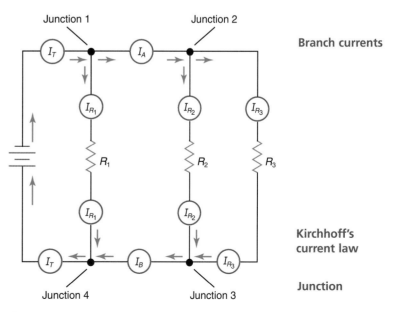

Kirchhoff's current law

Junction

Fig. 5-20 Current measurement in parallel circuits.

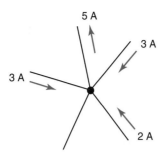

Total resistance

Lowest branch
resistance

Fig. 5-21 Current at a junction. The unmarked
line must carry 3 A of current away
from the junction.

As shown in Fig. 5-22, the branch of a parallel circuit with the lowest resistance dominates the circuit. That is, the lowest resistance takes the most current and power from the source. Ohm's law shows why this is so:

$$I = \frac{V}{R}$$

With V equal for all branches, a low value of R results in a high value of I. As R increases, I decreases. Since V is the same for all branches, the branch with the most current uses the most power ($P = IV$).

In Fig. 5-22, removing the 10-kΩ resistor would reduce the total power and current about 10 percent. Following the logic used for series circuits, engineers have developed a *rule of thumb for parallel circuits*. When *estimating* total current and power, ignore a parallel resistor whose resistance is 10 times higher than that of the other resistor. This rule assumes the resistors have a 10 percent tolerance. If the tolerance is 5 percent, the rule is modified to read "20 times higher." In Fig. 5-22, ignoring the 10-kΩ resistor results in an estimated total current of 10 mA. If both resistors were on the high side of their 10 percent tolerance, the actual total current would also be 10 mA. Of

**Rule of thumb
for parallel
circuits**

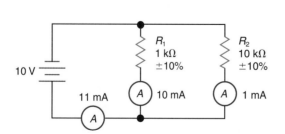

Fig. 5-22 Resistor tolerance and estimated current. Notice that the small resistance carries nearly all the circuit current.

course, if both resistors were on the low side of their tolerance, the actual current would then be 12.2 mA.

Resistance in Parallel Circuits

The *total resistance* of a parallel circuit is always less than the *lowest branch resistance*. It may seem illogical at first that adding more resistors to a parallel circuit decreases the total resistance. The logic of the above statements can be shown through the use of Ohm's law and reference to Fig. 5-23. In Fig. 5-23(*a*) the current in the circuit is

$$I = \frac{V}{R} = \frac{10 \text{ V}}{10,000 \ \Omega}$$
$$= 0.001 \text{ A} = 1 \text{ mA}$$

Adding R_2 in parallel, as in Fig. 5-23(*b*), does not change either the resistance of R_1 or the voltage across R_1. Therefore, R_1 will still draw 1 mA. The current drawn by R_2 can also be calculated:

$$I_{R_2} = \frac{V_2}{R_2} = \frac{10 \text{ V}}{100,000 \ \Omega}$$
$$= 0.0001 \text{ A} = 0.1 \text{ mA}$$

By Kirchhoff's current law, the total current in Fig. 5-23(*b*) is 1.1 mA ($I_T = I_{R_1} + I_{R_2}$). Now, if the total voltage is still 10 V and the total current has increased, the total resistance had to decrease. By Ohm's law, the total resistance of Fig. 5-23(*b*) is

$$R_T = \frac{V_T}{I_T} = \frac{10 \text{ V}}{0.0011 \text{ A}}$$
$$= 9091 \ \Omega = 9.09 \text{ k}\Omega$$

Notice that the addition of a 100-kΩ resistor in parallel with a 10-kΩ resistor *reduces* the total resistance. Also, notice that the total resistance is less than the lowest (10-kΩ) resistance. In Fig. 5-23(*c*), a 1-kΩ resistor has been added to the circuit of Fig. 5-23(*b*). The total resistance of Fig. 5-23(*c*) is

$$R_T = \frac{V_T}{I_T} = \frac{10 \text{ V}}{0.0111 \text{ A}}$$
$$= 900.9 \ \Omega = 0.901 \text{ k}\Omega$$

Again, notice that R_T has decreased and is less than the lowest resistance (1 kΩ).

In the above examples, the total resistance was calculated by using Ohm's law and the circuit voltage and current. A formula to deter-

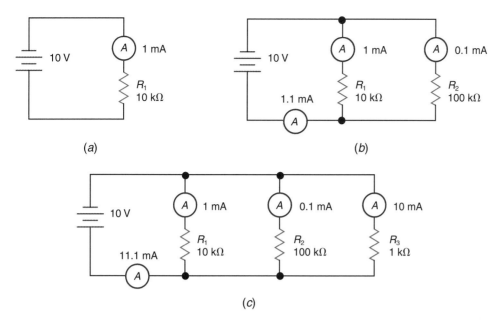

Fig. 5-23 Total resistance in parallel circuits. Adding resistors, as in circuits (b) and (c), increases total current and decreases total resistance.

mine the total resistance directly from the branch resistance can be developed using Ohm's law and the current and voltage relationships in a parallel circuit. From parallel-circuit relationships, $I_1 + I_2 + I_3$ + etc. can be substituted for I_T, and V_T can be substituted for V_{R_1} or V_{R_2}, etc. Since, by Ohm's law, V_{R_1}/R_1 can be substituted for I_{R_1}, and V_T can be substituted for V_{R_1}, we can substitute V_T/R_1 for I_{R_1}. Now, starting with Ohm's law for a parallel circuit, we can write

$$R_T = \frac{V_T}{I_T} = \frac{V_T}{I_{R_1} + I_{R_2} + I_{R_3} + \text{etc.}}$$

$$= \frac{V_T}{\dfrac{V_T}{R_1} + \dfrac{V_T}{R_2} + \dfrac{V_T}{R_3} + \text{etc.}}$$

Finally, both the numerator and the denominator of the right side of the equation can be divided by V_T to yield

$$R_T = \frac{1}{\dfrac{1}{R_1} + \dfrac{1}{R_2} + \dfrac{1}{R_3} + \text{etc.}}$$

This formula is often referred to as the *reciprocal formula* because the reciprocals of the branch resistances are added, and then the reciprocal of this sum is taken to obtain the total (equivalent) resistance.

Example 5-3

What is the total resistance of three resistors—20 Ω, 30 Ω, and 60 Ω—connected in parallel?

Given: $R_1 = 20 \ \Omega$
$R_2 = 30 \ \Omega$
$R_3 = 60 \ \Omega$

Find: R_T

Known: $R_T = \dfrac{1}{\dfrac{1}{R_1} + \dfrac{1}{R_2} + \dfrac{1}{R_3}}$

Solution: $R_T = \dfrac{1}{\dfrac{1}{20} + \dfrac{1}{30} + \dfrac{1}{60}}$

$= \dfrac{1}{\left(\dfrac{6}{60}\right)} = 10 \ \Omega$

Answer: The total resistance is 10 Ω.

Instead of using fractions to calculate the total resistance, you can convert the reciprocals to their *decimal equivalents*. Let us solve the problem in example 5-3 using the decimal equivalents:

Reciprocal formula

Decimal equivalents

$$R_T = \frac{R}{n}$$

$$R_T = \cfrac{1}{\cfrac{1}{20} + \cfrac{1}{30} + \cfrac{1}{60}}$$

$$= \frac{1}{0.05 + 0.033 + 0.017}$$

$$= \frac{1}{0.100} = 10 \ \Omega$$

Two resistors in parallel

When only *two resistors* are *in parallel,* a simplified formula can be used to solve parallel-resistance problems. This simplified formula is derived from the reciprocal formula. It is

$$R_T = \frac{R_1 \times R_2}{R_1 + R_2}$$

Example 5-4

What is the total resistance of a 27-Ω resistor in parallel with a 47-Ω resistor?

Given: $R_1 = 27 \ \Omega$
$R_2 = 47 \ \Omega$

Find: R_T

Known: $R_T = \dfrac{R_1 \times R_2}{R_1 + R_2}$

Solution: $R_T = \dfrac{27 \times 47}{27 + 47} = \dfrac{1269}{74}$
$= 17.1 \ \Omega$

Answer: The total resistance is 17.1 Ω.

Equivalent resistance

The simpler formula can also be used for circuits containing more than two resistors. The process is to find the *equivalent resistance* of R_1 and R_2 in parallel, label it $R_{1,2}$, and then use this equivalent resistance and R_3 in a second application of the formula to find R_T. As an illustration, let us solve example 5-3 using this two-step method.

$$R_{1,2} = \frac{R_1 \times R_2}{R_1 + R_2} = \frac{20 \times 30}{20 + 30} = \frac{600}{50} = 12 \ \Omega$$

$$R_T = \frac{R_{1,2} \times R_3}{R_{1,2} + R_3} = \frac{12 \times 60}{12 + 60} = \frac{720}{72} = 10 \ \Omega$$

When all the resistors in a parallel circuit have the same value, the total resistance can be found easily. Just divide the value of a resistor by the number of resistors. That is, $R_T = R/n$, where n is the number of resistors. For example, three 1000-Ω resistors in parallel have a total resistance of

$$R_T = \frac{R}{n} = \frac{1000}{3}$$
$$= 333.3 \ \Omega$$

Two 100-Ω parallel resistors have an equivalent resistance of

$$R_T = \frac{R}{n} = \frac{100}{2} = 50 \ \Omega$$

Measuring Resistance in Parallel

The total resistance of a parallel circuit is measured in the same way that it is measured in other types of circuits: the power source is disconnected and the resistance is measured across the points where the power was applied.

To measure an individual resistance in a parallel circuit, one end of the load must be disconnected from the circuit. The correct technique is shown in Fig. 5-24(*a*). When the load is not disconnected, as illustrated in Fig. 5-24(*b*), the meter again reads the total resistance.

Solving Parallel-Circuit Problems

Now that we know the relationship between individual and total resistance, current, voltage, and power, we can solve parallel-circuit problems. These relationships, plus Ohm's law and the power formula, allow us to solve most parallel-circuit problems. The formulas listed

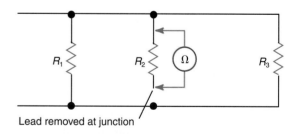

(*a*) Measuring R_2

(*b*) Measuring R_T

Fig. 5-24 Measurement of resistance in parallel.

below will be the "Known" for examples that follow.

Parallel-Circuit Formulas

$$I_T = I_{R_1} + I_{R_2} + I_{R_3} + \text{etc.}$$

$$V_T = V_{R_1} = V_{R_2} = V_{R_3} = \text{etc.}$$

$$R_T = \cfrac{1}{\dfrac{1}{R_1} + \dfrac{1}{R_2} + \dfrac{1}{R_3} + \text{etc.}}$$

$$P_T = P_{R_1} + P_{R_2} + P_{R_3} + \text{etc.}$$

$$I = \frac{V}{R}$$

$$P = IV = I^2R = \frac{V^2}{R}$$

Example 5-5

For the circuit in Fig. 5-25, find I_T, R_T, and P_T.

Given: $V_T = 10\ \text{V}$ $\quad R_1 = 100\ \Omega$

$P_{L_1} = 2\ \text{W}$ $\quad I_{R_2} = 0.5\ \text{A}$

Find: I_T, R_T, P_T

Known: Parallel-circuit formulas

Solution: For branch I:

$$I_{L_1} = \frac{P_{L_1}}{V_{L_1}} = \frac{2\ \text{W}}{10\ \text{V}} = 0.2\ \text{A}$$

For branch II:

$$I_{R_1} = \frac{V_{R_1}}{R_1} = \frac{10\ \text{V}}{100\ \Omega} = 0.1\ \text{A}$$

For the total circuit:

$$I_T = I_{L_1} + I_{R_1} + I_{R_2}$$
$$= 0.2\ \text{A} + 0.1\ \text{A} + 0.5\ \text{A}$$
$$= 0.8\ \text{A}$$

$$R_T = \frac{V_T}{I_T} = \frac{10\ \text{V}}{0.8\ \text{A}} = 12.5\ \Omega$$

$$P_T = I_T V_T = 0.8\ \text{A} \times 10\ \text{V}$$
$$= 8\ \text{W}$$

Answer: The total current is 0.8 A. The total resistance is 12.5 Ω. The total power is 8 W.

Example 5-6

With the data found in example 5-5, find the total resistance using the reciprocal formula.

Solution: $\quad R_{L_1} = \dfrac{V_{L_1}}{I_{L_1}} = \dfrac{10\ \text{V}}{0.2\ \text{A}} = 50\ \Omega$

$$R_2 = \frac{V_{R_2}}{I_{R_2}} = \frac{10\ \text{V}}{0.5\ \text{A}} = 20\ \Omega$$

$$R_T = \cfrac{1}{\dfrac{1}{50} + \dfrac{1}{100} + \dfrac{1}{20}}$$

$$= \frac{1}{0.02 + 0.01 + 0.05}$$

$$= \frac{1}{0.08} = 12.5\ \Omega$$

Answer: The total resistance is 12.5 Ω.

Using the reciprocal formula can be a laborious task—especially when the resistances do not have an easily determined common denominator. This task is greatly simplified by using one of the many computer programs available for analyzing or simulating electrical or electronic circuits.

Current-Divider Formula

When you are interested in finding the current through only one of two parallel resistors, you can use the *current-divider formula*. This formula is

$$I_{R_1} = \frac{I_T R_2}{R_1 + R_2} \qquad \text{or} \qquad I_{R_2} = \frac{I_T R_1}{R_1 + R_2}$$

Current-divider formula

The current-divider formula can easily be derived by substituting parallel relationships and formulas into the Ohm's law expression for I_{R_1}. The derivation is

$$I_{R_1} = \frac{V_{R1}}{R_1} = \frac{R_T I_T}{R_1} = \cfrac{\left(\dfrac{R_1 R_2}{R_1 + R_2}\right) I_T}{R_1}$$

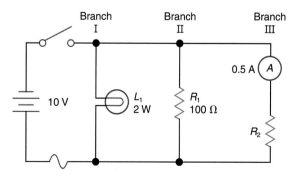

Fig. 5-25 Circuit for example 5-5.

$$= \frac{R_1 R_2 I_T}{(R_1 + R_2)R_1} = \frac{I_T R_2}{R_1 + R_2}$$

We can demonstrate the usefulness of the current-divider formula by solving for I_{R_1} in the circuit shown in Fig. 5-26:

$$I_{R_1} = \frac{I_T R_2}{R_1 + R_2} = \frac{2 \text{ A} \times 47 \text{ }\Omega}{22 \text{ }\Omega + 47 \text{ }\Omega} = 1.36 \text{ A}$$

Calculating I_{R_1} in Fig. 5-26 without the current divider formula involves the following steps:

1. Calculate R_T using the reciprocal formula.
2. Calculate V_T using Ohm's law.
3. Calculate I_{R_1} using Ohm's law.

Applications of Parallel Circuits

An electric system in which one section can fail and other sections continue to operate has parallel circuits. As previously mentioned, the electric system used in homes consists of many parallel circuits.

An automobile electric system uses parallel circuits for lights, heater motor, radio, etc. Each of these devices operate independently of the others.

Individual television circuits are quite complex. However, the complex circuits are connected in parallel to the main power source.

That is why the audio section of television receivers can still work when the video (picture) is inoperative.

5-6   Conductance

So far in this text, we have considered only a resistor's opposition to current. However, no resistor completely stops current. Therefore, we could just as well consider a resistor's ability to *conduct* current. Instead of considering a resistor's resistance, we could consider a resistor's conductance. *Conductance* refers to the ability to conduct current. It is symbolized by the letter G. The base unit for conductance is the *siemens,* abbreviated S, in honor of the inventor Ernst von Siemens.

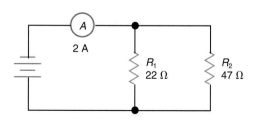

Fig. 5-26   The current-divider formula can solve for I_{R_1} without calculating R_T or V_T.

Conductance

Siemens

Conductance is the exact opposite of resistance. In fact, the two are mathematically defined as reciprocals of each other. That is,

$$G = \frac{1}{R} \quad \text{and} \quad R = \frac{1}{G}$$

Thus, a 100-Ω resistor has a conductance of $^1/_{100}$, or 0.01, siemens (S).

Using the relationship $R = 1/G$ and the series-resistance formula, we can determine total conductance of conductance in series.

$$R_T = R_1 + R_2 + R_3 + \text{etc.}$$

so

$$\frac{1}{G_T} = \frac{1}{G_1} + \frac{1}{G_2} + \frac{1}{G_3} + \text{etc.}$$

Taking the reciprocal of both sides yields

$$G_T = \frac{1}{\dfrac{1}{G_1} + \dfrac{1}{G_2} + \dfrac{1}{G_3} + \text{etc.}}$$

The formula for total conductance of parallel conductances can be found in a like manner.

$$R_T = \frac{1}{\dfrac{1}{R_1} + \dfrac{1}{R_2} + \dfrac{1}{R_3} + \text{etc.}}$$

Taking the reciprocal of both sides gives

$$\frac{1}{R_T} = \frac{1}{R_1} + \frac{1}{R_2} + \frac{1}{R_3} + \text{etc.}$$

so

$$\frac{1}{\dfrac{1}{G_T}} = \frac{1}{\dfrac{1}{G_1}} + \frac{1}{\dfrac{1}{G_2}} + \frac{1}{\dfrac{1}{G_3}} + \text{etc.}$$

This reduces to

$$G_T = G_1 + G_2 + G_3 + \text{etc.}$$

■ TEST

Answer the following questions.

19. Define *parallel circuit.*
20. How is the voltage distributed in a parallel circuit?
21. How is the current distributed in a parallel circuit?
22. Does the highest or the lowest resistance dominate a parallel circuit?
23. Give two formulas that could be used to find the total resistance of two parallel resistors.

Example 5-7

Determine the individual conductances and the total conductances of a 25-Ω resistor (R_1) and a 50-Ω resistor (R_2) connected in series.

$G = \dfrac{1}{R}$

$R = \dfrac{1}{G}$

Given: $R_1 = 25\ \Omega$

          $R_2 = 50\ \Omega$

Find: G_1, G_2, and G_T

Known: $G = 1/R$

$$G_T = \frac{1}{1/G_1 + 1/G_2}$$

Solution: $G_1 = 1/25 = 0.04$ S

            $G_2 = 1/50 = 0.02$ S

$$G_T = \frac{1}{1/0.04 + 1/0.02}$$

$$= \frac{1}{75} = 0.0133 \text{ S}$$

Answer: The conductances are as follows: 0.04 S, 0.02 S, and 0.0133 S. Notice that the total conductance in example 5-7 could also have been found by determining R_T and taking the reciprocal of it.

Example 5-8

Determine the total conductance for the resistors in example 5-7 when they are in parallel.

Given: $G_1 = 0.04$ S

          $G_2 = 0.02$ S

Find: G_T

Known: $G_T = G_1 + G_2$

Solution: $G_T = 0.04 + 0.02 = 0.06$ S

Answer: The total conductance is 0.06 S.

24. True or false. The base unit for conductance is the siemens.
25. True or false. Adding another resistor in parallel increases the total resistance.
26. True or false. The total resistance of a 15-Ω resistor in parallel with a 39-Ω resistor is less than 15 Ω.
27. True or false. The total resistance of two 100-Ω resistors in parallel is 200 Ω.

28. True or false. In a parallel circuit, a 50-Ω resistor dissipates more power than does a 150-Ω resistor.

29. True or false. The resistance of a parallel resistor can be measured while the resistor is connected in the circuit.

30. True or false. Voltage measurements are used to determine whether or not a load is open in a parallel circuit.

31. If one load in a parallel circuit opens, what happens to each of the following?
    a. Total resistance
    b. Total current
    c. Total power
    d. Total voltage

32. Refer to Fig. 5-27. What are the values of I_{R_1} and I_{R_2}?

33. Refer to Fig. 5-27. What is the voltage of B_1?

34. Refer to Fig. 5-27. What is the resistance of R_2?

35. Refer to Fig. 5-27. Determine the following:
    a. Total resistance
    b. Total conductance
    c. Current through R_2
    d. Power dissipated by R_1

5-7 Series-Parallel Circuits

Some of the features of both the series circuit and the parallel circuit are incorporated into *series-parallel* circuits. For example, R_2 and R_3 in Fig. 5-28(a) are in parallel. Everything that has been said about parallel circuits applies to these two resistors. In Fig. 5-28(d), R_7 and R_8 are in series. All the series-circuit relationships apply to these two resistors.

Resistors R_1 and R_2 in Fig. 5-28(a) are not directly in series because the same current does not flow through each. However, the equivalent of R_2 and R_3 in parallel [$R_{2,3}$ in Fig. 5-28(b)] is in series with R_1. Combining R_2

and R_3 is the first step in determining the total resistance of this circuit. The result is $R_{2,3}$. Of course, $R_{2,3}$ is not an actual resistor; it merely represents R_2 and R_3 in parallel. Resistor R_1 is in series with $R_{2,3}$. The final step in finding the total resistance is to combine R_1 and $R_{2,3}$, as shown in Fig. 5-28(c). Referring to Fig. 5-28(a), assume that $R_1 = 15\ \Omega$, $R_2 = 20\ \Omega$, and $R_3 = 30\ \Omega$. Combining R_2 and R_3 gives

$$R_{2,3} = \frac{R_2 \times R_3}{R_2 + R_3} = \frac{20 \times 30}{20 + 30} = \frac{600}{50} = 12\ \Omega$$

The total resistance is

$$R_T = R_1 + R_{2,3} = 15 + 12 = 27\ \Omega$$

Neither R_7 nor R_8 in Fig. 5-28(d) is in parallel with R_9 because neither is across the same voltage source as R_9. Therefore, the first step in reducing this circuit is to combine the series resistors R_7 and R_8. This results in the circuit shown in Fig. 5-28(e). Now the parallel circuit of Fig. 5-28(e) can be reduced, by a parallel-resistance formula, to the simple circuit of Fig. 5-28(f). Refer to Fig. 5-28(d) and let $R_7 = 40\ \Omega$, $R_8 = 60\ \Omega$, and $R_9 = 20\ \Omega$.

$$R_{7,8} = R_7 + R_8 = 40 + 60 = 100\ \Omega$$

$$\begin{aligned} R_T &= \frac{R_{7,8} \times R_9}{R_{7,8} + R_9} \\ &= \frac{100 \times 20}{100 + 20} = \frac{2000}{120} = 16.7\ \Omega \end{aligned}$$

A more complex series-parallel circuit is shown in Fig. 5-29(a). Determining the total resistance of the circuit is illustrated in Fig. 5-29(b) through (e). The calculations required to arrive at the total resistance are

$$R_{3,5} = R_3 + R_5 = 50 + 30 = 80\ \Omega$$

$$R_{3,4,5} = \frac{R_4\ \text{or}\ R_{3,5}}{2} = \frac{80}{2} = 40\ \Omega$$

$$R_{2,3,4,5} = R_2 + R_{3,4,5} = 60 + 40 = 100\ \Omega$$

$$\begin{aligned} R_T &= \frac{R_1 \times R_{2,3,4,5}}{R_1 + R_{2,3,4,5}} = \frac{200 \times 100}{200 + 100} = \frac{20{,}000}{300} \\ &= 66.7\ \Omega \end{aligned}$$

Using Kirchhoff's Laws in Series-Parallel Circuits

Individual currents and voltages in series-parallel circuits can often be determined by Kirchhoff's laws. In Fig. 5-30, current in some of the conductors is given. The rest of the cur-

Series-parallel

Kirchhoff's laws in series-parallel circuits

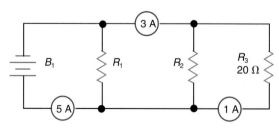

Fig. 5-27 Circuit test questions 32 to 35.

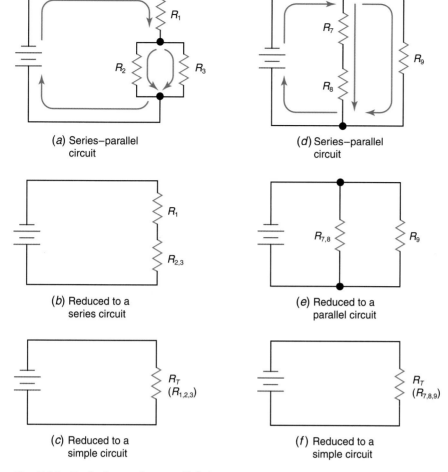

(a) Series–parallel circuit

(d) Series–parallel circuit

(b) Reduced to a series circuit

(e) Reduced to a parallel circuit

(c) Reduced to a simple circuit

(f) Reduced to a simple circuit

Fig. 5-28 Reducing series-parallel circuits to equivalent simple circuits.

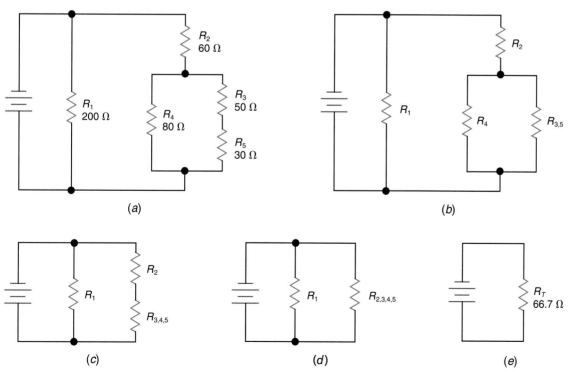

(a)

(b)

(c)

(d)

(e)

Fig. 5-29 Complex series-parallel circuit (a) can be simplified, as shown in (b) to (d).

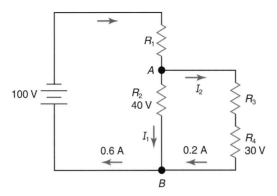

Fig. 5-30 Currents and voltages in series-parallel circuits. Kirchhoff's laws can be used to determine the unknown values.

The voltage across R_3 can be found by rearranging the second equation:

$$V_{R_3} = V_T - V_{R_1} - V_{R_4}$$
$$= 100 \text{ V} - 60 \text{ V} - 30 \text{ V}$$
$$= 10 \text{ V}$$

Also, the voltage across R_3 can be found without knowing the voltage across R_1:

$$V_{R_3} = V_{R_2} - V_{R_4} = 40 \text{ V} - 30 \text{ V} = 10 \text{ V}$$

Solving Series-Parallel Problems

Several problems are solved in the examples that follow. These problems illustrate how to use Ohm's law and Kirchhoff's two laws to solve series-parallel problems.

Example 5-9

For the circuit of Fig. 5-31(a), determine all unknown currents and voltages.

Given: $V_T = 60$ V

$V_{R_2} = 40$ V

$I_T = 4$ A

$R_3 = 20 \ \Omega$

Find: $V_{R_1}, V_{R_3}, I_{R_1}, I_{R_2}, I_{R_3}$

Known: Ohm's law and Kirchhoff's laws

Solution: $V_{R_3} = V_T - V_{R_2}$

$= 60 \text{ V} - 40 \text{ V}$

$= 20 \text{ V}$

$V_{R_1} = V_T = 60$ V

$I_{R_3} = \dfrac{V_{R_3}}{R_3}$

$= \dfrac{20 \text{ V}}{20 \ \Omega} = 1 \text{ A}$

$I_{R_2} = I_{R_3} = 1$ A

$I_{R_1} = I_T - I_{R_2}$

$= 4 \text{ A} - 1 \text{ A} = 3 \text{ A}$

Answer: $V_{R_1} = 60$ V, $V_{R_3} = 20$ V, I_{R_1} $= 3$ A, $I_{R_2} = 1$ A, $I_{R_3} = 1$ A

rents can be found using Kirchhoff's current law. Since 0.6 A enters the battery, the same amount must leave. This is the total current I_T. Thus $I_T = 0.6$ A. Since R_3 and R_4 are in series, the current entering R_3 is the same as the current leaving R_4. Therefore, $I_2 = 0.2$ A. The current entering point A is 0.6 A. Leaving that point are I_1 and I_2. From Kirchhoff's current law we know that $I_1 + I_2 = 0.6$ A. But $I_2 = 0.2$ A, and thus $I_1 + 0.2$ A $= 0.6$ A, and $I_1 = 0.4$ A. We can check this result by examining point B. Here $I_1 + 0.2$ A $= 0.6$ A. Since I_1 was found to be 0.4 A, we have 0.4 A $+ 0.2$ A $= 0.6$ A, which agrees with Kirchhoff's current law.

Some of the voltage drops in Fig. 5-30 are indicated beside the resistor symbols. The unspecified voltage drops can be found by using Kirchhoff's voltage law. In a series-parallel circuit, Kirchhoff's law applies to all loops, or current paths, in the circuit. Thus, for the circuit of Fig. 5-30, we can write two voltage relationships.

$$V_T = V_{R_1} + V_{R_2}$$
$$V_T = V_{R_1} + V_{R_3} + V_{R_4}$$

Studying the above two equations shows that

$$V_{R_2} = V_{R_3} + V_{R_4}$$

In other words, the voltage between points A and B is 40 V, no matter which path is taken. Rearranging the first equation, we can solve for the voltage drop across R_1:

$$V_{R_1} = V_T - V_{R_2} = 100 \text{ V} - 40 \text{ V} = 60 \text{ V}$$

All the currents and voltages for Fig. 5-31 (a) have now been determined. Using Ohm's law and the power formula, we can easily find the resistance and power of each resistor.

Example 5-10

Refer to Fig. 5-31(*b*). For this circuit compute the resistance of R_3, the power dissipation of R_4, and the voltage across R_1.

Given: $V_T = 100$ V

              $I_{R_1} = 0.8$ A

              $I_{R_3} = 0.3$ A

              $R_2 = 100$ Ω

              $V_{R_4} = 30$ V

Find: R_3, P_{R_4}, V_{R_1}

Known: Ohm's law and Kirchhoff's laws

Solution: $I_{R_2} = I_{R_1} - I_{R_3}$

                $= 0.8$ A $- 0.3$ A

                $= 0.5$ A

              $V_{R_2} = I_{R_2}R_2$

                $= 0.5$ A $\times 100$ Ω

                $= 50$ V

              $V_{R_1} = V_T - V_{R_2} - V_{R_4}$

                $= 100$ V $- 50$ V $- 30$ V

                $= 20$ V

              $V_{R_3} = V_{R_2} = 50$ V

              $R_3 = \dfrac{V_{R_3}}{I_{R_3}} = \dfrac{50 \text{ V}}{0.3 \text{ A}} = 166.7$ Ω

              $P_{R_4} = I_{R_4}V_{R_4}$

                $= 0.8$ A $\times 30$ V $= 24$ W

Answer: $R_3 = 166.7$ Ω, $P_{R_4} = 24$ W,

            $V_{R_1} = 20$ V

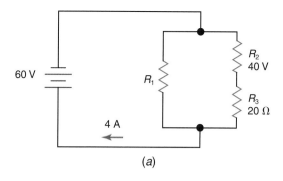

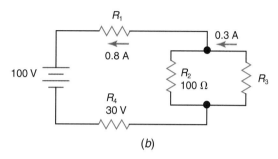

Fig. 5-31 Circuits for examples 5-9 and 5-10.

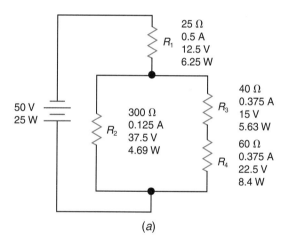

(*a*)

Relationships in Series-Parallel Circuits

As in series circuits, the current, the voltage, and the power in series-parallel circuits are dependent on one another. That is, changing any one resistance usually changes all currents, voltages, and powers except the source voltage. For example, increasing the resistance of R_3 in Fig. 5-32(*a*) from 40 to 90 Ω causes the changes listed below:

1. R_T increases (because $R_{3,4}$ increases).
2. I_T decreases (because R_T increases).
3. V_{R_1} decreases (because $I_T = I_{R_1}$).
4. V_{R_2} increases (because $V_{R_2} = V_T - V_{R_1}$).
5. I_{R_2} increases (because $I_{R_2} = V_{R_2}/R_2$).
6. I_{R_4} decreases (because $I_{R_4} = I_{R_1} - I_{R_2}$).
7. V_{R_4} decreases (because $V_{R_4} = I_{R_4}R_4$).

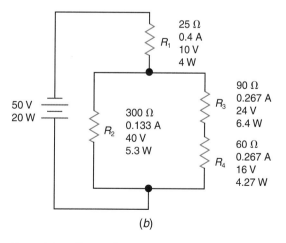

(*b*)

Fig. 5-32 Effects of changing R_3. (*a*) Original circuit values. (*b*) Values after R_3 increases.

8. V_{R_3} increases (because $V_{R_3} = V_{R_2} - V_{R_4}$).
9. P_{R_1} decreases, P_{R_2} increases, P_{R_4} decreases, and P_T decreases (because of the I and V changes specified above).
10. P_{R_3} increases (because V_{R_3} increases more than I_{R_3} decreases).

The magnitude of these changes is shown in Fig. 5-32(b).

The changes detailed above occur when any resistor is changed, except in circuits like the one in Fig. 5-29(a). In this circuit, changing R_1 affects only the current and power of R_1 and the battery. This is because R_1 is in parallel with the combination of R_2, R_3, R_4, and R_5 [Fig. 5-29(d)].

TEST

Answer the following questions.

36. Refer to Fig. 5-29(a). Which resistors, if any, are
    a. Directly in series?
    b. Directly in parallel?
37. Referring to Fig. 5-29(a), determine whether each of the following statements is true or false.
    a. $I_{R_2} = I_{R_3} + I_{R_5}$
    b. $I_{R_2} = I_{R_4} + I_{R_5}$
    c. $I_{R_1} = I_T - I_{R_2}$
    d. $V_{R_3} = V_{R_5}$
    e. $V_{R_4} = V_{R_1} - V_{R_2}$
38. Refer to Fig. 5-31(a) and compute the following:
    a. R_1

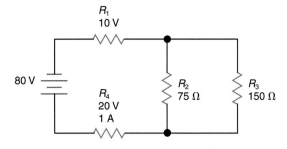

Fig. 5-33 Circuit diagram for test question 41.

    b. P_{R_2}
    c. P_T
39. Refer to Fig. 5-28(a). If R_2 is decreased, indicate whether each of the following increases or decreases:
    a. I_{R_1}
    b. V_{R_3}
    c. P_{R_3}
    d. V_{R_1}
40. Refer to Fig. 5-32(a). Change the value of R_1 to 50 Ω. Then determine the value of each of the following:
    a. R_T
    b. V_{R_1}
    c. I_{R_4}
    d. P_{R_3}
41. For Fig. 5-33, determine the following:
    a. R_1
    b. I_{R_2}
    c. P_{R_3}
    d. I_T
    e. R_T

Summary

1. Multiple-load circuits include series, parallel, and series-parallel circuits.
2. Series circuits are single-path circuits.
3. The same current flows throughout a series circuit.
4. The total resistance equals the sum of the individual resistances in a series circuit.
5. The sum of the voltage drops around a circuit equals the toal source voltage (Kirchhoff's voltage law).
6. A voltage drop (voltage across a load) indicates that electric energy is being converted to another form.
7. The polarity of a voltage drop indicates the direction of current flow.
8. The voltage across an open series load is equal to the source voltage.
9. A shorted load in a series circuit increases the current, voltage, and power of the other loads.
10. When one resistance in a series circuit is smaller than the tolerance of another, the smaller resistance has little effect on the circuit current and power.
11. The highest resistance in a series circuit drops the most voltage.
12. Maximum power transfer occurs when the source resistance equals the load resistance.
13. Conductance is the ability to conduct current.
14. Parallel circuits are multiple-path circuits.
15. Each branch of a parallel circuit is independent of the other branches.
16. The same voltage appears across each branch of a parallel circuit.
17. The currents entering a junction must equal the currents leaving a junction (Kirchhoff's current law).
18. The total current in a parallel circuit equals the sum of the branch currents.

19. Adding more resistance in parallel decreases the total resistance.
20. The total resistance in a parallel circuit is always less than the lowest branch resistance.
21. The relationships of both series and parallel circuits are applicable to parts of series-parallel circuits.
22. In all circuits:

$$P_T = P_{R_1} + P_{R_2} + P_{R_3} + \text{etc.}$$

$$R = \frac{1}{G}$$

23. In series circuits:

$$I_T = I_{R_1} = I_{R_2} = I_{R_3} = \text{etc.}$$
$$V_T = V_{R_1} + V_{R_2} + V_{R_3} + \text{etc.}$$
$$R_T = R_1 + R_2 + R_3 + \text{etc.}$$

$$G_T = \frac{1}{\dfrac{1}{G_1} + \dfrac{1}{G_2} + \dfrac{1}{G_3} + \text{etc.}}$$

24. In parallel circuits:

$$I_T = I_{R_1} + I_{R_2} + I_{R_3} + \text{etc.}$$
$$V_T = V_{R_1} = V_{R_2} = V_{R_3} = \text{etc.}$$

$$R_T = \frac{1}{\dfrac{1}{R_1} + \dfrac{1}{R_2} + \dfrac{1}{R_3} + \text{etc.}}$$

$$G_T = G_1 + G_2 + G_3 + \text{etc.}$$

For two parallel branches:

$$R_T = \frac{R_1 \times R_2}{R_1 + R_2}$$

For n parallel resistors, R ohms each:

$$R_T = \frac{R}{n}$$

Chapter Review Questions

For questions 5-1 to 5-5. Supply the missing word or phrase in each statement.

5-1. The total current is equal to the sum of the individual currents in a _____ circuit.

5-2. The highest resistance dissipates the least power in a _____ circuit.

5-3. The highest resistance drops the _____ voltage in a series circuit.

5-4. Adding another parallel resistor _____ the total resistance.

5-5. The total voltage is equal to the sum of the voltage drops in a _____ circuit.

For questions 5-6 to 5-20, determine whether each statement is true or false.

5-6. To measure the resistance of an individual resistor in a parallel circuit, one end of the resistor must be disconnected from the circuit.

5-7. Maximum power transfer occurs when the source resistance is very low compared with the load resistance.

5-8. The negative polarity marking on a resistor in a schematic diagram indicates an excess of electrons at that point.

5-9. Voltmeters have a very low internal resistance.

5-10. In a series circuit, an open load drops no voltage.

5-11. If one load in a parallel circuit shorts out, all the other loads will use more power.

5-12. If one load in a series circuit shorts out, all the other loads will use more power.

5-13. Adding more resistors to a parallel circuit increases the total power used by the circuit.

5-14. The total resistance of parallel resistances is always less than the value of the lowest resistance.

5-15. Changing the value of one resistor in a parallel circuit changes the current through all other resistors in that circuit.

5-16. The unit of conductance is the siemen.

5-17. A voltage drop indicates that some other form of energy is being converted to electric energy.

5-18. The direction of current flow determines the polarity of the voltage drop across the resistor.

5-19. The lowest resistance in a series circuit dominates the circuit current and power.

5-20. Most circuits in a home are series circuits.

For questions 5-21 and 5-22, choose the letter that best completes each sentence.

5-21. The total resistance of a 45-Ω resistor and a 90-Ω resistor connected in series is
   a. 30 Ω
   b. 45 Ω
   c. 67.5 Ω
   d. 135 Ω

5-22. The total resistance of a 30-Ω resistor and a 60-Ω resistor connected in parallel is
   a. 20 Ω
   b. 30 Ω
   c. 45 Ω
   d. 90 Ω

Solve the following problems.

5-23. For the circuit in Fig. 5-34, compute the following:
   a. V_{R_1}
   b. R_2
   c. P_T
   d. G_3

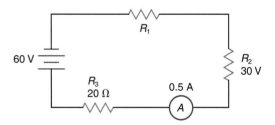

Fig. 5-34 Circuit for chapter review question 5-23.

5-24. For the circuit in Fig. 5-35, compute the following:
   a. I_{R_1}
   b. R_1
   c. I_{R_2}
   d. G_T

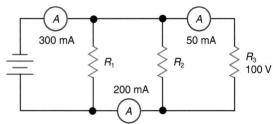

Fig. 5-35 Circuit for chapter review question 5-24.

5-25. For the circuit in Fig. 5-36, compute the following:
   a. V_{R_3}
   b. I_{R_1}
   c. V_{R_1}

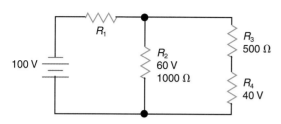

Fig. 5-36 Circuit for chapter review question 5-25.

Answer the following questions.

5-26. If R_2 in Fig. 5-36 is increased to 1200 Ω, does V_{R_1} increase, decrease, or remain the same?

5-27. What would happen to G_T in Fig. 5-35 if a fourth parallel resistor were added?

5-28. What would happen to G_T in Fig. 5-34 if a fourth series resistor were added?

Critical Thinking Questions

5-1. List and explain several applications of parallel circuits not mentioned in this chapter.

5-2. For Fig. 5-11, assume that all resistors have a tolerance of ±5 percent and that the source has an internal resistance of 9.2 Ω. Determine the maximum total power dissipated by the loads.

5-3. Is it desirable to have an amplifier's output impedance (opposition) equal to the speaker's impedance? Why?

5-4. Is it desirable to have a 100-kW generator's internal resistance equal to the resistance of its load? Why?

5-5. Derive the reciprocal formula for parallel resistances from the formula for parallel conductances.

5-6. Are measured circuit voltages more likely to be inaccurate in a series or a parallel circuit? Why?

5-7. Discuss the results of shorting out R_2 in Fig. 5-11 when the power ratings for R_1, R_2, and R_3 are 25 W, 50 W, and 25 W respectively.

5-8. Without disconnecting either end of any of the loads in Fig. 5-37, how could you determine whether one or more of the loads are outside its tolerance range?

5-9. Without disconnecting either end of any of the loads in Fig. 5-37, how could you determine whether any of the loads were open?

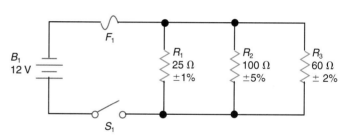

Fig. 5-37 Circuit for critical thinking questions 5-8 and 5-9.

5-10. Determine the efficiency of a 24-V battery with 0.04 Ω of internal resistance when it is supplying 150 A to a load.

Answers to Tests

1. a. V_{R_4}
   b. I_T
   c. I_{R_2}
2. T
3. T
4. T

5. $R_T = R_1 + R_2 +$ R_3 + etc.
   $R_T = V_T / I_T$
6. Yes.
7. $V_T = V_{R_1} + V_{R_2} +$ R_3 + etc.

8. 100-Ω resistor
9. No.
10. across the open resistor
11. a. increases
    b. increases
    c. increases

12. a. 500 Ω
    b. 0.2 A
    c. 25 V
    d. 15 W
13. R_1 is 37.3 Ω. The battery has to furnish 1.89 W.

14. It would increase.
15. 0 V
16. 9 mA
17. when load resistance is equal to source resistance
18. greater than
19. A parallel circuit is one that has two or more loads and two or more independent current paths.
20. The total (source) voltage is applied to each load:

$$V_T = V_{R_1} = V_{R_2} = V_{R_3}$$

21. The total current divides among the branches of the circuit:

$$I_T = I_{R_1} + I_{R_2} + I_{R_3} + \text{etc.}$$

22. lowest
23. $R_T = \dfrac{1}{\dfrac{1}{R_1} + \dfrac{1}{R_2}}$

$R_T = \dfrac{R_1 \times R_2}{R_1 + R_2}$

24. T
25. F
26. T
27. F
28. T
29. F
30. F
31. a. increases
   b. decreases
   c. decreases
   d. remains the same
32. $I_{R_1} = I_T - (I_{R_2} + I_{R_3})$
     $= 5\,A - 3\,A = 2\,A$
   $I_{R_2} = (I_{R_2} + I_{R_3}) - I_{R_3}$
     $= 3\,A - 1\,A = 2\,A$
33. $V_{B_1} = V_{R_3} = I_{R_3}R_3 = 1\,A$
   $\times\ 20\,\Omega = 20\,V$
34. $V_{R_2} = V_{R_3} = 20\,V$

$$R_2 = \dfrac{V_{R_2}}{I_{R_2}} = \dfrac{20\,V}{2\,A} = 10\,\Omega$$

35. a. 4 Ω
   b. 0.25 S
   c. 2 A
   d. 40 W

36. a. R_3 and R_5
   b. none
37. a. F
   b. T
   c. T
   d. F
   e. T
38. a. 20 Ω
   b. 40 W
   c. 240 W
39. a. increases
   b. decreases
   c. decreases
   d. increases
40. a. 125 Ω
   b. 20 V
   c. 0.3 A
   d. 3.6 W
41. a. 10 Ω
   b. 0.67 A
   c. 16.7 W
   d. 1 A
   e. 80 Ω

Complex-Circuit Analysis

Chapter Objectives

This chapter will help you to:

1. *Use* simultaneous equations to solve equations with more than one unknown variable.
2. *Write* loop equations using Kirchhoff's voltage law.
3. *Determine* the values of electrical quantities of either single-source or multiple-source complex circuits using a variety of techniques.
4. *Use* the superposition theorem to solve multiple-source complex circuits.
5. *Understand* the advantages of viewing a circuit as a two-terminal network.
6. *Apply* Thevenin's theorem and Norton's theorem to reduce complex circuits.

S ome circuits are arranged so that the loads are neither directly in series nor directly in parallel. Such circuits are called *complex circuits or networks.* Some complex circuits (networks) include two or more voltage sources. Reducing complex circuits to a single equivalent resistance, or determining individual currents and voltages for the circuit, requires techniques beyond those covered in the previous chapter.

6-1 Simultaneous Equations

Equations with two or more unknown variables are readily solvable if there are at least as many independent equations relating the variables as there are variables. Such sets, or groups, of equations are known as *simultaneous equations.* We will use the *variable elimination* method for solving these types of equations. This method is simple and straightforward in that it involves only three algebraic operations: (1) multiplying by a constant, (2) adding equations, and (3) substituting numerical value for variables.

YOU MAY RECALL

. . . that in solving problems in previous chapters, you have been performing all these operations except the second.

Adding Equations

The addition of equations can most readily be illustrated with a few examples. Suppose you want to add the equation $A = 4$ to the equation $B = 7$. The first step is to align the variables A and B vertically so that A is in one column and B in another column:

$$\begin{aligned} A \quad\;\; &= 4 \\ B &= 7 \\ \hline A + B &= 11 \end{aligned}$$

Notice that the addition of two unequal variables yields A plus B, not A times B.

Next, let us add $2A + B = 13$ to $2B = 6$. Again, like variables are aligned vertically:

$$\begin{aligned} 2A + \;\; B &= 13 \\ 2B &= 6 \\ \hline 2A + 3B &= 19 \end{aligned}$$

Simultaneous equations

Variable-elimination

Finally, let us add $2A - 6B + 3C = -25$ to $A + B - 2C = 2$:

$$
\begin{array}{r}
2A - 6B + 3C = -25 \\
A + B - 2C = 2 \\
\hline
3A - 5B + C = -23
\end{array}
$$

■ TEST

Answer the following questions:

1. Add $2A + 5B = 43.2$ to $4A + B = 16.4$
2. Add $3R - 2S - 4T = -23$ to $R + 3S - 2T = 48$
3. Add $-2A + 3B = 10$ to $-3B + 4C = -2$

Variable Elimination with Two Variables

Two unknown variables

Check for error

Constant

Solving simultaneous equations with *two unknown variables* requires two independent equations, each of which shows a relationship between the unknown variables. The procedure involves multiplying each of the terms of one of the equations by a *constant* to obtain a new, third, equation. The new equation and the equation that was not multiplied are then added together. The constant is selected so that when the two equations are added, one of the variables cancels out and does not appear in the sum. For example, $2A$ in one equation cancels $-2A$ in the other equation. The constant is obtained by dividing a term in the first equation by the corresponding term in the second equation and then multiplying the answer by -1. If each of the terms in the second equation is then multiplied by the constant, a term in the new equation will be equal in magnitude but opposite in sign to the corresponding term in the first equation. When the two equations are added, the variables will cancel each other. Only one variable will remain, and its value can then be calculated.

Let us try this process by solving the two-variable simultaneous equations $6A + 5B = 45$ and $2A + 3B = 19$. First divide $6A$ by $2A$ and multiply by -1 to get the constant:

$$
\frac{6A}{2A} \times (-1) = 3 \times (-1) = -3
$$

Next, multiply the second equation by the constant -3:

$$
-3(2A + 3B) = -3(19)
$$
$$
-6A - 9B = -57
$$

Then add the first equation to the new equation that is the result of the above multiplication:

First equation:	$6A + 5B = 45$
New equation:	$-6A - 9B = -57$
Sum of the two equations:	$-4B = -12$

Solving for B yields

$$
B = \frac{-12}{-4} = 3
$$

Now we can go back to either the first equation or the second and solve for A by substituting the value of B just found:

$$
\begin{align}
6A + 5B &= 45 \\
6A + 5(3) &= 45 \\
6A + 15 &= 45 \\
6A &= 30 \\
A &= 5
\end{align}
$$

Finally, we can *check for* any arithmetic *error* by substituting the values for A and B into the equation we did not use in the above step:

$$
\begin{align}
2A + 3B &= 19 \\
2(5) + 3(3) &= 19 \\
10 + 9 &= 19 \\
19 &= 19 \text{ check}
\end{align}
$$

Example 6-1

Determine the values of A and B for $4A + 3B = 36.9$ and $A + 2B = 17.1$.

Given:	$4A + 3B = 36.9$
	$A + 2B = 17.1$
Find:	A and B
Known:	Variable-elimination method
Solution:	$\dfrac{4A}{A} \times (-1) = -4$
	$-4(A + 2B) = -4(17.1)$
New equation:	$-4A - 8B = -68.4$
First equation:	$4A + 3B = 36.9$
New equation:	$-4A - 8B = -68.4$
	$\overline{-5B = -31.5}$
	$B = 6.3$
	$4A + 3(6.3) = 36.9$
	$4A + 18.9 = 36.9$
	$4A = 18$
	$A = 4.5$
Answer:	A equals 4.5, and B equals 6.3.

Answer the following questions.

4. Solve for X and Y when $3X + 4.2Y = 38.1$ and $2.5X + 3Y = 28.75$.
5. Solve for C and D when $4C + 5D = 42$ and $-3C + 7D = 26$.
6. Solve for A and B when $3A - 4B = 6$ and $-2A - 2B = -18$.

Variable Elimination with Three Variables

Solving for three unknown variables requires three *independent equations* involving the *three variables*. Suppose the three equations are

$$
\begin{array}{ll}
(1) & 2X + 4Y - 5Z = -1 \\
(2) & -4X - 2Y + 6Z = -2 \\
(3) & 3X + 5Y + 7Z = 96
\end{array}
$$

The step-by-step procedure for solving for X, Y, and Z is as follows:

1. Combine any two of the three equations so that any one of the three variables is eliminated. For example, let us combine Eqs. 1 and 3 to eliminate variable Y:

Find the constant: $\dfrac{4Y}{5Y} \times (-1) = -0.8$

Multiply: $-0.8(3X + 5Y + 7Z) = -0.8(96)$
$$-2.4X - 4Y - 5.6Z = -76.8$$

Add:
Eq. 1: $\qquad\quad 2X + 4Y - 5Z = -1$
New equation: $-2.4X - 4Y - 5.6Z = -76.8$
$$\overline{-0.4X - 10.6Z = -77.8}$$

To simplify this equation and eliminate the decimal numbers, we multiply the equation by 5:

$$-2X - 53Z = -389$$

2. Combine any other two of the three original equations to eliminate the same variable as eliminated in step 1. Let us use Eqs. 1 and 2 and again eliminate Y by the same procedure:

Constant: $\dfrac{4Y}{-2Y} \times (-1) = 2$

Multiply: $2(-4X - 2Y + 6Z) = 2(-2)$
$$-8X - 4Y + 12Z = -4$$

Add:
Eq. 1: $\qquad\quad 2X + 4Y - 5Z = -1$
New Eq. 2: $\quad -8X - 4Y + 12Z = -4$
$$\overline{-6X +7Z = -5}$$

3. Combine the equations derived in steps 1 and 2 to eliminate one of the two remaining variables. Let us eliminate X and solve for Z:

Constant: $\dfrac{-6X}{-2X} \times (-1) = -3$

Multiply: $-3(-2X - 53Z) = -3(-389)$
$$6X + 159Z = 1167$$

Add: $\qquad -6X + 7Z = -5$
$$\underline{6X + 159Z = 1167}$$
$$166Z = 1162$$
$$Z = 7$$

4. Substitute the value for Z found in step 3 into the equation derived in either step 1 or step 2 to find the value of X. Using the equation of step 1 yields:

$$
\begin{aligned}
-2X - 53Z &= -389 \\
-2X - 53(7) &= -389 \\
-2X &= -389 + 371 \\
2X &= 18 \\
X &= 9
\end{aligned}
$$

5. Substitute the values of Z and X from steps 3 and 4 into any one of the three original equations and solve for Y. Let us use Eq. 1:

$$
\begin{aligned}
2(9) + 4Y - 5(7) &= -1 \\
18 + 4Y - 35 &= -1 \\
4Y - 17 &= -1 \\
4Y &= 16 \\
Y &= 4
\end{aligned}
$$

All five steps are needed only if all three equations contain all three variables. If either one or two of the equations contain only two of the three variables, then step 2 can be omitted.

Independent equations

Three variables

Example 6-2

Determine the values of A, B, and C for the given equations:

Given: $\quad 2A + 4C = 40$
$\qquad\qquad -4B + 3C = -4$
$\qquad\qquad 5A + 3B - 7C = -15$
Find: $\quad A$, B, and C
Known: Variable-elimination technique
Solution: $\dfrac{-4B}{3B} \times (-1) = \dfrac{4}{3}$

$$\dfrac{4}{3}(5A + 3B - 7C) = \dfrac{4}{3}(-15)$$

Multiple-source
circuits

Single-source
circuits

Loop equations

Bridge circuits

Negative

$$6.6A + 4B - 9.3C = -20$$
$$-4B + 3C = -4$$
$$\underline{6.6A + 4B - 9.3C = -20}$$
$$\overline{6.6A \qquad - 6.3C = -24}$$
$$10(6.6A) - 10(6.3C) = 10(-24)$$
$$66A - 63C = -240$$
$$\frac{66A}{2A} \times (-1) = -33$$
$$-33(2A + 4C) = 33(40)$$
$$-66A - 132C = 1320$$
$$66A - 63C = -240$$
$$\underline{-66A - 132C = -1320}$$
$$\overline{-195C = -1560}$$
$$C = 8$$
$$2A + 4(8) = 40$$
$$2A + 32 = 40$$
$$2A = 8$$
$$A = 4$$
$$-4B + 3(8) = -4$$
$$-4B + 24 = -4$$
$$-4B = -28$$
$$B = 7$$

Answer: *A* equals 4, *B* equals 7, and *C* equals 8.

▪ TEST

Answer the following questions.

7. Solve for *R*, *S*, and *T* when

$$2R - 4S + 3T = 49$$
$$-3R + 5S = -52$$
$$6R - 2S + 4T = 52$$

8. Solve for *X*, *Y*, and *Z* when

$$2X - Y + 2Z = 5.6$$
$$2X + 3Y - Z = 17.1$$
$$-3X + 4Y + 5Z = 29.4$$

9. Solve for *C*, *D*, and *E* when

$$4C - 5D + 3E = -9$$
$$2C - 3D = 4$$
$$8D - 5E = -18$$

6-2 Loop-Equations Technique

Now that you are familiar with simultaneous-equation techniques, you can use them to solve complex electric networks. The equations you will be working with are called *loop equations*. They are derived by applying Kirchhoff's voltage law to the various loops of an electric circuit.

Either *multiple source* or *single-source circuits* can be solved by the *loop equations* technique. This technique allows one to find every voltage and every current in the circuit.

Single-Source Circuits

A complex circuit with a single-source voltage is shown in Fig. 6-1. This circuit configuration is called a *bridge circuit*. It is a popular configuration for measuring temperature indirectly. In such an application R_5 represents the internal resistance of an electric meter and R_4 is a thermistor. The scale of the meter is calibrated in temperature units rather than in current or voltage units.

A study of Fig. 6-1 shows that none of the resistors are directly in series or directly in parallel. Thus, there is no place to start using series rules, parallel rules, or Ohm's law. This complex circuit must be solved by developing loop equations using Kirchhoff's voltage law.

A logical way to determine the loops in a circuit is to trace possible paths of current flow, as shown in Fig. 6-2. Without knowing resistor values, it is impossible to determine whether I_1 (loop 1) is in the direction shown in Fig. 6-2(*a*) or in that shown in Fig. 6-2(*b*). The direction of the current depends upon the polarity of the voltage between *A* and *B*. It does not matter which direction one selects when analyzing circuits by loop equations. If the wrong direction is selected, then the value of I_1 will be *negative*. If I_1 turns out to be negative, then we know our assumption about current direction was wrong and we have to reverse the polarity markings of the voltage across R_5. Notice that it requires three loops (current paths) in

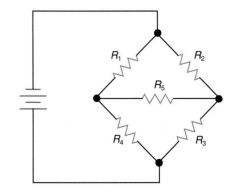

Fig. 6-1 Bridge circuit. None of the resistors are in series or parallel.

Fig. 6-2 to include every resistor in one or more loops. Thus, three loop equations will have to be derived and then solved simultaneously. To aid in deriving the loop equations, the three loops of Fig. 6-2(a), (c), and (d) have been combined on a single diagram in Fig. 6-3. Also, values have been assigned to the resistors and the voltage source in Fig. 6-3 so that we can calculate the values for all currents and voltages. Notice in Fig. 6-3 that R_1 has two currents (I_1 and I_2) flowing through it. Both currents are flowing in the same direction; therefore, the voltage drop across R_1 is

$$V_{R_1} = R_1(I_1 + I_2) = R_1 I_1 + R_1 I_2$$
$$= 2 \text{ k}\Omega \ (I_1) + 2\text{k}\Omega \ (I_2)$$

The same procedure is applicable to V_{R_3}.

We are now ready to derive the loop equations for the *three loops* of Fig. 6-3. We use Kirchhoff's voltage law, which states that "the algebraic sum of the voltage drops around a loop of a circuit is equal to the source voltage." For loop 1 we can write

$$V_T = V_{R_1} + V_{R_5} + V_{R_3}$$

By substituting known and equivalent values, we get

$$9 \text{ V} = 2\text{k}\Omega \ (I_1 + I_2) + 4\text{k}\Omega \ (I_1) + 5\text{k}\Omega \ (I_1 + I_3)$$

By expanding we get

$$9 \text{ V} = 2\text{k}\Omega \ (I_1) + 2\text{k}\Omega \ (I_2) + 4 \text{ k}\Omega \ (I_1)$$
$$+ 5\text{k}\Omega \ (I_1) + 5 \text{ k}\Omega \ (I_3)$$

By collecting terms, the equation for loop 1 becomes

$$9 \text{ V} = 11 \text{ k}\Omega \ (I_1) + 2 \text{ k}\Omega \ (I_2) + 5 \text{ k}\Omega \ (I_3)$$
$$\text{(loop 1)}$$

Three loops

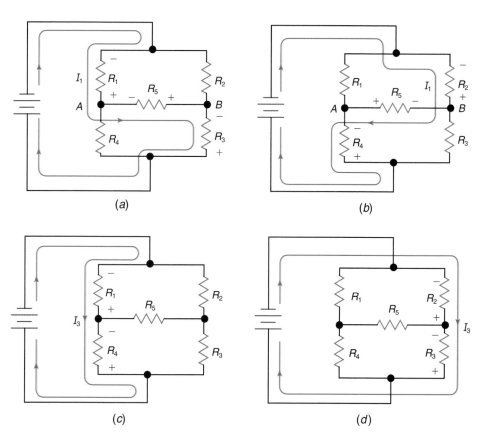

(a)

(b)

(c)

(d)

Fig. 6-2 Three loops for analyzing a bridge circuit using Kirchhoff's voltage law. (a) Loop 1—assuming point A is negative with respect to point B. (b) Loop 1—assuming point A is positive with respect to point B. (c) Loop 2. (d) Loop 3.

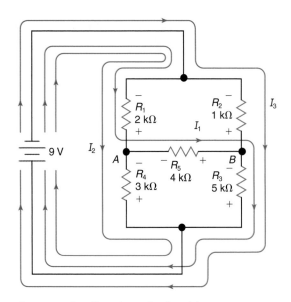

Fig. 6-3 The direction of I_1 is arbitrary. Two currents must be used in calculating the voltage drops across R_1 and R_3.

Using the same procedure for loop 2 yields

$$V_T = V_{R_1} + V_{R_4}$$
$$9 \text{ V} = 2 \text{ k}\Omega \ (I_1 + I_2) + 3 \text{ k}\Omega \ (I_2)$$
$$= 2 \text{ k}\Omega \ (I_1) + 2 \text{ k}\Omega \ (I_2) + 3 \text{ k}\Omega \ (I_2)$$
$$= 2 \text{ k}\Omega \ (I_1) + 5 \text{ k}\Omega \ (I_2)$$

(loop 2)

And for loop 3 the result is

$$V_T = V_{R_2} + V_{R_3}$$
$$9 \text{ V} = 1 \text{ k}\Omega \ (I_3) + 5 \text{ k}\Omega \ (I_1 + I_3)$$
$$= 1 \text{ k}\Omega \ (I_3) + 5 \text{ k}\Omega \ (I_1) + 5 \text{ k}\Omega \ (I_3)$$
$$= 5 \text{ k}\Omega \ (I_1) + 6 \text{ k}\Omega \ (I_3)$$

(loop 3)

Our next step is to solve the three equations simultaneously using the variable-elimination technique we learned in the preceding section. We start by adding the equations for loops 1 and 2 to eliminate the variable I_2:

$$\frac{5 \text{ k}\Omega \ (I_2)}{2 \text{ k}\Omega \ (I_2)} \times (-1) = -2.5$$

$$-2.5(9 \text{ V}) = -2.5 \ [11 \text{ k}\Omega \ (I_1) + 2 \text{ k}\Omega \ (I_2) \\ + 5 \text{ k}\Omega \ (I_3)]$$
$$-22.5 \text{ V} = -27.5 \text{ k}\Omega \ (I_1) - 5 \text{ k}\Omega \ (I_2) \\ - 12.5 \text{ k}\Omega \ (I_3)$$
$$\underline{\phantom{-22.5 \text{ V} = } 9 \text{ V} = 2 \text{ k}\Omega \ (I_1) + 5 \text{ k}\Omega \ (I_2)}$$
$$-13.5 \text{ V} = -25.5 \text{ k}\Omega \ (I_1) - 12.5 \text{ k}\Omega \ (I_3)$$

(loops 1 and 2)

Next, let us add loops 1 and 2 to loop 3 so that we can eliminate the variable I_3 and solve for I_1:

$$\frac{-12.5 \text{ k}\Omega \ (I_3)}{6 \text{ k}\Omega \ (I_3)} \times (-1) = 2.083$$

$$2.083 \ (9 \text{ V}) = 2.083 \ [5 \text{ k}\Omega \ (I_1) + 6 \text{ k}\Omega \ (I_3)]$$
$$18.75 \text{ V} = 10.416 \text{ k}\Omega \ (I_1) + 12.5 \text{ k}\Omega \ (I_3)$$
$$\underline{-13.5 \text{ V} = -25.5 \text{ k}\Omega \ (I_1) - 12.5 \text{ k}\Omega \ (I_3)}$$
$$5.25 \text{ V} = -15.083 \text{ k}\Omega \ (I_1)$$

$$I_1 = \frac{-5.25 \text{ V}}{15.083 \text{ k}\Omega} = -0.348 \text{ mA}$$

Now we can solve for I_3 by substituting the value of I_1 into the equation for loop 3:

$$9 \text{ V} = (5 \text{ k}\Omega) \ (-0.348 \text{ mA}) + 6 \text{ k}\Omega \ (I_3)$$
$$I_3 = \frac{9 \text{ V} + 1.74 \text{ V}}{6 \text{ k}\Omega} = 1.79 \text{ mA}$$

Finally, we solve for I_2 by substituting the value of I_1 into the equation for loop 2:

$$9 \text{ V} = (2 \text{ k}\Omega) \ (-0.348 \text{ mA}) + 5 \text{ k}\Omega \ (I_2)$$
$$I_2 = \frac{9 \text{ V} + 0.696 \text{ V}}{5 \text{ k}\Omega} = 1.939 \text{ mA}$$

Notice that the value of I_1 is negative. Thus, we assumed the wrong direction for I_1. Therefore, we must change the polarity of the voltage drop across R_5.

Now that we know the values of I_1, I_2, and I_3, we can calculate the current through and the voltage across each resistor. Referring to Fig. 6-3 and our values of I_1, I_2, and I_3, we can write

$$I_{R_1} = I_1 + I_2 = -0.348 \text{ mA} + 1.939 \text{ mA}$$
$$= 1.591 \text{ mA}$$
$$I_{R_2} = I_3 = 1.79 \text{ mA}$$
$$I_{R_3} = I_1 + I_3 = -0.348 \text{ mA} + 1.79 \text{ mA}$$
$$= 1.442 \text{ mA}$$
$$I_{R_4} = I_2 = 1.939 \text{ mA}$$
$$I_{R_5} = I_1 = -0.348 \text{ mA}$$
$$I_T = I_1 + I_2 + I_3 = -0.348 \text{ mA}$$
$$+ 1.939 \text{ mA} + 1.79 \text{ mA} = 3.381 \text{ mA}$$
$$V_{R_1} = I_{R_1} R_1 = (1.591 \text{ mA})(2 \text{ k}\Omega) = 3.182 \text{ V}$$
$$V_{R_2} = I_{R_2} R_2 = (1.79 \text{ mA})(1 \text{ k}\Omega) = 1.79 \text{ V}$$
$$V_{R_3} = I_{R_3} R_3 = (1.442 \text{ mA})(5 \text{ k}\Omega) = 7.21 \text{ V}$$
$$V_{R_4} = I_{R_4} R_4 = (1.939 \text{ mA})(3 \text{ k}\Omega) = 5.817 \text{ V}$$
$$V_{R_5} = I_{R_5} R_5 = (-0.348 \text{ mA})(4 \text{ k}\Omega) = -1.392 \text{ V}$$

$$R_T = \frac{V_T}{I_T} = \frac{9 \text{ V}}{3.381 \text{ mA}} = 2.662 \text{ k}\Omega$$

The negative sign for V_{R_5} again tells us that the *assumed polarity* of the voltage across R_5 in Fig. 6-3 is incorrect.

After finishing an involved problem like the one above, it is worthwhile to check for errors. One way to do this is to substitute the voltages calculated into the original loop equations and see if the equations balance. For instance,

$$V_T = V_{R_1} + V_{R_5} + V_{R_3}$$
$$= 3.182 \text{ V} - 1.392 \text{ V} + 7.21 \text{ V}$$
$$= 9 \text{ V}$$

In analyzing the bridge circuit, we arbitrarily chose the direction of the current through the center resistor (R_5). Although choosing the wrong direction makes no difference in the final results, such an error can be easily avoided. Suppose we wish to know the direction of current through R_5 in Fig. 6-4(a). The direction can be found by temporarily removing R_5, as shown in Fig. 6-4(b), and determining the polarity of point A with respect to point B. Removal of R_5 makes the circuit a series-parallel circuit with two series resistors in each parallel branch. Inspection of Fig. 6-4(b) shows that, with respect to the negative terminal of the power source, point A must be more positive than point B because the ratio R_1/R_4 is greater than the ratio R_2/R_3. Thus, point A is positive with respect to point B to point A. The calculations used to determine the values shown in Fig. 6-4(b) are

Assumed polarity

$$V_{R_1} = \frac{R_1 V_T}{R_1 + R_4} = \frac{(10 \ \Omega)(18 \text{ V})}{10 \ \Omega + 15 \ \Omega}$$
$$= \frac{180 \ \Omega \cdot \text{V}}{25 \ \Omega} = 7.2 \text{ V}$$

$$V_{R_2} = \frac{R_2 V_T}{R_2 + R_3} = \frac{(50 \ \Omega)(18 \text{ V})}{50 \ \Omega + 100 \ \Omega}$$
$$= \frac{900 \ \Omega \cdot \text{V}}{150 \ \Omega} = 6 \text{ V}$$

Although connecting R_5 back into the circuit of Fig. 6-4(b) changes the voltage distributions, the relative polarities remain the same. Current flows from point B to point A in Fig. 6-4(a).

■ TEST _____

Answer the following questions.

10. Write the three loop equations for Fig. 6-3 assuming the direction of I_1 shown in Fig. 6-2(b).
11. Do the equations you wrote for question 10 yield the same absolute values of I_1, I_2, and I_3 as those we obtained using Fig. 6-3?
12. Do the equations of question 10 yield the same values for the current through and voltage across each of the resistors as those we obtained using Fig. 6-3?
13. Change the values in Fig. 6-4(a) to $R_1 = 40 \ \Omega$, $R_2 = 70 \ \Omega$, $R_3 = 80 \ \Omega$, $R_4 = 80 \ \Omega$, and $R_5 = 500 \ \Omega$. Will the current now flow from point A to point B or from point B to point A?
14. Does the value of R_5 in Fig. 6-4(a) have any effect on the polarity of the voltage between points A and B?

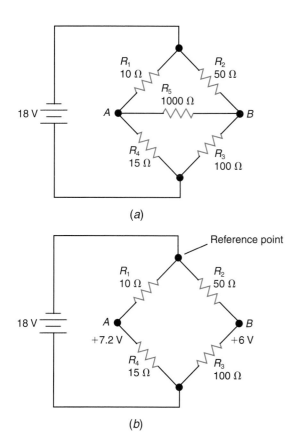

(a)

(b)

Fig. 6-4 Determining the direction of current through R_5. (a) Current direction through R_5 unspecified. (b) Point A is positive with respect to point B.

Multiple-Source Circuits

Multiple-source circuits

Several voltage sources can be applied to the same circuit components. When some, or all, of the components have no series or parallel relationship to other components, a complex multiple-source circuit exists. Such circuits are encountered in electronic systems and must be analyzed.

Look at the circuit of Fig. 6-5(*a*) and note that the resistors are neither in series nor in parallel. Therefore, we must use loop equations to solve the circuit. Two loops are required to include all the resistors in at least one loop. The traditional way to *draw the loops* is shown in Fig. 6-5(*b*). Although the traditional way is correct, the true direction of current flow is as shown in either Fig. 6-5(*c*) or Fig. 6-5(*d*). Which direction the current takes through R_3 depends on whether it is B_1 or B_2 that dominates the circuit. We could easily find the *direction of current* through R_3 by temporarily removing R_3 from the circuit and determining the polarity of the voltage between points A and B. However, there is no need to determine the true direction because the final results using any set of loops will give the current direction. If the loops of Fig. 6-5(*b*) are used, the direction will be that of the highest current (I_1 or I_2) and the magnitude will be the difference between the two currents. If the loops of Fig. 6-5(*c*) are used, the direction will be as indicated if the calculated value of I_1 is positive. If I_1 turns out negative when using the loops of Fig. 6-5(*c*), then the true current paths are those of Fig. 6-5(*d*). If the Fig. 6-5(*d*) loops are used, the polarity of I_2 determines the direction of the current.

Let us verify the equivalency of the three sets of loops by determining the circuit values using each set. For Fig. 6-5(*b*) the equations and calculations are

$$85\ \text{V} = 40\ \Omega\ (I_1) + 10\ \Omega\ (I_1) - 10\ \Omega\ (I_2)$$
(loop 1)

$$25\ \text{V} = -10\ \Omega\ (I_1) + 10\ \Omega\ (I_2) + 20\ \Omega\ (I_2)$$
(loop 2)

Collect terms:

$$85\ \text{V} = 50\ \Omega\ (I_1) - 10\ \Omega\ (I_2) \quad \text{(loop 1)}$$
$$25\ \text{V} = -10\ \Omega\ (I_1) + 30\ \Omega\ (I_2) \quad \text{(loop 2)}$$

Draw the loops

Current direction

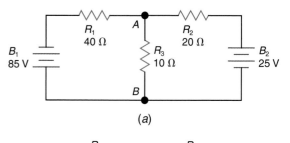

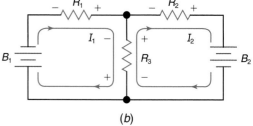

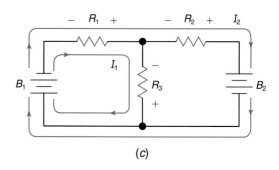

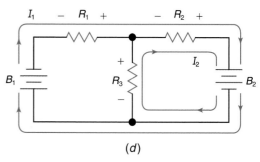

Fig. 6-5 Alternate sets of loops for a complex circuit. Any set will lead to the correct solution of circuit values. (*a*) Circuit to be analyzed. (*b*) Traditional loops. (*c*) Current paths if B_1 dominates. (*d*) Current path if B_2 dominates.

Determine the constant and eliminate I_1:

$$\frac{50\ \Omega\ (I_1)}{-10\ \Omega\ (I_1)} \times (-1) = 5$$

$$125\ \text{V} = -50\ \Omega\ (I_1) + 150\ \Omega\ (I_2)$$
$$\underline{85\ \text{V} = 50\ \Omega\ (I_1) - 10\ \Omega\ (I_2)}$$
$$210\ \text{V} = 140\ \Omega\ (I_2)$$

Determine I_2:

$$I_2 = \frac{210\ \text{V}}{140\ \Omega} = 1.5\ \text{A}$$

JOB TIP

People who seek technical jobs must also seek lifelong learning.

124   **Chapter 6**   Complex-Circuit Analysis

Determine I_1:

$$25 \text{ V} = -10 \text{ }\Omega \text{ }(I_1) + (30 \text{ }\Omega)(1.5 \text{ A})$$
$$= -10 \text{ }\Omega \text{ }(I_1) + 45 \text{ V}$$
$$-20 \text{ V} = -10 \text{ }\Omega \text{ }(I_1)$$
$$I_1 = \frac{20 \text{ V}}{10 \text{ }\Omega} = 2 \text{ A}$$

Determine the current through each resistor:

$$I_{R_3} = I_1 - I_2 = 2 \text{ A} - 1.5 \text{ A} - = 0.5 \text{ A}$$

(Note that I_{R_3} flows from point A to point B and that point A is negative with respect to point B.)

$$I_{R_2} = I_2 = 1.5 \text{ A}$$
$$I_{R_1} = I_1 = 2 \text{ A}$$

For Fig. 6-5(c) the procedure is

$$85 \text{ V} = 40 \text{ }\Omega \text{ }(I_1) + 40 \text{ }\Omega \text{ }(I_2) + 10 \text{ }\Omega \text{ }(I_1)$$
$$\text{(loop 1)}$$
$$85 \text{ V} = 40 \text{ }\Omega \text{ }(I_1) + 40 \text{ }\Omega \text{ }(I_2)$$
$$+ 20 \text{ }\Omega \text{ }(I_2) - 25 \text{ V} \quad \text{(loop 2)}$$

(Note that the B_2 voltage is negative because its polarity is the reverse of the voltage drops across R_1 and R_2). Collect terms:

$$85 \text{ V} = 50 \text{ }\Omega \text{ }(I_1) + 40 \text{ }\Omega \text{ }(I_2) \quad \text{(loop 1)}$$
$$110 \text{ V} = 40 \text{ }\Omega \text{ }(I_1) + 60 \text{ }\Omega \text{ }(I_2) \quad \text{(loop 2)}$$

Determine the constant and eliminate I_2:

$$\frac{60 \text{ }\Omega \text{ }(I_2)}{40 \text{ }\Omega \text{ }(I_2)} \times (-1) = -1.5$$
$$-127.5 \text{ V} = -75 \text{ }\Omega \text{ }(I_1) - 60 \text{ }\Omega \text{ }(I_2)$$
$$\underline{110 \text{ V} = \quad 40 \text{ }\Omega \text{ }(I_1) + 60 \text{ }\Omega \text{ }(I_2)}$$
$$-17.5 \text{ V} = -35 \text{ }\Omega \text{ }(I_1)$$

Determine I_1:

$$I_1 = \frac{17.5 \text{ V}}{35 \text{ }\Omega} = 0.5 \text{ A}$$

Determine I_2:

$$85 \text{ V} = 50 \text{ }\Omega \text{ }(0.5 \text{ A}) + 40 \text{ }\Omega \text{ }(I_2)$$
$$60 \text{ V} = 40 \text{ }\Omega \text{ }(I_2)$$
$$I_2 = \frac{60 \text{ V}}{40 \text{ }\Omega} = 1.5 \text{ A}$$

Determine the current through each resistor:

$$I_{R_3} = I_1 = 0.5 \text{ A}$$
$$I_{R_2} = I_2 = 1.5 \text{ A}$$
$$I_{R_1} = I_1 + I_2 = 0.5 \text{ A} + 1.5 \text{ A} = 2 \text{ A}$$

These values agree with those for Fig. 6-5(b). For Fig. 6-5(d) the calculations and results are

$$85 \text{ V} = 40 \text{ }\Omega \text{ }(I_1) + 20 \text{ }\Omega \text{ }(I_1)$$
$$+ 20 \text{ }\Omega \text{ }(I_2) - 25 \text{ V} \quad \text{(loop 1)}$$
$$25 \text{ V} = 10 \text{ }\Omega \text{ }(I_2) + 20 \text{ }\Omega \text{ }(I_1)$$
$$+ 20 \text{ }\Omega \text{ }(I_2) \quad \text{(loop 2)}$$

Collect terms:

$$110 \text{ V} = 60 \text{ }\Omega \text{ }(I_1) + 20 \text{ }\Omega \text{ }(I_2)$$
$$25 \text{ V} = 20 \text{ }\Omega \text{ }(I_1) + 30 \text{ }\Omega \text{ }(I_2)$$

Determine the constant and eliminate I_1:

$$\frac{60 \text{ }\Omega \text{ }(I_1)}{20 \text{ }\Omega \text{ }(I_1)} \times (-1) = -3$$
$$-75 \text{ V} = -60 \text{ }\Omega \text{ }(I_1) - 90 \text{ }\Omega \text{ }(I_2)$$
$$\underline{110 \text{ V} = 60 \text{ }\Omega \text{ }(I_1) + 20 \text{ }\Omega \text{ }(I_2)}$$
$$35 \text{ V} = -70 \text{ }\Omega \text{ }(I_2)$$

Determine I_2:

$$I_2 = \frac{-35 \text{ V}}{70 \text{ }\Omega} = -0.5 \text{ A}$$

Thus, Fig. 6-5(d) shows the wrong direction for I_{R_3}. Determine I_1:

$$25 \text{ V} = 20 \text{ }\Omega \text{ }(I_1) + 30 \text{ }\Omega(-0.5 \text{ A})$$
$$40 \text{ V} = 20 \text{ }\Omega \text{ }(I_1)$$
$$I_1 = \frac{40 \text{ V}}{20 \text{ }\Omega} = 2 \text{ A}$$

Determine the current through each resistor:

$$I_{R_3} = I_2 = -0.5 \text{ A}$$
$$I_{R_2} = I_1 + I_2 = 2 \text{ A} - 0.5 \text{ A} = 1.5 \text{ A}$$
$$I_{R_1} = I_1 = 2 \text{ A}$$

Again, these current values agree with the previously calculated values.

The voltage sources in Fig. 6-5 are connected so that they are *series-aiding*. That is, the negative terminal of B_2 is connected to the positive terminal of B_1. Multiple-source circuits can also be connected so that the sources are *series-opposing*. Furthermore, both the negative and positive source terminals can be separated by one or more resistors. The possible variations are almost limitless; yet the procedures and rules for working with the circuits remain the same.

Series-aiding

Series-opposing

Example 6-3

Determine the voltages across R_1, R_2, and R_3 in Fig. 6-6(a) using the loop-equations technique.

Given: Circuit diagram and values of Fig. 6-6(a).

Find: V_{R_1}, V_{R_2}, and V_{R_3}

Known: Kirchhoff's voltage law and simultaneous equations

Solution: Determine loops and polarities as shown in Fig. 6-6(b). Derive loop equations:

$$3\text{ V} = 4\ \Omega\ (I_1) + 9\ \Omega\ (I_1) + 9\ \Omega\ (I_2)$$
(loop 1)

$$24\text{ V} = 9\ \Omega\ (I_1) + 9\ \Omega\ (I_2) + 6\ \Omega\ (I_2)$$
(loop 2)

Collect terms:

$$3\text{ V} = 13\ \Omega\ (I_1) + 9\ \Omega\ (I_2) \quad \text{(loop 1)}$$
$$24\text{ V} = 9\ \Omega\ (I_1) + 15\ \Omega\ (I_2) \quad \text{(loop 2)}$$

Add, eliminate I_2, and solve for I_1:

$$\frac{9\ \Omega\ (I_2)}{15\ \Omega\ (I_2)} \times (-1) = -0.6$$

$$-14.4\text{ V} = -5.4\ \Omega\ (I_1) - 9\ \Omega\ (I_2)$$
$$\underline{3\text{ V} = 13\ \Omega\ (I_1) + 9\ \Omega\ (I_2)}$$
$$-11.4\text{ V} = 7.6\ \Omega\ (I_1)$$

$$I_1 = \frac{-11.4\text{ V}}{7.6\ \Omega} = -1.5\text{ A}$$

Solve for I_2:

$$24\text{ V} = 9\ \Omega\ (-1.5\text{ A}) + 15\ \Omega\ (I_2)$$

$$I_2 = \frac{37.5\text{ V}}{15\ \Omega} = 2.5\text{ A}$$

Solve for the currents through the resistors:

$$I_{R_1} = I_1 = -1.5\text{ A}$$
$$I_{R_2} = I_1 + I_2 = -1.5\text{ A} + 2.5\text{ A} = 1\text{ A}$$
$$I_{R_3} = I_2 = 2.5\text{ A}$$

Solve for the voltages across the resistors:

$$V_{R_1} = I_{R_1}R_1$$
$$= (-1.5\text{ A})(4\ \Omega) = -6\text{ V}$$
$$V_{R_2} = I_{R_2}R_2$$
$$= (1\text{ A})(9\ \Omega) = 9\text{ V}$$
$$V_{R_3} = I_{R_3}R_3 = (2.5\text{ A})(6\ \Omega) = 15\text{ V}$$

Answer: $V_{R_1} = -6\text{ V}$
$$V_{R_2} = 9\text{ V}$$
$$V_{R_3} = 15\text{ V}$$

In the above example, the negative sign for V_{R_1} tells us that the polarity for R_1 in Fig. 6-6(b) is incorrect. The correct polarities and

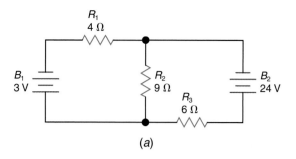

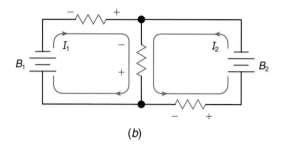

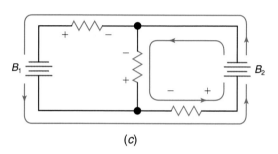

Fig. 6-6 Circuits for example 6-3. Battery B_1 is being charged. (a) Given circuit and values. (b) Loops used in circuit solution. (c) True polarities and current directions.

current directions are shown in Fig. 6-6(c). Note that B_1 is being charged because electrons are entering its negative terminal. Battery B_2 is furnishing all the power for the circuit.

As a final illustration of a multiple-source circuit, let us solve one that involves three loops. The general procedures are the same as those used for the bridge circuits except that this time we have two voltage sources. In this example, let us omit all units from the formulas for the sake of simplicity. Our answers for I will still be in amperes as long as circuit resistances are in ohms and circuit voltages are in volts.

Example 6-4

Determine the current through each of the resistors in Fig. 6-7(a).

Given: Circuit diagram and values of Fig. 6-7(a).

Find: $I_{R_1}, I_{R_2}, I_{R_3}, I_{R_4}, I_{R_5}$

Known: Kirchhoff's laws and three-loop techniques

Solution: Specify loops and polarities [Fig. 6-7(b)]. Write loop equations:

$$44 = 10I_1 + 10I_2 + 20I_1 \quad \text{(loop 1)}$$
$$44 = 10I_1 + 10I_2 + 15I_2$$
$$\quad + 12I_2 + 12I_3 \quad \text{(loop 2)}$$
$$18 = 30I_3 + 12I_3 + 12I_2 \quad \text{(loop 3)}$$

Collect terms:

$$44 = 30I_1 + 10I_2 \quad \text{(loop 1)}$$
$$44 = 10I_1 + 37I_2 + 12I_3 \quad \text{(loop 2)}$$
$$18 = 12I_2 + 42I_3 \quad \text{(loop 3)}$$

Add loops 1 and 2 and eliminate I_1:

$$\frac{30I_1}{10I_1} \times (-1) = -3$$
$$-132 = -30I_1 - 111I_2 - 36I_3$$

$$44 = \quad 30I_1 + 10I_2$$
$$\underline{-88 = -101I_2 - 36I_3} \quad \text{(loops 1 and 2)}$$

Add loops 1 and 2 to loop 3, eliminate I_3, and solve for I_2:

$$\frac{42I_3}{-36I_3} \times (-1) = 1.16$$
$$-102.6 = -117.83I_2 - 42I_3$$
$$\underline{18 = \quad 12I_2 + 42I_3}$$
$$-84.6 = -105.83I_2$$
$$I_2 = 0.8 \text{ A}$$

Solve for I_1 and I_3:

(from loop 1 equation) $44 = 30I_1 + 10(0.8)$
$$I_1 = 1.2 \text{ A}$$

(from loop 3 equation) $18 = 12(0.8) + 42I_3$
$$I_3 = 0.2 \text{ A}$$

Solve for currents through each resistor:

$$I_{R_1} = I_1 + I_2 = 1.2 \text{ A} + 0.8 \text{ A} = 2 \text{ A}$$
$$I_{R_2} = I_1 = 1.2 \text{ A}$$

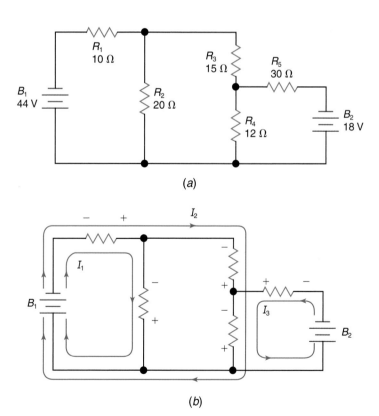

(a)

(b)

Fig. 6-7 Three-loop circuit analyzed in example 6-4. The selected loops turned out also to be the true current paths. (a) Given circuit and values. (b) Assumed loops and resulting polarities.

$$I_{R_3} = I_2 = 0.8 \text{ A}$$

$$I_{R_4} = I_2 + I_3 = 0.8 \text{ A} + 0.2 \text{ A} = 1 \text{ A}$$

$$I_{R_5} = I_3 = 0.2 \text{ A}$$

Answer: $I_{R_1} = 2 \text{ A}$,
$I_{R_2} = 1.2 \text{ A}$,
$I_{R_3} = 0.8 \text{ A}$,
$I_{R_4} = 1 \text{ A}$, $I_{R_5} = 0.2 \text{ A}$

Short circuit

Note in example 6-4 that the true direction of the currents is as shown in Fig. 6-7(*b*) because I_1, I_2 and I_3 all came out positive. Thus, the polarity of the voltage drops in Fig. 6-7(*b*) is also correct.

■ TEST

Answer the following questions.

15. True or false. There can be more than one correct set of loop equations for a multiple-source circuit.

16. True or false. When the calculated value of a current is negative, the wrong loops have been used in analyzing the circuit.

17. Determine I_{R_1}, I_{R_2}, I_{R_3}, and V_{R_3} for the circuit shown in Fig. 6-8.

Active source

Algebraically add

6-3 Superposition Theorem

Superposition theorem

Another approach to analyzing multiple-source circuits involves the use of the *superposition theorem*. The general idea of this theorem is that the overall effect of all sources on a circuit is the sum of effects of the individual sources. Since each voltage source has an effect on the circuit currents, the circuit currents are analyzed once for each voltage source. Then, the currents resulting from each source are added together to arrive at the final circuit currents.

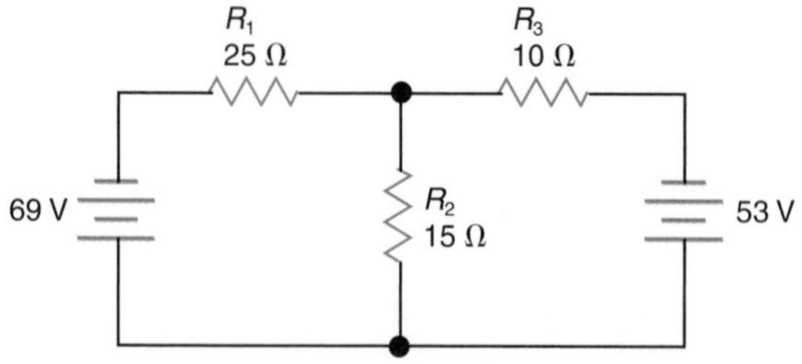

Fig. 6-8   Multiple-source circuit for test question 15.

Once the currents are known, it is a simple matter of applying Ohm's law to find the voltage drops in the circuit.

The specific steps used in applying the superposition theorem are as follows:

1. If the internal resistances of the voltage sources are very small compared with the external resistance in the circuit, which is the case in most electronic circuits, replace every voltage source but one with a *short circuit* (conductor). If the internal resistance is a significant part (1 percent or more) of the total circuit resistance, replace the source with a resistor equal to its internal resistance.

2. Calculate the magnitude and direction of the current through each resistor in the temporary circuit produced by step 1. (Use the procedures developed in Chap. 5 for series-parallel circuits when making these calculations.)

3. Repeat steps 1 and 2 until each voltage source has been used as the *active source*.

4. *Algebraically add* all the currents (from step 2) through a specific resistor to determine the magnitude of the current through that resistor in the original circuit. The direction of the current is the direction of the dominant current through the resistor.

Now, let us apply these four steps to the circuit of Fig. 6-9(*a*) so that we can observe in detail how the superposition theorem works. Applying step 1 by shorting out B_1 yields the temporary circuit shown in Fig. 6-9(*b*). The currents shown in Fig. 6-9(*b*) result from following step 2. The calculations involved are

$$\begin{aligned}
R_{1,2} &= \frac{R_1 R_2}{R_1 + R_2} \\
&= \frac{(40 \ \Omega)(10 \ \Omega)}{40 \ \Omega + 10 \ \Omega} = 8 \ \Omega \\
R_T &= R_{1,2} + R_3 \\
&= 8 \ \Omega + 30 \ \Omega = 38 \ \Omega \\
I_{R_3} &= I_T \\
&= \frac{V_T}{R_T} \\
&= \frac{38 \text{ V}}{38 \ \Omega} = 1 \text{ A} \\
I_{R_2} &= \frac{I_T R_1}{R_1 + R_2}
\end{aligned}$$

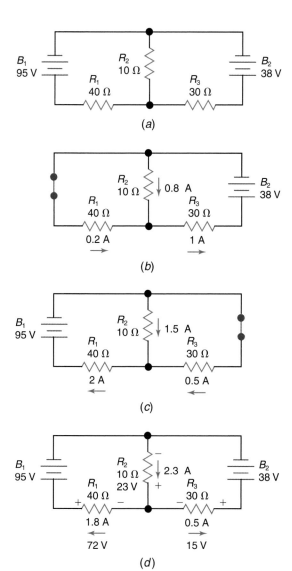

(a)

(b)

(c)

(d)

Fig. 6-9 Using the superposition theorem to analyze a multiple-source circuit. Currents and voltage drops for the original circuit (*a*) are shown in (*d*). (*a*) Original circuit. (*b*) Shorting B_1 puts R_1 and R_2 in parallel. (*c*) Shorting B_2 puts R_2 and R_3 in parallel. (*d*) Final results.

$$= \frac{(1 \text{ A})(40 \text{ } \Omega)}{40 \text{ } \Omega + 10 \text{ } \Omega}$$
$$= 0.8 \text{ A}$$
$$I_{R_1} = I_{R_3} - I_{R_2}$$
$$= 1 \text{ A} - 0.8 \text{ A} = 0.2 \text{ A}$$

Going on to step 3 leads us back to steps 1 and 2 for the second time. The second time through steps 1 and 2 results in the circuit and current values shown in Fig. 6-9(*c*). The currents were calculated by using the procedure outlined for Fig. 6-9(*b*). We have now used each of the

sources as the active source, and so we move on to step 4. Following step 4 leads to the currents shown in Fig. 6-9(*d*). These currents show the magnitude and direction of the currents in the original circuit. They were obtained by algebraically adding the appropriate currents from Fig. 6-9(*b*) and (*c*). When the currents are in opposite directions, the smaller current is assigned a minus sign because the larger current shows the direction of current in the original circuit. The calculations are:

$$I_{R_1} = 2 \text{ A} - 0.2 \text{ A} = 1.8 \text{ A}$$
$$I_{R_2} = 1.5 \text{ A} + 0.8 \text{ A} = 2.3 \text{ A}$$
$$I_{R_3} = 1 \text{ A} - 0.5 \text{ A} = 0.5 \text{ A}$$

Also shown in Fig. 6-9(*d*) are the voltage drops, which were calculated with Ohm's law. We can use these voltage drops and Kirchhoff's voltage law to verify that our analysis is correct. For example,

$$95 \text{ V} = 23 \text{ V} + 72 \text{ V} = 38 \text{ V} - 15 \text{ V} + 72 \text{ V}$$
$$38 \text{ V} = 23 \text{ V} + 15 \text{ V} = 95 \text{ V} - 72 \text{ V} + 15 \text{ V}$$

The use of the superposition theorem allows us to analyze Fig. 6-9(*a*) without solving simultaneous equations. However, it does require repeated use of series-parallel techniques.

TEST

Answer the following questions.

18. Change R_1 in Fig. 6-6(*a*) to 8 Ω and then use the superposition theorem to find I_{R_1}, I_{R_2}, and I_{R_3}.
19. Can the superposition theorem be used with single-source circuits?
20. True or false. Application of the superposition theorem always leads to the correct direction of currents through all resistors.

Three-Loop Circuits

The superposition theorem can be applied to *three-loop circuits* like the one shown in Fig. 6-7. For example, shorting out B_1 sets up a series-parallel circuit where R_T can be found by applying these formulas:

$$R_{1,2} = \frac{R_1 R_2}{R_1 + R_2}$$

$$R_{1,2,3} = R_{1,2} + R_3$$

$$R_{1,2,3,4} = \frac{(R_{1,2,3})(R_4)}{R_{1,2,3} + R_4}$$

$$R_T = R_{1,2,3,4} + R_5$$

The remainder of the analysis of Fig. 6-7 by the superposition theorem is left up to you. You can check your answers against the answers given in example 6-4.

So far we have not examined any circuits that involve three sources. Such circuits are easily solved using either the loop techniques of Sec. 6-2 or the superposition theorem. We will analyze a *three-source circuit* using the superposition theorem.

Example 6-5

Using the superposition theorem, calculate the current through every resistor in Fig. 6-10(*a*).

Given: Circuit diagram and values of Fig. 6-10(*a*).

Find: I_{R_1}, I_{R_2}, I_{R_3}, and I_{R_4}

Known: Superposition theorem, series-parallel circuit techniques

Solution: Shorts out B_2 and B_3 and solve for all currents:

$$R_{1,2} = \frac{R_1 R_2}{R_1 + R_2}$$

$$= \frac{(10)(15)}{10 + 15} = 6\ \Omega$$

$$R_{3,4} = \frac{R_3 R_4}{R_3 + R_4}$$

$$= \frac{(12)(20)}{12 + 20} = 7.5\ \Omega$$

$$R_T = R_{1,2} + R_{3,4}$$

$$= 6 + 7.5 = 13.5\ \Omega$$

$$I_T = \frac{V_T}{R_T} = \frac{55\ \text{V}}{13.5\ \Omega}$$

$$= 4.074\ \text{A}$$

$$I_{R_1} = \frac{I_T R_2}{R_1 + R_2}$$

$$= \frac{(4.074)(15)}{10 + 15} = 2.444\ \text{A} \downarrow$$

The arrow indicates current direction.

$$I_{R_2} = I_T - I_{R_1}$$

$$= 4.074 - 2.444 = 1.630\ \text{A} \downarrow$$

$$I_{R_3} = \frac{I_T R_4}{R_3 + R_4}$$

$$= \frac{(4.074)(20)}{12 + 20} = 2.546\ \text{A} \downarrow$$

$$I_{R_4} = I_T - I_{R_3}$$

$$= 4.074 - 2.546 = 1.528\ \text{A} \downarrow$$

Next, short out B_1 and B_2 and solve all for currents:

$$R_{1,2,4} = \frac{1}{\dfrac{1}{R_1} + \dfrac{1}{R_2} + \dfrac{1}{R_4}}$$

$$R_{1,2,4} = \frac{1}{\dfrac{1}{10} + \dfrac{1}{15} + \dfrac{1}{20}} = 4.615\ \Omega$$

$$R_T = R_{1,2,4} + R_3 = 4.615 + 12$$

$$= 16.615\ \Omega$$

$$I_T = \frac{V_T}{R_T} = \frac{58\ \text{V}}{16.615\ \Omega} = 3.491\ \text{A}$$

$$V_{R_3} = I_{R_3} R_3 = I_T R_3 = (3.491\ \text{A})(12\ \Omega)$$

$$= 41.892\ \text{V}$$

$$V_{R_1} = V_{R_1} = V_{R_4} = V_T - V_{R_3}$$

$$= 58 - 41.892 = 16.108\ \text{V}$$

$$I_{R_1} = \frac{V_{R_1}}{R_1} = \frac{16.108\ \text{V}}{10\ \Omega} = 1.611\ \text{A} \uparrow$$

$$I_{R_2} = \frac{V_{R_2}}{R_2} = \frac{16.108\ \text{V}}{15\ \Omega} = 1.074\ \text{A} \uparrow$$

$$I_{R_3} = I_T = 3.491\ \text{A} \uparrow$$

$$I_{R_4} = \frac{V_{R_4}}{R_4} = \frac{16.108\ \text{V}}{20\ \Omega} = 0.805\ \text{A} \downarrow$$

Next, short out B_1 and B_3 and again solve for all currents:

$$R_{1,3,4} = \frac{1}{\dfrac{1}{R_1} + \dfrac{1}{R_3} + \dfrac{1}{R_4}} = \frac{1}{\dfrac{1}{10} + \dfrac{1}{12} + \dfrac{1}{20}}$$

$$= 4.286\ \Omega$$

$$R_T = R_{1,3,4} + R_2 = 19.286\ \Omega$$

$$I_T = \frac{V_T}{R_T} = \frac{30\ \text{V}}{19.286\ \Omega} = 1.556\ \text{A}$$

$$V_{R_2} = I_{R_2}R_2 = I_T R_2 = (1.556\ \text{A})(15\ \Omega)$$
$$= 23.34\ \text{V}$$
$$V_{R_1} = V_{R_3} = V_{R_4} = V_T - V_{R_2} = 30 - 23.34$$
$$= 6.66\ \text{V}$$
$$I_{R_1} = \frac{V_{R_1}}{R_1} = \frac{6.66\ \text{V}}{10\ \Omega} = 0.666\ \text{A} \downarrow$$
$$I_{R_2} = I_T = 1.556\ \text{A} \uparrow$$
$$I_{R_3} = \frac{V_{R_3}}{R_3} = \frac{6.66\ \text{V}}{12\ \Omega} = 0.555\ \text{A} \uparrow$$
$$I_{R_4} = \frac{V_{R_4}}{R_4} = \frac{6.66\ \text{V}}{20\ \Omega} = 0.333\ \text{A} \uparrow$$

Finally, we must algebraically add the individual currents:

$$I_{R_1} = \quad 2.444 - 1.611 + 0.666 = 1.5\ \text{A} \downarrow$$
$$I_{R_2} = -1.630 + 1.074 + 1.556 = 1\ \text{A} \uparrow$$
$$I_{R_3} = -2.546 + 3.491 + 0.555 = 1.5\ \text{A} \uparrow$$
$$I_{R_4} = \quad 1.528 + 0.805 - 0.333 = 2\ \text{A} \downarrow$$

Answers: $I_{R_1} = 1.5\ \text{A}$, $I_{R_2} = 1\ \text{A}$
$\qquad\quad I_{R_3} = 1.5\ \text{A}$, $I_{R_4} = 2\ \text{A}$

We could cross-check the answers to example 6-5 with Kirchhoff's current law and/or Kirchhoff's voltage law. The cross-check is left up to you.

◼ TEST

Answer the following questions.

21. Apply Kirchhoff's current law to the junction of R_1, R_3, R_4, and B_2 in Fig. 6-10.
22. How much current does B_1 provide in Fig. 6-10?

6-4 Voltage Sources

There are two ways of looking at a voltage source. It can be considered a *constant voltage source* (also called an *ideal voltage source*), or it can be considered a constant voltage source in series with some internal resistance. This latter approach is referred to as an *equivalent-circuit voltage source*. The way in which we view a voltage source depends on both the na-

Constant voltage source

Ideal voltage source

Equivalent-circuit voltage source

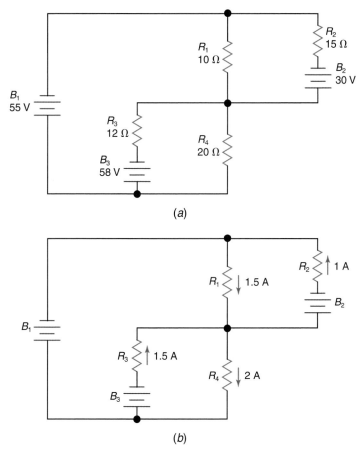

(a)

(b)

Fig. 6-10 Three-source circuit for example 6-5. Magnitude and direction of current are shown in (b). (a) Given circuit and values. (b) Final results of analysis.

ture of the source and the nature of the circuit to which the source is connected. Actual voltage sources are not true constant voltage sources; however, for practical applications, many voltage sources are assumed to be constant voltage sources.

If a constant (ideal) voltage source were possible, it would have no internal resistance; therefore, the terminal voltage would be constant regardless of the amount of current drawn from the source. The *open-circuit terminal voltage* (V_{oc}) of Fig. 6-11(a) would be the same as the closed-circuit terminal voltage of Fig. 6-11(b). Of course, such a voltage source does not exist because all voltage sources have some internal resistance. The conductors in the coils of a generator have some resistance, the chemicals in the lead-acid cell have some resistance, and so forth. Thus, the terminal voltage of all voltage sources must decrease when the current drawn from the source is increased. Viewing a battery as a constant voltage source when analyzing circuits introduces some error. In most cases, however, the error is less than that caused by rounding off numbers when doing the calculations. With but one exception (Sec. 5-4), we have assumed all of our circuits to be powered by constant voltage sources. (The battery symbol, with a voltage specified, implies a constant voltage source.) This is stan-

Open-circuit terminal voltage

Internal resistance

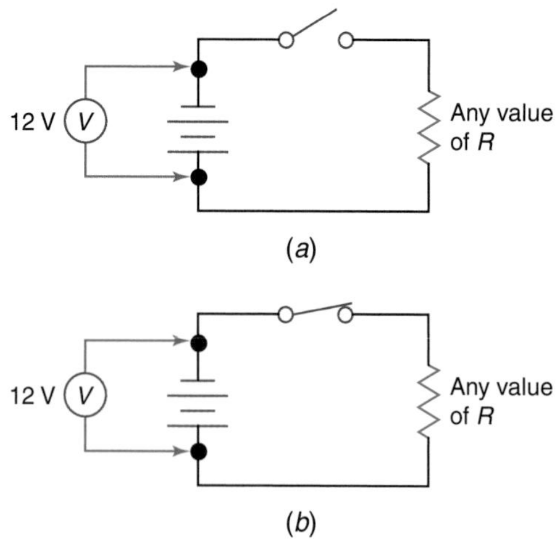

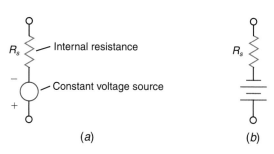

Fig. 6-12 Equivalent circuit of a voltage source. (a) Equivalent circuit using a constant voltage source symbol. (b) Alternative circuit using a battery symbol.

dard practice as long as the internal resistance of the source is less than 1 percent of the total circuit resistance.

When the *internal resistance* of a voltage source becomes a significant part of the total circuit resistance, we must take it into account when analyzing the circuit. To do this, we replace the simple battery symbol with its true equivalent circuit. That is, we view the source as an equivalent-circuit voltage source. The equivalent circuit of a voltage source is shown in Fig. 6-12(a). In this figure, the circle with the polarity markings represents the constant voltage source. Its value is the open-circuit voltage (V_{oc}) of the voltage source. It can be thought of as the voltage produced by the chemicals from which a cell is constructed. The voltage produced by the chemical reaction in the cell is indeed independent of the current. The internal resistance (R_S) represents the resistance of the chemicals in the cell. When current flows through the cell, this resistance drops part of the voltage produced by the chemical reaction. When dc circuits are being analyzed, the polarized circle of Fig. 6-12(a) is sometimes replaced by a battery symbol, as shown in Fig. 6-12(b). The small circles at the end of the equivalent circuits in Fig. 6-12 represent the terminals of the voltage source. Sometimes they are omitted.

Fig. 6-11 An ideal voltage source. Such voltage sources do not actually exist. (a) Open-circuit terminal voltage (V_{oc}) equals 12 V. (b) Closed-circuit (loaded) terminal voltage equals 12 V.

■ **TEST** _____

Answer the following questions.

23. What determines whether or not a voltage source should be assumed to be constant?

24. Whether we view a voltage source as constant or as an equivalent circuit (constant

voltage source in series with its internal resistance), the voltage of the constant voltage source is equal to the _____ .

25. True or false. Batteries are never assumed to be a constant voltage source.

Example 6-6

A voltage source has an open-circuit voltage of 30 V. Its internal resistance is 1 Ω. Calculate the terminal voltage when this source is connected to a 10-Ω load.

Given: $V_{oc} = 30$ V, $R_S = 1$ Ω, $R_L = 10$ Ω ("R_L" stands for "load resistor")

Find: Closed-circuit terminal voltage for a specified load resistance

Known: Equivalent circuit for a source and voltage-divider formula

Solution: Draw the schematic diagram as shown in Fig. 6-13. Then use the voltage divider equation to determine V_{RL}:

$$V_{RL} = \frac{(30 \text{ V})(R_L)}{R_L + R_S} = \frac{(30 \text{ V})(10 \ \Omega)}{10 \ \Omega + 1 \ \Omega} = 27.27 \text{ V}$$

Since R_L is connected directly to the source terminals, the closed-circuit terminal voltage equals V_{RL}.

Answer: The terminal voltage with a 10-Ω load is 27.27 V.

6-5 Thevenin's Theorem

Any of the circuits we have been working with can be viewed as a *two-terminal network*. For example, the circuit in Fig. 6-14(*a*) can be as-signed two terminal locations as shown in Fig. 6-14(*b*). Between these two terminals a voltage exists; however, the amount of voltage depends on the amount of current we draw from the terminals. Thus, we can view everything to the left of the terminals, as shown in Fig. 6-14(*c*), as a voltage source. Of course, this voltage source must be viewed as an equivalent-circuit voltage source because its internal resistance is quite large. The location of the two terminals in a circuit is entirely up to the person analyzing the circuit. The location of the terminals for the circuit of Fig. 6-14(*a*) could just as well have been as shown in Fig. 6-14(*d*).

Thevenin's theorem provides us with an easy way to develop an equivalent circuit of a two-terminal network. This equivalent circuit will, of course, be an equivalent-circuit voltage source like that shown in Fig. 6-12(*a*). Applying Thevenin's theorem to a two-terminal network requires only two steps:

1. Determine the open-circuit voltage between the terminals. This can be done either by calculating the voltage using the schematic diagram or by measuring the voltage if the circuit actually exists.
2. Determine the internal resistance of the equivalent-circuit voltage source after replacing all voltage sources in the original circuit with resistances equal to their internal resistance. When the original voltage sources are viewed as ideal sources, they are short-circuited. After the original voltage sources are replaced by conductors, the internal resistance (total resistance between the terminals) can be either calculated or measured.

Equivalent circuits developed by Thevenin's theorem are often called *Thevenin equivalent circuits;* the original circuit is said to have been *thevenized.* The constant voltage source determined in step 1 is usually labeled V_{TH} instead of V_{oc}, and the internal resistance of step 2 is R_{TH} instead of R_S.

Let us try an example using Thevenin's theorem.

Closed-circuit terminal voltage

Thevenin's theorem

Two-terminal network

Thevenin equivalent circuit

Thevenized

V_{TH}

R_{TH}

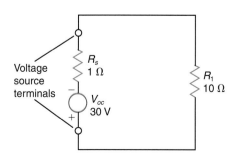

Fig. 6-13 Circuit diagram for example 6-6.

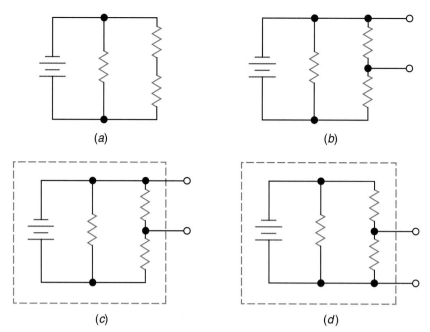

Fig. 6-14 Viewing a circuit as a two-terminal voltage source. (*a*) Original circuit. (*b*) Two terminals added. (*c*) Voltage source with large internal resistance. (*d*) Alternative location of terminals.

Example 6-7

Determine the Thevenin equivalent circuit for the circuit in Fig. 6-15(*a*) when the terminals are considered to be the two ends of R_2.

Given: Circuit diagram and values in Fig. 6-15(*a*)

Find: V_{TH} and R_{TH}

Known: Thevenin's theorem

Solution: Calculate V_{TH} (note that V_{TH} is the same as V_{R_2}):

$$V_{TH} = \frac{V_{B_1}R_2}{R_1 + R_2}$$

$$= \frac{(30\text{ V})(1000\ \Omega)}{500\ \Omega + 1000\ \Omega} = 20\text{ V}$$

Calculate R_{TH} (note that shorting out V_{B_1} puts R_1 in parallel with R_2):

$$R_{TH} = \frac{R_1 R_2}{R_1 + R_2}$$

$$= \frac{(500\ \Omega)(1000\ \Omega)}{500\ \Omega + 1000\ \Omega} = 333.3\ \Omega$$

Answer: The Thevenin equivalent circuit is drawn in Fig. 6-15(*b*).

$$V_{TH} = V_{oc} = 20\text{ V}$$

$$R_{TH} = R_S = 333.3\ \Omega$$

Any load connected to the terminals of the original circuit in Fig. 6-15(*a*) have the same current, voltage, and power as the same load connected to the Thevenin equivalent circuit of Fig. 6-15(*b*). We can prove this statement by applying a 1000-Ω load to each circuit, as shown in Fig. 6-15(*c*) and (*d*). For the loaded Thevenin equivalent circuit [Fig. 6-15(*d*)], the calculations are:

$$V_{RL} = \frac{V_{TH}R_L}{R_{TH} + R_L} = \frac{(20\text{ V})(1000\ \Omega)}{333.3\ \Omega + 1000\ \Omega} = 15\text{ V}$$

$$I_{RL} = \frac{V_{RL}}{R_L} = \frac{15\text{ V}}{1000\ \Omega} = 15\text{ mA}$$

For the loaded original circuit shown in Fig. 6-15(*c*), the calculations are

$$R_{2,L} = \frac{R_2 R_L}{R_2 + R_L} = \frac{(1000\ \Omega)(1000\ \Omega)}{1000\ \Omega + 1000\ \Omega} = 500\ \Omega$$

$$V_{RL} = V_{R2,L} = \frac{V_{B_1}R_{2,L}}{R_1 + R_{2,L}} = \frac{(30\text{ V})(500\ \Omega)}{500\ \Omega + 500\ \Omega}$$

$$= 15\text{ V}$$

$$I_{RL} = \frac{V_{RL}}{R_L} = \frac{15\text{ V}}{1000\ \Omega} = 15\text{ mA}$$

From the above calculations, you can see that the calculations for solving the loaded original circuit are more difficult than for solving the loaded Thevenin equivalent circuit. The level of difficulty becomes more pronounced

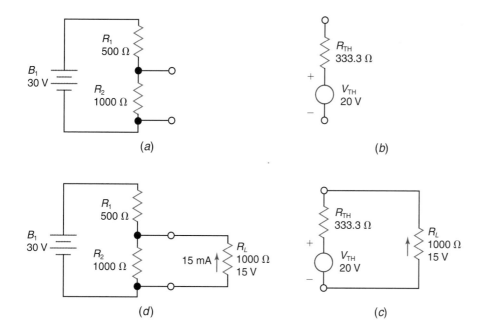

Fig. 6-15 Thevenin equivalent circuit without and with a load. (*a*) Original circuit (voltage divider). (*b*) Thevenin equivalent circuit. (*c*) Original circuit with a load added. (*d*) Thevenin equivalent circuit with a load added.

as the original circuit becomes more complex. Now you can see the advantage of thevenizing a circuit if you need to know the load voltage for many different values of R_L.

The preceding illustration shows how to thevenize a circuit and then determine the effects of adding a load resistance. Next, we will see how to use a Thevenin equivalent circuit to analyze a complex circuit. The general procedure involves selecting one resistor in the complex circuit to act as a load and then thevenizing the remainder of the circuit. The specific steps for using Thevenin's theorem to analyze a complex circuit are as follows:

1. Select one of the resistors in the complex circuit to be removed from the circuit. This resistor will later be added back to the Thevenin equivalent circuit and treated like the load resistor in the above example. If possible, select this resistor so that the remainder of the original circuit can be analyzed using series, parallel, or series-parallel rules.
2. Analyze the remainder of the circuit to determine the open-circuit voltage (V_{TH}) at the terminals created by step 1.
3. Analyze the remainder of the circuit to determine the internal resistance (R_{TH}).
4. Draw the Thevenin equivalent circuit (using the values obtained in steps 2 and

3) and load it with the resistor removed in step 1.
5. Determine the voltage across and the current through the resistor added in step 4.
6. Insert the resistor removed in step 1 back into the original circuit. The current and voltage of the reinserted resistor are the same in the original circuit as in the loaded Thevenin equivalent circuit of step 5.
7. If possible, use Kirchhoff's current and voltage laws to calculate the other currents and voltages of the complex circuit. Steps 6 and 7 may be omitted if you are interested only in the current and voltage of the resistor removed in step 1.

Let us illustrate these steps by evaluating the complex circuit of Fig. 6-16(*a*). The result of applying step 1 is shown in Fig. 6-16(*b*), where terminals have been added to the points from which R_3 was removed. Removal of R_3 results in a simple series circuit with two opposing source voltages. Of course, the larger voltage (B_2) dominates the circuit and forces current through B_1 in a reverse direction. Applying step 2 to the circuit in Fig. 6-16(*b*) leads to the voltages specified in Fig. 6-16(*c*), which were calculated as follows:

$$V_T = B_2 - B_1 = 19 \text{ V} - 10 \text{ V} = 9 \text{ V}$$

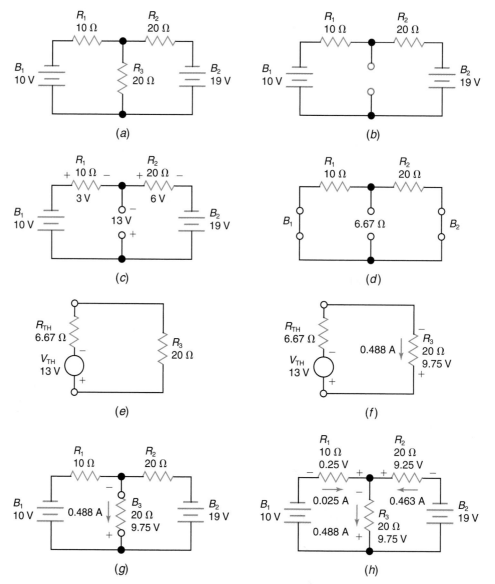

Fig. 6-16 Analyzing a complex circuit by thevenizing the circuit. (*a*) Original complex circuit. (*b*) Step 1—remove R_3. (*c*) Step 2—calculate V_{TH}. (*d*) Step 3—calculate R_{TH}. (*e*) Step 4—loaded Thevenin equivalent circuit. (*f*) Step 5—calculate V_{R_3} and I_{R_3}. (*g*) Step 6—replace R_3. (*h*) Step 7—calculate other currents and voltages.

$$V_{R_2} = \frac{V_T R_2}{R_1 + R_2} = \frac{(9 \text{ V})(20 \text{ }\Omega)}{10 \text{ }\Omega + 20 \text{ }\Omega} = 6 \text{ V}$$

$$V_{TH} = B_2 - V_{R_2} = 19 \text{ V} - 6 \text{ V} = 13 \text{ V}$$

As called for in step 3, the circuit of Fig. 6-16(*b*) was modified to arrive at the circuit in Fig. 6-16(*d*) so that R_{TH} could be calculated. As seen from the two terminals, R_1 and R_2 in Fig. 6-16(*d*) are in parallel. Thus, R_{TH} is

$$R_{TH} = \frac{R_1 R_2}{R_1 + R_2} = \frac{(10 \text{ }\Omega)(20 \text{ }\Omega)}{10 \text{ }\Omega + 20 \text{ }\Omega} = 6.67 \text{ }\Omega$$

Next, step 4 tells us to draw the Thevenin equivalent circuit for the circuit of Fig. 6-16(*b*) and load it with R_3. This step is illustrated in Fig. 6-16(*e*). Then, as called for in step 5 and illustrated in Fig. 6-16(*f*), the voltage and current associated with R_3 are calculated:

$$V_{R_3} = \frac{V_{TH} R_3}{R_{TH} + R_3} = \frac{(13 \text{ V})(20 \text{ }\Omega)}{6.67 \text{ }\Omega + 20 \text{ }\Omega} = 9.75 \text{ V}$$

$$I_{R_3} = \frac{V_{R_3}}{R_3} = \frac{9.75 \text{ V}}{20 \text{ }\Omega} = 0.488 \text{ A}$$

Following through with step 6 yields the conditions shown in Fig. 6-16(g). Step 7 yields the values of additional voltages and currents shown in Fig. 6-16(h), calculated as follows:

$$V_{R_1} = B_1 - V_{R_3} = 10 \text{ V} - 9.75 \text{ V} = 0.25 \text{ V}$$

$$I_{R_1} = \frac{V_{R_1}}{R_1} = \frac{0.25 \text{ V}}{10 \text{ }\Omega} = 0.025 \text{ A}$$

$$V_{R_2} = B_2 - V_{R_3} = 19 \text{ V} - 9.75 \text{ V} = 9.25 \text{ V}$$

$$I_{R_2} = I_{R_3} - I_{R_1} = 0.488 \text{ A} - 0.025 \text{ A}$$
$$= 0.463 \text{ A}$$

You may wonder why the last sentence in step 1 starts out, "If possible." Studying Fig. 6-17 should reveal that the removal of any one resistor still leaves a complex circuit. Such circuits can still be thevenized, but one of the methods we have learned for analyzing a complex circuit must be used to calculate R_{TH} and V_{TH}.

Note that step 7 also starts out, "If possible." The bridge circuit of Fig. 6-4(a) shows that knowing the current and voltage for only one of five resistors does not allow solving the other currents and voltages by using series-parallel rules or Kirchhoff's laws. All currents and voltages for the bridge circuit could be found by applying Thevenin's theorem to the circuit two times. With the first application, find V_{R_5} and I_{R_5}. With the second application, find the current and voltage for any one of the other four resistors. Then Kirchhoff's laws can be used to find the remaining currents and voltages.

As mentioned above, thevenizing a circuit sometimes requires the use of other techniques, such as loop equations, to determine V_{TH}.

$$58 = 12I_1 + 10I_1 - 10I_2 + 55 \quad \text{(loop 1)}$$
$$3 = 22I_1 - 10I_2 \quad \text{(loop 1)}$$
$$30 = 15I_2 + 10I_2 - 10I_1 \quad \text{(loop 2)}$$
$$30 = -10I_1 + 25I_2 \quad \text{(loop 2)}$$

$$\frac{25I_2}{-10I_2} \times (-1) = 2.5 \quad \text{(constant)}$$

$$7.5 = 55I_1 - 25I_2 \quad \text{(loop 1)}$$
$$30 = -10I_1 + 25I_2 \quad \text{(loop 2)}$$
$$\overline{37.5 = 45I_1} \quad \text{(loops 1 and 2)}$$
$$I_1 = 0.83 \text{ A}$$

Find V_{TH}:

$$V_{R_3} = I_1 R_3 = (0.83 \text{ A})(12 \text{ }\Omega) = 10 \text{ V}$$
$$V_{TH} = B_3 - V_{R_3} = 58 \text{ V} - 10 \text{ V} = 48 \text{ V}$$

Find R_{TH} [Fig. 6-18(d) shows that R_1, R_2, and R_3 are in parallel]:

$$R_{TH} = \frac{1}{\dfrac{1}{10} + \dfrac{1}{15} + \dfrac{1}{12}} = \frac{60}{15} = 4 \text{ }\Omega$$

Construct the Thevenin equivalent circuit. Load it with R_4, as in Fig. 6-18(e), and solve for V_{R_4}:

$$V_{R_4} = \frac{V_{TH} R_4}{R_{TH} + R_4} = \frac{(48 \text{ V})(20 \text{ }\Omega)}{4 \text{ }\Omega + 20 \text{ }\Omega} = 40 \text{ V}$$

Finally, solve for other voltage drops. The polarities of these voltage drops are shown in Fig. 6-18(f):

$$V_{R_3} = B_3 - V_{R_4} = 58 \text{ V} - 40 \text{ V} = 18 \text{ V}$$
$$V_{R_1} = B_1 - V_{R_4} = 55 \text{ V} - 40 \text{ V} = 15 \text{ V}$$
$$V_{R_2} = B_2 - V_{R_1} = 30 \text{ V} - 15 \text{ V} = 15 \text{ V}$$

Answer: $V_{R_1} = 15 \text{ V}$, $V_{R_2} = 15 \text{ V}$,
$V_{R_3} = 18 \text{ V}$, $V_{R_4} = 40 \text{ V}$

Example 6-8

Use Thevenin's theorem to find the voltage across each resistor in Fig. 6-18(a).

Given: Complex circuit and values of Fig. 6-18(a)

Find: V_{R_1}, V_{R_2}, V_{R_3}, V_{R_4}

Known: Thevenin's theorem and loop-equations technique

Solution: Remove R_4 as in Fig. 6-18(b). Define loops [Fig. 6-18(c)] and solve for I_1:

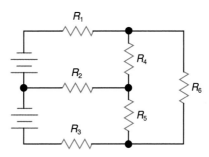

Fig. 6-17 Removal of any one resistor still leaves a complex circuit for which R_S and V_{oc} cannot be determined by applying series-parallel rules.

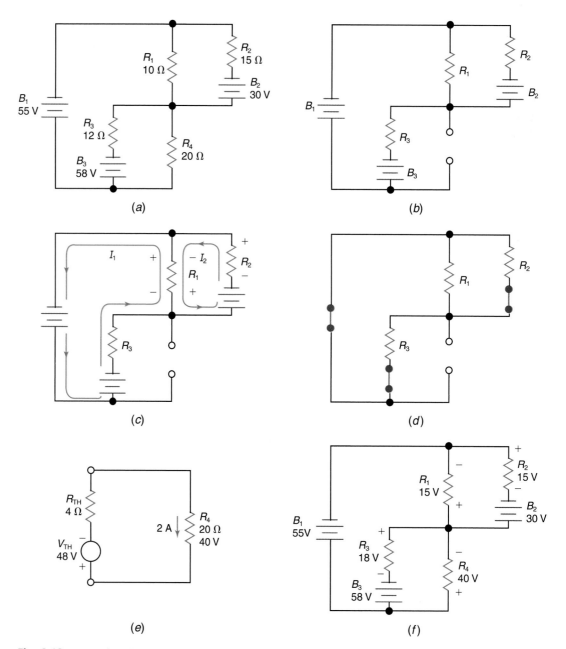

Fig. 6-18 Complex circuit analyzed in example 6-8. (*a*) Original circuit. (*b*) R_4 removed. (*c*) Loops used to find V_{TH}. (*d*) Finding R_{TH}. (*e*) Equivalent circuit with load. (*f*) Final results.

The circuit analyzed in example 6-8 is the same circuit analyzed with the superposition theorem in example 6-5. In example 6-5 we find the individual currents, whereas in example 6-8 we find the individual voltages. Of course, application of Ohm's law solves the voltages in example 6-5 and the currents in example 6-8. Note that the number of calculations involved is fewer with the Thevenin method than with the superposition method. Once we have the Thevenin equivalent circuit of Fig. 6-18(*e*), we can easily find the voltage distribution for any value of R_4. By contrast, changing the value of R_4 when using the superposition method requires that the complete procedure be repeated to find the new currents.

■ TEST_____

Answer the following questions.

26. Change the value of R_1 in Fig. 6-9(*a*) to 15 Ω and calculate all voltage drops using Thevenin's theorem.

27. If you remove R_3 from Fig. 6-7(*a*), can you

a. Find R_{TH} without using the superposition theorem or loop analysis

b. Find V_{TH} without using the superposition theorem or loop analysis

28. If you use Thevenin's theorem to find I_{R_3} and V_{R_3} in Fig. 6-7(a), can you then find the other voltage drops with Kirchhoff's voltage law?

29. Which resistor(s) can be removed in Fig. 6-7(a) so that all voltages can be found without using the superposition theorem or loop analysis and by using Thevenin's theorem only once?

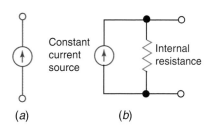

(a)   (b)

Fig. 6-19 Current sources. (a) Ideal constant current source. (b) Equivalent circuit of a current source.

6-6   Current Source

Up to this point we have viewed the source that powers a circuit as a voltage source—either as a constant (ideal) voltage source or as an equivalent-circuit voltage source. We can also consider the source that powers a circuit to be a *current source*. The current source, too, can be viewed as a *constant current source* or as an equivalent-circuit current source.

By definition, a constant current source is a source which supplies the same amount of current to its terminals regardless of the terminal voltage. This definition implies that the parallel internal resistance of the current source is infinite. The symbol for a constant current source is drawn in Fig. 6-19(a). The arrow in the symbol indicates the direction of the constant current.

Like the constant voltage source, the constant current source does not really exist. All current sources have some finite internal resistance. However, it is possible to construct a current source with such a large internal resistance that it can be treated as a constant current source for many practical applications. In fact, a transistor is sometimes viewed as a constant current source.

The *equivalent-circuit model* of a current source is shown in Fig. 6-19(b). It consists of a constant current source in parallel with the internal resistance of the source. This model of a current source is the one we are interested in because it is very useful in analyzing complex circuits. Therefore, let us see how we can develop an equivalent-circuit current source.

Figure 6-20 illustrates why we can view a source as either a voltage source or a current source. Figure 6-20(a), (b), and (c) shows a

voltage source under three conditions: open-circuited, short-circuited, and loaded. Figure 6-20(d), (e), and (f) shows a current source under the same three conditions. Notice that the end results are the same whether we view the source as a current source or as a voltage source. In Fig. 6-20(d), the constant current source provides 1 A of current which must all flow through the internal resistance of the source since there is no load. This 1 A of current through the 12 Ω of resistance provides 12 V of open-circuit voltage. In Fig. 6-20(e), the 1 A of the constant current is routed around the 12 Ω of internal resistance R_S and through the external short circuit. In Fig. 6-20(f) the 1 A from the constant current source is split between the internal resistance and the external resistance. When the ratio of the current split is not obvious, use the current-divider formula for parallel circuits:

$$I_{RL} = \frac{I_{constant}R_S}{R_L + R_S}$$

Three very important points should be noted from Fig. 6-20.

1. The internal resistance of a source is the same whether the source is treated as a voltage source or as a current source.
2. The *short-circuit current* of a voltage source is the same as the constant current of the current source. Thus, the abbreviation I_{sc} is sometimes used for the constant current source.
3. The open-circuit voltage of a current source is the same as the constant voltage of the voltage source.

Keeping these three points in mind, it is very easy to convert a voltage source to a current source and vice versa.

Example 6-9

Convert the voltage source shown in Fig. 6-21(a) to a current source.

Given: Voltage source and values of Fig. 6-21(a)

Find: Constant current and equivalent-circuit current source

Known: Relationships between voltage source and current source

Solution: Short-circuit the voltage source to find the constant current [Fig. 6-21(b)]:

$$I_{sc} = \frac{V_{oc}}{R_S} = \frac{24\ V}{4\ \Omega} = 6\ A$$

The R_S of the two sources will be the same.

Answer: The constant current is 6 A, and the current source is as shown in Fig. 6-21(c).

Norton's theorem

Equivalent-circuit current source

■ TEST_____

Answer the following questions.

30. A current source has an internal resistance of 2 Ω and a constant current of 7 A.

a. If this current source were converted to a voltage source, what would the value of the constant voltage be?

b. If this current source were loaded with a 12-Ω resistor, how much current would flow through the resistor?

31. Determine the constant current for the voltage source in Fig. 6-15(b).

6-7 Norton's Theorem

Norton's theorem provides a way of converting any two-terminal circuit into an *equivalent-circuit current source*. The conversion can be accomplished by following the four steps listed below and illustrated in Fig. 6-22:

1. Select two points on the circuit to be used as the terminals of the current source. This is done in Fig. 6-22(b).

2. Calculate or measure the current available from the terminals when the terminals are short-circuited as shown in Fig. 6-22(c). This short-circuit current is labeled I_{sc} in Fig. 6-22(c), and it is calculated

$$I_{sc} = \frac{B_1}{R_1} = \frac{16\ V}{10\ \Omega} = 1.6\ A$$

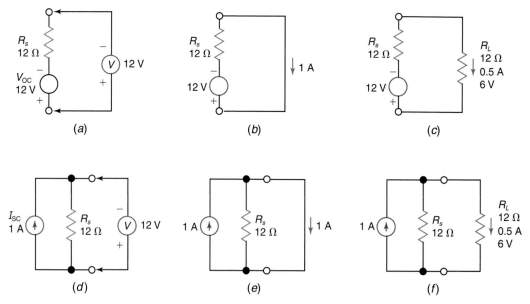

Fig. 6-20 Equivalency of voltage and current sources. R_S is in parallel for a current source and in series for a voltage source. (a) Voltage source—open-circuit voltage. (b) Voltage source—short-circuit current. (c) Voltage source—load current and voltage. (d) Current source—open-circuit voltage. (e) Current source—short-circuit current. (f) Current source—load current and voltage.

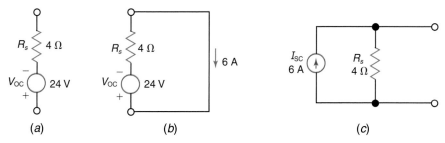

Fig. 6-21 Circuit for example 6-9—converting a voltage source to a current source. (a) Voltage source. (b) Short-circuit current. (c) Current source.

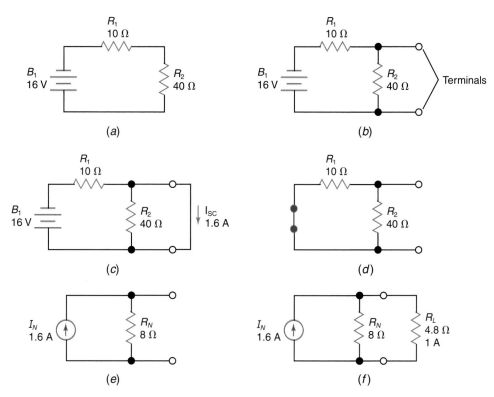

Fig. 6-22 Converting a circuit to an equivalent-circuit current source. The equivalent circuit is a constant current source in parallel with an internal resistance. (a) Original circuit. (b) Source terminals added. (c) Short-circuit current. (d) Internal resistance ($R_S = 8\ \Omega$). (e) Norton equivalent circuit (equivalent-circuit current source). (f) Norton equivalent circuit with a load connected.

3. Calculate R_S (the internal resistance) using the technique used for Thevenin's theorem. As shown in Fig. 6-22(d), R_1 and R_2 are in parallel, and so R_S is 8 Ω.

4. Using I_{sc} from step 2 and R_S from step 3, draw the equivalent-circuit current source shown in Fig. 6-22(e). This current source is called a Norton equivalent circuit. Thus, I_{sc} is labeled I_N and R_S is labeled R_N in Fig. 6-22(e).

Figure 6-22(f) illustrates the results of connecting a 4.8-Ω load to the Norton equivalent circuit of Fig. 6-22(e). The load current I_{RL} is calculated using the current-divider equation:

$$I_{RL} = \frac{I_N R_N}{R_L + R_N} = \frac{(1.6\ \text{A})(8\ \Omega)}{4.8\ \Omega + 8\ \Omega} = 1\ \text{A}$$

It is left to you to verify that connecting a 4.8-Ω load in parallel with R_2 in Fig. 6-22(a) yields the same value for I_{RL} and V_{RL}.

Norton's theorem can also be used to analyze complex circuits. The procedure is the same as with Thevenin's theorem except that the circuit is reduced to a current source after a resistor is removed. Thus, instead of determining the open-circuit voltage (V_{TH}) at the terminals, we determine the short-circuit current (I_N) at the terminals. Let us try an example to illustrate the use of Norton's theorem in analyzing a complex circuit.

Example 6-10

Using Norton's theorem, find the voltage across and current through R_2 in Fig. 6-23(a).

Given: Values and diagram of Fig. 6-23(a)

Find: V_{R_2} and I_{R_2}

Known: Norton's theorem

Solution: Remove R_2 and calculate I_N as indicated in Fig. 6-23(b):

$$I_1 = \frac{B_1}{R_1} = \frac{3 \text{ V}}{4 \text{ }\Omega} = 0.75 \text{ A}$$

$$I_2 = \frac{B_2}{R_3} = \frac{24 \text{ V}}{6 \text{ }\Omega} = 4 \text{ A}$$

$$I_N = I_1 + I_2 = 0.75 \text{ A} + 4 \text{ A} + 4.75 \text{ A}$$

Replace B_1 and B_2 with short circuits as in Fig. 6-23(c) and calculate R_N:

$$R_N = \frac{R_1 R_2}{R_1 + R_2} = \frac{24}{10} = 2.4 \text{ }\Omega$$

Draw the Norton equivalent, load it with R_2 as in Fig. 6-23(d), and calculate V_{R_2} and I_{R_2}:

$$I_{R2} = \frac{I_N R_N}{R_N + R_2} = \frac{(4.75 \text{ A})(2.4 \text{ }\Omega)}{2.4 \text{ }\Omega + 9 \text{ }\Omega} = 1 \text{ A}$$

$$V_{R_2} = I_{R_2} R_2 = (1 \text{ A})(9 \text{ }\Omega) = 9 \text{ V}$$

Answer: $I_{R_2} = 1$ A, and $V_{R_2} = 9$ V

The circuit used in example 6-10 is the same circuit used in the example 6-3 and solved with loop equations. Notice that the results for V_{R_2} and I_{R_2} are the same in both cases. Of course, we can use Kirchhoff's laws to find the remaining currents and voltages in Fig. 6-23(a) after we know V_{R_2} and I_{R_2}.

■ TEST

Answer the following questions.

32. Using Norton's theorem, calculate I_{R_2} and V_{R_2} for the circuit of Fig. 6-24.

33. Can you solve for the other values of current and voltage in Fig. 6-24 using only

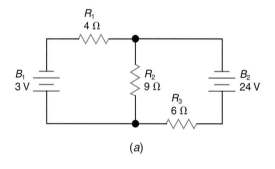

(a)

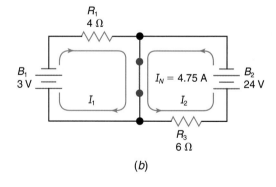

(b)

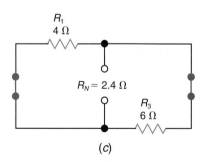

(c)

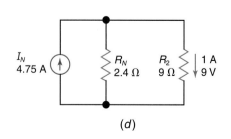

(d)

Fig. 6-23 Circuits for example 6-10.

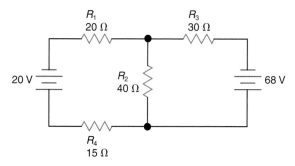

Fig. 6-24 Complex circuit for test questions 27, 28, and 29.

Kirchhoff's laws, Ohm's law, and the values of I_{R_2} and V_{R_2} found in question 32?

34. Change the value of R_3 in Fig. 6-24 to 60 Ω. Then use Norton's theorem to find I_{R_3} and V_{R_3}. Finally, determine I_{R_4} and V_{R_1} using Kirchhoff's and Ohm's laws.

6-8 Comparison of Techniques

Most circuits can be analyzed by two or more of the techniques covered in this chapter. In deciding which techniques to use, ask yourself two questions.

1. Which technique will work?
2. Which technique will involve the fewest and simplest calculations to obtain the desired information?

Answering these two questions is not too difficult if you keep in mind the major characteristics of the various techniques, which are as follows:

Loop equations:

1. Can be used with either single-source or multiple-source circuits
2. Solve for all currents and voltages in the circuit

Superposition theorem:

1. Applicable only to multiple-source circuits
2. Solves for all currents and voltages in the circuit
3. Uses only series-parallel rules and procedures

Thevenin's theorem:

1. Can be used with either single-source or multiple-source circuits
2. May not solve for all values of current and voltage
3. May require other techniques to develop the Thevenin equivalent circuit
4. Requires only a single calculation using the voltage division formula to determine the voltage across any value of load resistor

Norton's theorem:

1. Can be used with either single-source or multiple-source circuits
2. May not solve for all values of current and voltage

Loop equations

Superposition theorem

Thevenin's theorem

Comparison of techniques

Norton's theorem

? Did You Know?

A New Buzz in Pollination Farmers trying to raise crops in regions without bees often rent bees to carry out pollination—which is expensive. A new, "beeless" pollination method allows farmers to apply a pollen mixture to the plants. This mixture is electrically charged to help pull the pollen to the stigma. Charged grains "stick" to plants five times better than uncharged pollen.

3. May require other techniques to develop the Norton equivalent circuit
4. Requires only one application of the current divider formula to determine the current through any value of load resistor

From the above summary you can tell that Thevenin's and Norton's theorems have much in common.

YOU MAY RECALL . . . that Sec. 6-6 points out the relationships between current sources and voltage sources and illustrates how to convert from one to the other.

Since Thevenin's equivalent circuit is a voltage source and Norton's equivalent circuit is a current source, we can also convert one to the other. The appropriate formulas are

$$R_N = R_{TH}$$

$$I_N = \frac{V_{TH}}{R_{TH}}$$

$$V_{TH} = I_N R_N$$

An example will illustrate why we may want to convert from a Thevenin equivalent circuit to a Norton equivalent circuit. Suppose we want to know the current through R_5 in Fig. 6-4(a) when R_5 is any one of 10 different values. This is an easy task if we have the Norton equivalent circuit derived from viewing R_5 as the load resistor. But try calculating I_N for Fig. 6-4(a); it is not easy. However, calculate V_{TH} and R_{TH} requires only series-parallel procedures. Thus, for this circuit, it is best to find the Thevenin equivalent circuit and then convert it to a Norton equivalent circuit.

■ TEST

Answer the following questions.
35. What is the value of the Thevenin resistance in Fig. 6-23(d)?
36. Which theorem cannot be used with a single-source circuit?

Summary

1. Complex circuits (networks) cannot be analyzed using only series-parallel rules and procedures.
2. Equations with more than one variable can be solved by simultaneous-equation techniques.
3. To solve simultaneous equations, there must be as many independent equations as there are unknown variables.
4. Loop equations are derived by applying Kirchhoff's voltage law to each loop of a circuit.
5. The loop-equations technique can solve for all voltages and currents in a circuit.
6. The loop-equations technique can be applied to either single-source or multiple-source circuits.
7. When the calculated value of current has a negative sign, the direction of current originally assumed is incorrect.
8. The superposition theorem can be used to analyze complex multiple-source circuits without using simultaneous equations. This technique can solve for all currents and voltages in the circuit.
9. When the superposition theorem is used, all but one voltage source is replaced by its internal resistance.
10. In many circuits the internal resistance of the voltage source can be assumed to be 0 Ω, that is, the voltage source can be viewed as a constant (ideal) voltage source.
11. An equivalent-circuit voltage source consists of a constant voltage source (V_{oc} or V_{TH}) in series with an internal resistance (R_S or R_{TH}).

12. If an electric circuit is viewed as a two-terminal network, Thevenin's theorem provides a way of reducing the circuit to an equivalent-circuit voltage source.
13. Thevenin's theorem can be used to find the current and voltage associated with a single resistor in a complex circuit.
14. A constant current source assumes that current from the terminals of the source remains the same for all values of load resistance. Furthermore, the constant current source is assumed to have an infinite internal resistance.
15. An equivalent-circuit current source is a constant current source (I_{sc}) in parallel with an internal resistance (R_S).
16. For a given circuit, the internal resistances of the equivalent-circuit current source and the equivalent-circuit voltage source are the same.
17. In a circuit, $I_{sc} = V_{oc}/R_S$ and $V_{oc} = I_{sc}R_S$; therefore, $I_N = V_{TH}/R_{TH}$ and $V_{TH} = I_N R_N$.
18. Norton's theorem provides a way of reducing any two-terminal circuit to an equivalent-circuit current source.
19. Norton's theorem can be used to determine the voltage across and the current through an individual resistor in a complex circuit.
20. Neither Thevenin's nor Norton's theorem always leads to the easy solution of all voltages and currents in a circuit.
21. Either Thevenin's or Norton's theorem can lead to easy determination of I_{RL} and V_{RL} when the value of R_L is changed.

For questions 6-1 to 6-4, supply the missing word or phrase in each statement.

6-1. A complex circuit can also be referred to as a _____ .

6-2. A loop equation is written by using _____ law.

6-3. A negative value of calculated current indicates that the assumed current direction was _____ .

6-4. _____ independent equations are required to solve for two unknown variables.

For questions 6-5 to 6-9, determine whether each statement is true or false.

6-5. Each equation in a set of equations must contain all unknown variables.

6-6. Application of the superposition theorem allows you to determine all voltages and currents in a dual-source complex circuit without using simultaneous equations.

6-7. A constant current source provides a constant voltage at its terminals under all load conditions.

6-8. For a given circuit, R_N of the Norton equivalent circuit always equals R_{TH} of the Thevenin equivalent circuit.

6-9. Norton's theorem reduces a circuit to an equivalent-circuit voltage source.

Answer the following questions.

6-10. In a particular complex circuit, three loops are needed to include all resistors in at least one loop. How many loop equations are required to analyze the circuit?

6-11. Refer to question 6-10. How many unknown variables will there be in the set of equations?

6-12. Which theorem cannot be used on a single-source circuit?

6-13. A power source is represented by an equivalent-circuit current source and an equivalent-circuit voltage source. Will the two equivalent circuits have the same value of
a. Internal resistance
b. Open-circuit voltage
c. Short-circuit current

6-14. Add $3A + 5B = 14$ to $-A + 2B - 3C = -7$.

6-15. Solve for A and B when $3A + 8B = 30$ and $2A - 4B = -8$.

6-16. A current source has an internal resistance of $4\ \Omega$ and a constant current of 9 A.
a. What is its open-circuit voltage?
b. What is its terminal voltage when loaded with an 8-Ω resistor?

6-17. If you convert the current source in question 6-16 to a voltage source, what will the value of the constant voltage source be?

6-18. If you convert a Thevenin equivalent circuit with $V_{TH} = 16$ V and $R_{TH} = 2$ kΩ to a Norton equivalent circuit, what will the values of I_N and R_N be?

6-19. Solve for A, B, and C when $2A + 4B - 3C = 10$, $-A - 3B + 2C = 12$, and $3A + 2B + 4C = 20$.

Solve the following problems.

6-20. For the circuit in Fig. 6-25(a), compute the following by applying Thevenin's theorem:
a. V_{R_3}
b. I_{R_3}

6-21. For the circuit in Fig. 6-25(b), compute the following by using loop equations:
a. V_{R_3}
b. V_{R_4}
c. I_{R_1}
d. I_{R_2}

6-22. For the circuit in Fig. 6-25(c), compute the following by using the superposition theorem:
a. I_{R_1}
b. I_{R_2}
c. I_{R_3}
d. V_{R_4}

6-23. For the circuit in Fig. 6-25(a), compute the following by applying Norton's theorem:
a. I_{R_2}
b. V_{R_2}

6-24. Write the loop equations for Fig. 6-25(a).

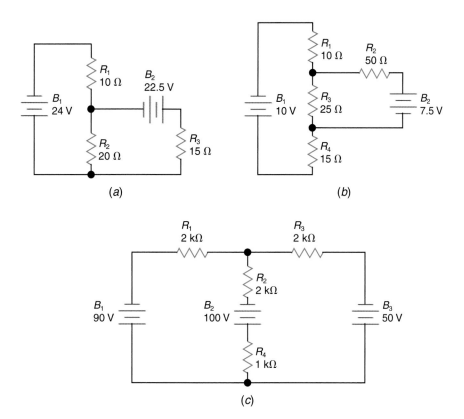

Fig. 6-25 Circuits for chapter review questions 6-19 to 6-22.

Critical Thinking Questions

6-1. When loop techniques are used on a two-source complex circuit, is it possible for all calculated currents to be negative? Explain.

6-2. Is it possible to have a complex circuit in which no current flows in one of the resistors? Explain.

6-3. Would a 6-V carbon-zinc battery rated at 1 Ah be more like a constant voltage source than a 6-V carbon-zinc battery rated at 10 Ah? Why?

6-4. Compare the efficiency of an equivalent-circuit voltage source to an equivalent-circuit current source.

6-5. Using the values from example 6-4, prove that $V_{R_1} + V_{R_2} = B_1$ in Fig. 6-7.

6-6. Using the values from example 6-4, prove that $B_1 = V_{R_1} + V_{R_3} + V_{R_5} + B_2$.

Answers to Tests

1. $6A + 6B = 59.6$
2. $4R + S - 6T = 25$
3. $-2A + 4C = 8$
4. $X = 4.3, Y = 6$
5. $C = 3.81, D = 5.35$
6. $A = 30, B = 21$

7. $R = 4, S = -8,$
   $T = 3$
8. $X = 2.3, Y = 5.2,$
   $Z = 3.1$
9. $C = -5.19, D = -4.79, E = -4.07$

10. $9 V = 8 k\Omega(I_1) + 3k\Omega(I_2) + 1 k\Omega(I_3)$
              (loop 1)
    $9 V = 3 k\Omega(I_1) + 5 k\Omega(I_2)$ (loop 2)
    $9 V = 1 k\Omega(I_1) + 6 k\Omega(I_3)$ (loop 3)

11. No; I_2 and I_3 are different.
12. Yes; however, V_{R_5} is positive with the equations from question 10.

13. point A to point B

14. no

15. T

16. F

17. $I_{R_1} = 1.2$ A, $I_{R_2} = 2.6$ A, $I_{R_3} = 1.4$ A, $V_{R_3} = 14$ V

18. $I_{R_1} = 0.983$ A, $I_{R_2} = 1.207$ A, $I_{R_3} = 2.190$ A

19. no

20. T

21. $I_{R_3} + I_{R_1} = I_{R_4} + I_{B_2}$, or 1.5 A + 1.5 A = 2 A + 1 A

22. 0.5 A

23. the amount of internal resistance R_S relative to the external-circuit resistance R_L

24. open-circuit voltage V_{oc}

25. F

26. $V_{R_1} = 57$ V, $V_{R_2} = 38$ V, $V_{R_3} = 0$ V

27. a. yes
    b. yes

28. no

29. R_1 or R_5

30. a. 14 V
    b. 1 A

31. 0.06 A

32. $I_{R_2} = 0.488$ A, $V_{R_2} = 19.5$ V

33. yes

34. $I_{R_3} = 1$ A, $V_{R_3} = 60$ V, $I_{R_4} = 0.8$ A, $V_{R_1} = 16$ V

35. 2.4 Ω

36. superposition theorem

Chapter 7

Magnetism and Electromagnetism

Chapter Objectives

This chapter will help you to:

1. *Visualize* magnetic fields, flux, and forces.
2. *Determine* the direction of the magnetic flux created by a current-carrying conductor.
3. *Predict* the direction of the force between current-carrying conductors.
4. *Explain* why some magnetic materials make permanent magnets and others make temporary magnets.

5. *Understand and properly use* the many terms needed to describe magnetism and magnetic circuits.
6. *Use* magnetic quantities and units to solve magnetic circuit problems.
7. *Understand* the basic principle of operation of a motor, generator, transformer, solenoid, and relay.

lectricity and magnetism cannot be separated. Wherever an electric current exists, a magnetic field also exists. Magnetism, created by an electric current, operates many devices, such as transformers, motors, and loudspeakers.

7-1   Magnetism and Magnets

Magnetism is a force field that acts on some materials but not on other materials. Physical devices which possess this force are called *magnets. Lodestone* (an iron compound) is a natural magnet which was discovered centuries ago.

The magnets we use today are all manufactured. They are made from various alloys containing elements like copper, nickel, aluminum, iron, and cobalt. These magnets are much, much stronger than the natural lodestone magnet.

7-2   Magnetic Fields, Flux, and Poles

The force of magnetism is referred to as a *magnetic field.* This field extends out from the magnet in all directs, as illustrated in Fig. 7-1. In this figure the lines extending from the magnet represent the magnetic field.

The invisible lines of force that make up the magnetic field are known as magnetic *flux* (ϕ). The lines of force in Fig. 7-1 represent the flux. Where the lines of flux are dense, the magnetic field is strong. Where the lines of flux are sparse, the field is weak. The lines of flux are most dense at the ends of the magnet; there-

Flux

Magnetism

Lodestone

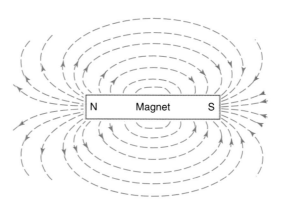

Magnetic field

Fig. 7-1   Magnetic field of a magnet. The flux is most dense at the poles.

fore, the magnetic field is strongest at the ends of the magnet.

The arrows on the lines in Fig. 7-1 indicate the direction of the flux. Lines of force are always assumed to leave the *north pole* (N) and enter the *south pole* (S) of a magnet. *North pole* and *south pole* refer to the polarity of the ends of a magnet. When a magnet is suspended on a string and allowed to rotate, its ends will point north and south. The end of the magnet which seeks (or points to) the earth's north magnetic pole is the magnet's north pole. Assuming that flux (lines of force) leaves the north pole and enters the south pole was an arbitrary decision. However, assigning direction to the lines aids in understanding the behavior of magnetism.

Like magnetic poles repel each other. The two north poles of Fig. 7-2(*a*) and the two south poles of Fig. 7-2(*b*) create a repelling force. The closer the poles are, the more they repel each other. The force of repulsion between magnetic poles varies inversely as the square of the distance between them. That is, if the distance is doubled, the force becomes one-fourth as great. Or, if the distance is halved, the force becomes four times as great.

Unlike magnetic poles (Fig. 7-3) create a force of attraction. This force also varies inversely as the square of the distance between the poles. As shown in Fig. 7-3(*a*), much of the

North pole

South pole

Like magnetic poles

Unlike magnetic poles

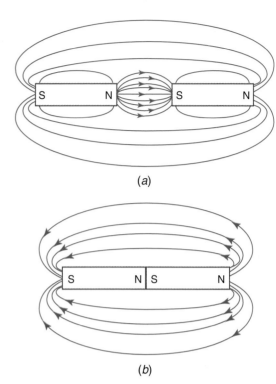

(*a*)

(*b*)

Fig. 7-3 Attraction of unlike poles. The force of attraction is greatest when the poles are touching, as in (*b*).

flux of the two magnets joins together to form the force of attraction. When the two poles touch, essentially all the flux joins together [Fig. 7-3(*b*)]. When joined together, the two magnets behave as a single magnet. They create a single magnetic field.

Magnetic lines of force (flux) are assumed to be continuous loops. Although not shown in Fig. 7-3, the flux lines continue on through the magnet. They do not stop at the poles. In fact, the poles of the magnet are merely the areas where most of the flux leaves the magnet and enters the air. If a magnet is broken in half (Fig. 7-4), two new poles are created. The one magnet becomes two magnets. A magnet can be broken into many pieces, and each piece be-

Flux lines

(*a*)

(*b*)

Fig. 7-2 Repulsion of like poles. The repelling poles can both be north poles, as in (*a*), or south poles, as in (*b*).

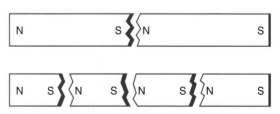

Fig. 7-4 Creation of poles. Each time the magnet is broken, a new pair of poles is created.

comes a new magnet with its own north pole and south pole.

A magnet does not have to have poles. All the flux lines can be within the magnet. This idea is illustrated in Fig. 7-5. In Fig. 7-5(*a*), a typical horseshoe magnet is shown. The dark lines represent the flux within the magnet. In Fig. 7-5(*b*) the horseshoe magnet has been bent around to form a circle with a small gap. And finally, in Fig. 7-5(*c*) the gap has been closed to form a circle. Now all the flux (and the magnetic field) is confined within the magnet. The magnet has no poles.

It should be noted that a circular magnet can be (and often is) made with poles. In this case, the flux runs parallel to the hollow core of the magnet, and the flat surfaces of the magnet are the north and south poles.

◀ TEST_____

Answer the following questions.

1. Define *magnetism.*
2. Define *magnetic field.*
3. The invisible lines of force in a magnetic field are called _____.
4. A magnetic north pole is repelled by a magnetic _____ pole.
5. Magnetic flux flows from the _____ pole to the _____ pole.
6. Halving the distance between two poles will _____ the force between the poles.
7. The flux from a north pole will join the flux from a _____ pole.
8. Are lines of flux continuous loops?
9. Does breaking a magnet in half destroy its magnetic field?
10. Do all magnets have a north pole?

7-3 Electromagnetism

So far our discussion has centered around the magnetic field and flux possessed by a magnet. However, magnetic fields are also created by electric current. The current-carrying conductor in Fig. 7-6 has a magnetic field around it. The field is always at right angles (perpendicular) to the direction of current. Although the field is shown in only five places along the conductor, it actually exists as a continuous field for the entire length. Notice in Fig. 7-6 that the magnetic field has no poles. The flux exists only in the air. However, the flux still has an assumed direction, just as it does in the circular magnet of Fig. 7-5(*c*). The *direction of the flux* around a conductor can be determined by using what is called the *left-hand rule.* Grasp the conductor with your left hand so that your thumb points in the direction of current. Your fingers then indicate the direction of the flux.

Direction of flux

Left-hand rule

Often the direction of current and flux is indicated as shown in Fig. 7-7. In this figure, you are looking at the ends of current-carrying conductors. Of course, the end of a round conductor looks like a circle. An **x** in the circle, as in

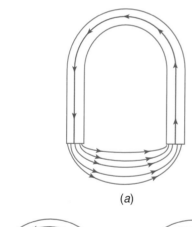
(*a*)

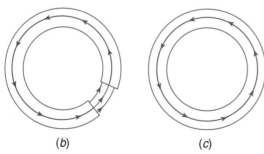
(*b*)   (*c*)

Fig. 7-5 Magnet without poles. Forming a horseshoe magnet (*a*) into a circle (*c*) eliminates the poles.

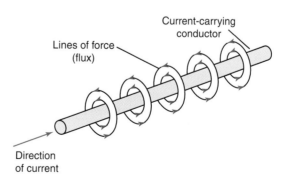
Direction of current

Fig. 7-6 Magnetic field around a conductor. The flux is perpendicular to the direction of current.

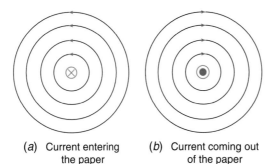

(a) Current entering
the paper

(b) Current coming out
of the paper

Fig. 7-7 Indicating direction of current.

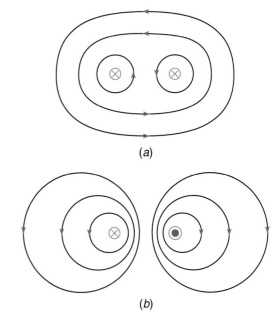

(a)

(b)

Fig. 7-8 Forces between parallel conductors. (a) Attraction of conductors. (b) Repulsion of conductors.

Fig. 7-7(*a*), means that the current is going into the paper. A way to remember this convention is to visualize an arrow or a dart as indicating the direction of current. If the dart were going into the paper, you would see the tail end with feathers arranged in an **x** shape. If the dart were coming toward you (out of the paper), you would see the front, pointed end. Thus, the dot in a circle represents current flowing toward you.

The strength of the magnetic field around a conductor is determined by the amount of current flowing through the conductor. The strength at some fixed distance from the conductor is directly proportional to the current. Doubling the current doubles the strength of the magnetic field.

Force between Conductors

Two *parallel,* current-carrying *conductors* attract each other if the currents in the conductors are in the same direction. This is because the fields of the two conductors join together, as shown in Fig. 7-8(*a*). When fields join together, there is a force of attraction. This is the same type of attraction that occurs between unlike poles (Fig. 7-3).

When the currents in parallel conductors are not in the same direction, a force of repulsion exists. The magnetic fields in Fig. 7-8(*b*) are not able to join together because they are opposing each other. Since lines of flux cannot cross each other and cannot share the same space, they repel each other.

Coils

The magnetic field of a single conductor is too weak for many applications. A stronger field can be created by combining the fields associ-

ated with two or more conductors. This is done by coiling a conductor as shown in Fig. 7-9. A conductor formed into this shape is called a *coil.* Forming a coil out of a conductor creates an *electromagnet.* Notice that the coils in Fig. 7-9 have poles at the ends of the coil where the flux enters and leaves the center of the coil.

The *polarity of a coil* can be determined by again applying the left-hand rule. This time, wrap your fingers around the coil in the direction current flows through the turns of the coil. Your thumb will then point to the north pole (it also indicates direction of the flux). Reversing the direction of the current reverses the polarity of an electromagnet.

In Fig. 7-9(*a*) the three turns of the coil have considerable space between them. This spacing of the turns allows some of the flux to loop only a single conductor. The result is a weakening of the magnetic field at the poles of the magnet. This loss of flux at the poles can be minimized by *close-winding* the coil, as shown in Fig. 7-9(*b*). In this figure all the flux of the individual turns combines to produce one stronger magnetic field.

▪ TEST

Answer the following questions.

11. Magnetic fields around a conductor are
_____ to the conductor.

Coil

Electromagnet

Polarity of a coil

Parallel conductors

Close-winding

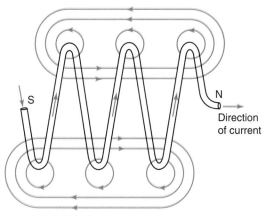

(a) Coil with space between turns

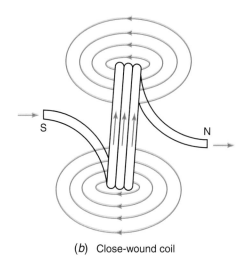

(b) Close-wound coil

Fig. 7-9 Magnetic fields of coils.

12. True or false. The magnetic field around a straight conductor has poles.
13. Is the direction of the flux around a conductor clockwise or counterclockwise when the current is flowing away from you?
14. When a wire is twisted into circular turns, it is called a _____ .
15. Refer to Fig. 7-8(b). Is the north pole above or below the two wires?
16. Increasing the space between the turns of a coil _____ the pole flux.
17. True or false. Lines of flux can cross each other.

7-4   Magnetic Materials

Materials that are attracted by magnetic fields (and materials from which magnets can be made) are called *magnetic materials.* The most common magnetic materials are iron, iron compounds, and alloys containing iron or steel. These magnetic materials are also called *ferromagnetic materials.* (*Ferro* is a prefix that means "iron.") A few materials, such as nickel and cobalt, are slightly magnetic. They are attracted by strong magnets. Compared with iron, however, they are only weakly magnetic.

Materials that are not attracted by magnets are called *nonmagnetic materials.* Most materials, both metallic and nonmetallic, are in this category. A magnet does not attract metals like copper, brass, aluminum, silver, zinc, and tin. Neither does a magnet attract nonmetals like wood, paper, leather, plastic, and rubber. A nonmagnetic material does not stop magnetic flux. Flux goes through nonmagnetic materials about as readily as it goes through air.

Theory of Magnetism

It is well known, and easily demonstrated, that an electric current produces a magnetic field. Since current is nothing but the movement of electric charges, any moving charge should create a magnetic field. The electrons of an atom or molecule possess electric charges, and they are in motion. Therefore, they should have a magnetic field associated with them. In the molecules of nonmagnetic materials, each electron is paired with another electron spinning in the opposite direction. This causes the magnetic field of one electron to be canceled by the opposite field of the other electron. The result is that the molecule ends up with no overall magnetic field. In magnetic materials, more electrons spin in one direction than in the other. Thus, not all the magnetic fields are canceled, and a molecule of the material possesses a weak magnetic field. These molecules then group together to form small *domains* that have a magnetic field with definite poles.

Figure 7-10 represents the theoretical internal structure of a piece of magnetic material. In this figure, magnetic domains are represented by small rectangles. The domains are arranged randomly. They are also moving about randomly in the material. The magnetic fields of the randomly arranged domains cancel each other. Overall, the piece of material in Fig. 7-10 has no magnetic field or poles. It is an *unmagnetized* piece of magnetic material.

Domains

Magnetic materials

Unmagnetized

Silicon steel

Temporary magnet

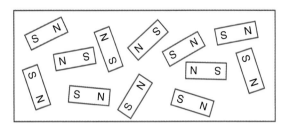

Fig. 7-10 Random arrangement of magnetic domains in an unmagnetized material.

Temporary and Permanent Magnets

Magnetized

When a magnetic material is put in the magnetic field of a magnet, it becomes *magnetized.* As shown in Fig. 7-11, all the domains are aligned in an orderly fashion. The magnetic field of one domain supports the field of the next domain. The magnetic material in Fig. 7-11 became a magnet when it was magnetized. Whether it is a *temporary* or a *permanent* magnet depends on how it reacts when removed from the original magnetic field. If most of the domains remain aligned, as shown in Fig. 7-11, the magnetic material becomes a *permanent magnet.* Many alloys of iron, especially those that contain more than 0.8 percent carbon, become permanent magnets. Most tools, such as screwdrivers, pliers, and hacksaw blades, contain more than 0.8 percent carbon. They can become permanent magnets capable of attracting other magnetic materials. Most permanent magnets are made of alloys (such as alnico) which can be highly magnetized. *Alnico* magnets are composed of iron, cobalt, nickel, aluminum, and copper. Ceramic materials also make strong permanent mag-

Permanent magnet

Ferrite

Alnico

nets. Permanent magnets are used to make door catches and the magnetic poles for loudspeakers, electric meters, and motors.

Materials such as pure iron, ferrite, and *silicon steel* make *temporary magnets.* When these materials are removed from a magnetic field, their domains immediately revert to the random arrangement of Fig. 7-10. They lose almost all their magnetism. They no longer attract other magnetic materials. Temporary magnetic materials are used in great quantities in motors, generators, transformers, and electromagnets.

7-5 Magnetizing Magnetic Materials

Magnetic materials can be magnetized by the magnetic fields of either a permanent magnet or an electromagnet. Magnetizing with an electromagnet is shown in Fig. 7-12(*a*). When the switch is closed, the field from the coil magnetizes the tool steel in the screwdriver. Since tool steel makes a permanent magnet, the screwdriver remains magnetized after the switch is opened. Now the screwdriver can be used to attract other magnetic materials, such as steel screws [Fig. 7-12(*b*)].

When the material in the center of a coil is silicon steel or *ferrite*, a temporary magnet is made. Refer to Fig. 7-13(*a*), where a tempo-

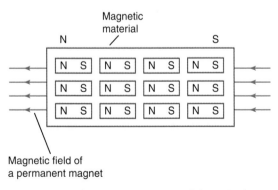

Fig. 7-11 Orderly arrangement of domains in a magnetized material.

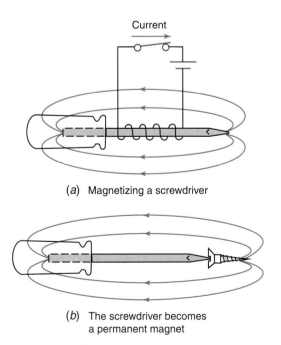

(*a*) Magnetizing a screwdriver

(*b*) The screwdriver becomes a permanent magnet

Fig. 7-12 Making a permanent magnet.

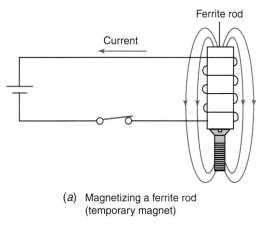

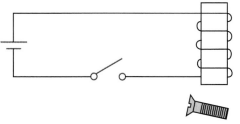

(a) Magnetizing a ferrite rod
(temporary magnet)

(b) When current stops, ferrite rod loses its magnetism

Fig. 7-13 Making a temporary magnet.

rary magnet is shown attracting a magnetic material. It continues to attract the material as long as current flows. However, as shown in Fig. 7-13(b), the magnetic field disappears the instant the switch is opened. The screw is no longer attracted to the ferrite rod.

■ TEST

Answer the following questions.

18. The most strongly magnetic element is _____.

19. _____ steel is used for temporary magnets.

20. A cluster of molecules that has a small magnetic field is called a _____.

21. Transformers and motors use _____ steel.

22. True or false. Most permanent magnets are made from an alloy rather than from a single element.

23. What happens to the flux in a ferrite rod in a coil when the electric circuit is opened?

24. What happens to the magnetic flux in a carbon-steel rod which is in a coil when the circuit is opened?

7-6 Magnetomotive Force

The effort exerted in creating a magnetic field (and flux) is called *magnetomotive force (mmf)*. Increasing either the number of turns or the current in the coil of Fig. 7-13 increases the mmf. It also increases the flux *if* the ferrite rod can support more flux.

Magnetomotive force (mmf)

7-7 Saturation

A magnetic material is saturated when an increase in mmf no longer increases the flux in the material. Increasing the coil current (or turns) in the coil shown in Fig. 7-13(a) increases the flux until the ferrite rod saturates. Once the rod has saturated, increasing either the current or the number of turns no longer increases the flux.

7-8 Demagnetizing

A permanent magnet can be partially demagnetized by hammering on it. It can also be demagnetized by heating it to a very high temperature. However, neither of these methods is very practical for demagnetizing many materials and devices. Most demagnetizing is done with a coil and an ac source, as shown in Fig. 7-14(a). Alternating current periodically reverses its direction of flow. It is also constantly

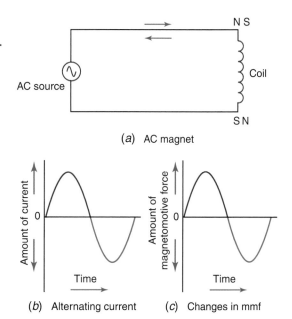

(a) AC magnet

(b) Alternating current

(c) Changes in mmf

Fig. 7-14 Electromagnetism with alternating current. The polarity of the magnetic field periodically reverses.

varying in value [Fig. 7-14(*b*)]. In other words, alternating current starts at zero current, slowly increases to maximum current, and then decreases back to zero current. At this time it reverses direction, increases to a maximum, and decays back to zero. This sequence is repeated rapidly (every $\frac{1}{60}$ s in the ac system in your home) as long as the circuit is connected. This change in value and direction of the current creates a magnetic field that changes in mmf and polarity. The changes in mmf [Fig. 7-14(*c*)] follow the changes in current. Both changes (current and mmf) are indicated in Fig. 7-14(*a*) by the arrows and the pole markings.

When a permanent-magnet material is put into the coil of Fig. 7-14(*a*), it is magnetized. It is magnetized first in one polarity and then in the reverse polarity. Just as the polarity is reversing, there is an instant when no magnetic field exists. If the switch were opened at that exact instant, the magnet would be demagnetized. However, the chances of opening the switch at that instant are very small. It is very likely that the switch would be opened at some other instant, when a magnetic field was present. Then the magnet would still be magnetized. The strength of the magnet depends on the strength of the field at the instant the switch is opened.

To demagnetize with the circuit of Fig. 7-14, the switch must remain closed as the magnet is removed. As the magnet is removed, it is being magnetized first in one polarity and then in the other. However, each reversal of polarity yields a weaker magnet. This is because the field of the coil gets weaker as the magnet is moved farther away. When the permanent-magnet material has been moved several feet away from the coil, it is completely demagnetized. The switch to the coil can then be opened.

A soldering gun like the one shown in Fig. 7-15 can be used to either magnetize or demagnetize. The soldering tip and connecting leads are a single-turn coil. The coil has a large alternating current flowing in it when the gun is on. To magnetize the screwdriver shown in Fig. 7-15, just turn the gun on and off with the screwdriver located as shown. To

Residual magnetism

Reluctance

JOB TIP

Don't overlook related areas when assembling your portfolio. If your hobby is building models, some photographs could be helpful.

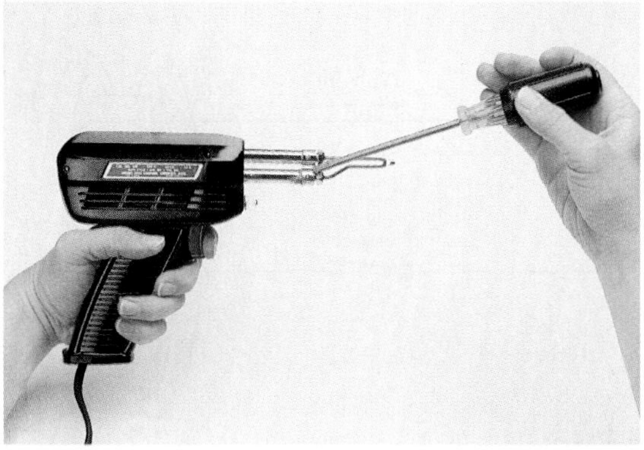

Fig. 7-15 Magnetizing and demagnetizing with a soldering gun. The soldering gun tip is a one-turn coil. *(Mark Steinmetz)*

demagnetize, turn the gun on and pull the screwdriver away from the gun. Then turn off the gun.

7-9 Residual Magnetism

The flux that remains in a temporary magnet after it is removed from a magnetic field is called *residual magnetism*. All magnetic materials retain some flux after they have been exposed to a magnetic field. Permanent-magnet materials retain a high magnetism. The ideal permanent magnet would retain all the flux.

Temporary-magnet materials have very low residual magnetism. In most temporary magnets, the magnetic field is created by current-carrying conductors. When the current stops, the temporary-magnet material does not become completely unmagnetized. A small residual magnetism always remains. However, the magnetic field of the residual magnetism is very weak.

Residual magnetism causes ac-operated magnetic devices to heat up. It is one of the reasons devices like motors and transformers are not 100 percent efficient.

7-10 Reluctance

The opposition to magnetic flux is called *reluctance*. The reluctance of an object depends upon both the material and the dimensions of the material. Nonmagnetic materials have about the same (within 1 percent) reluctance as air does. On the other hand, air has 50 to 5000 times as much reluctance as common magnetic materials do. For example, silicon steel used in motors and transformers has about ⅓₀₀₀th as much reluctance as air.

When offered a choice, flux takes the lowest-reluctance path. In Fig. 7-16(*a*), the flux extends away from the poles into the surrounding air. However, in Fig. 7-16(*b*), the flux concentrates in the small piece of iron. This is because the iron has much, much less reluctance. The insertion of the iron between the poles concentrates the flux into a smaller volume. If the iron in Fig. 7-16(*b*) were replaced by a piece of lead, the flux pattern would be the same as it is in Fig. 7-16(*a*).

Magnetic materials are attracted to a magnet because of their low reluctance. Magnetic flux lines try to take the shortest path between the poles of a magnet. They also try to take the path of lowest reluctance. The flux lines can do both if they can pull the surrounding magnetic material to the magnet. This idea is illustrated in Fig. 7-17. The normal lines of force around the poles of a horseshoe magnet are shown in Fig. 7-17(*a*). In Fig. 7-17(*b*), the flux lines have bent to take the path of low reluctance through the iron. There is a force trying to shorten the lines of force. This resembles the force exerted by a stretched rubber band. If either the bar or the magnet is free to move, the two will come together, as shown in Fig. 7-17(*c*). Now the flux has both the shortest path and the lowest-reluctance path.

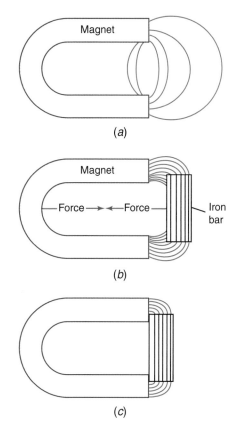

Fig. 7-17 Attraction of magnetic materials. The flux in (*a*) is concentrated and shortened by the magnetic material in (*b*) and (*c*).

7-11 Magnetic Shields

Magnetic flux can be bent, distorted, and guided by low-reluctance materials inserted in the magnetic field. For example, most of the flux in Fig. 7-18 is being bent and guided through the crooked iron bar. *Magnetic shields* make use of the tendency of flux to distort and follow the path of lowest reluctance. A magnetic shield is just a material of very low reluctance. The material is put around the object which is to be protected from any magnetic field in its vicinity. Figure 7-19 illustrates how a watch movement can be protected by a shield.

Magnetic shields

■ TEST _____

Answer the following questions.

25. The force that creates magnetic flux is a _____ .

26. When an increase in mmf causes no increase in the flux in a material, the material is _____ .

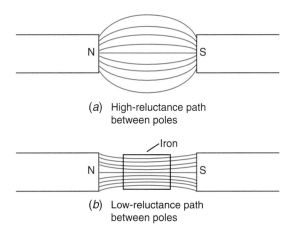

(*a*) High-reluctance path between poles

(*b*) Low-reluctance path between poles

Fig. 7-16 Reluctance. Iron has less reluctance than air does.

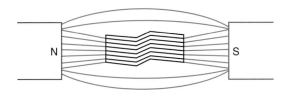

Fig. 7-18 Guiding flux through a low-reluctance path.

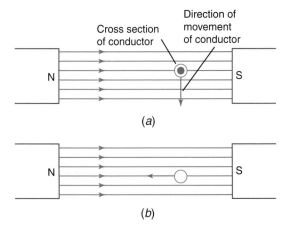

(a)

(b)

Fig. 7-20 (a) Voltage induced in a moving conductor. The conductor must cut the lines of flux. (b) No voltage induced in the conductor.

27. True or false. Opening and closing the switch controlling an ac coil will demagnetize the material inserted in the coil.
28. True or false. A dc coil can be used to demagnetize a permanent-magnet material.
29. Magnetism remaining in a material after the material has been removed from a magnetic field is called _____.
30. Which will have more residual magnetism, a permanent-magnet material or a temporary-magnet material?
31. Why are magnetic materials attracted to a magnet?
32. True or false. Magnetic shields are made from nonmagnetic materials.

Electric
generator

Commutator

Induced voltage

Brushes

7-12 Induced Voltage

We have seen that a current-carrying conductor produces a magnetic field. It is equally true that a magnetic field can induce a voltage (and thus a current) in a conductor. When a conductor moves across lines of flux (Fig. 7-20), a voltage is induced in the conductor. The polarity of the *induced voltage* depends on the direction of motion of the conductor and on the

direction of the flux. Changing the direction of either changes the polarity of the induced voltage.

Generator Action

Induced voltage is the principle on which an *electric generator* operates. When the shaft of a generator is turned, a conductor loop is forced through a magnetic field. A simplified dc generator is shown in Fig. 7-21. As illustrated, the ends of the loop are connected to *commutator* bars and *brushes*. The commutator and brushes serve two purposes. First, they make an electric connection to the rotating loop of wire.

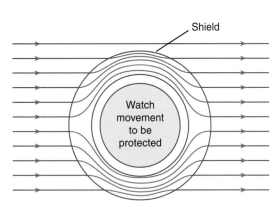

Fig. 7-19 Magnetic shield. The shield provides a low-reluctance path around the protected area.

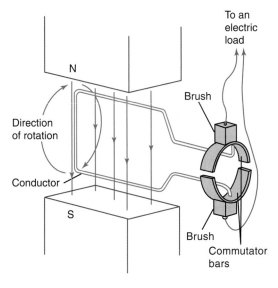

Fig. 7-21 Generator principle.

The ends of the loop are permanently attached to the commutator bars. As the loop turns, the commutator turns with it. The commutator bars touch the brushes and slide against them as the loop rotates. Second, the commutator and brushes reverse connections to the rotating loop every time the polarity of the induced voltage changes. The polarity of the induced voltage changes because the relative direction of motion of the conductor reverses. The top part of the rotating loop in Fig. 7-21 is shown moving from left to right. Ninety degrees later it will start moving from right to left and its polarity will reverse. At the same time, the brushes will change commutator bars. Thus, one brush is always negative with respect to the other brush. This provides a *dc output* from the generator.

Transformer Action

Induced voltage is also created when a conductor is in the immediate vicinity of a *changing* magnetic *flux*. The conductor must be perpendicular to the lines of flux (Fig. 7-22). When the switch in Fig. 7-22 is closed and opened, the current in the circuit starts to flow and then stops. The changing current in the top coil produces a changing flux in the iron core. This changing flux in the core induces a voltage in the bottom coil. The magnitude of the induced voltage is a direct function of both the amount of flux change and the rate of flux change.

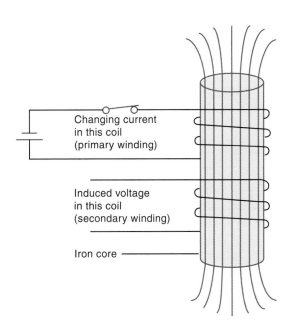

Fig. 7-22 Voltage induced by a changing flux.

Voltage induced by changing flux is the principle on which *transformers* and other devices operate. The ignition coil (which is a transformer) of an automobile operates similarly to the circuit shown in Fig. 7-22.

◼ TEST

Answer the following questions.

33. List the factors which determine the amount of voltage induced in a conductor by transformer action.
34. What are the functions of the commutator and brushes in a generator?
35. What is the principle on which an automotive ignition coil operates?

DC output

7-13 Magnetic Quantities and Units

A number of magnetic quantities, such as flux, have already been discussed. However, we need to know the units in which these quantities are specified and measured.

Changing flux

Magnetomotive Force—The Ampere-Turn

 . . . that in Sec. 7-6 we defined magnetomotive force (mmf) as the force that creates flux.

We will use the *ampere-turn* (A · t) for the base unit of mmf. One ampere-turn is the mmf created by 1 A flowing through one turn of a coil. Three ampere-turns of mmf is created by 1 A moving through three turns. Three ampere-turns is also equal to 3 A through one turn.

Ampere-turn (A · t)

Example 7-1

What is the mmf of the coil in Fig. 7-22 if 2.8 A of current is flowing in the circuit?

Given:	Number of turns (N) = 3 t
	Current (I) = 2.8 A
Find:	mmf
Known:	mmf = turns × current
Solution:	mmf = 3 t × 2.8 A = 8.4 A · t
Answer:	The mmf of the coil is 8.4 A · t.

Weber (Wb)

Magnetizing force (H)

Field strength

Ampere-turn per meter (A · t/m)

Three ampere-turns can be created by any combination in which the product of amperes times turns equals 3.

Strictly speaking, the base unit of mmf in the metric system is just the ampere. At first, this may seem confusing because the ampere is also the base unit of current in the metric system. The reasonableness of using *ampere* as the unit of mmf is illustrated in Fig. 7-23. In 7-23(*a*), 4 A of current is moving around the iron core and creating flux. In Fig. 7-23(*b*), 4 A is again moving around the iron core. The fact that the 4 A is split into two paths does not change the total force exerted in creating flux. The same reasoning extends to Fig. 7-23(*c*). The four 1-A currents across the iron produce the same mmf as one 4-A current. Thus, it is the total current flowing around the iron that determines the mmf. The number of turns carrying the current really does not matter. However, the easiest way to determine the total current flowing by the iron is to multiply the coil current by the number of turns. Thus, the ampere-turn is a more descriptive unit than ampere. Therefore, we will use ampere-turn as the unit of mmf.

Flux—The Weber

The base unit of magnetic flux (ϕ) is the weber. A *weber* (Wb) can be defined only in terms of a change in the flux in a magnetic circuit. One weber is the amount of flux change required in 1 s to induce 1 V in a single conductor. Refer to Fig. 7-22 for an example of the use of the weber. The bottom coil has a voltage induced in it when the flux changes. Note that this coil is a three-turn coil. When the flux in the core changes 1 Wb in 1 s, 3 V is induced in the coil.

Magnetic Field Strength—Ampere-Turn per Meter

Magnetizing force, field intensity, and *magnetic field strength* all mean the same thing. They refer to the amount of mmf available to create a magnetic field for each unit length of a magnetic circuit. The symbol for magnetic field strength is H. The base unit of magnetic field strength is the *ampere-turn per meter* (A · t/m). In the magnetic circuit of Fig. 7-24(*a*), a four-turn coil carries 3 A of current. Therefore, the mmf is 12 A · t (3 A × 4 t = 12 A · t). The average length of the magnetic circuit (iron core) of Fig. 7-24(*a*) is 0.25 m; therefore, the magnetic field strength is the mmf divided by the circuit length:

$$H = \frac{\text{mmf}}{\text{length}}$$
$$= \frac{12 \text{ A} \cdot \text{t}}{0.25 \text{ m}} = 48 \text{ A} \cdot \text{t/m}$$

In Fig. 7-24(*b*) the mmf is

$$\text{mmf} = 4.8 \text{ A} \times 6 \text{ t} = 28.8 \text{ A} \cdot \text{t}$$

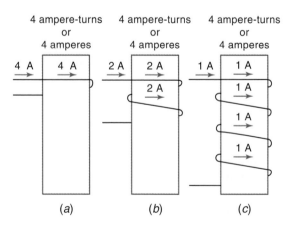

Fig. 7-23 Ampere and ampere-turn.

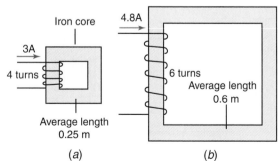

Fig. 7-24 Comparing magnetic circuits. The circuits in (*a*) and (*b*) have the same magnetic field strength (48 A · t/m).

and the magnetic field strength is

$$H = \frac{mmf}{l} = \frac{28.8\ A \cdot t}{0.6\ m} = 48\ A \cdot t/m$$

where l is the length.

Notice that the magnetic field strength is the same for both circuits in Fig. 7-24. Yet the two circuits have different currents, different numbers of turns, different mmf's, and different lengths. Specifying magnetic field strength rather than mmf makes it possible to compare magnetic circuits without specifying physical dimensions.

Example 7-2

What is the magnetizing force of a magnetic circuit with 150 turns on a core with an average length of 0.3 meters if the current is 0.4 amperes?

Given: Number of turns $(N) = 150\ t$
Current $= 0.4\ A$
Length $(l) = 0.3\ m$
Find: Magnetizing force (H)
Known: $mmf = IN$

$$H = \frac{mmf}{l}$$

Solution: $mmf = 0.4\ A \times 150\ t = 60\ A \cdot t$

$$H = \frac{60\ A \cdot t}{0.3\ m} = 200\ A \cdot t/m$$

Answer: The magnetizing force is 200 ampere-turns per meter.

Flux Density—The Tesla

The amount of flux per unit cross-sectional area is called *flux density* (B). The base unit of flux density is the *tesla* (T). One tesla is equal to one weber per square meter. Specifying flux density rather than flux also makes it easier to compare magnetic circuits of unequal size. For example, the iron cores in Fig. 7-25 have equal flux density. Yet they have different cross-sectional areas. Flux density can be calculated by dividing the total flux by the cross-sectional area. In Fig. 7-25(*a*), the cross-sectional area is

$$\text{Area} = 0.02\ m \times 0.02\ m = 0.0004\ m^2$$

The flux density therefore is

$$\text{Flux density} = \frac{\text{flux}}{\text{area}} = \frac{2\ Wb}{0.0004\ m^2}$$
$$= 5000\ T$$

If you calculate the flux density for Fig. 7-25(*b*), you will find that it is also 5000 T.

Example 7-3

How much flux will there be in a core that is 0.03 m × 0.05 m and has a flux density of 400 teslas?

Given: Dimensions $= 0.03\ m \times 0.05\ m$
Flux density $(B) = 400\ T$
Find: Flux (ϕ)

Known: $B = \dfrac{\phi}{\text{area}}$; therefore, $\phi = B$ area

Solution: $\phi = 400\ T \times 0.03\ m \times 0.05\ m$
$= 0.6\ Wb$
Answer: The flux is 0.6 weber.

Permeability and Relative Permeability

Permeability refers to the ability of a material to conduct flux. Its symbol is μ. Permeability is defined as the ratio of flux density to magnetic field strength:

$$\mu = \frac{B}{H}$$

 . . . that both flux density and field strength are independent of size.

Permeability (μ)

$$\mu = \frac{B}{H}$$

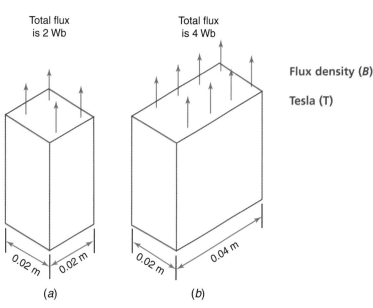

Flux density (B)

Tesla (T)

Fig. 7-25 Comparing flux density. Both (*a*) and (*b*) have a flux density of 5000 T.

About ⬅💠➡ Electronics

Relative permeability (μ_r)

Therefore, permeability is also independent of size. Substituting the base units for the flux density and magnetizing force into the above formula shows that the base unit for permeability is the weber per ampere-turn-meter [Wb/(A·t·m)].

Often the permeability of a material is specified as relative permeability (μ_r). *Relative permeability* compares the permeability of the material with that of air. Suppose a piece of iron has a relative permeability of 600. This means the iron carries 600 times as much flux as an equal amount of air. Since relative permeability is a ratio of two permeabilities, the units cancel out. Thus, relative permeability is a unitless (pure) number.

The relative permeability of all nonmagnetic materials is very close to 1. Relative permeabilities of magnetic materials range from about 30 to more than 6000.

Strength of an electromagnet

The permeability of a magnetic material decreases as the flux density increases. As a material approaches saturation, its permeability decreases rapidly. When the material is fully saturated, the permeability is only a small fraction of its value at lower flux densities.

Example 7-4

What is the permeability of a material when a magnetizing force of 100 A · t/m produces a flux density of 0.2 T?

Given: $H = 100$ A · t/m, $B = 0.2$ T
Find: μ
Known: $\mu = B/H$, T = Wb/m²

Solution: $\mu = \dfrac{0.2 \text{ Wb/m}^2}{100 \text{ A} \cdot \text{t/m}}$

$= 0.002$ Wb/(A·t·m)

Answer: The permeability is 0.002 weber per ampere-turn-meter.

Answer the following questions.

36. What is the mmf of a 200-t coil when the coil current is 0.4 A?
37. Magnetic field strength is also called _____ or _____.
38. The base unit for magnetic field strength is _____.
39. The base unit for flux is _____.
40. The base unit for flux density is _____.
41. The ratio of B to H is called _____.
42. What is the advantage of specifying flux density rather than flux?
43. Define relative permeability.
44. True or false. One tesla equals one weber per meter.

7-14 Electromagnets

. . . that we have already discussed many of the ideas of an electromagnet in Secs. 7-3 and 7-5.

A coil of wire with a current through it is an electromagnet. However, most useful electromagnets are wound on, and partially encased in, a material that can become a temporary magnet.

The *strength of an electromagnet* depends upon four factors:

1. The type of core material (the magnetic material)
2. The size and shape of the core material
3. The number of turns on the coil
4. The amount of current in the coil

In general, an electromagnet is strongest when

1. The core material has the highest permeability
2. The core material has the largest cross-sectional area and the shortest length for flux lines. This results in a low reluctance.
3. The number of turns and amount of current are greatest. This provides a large mmf.

Electromagnets are used in industry for a variety of jobs. They are used to hold steel while it is being machined, sort magnetic from nonmagnetic materials, lift and move heavy iron and steel products. Figure 7-26 shows a large electromagnet lifting heavy iron pipes.

Fig. 7-26 Large industrial electromagnet.
(Courtesy of O. S. Walker Company.)

7-15 DC Motors

Physically and electrically, a *dc motor* resembles a dc generator. In fact, in some cases a single machine may be used as either a generator or a motor. Whereas the shaft of a generator turns because of some outside mechanical force, the shaft of a motor turns because of the interaction of two magnetic fields within the motor. The principle of a simple dc motor is illustrated in Fig. 7-27. Current from an external source flows through the brushes, commutator, and *armature coil*. This produces a magnetic field in the armature iron. The armature poles are attracted by the *field poles* of the permanent magnet. The result is a rotational force called *torque,* as shown in Fig. 7-27(*a*).

The *commutator* and *brushes* change the direction of current in the armature coil every 180°. This change reverses the magnetic polarity of the armature so that the direction of the torque (rotational force) is always the same. When the armature in Fig. 7-27(*a*) is rotated 90°, its south pole is in line with the field's north pole. However, the inertia (which is the tendency of a moving body to keep moving) of the moving armature carries it to just past the vertical position. At this time, the current in the armature coil changes and the armature field reverses. Now the north pole of the armature is repelled by the north pole of the field [Fig. 7-27(*b*)].

In general, motors have more than two commutator segments and many more armature coils. With more commutator segments, it is not necessary to rely on inertia as the armature polarity is reversed. As seen in Fig. 7-28, the segments change an instant before the poles of

the fields are aligned. During the time that each brush is in contact with two segments of the commutator, four poles are created. However, as shown in Fig. 7-28(*b*), all four poles provide torque in the same direction.

The field poles in a dc motor can be either permanent magnets or electromagnets. When electromagnets are used, the current in the field never changes direction. Thus dc motors with electromagnetic fields operate in the same way as those with permanent-magnet fields.

7-16 Solenoids

A *solenoid* is an electromagnetic device that enables an electric current to control a me-

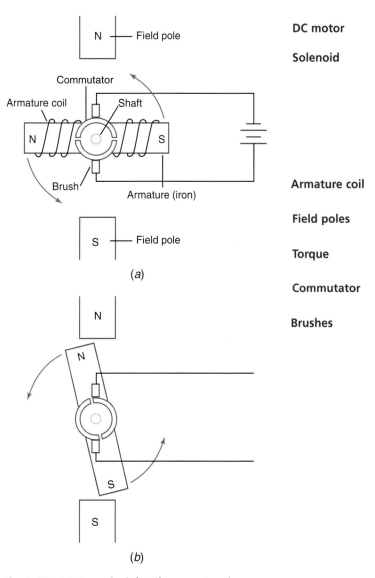

Fig. 7-27 Motor principle. The armature is rotated by alternate attraction (*a*) and repulsion (*b*) of magnetic poles.

DC motor

Solenoid

Armature coil

Field poles

Torque

Commutator

Brushes

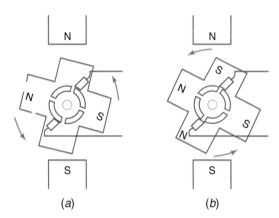

Fig. 7-28 Four-segment commutator.

of ratings for both the coil and the contacts. Coils are rated for both the current required to energize the relay and the voltage required to produce that current. The contacts are also rated for both current and voltage, just as any switch is rated. Relays often have more than one set of contacts. The contacts may be either normally open or normally closed or a mixture of both. The relay may be energized by either alternating or direct current. However, ac and dc relays are not interchangeable.

One of the chief advantages of a relay over a simple switch is that it allows *remote operation*. A low-voltage, low-current supply can control the relay coil. Then the relay contacts can control a high-voltage and/or high-current circuit. The switch that operates the coil can be located in a remote place. Only a small two-conductor cable has to run between the switch and the relay.

Plunger

Remote operation

chanical mechanism. The movable *plunger* in Fig. 7-29 can be used to operate, for example, a mechanical brake, a clutch, or a valve to control water flow.

When the solenoid is deenergized [Fig. 7-29(b)], there is no magnetic force to hold the plunger in the center of the coil. When current is supplied to the coil, the solenoid is energized and the plunger is pulled into the coil because of its high permeability. The plunger has much less reluctance than the air it replaces. As shown in Fig. 7-29(c), the flux path is almost entirely through iron when the solenoid is energized. The pull of the plunger of a solenoid is dependent on its magnetic properties. In general, increasing the mmf or decreasing the reluctance increases the pull.

Most solenoids are made to operate from an ac supply. However, dc solenoids are also available. The two types (ac and dc) are not interchangeable.

7-17 Relays

Magnetic relay

Armature

A *magnetic relay* uses the attraction between an iron *armature* and an energized coil to operate a pair of electric contacts. Figure 7-30 shows the cross section of a relay in the deenergized position. When current flows through the coil, the armature is pulled down against the iron core. This closes the set of contacts that can complete another electric circuit. When current stops flowing in the coil, the spring pulls the armature up and opens the contacts.

Relays are manufactured in a wide variety

■ TEST

Answer the following questions.

45. True or false. The core of an electromagnet should be made from a material with high reluctance.

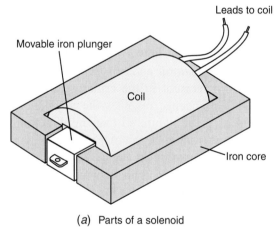

(a) Parts of a solenoid

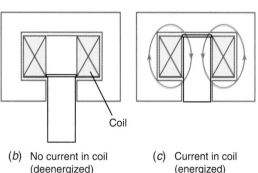

(b) No current in coil (deenergized) (c) Current in coil (energized)

Fig. 7-29 Movable plunger.

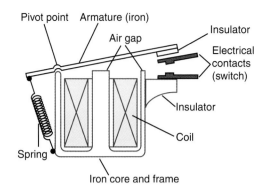

Pivot point   Armature (iron)

Air gap

Insulator

Electrical contacts (switch)

Insulator

Coil

Spring

Iron core and frame

Fig. 7-30 Relay. The armature is attracted to the iron core when current flows through the coil.

46. True or false. Changing the core of an electromagnet from a low-permeability material to a high-permeability material will increase the strength of the magnet.
47. True or false. The field in a dc motor periodically changes magnetic polarity.
48. True or false. AC and dc solenoids are interchangeable.
49. True or false. A single relay can control power to two circuits even if the circuits operate from different power sources.
50. What is the purpose of the commutator and brushes in a dc motor?
51. Why is the plunger of a solenoid pulled into the center of the coil?
52. List the major electrical ratings needed to specify a relay.

Relay armature

Summary

1. Magnetism is an invisible force field.
2. A magnetic field exists around a magnet.
3. The lines of force of a magnetic field are called magnetic flux.
4. The denser the flux, the stronger the magnetic field.
5. The magnetic field is strongest at the poles of a magnet.
6. The direction of flux is from the north pole to the south pole.
7. Like magnetic poles repel; unlike poles attract.
8. When magnetic poles attract each other, their flux lines join together.
9. Lines of flux are continuous loops.
10. Magnetic fields exist around all current-carrying conductors.
11. The magnetic field around a straight conductor has no poles.
12. The left-hand rule is used to determine the direction of flux around a conductor.
13. The strength of the magnetic field around a conductor is directly proportional to the current in the conductor.
14. Two parallel conductors attract each other if their currents are in the same direction. They repel if the currents are in opposite directions.
15. A coil of wire carrying a current forms magnetic poles.
16. The north pole of a coil can be determined by use of the left-hand rule.
17. Decreasing the space between turns of a coil increases the flux at the poles of a coil.
18. Iron, nickel, and cobalt are all magnetic materials.
19. Iron is the most magnetic of the elements.
20. Nonmagnetic materials do not prohibit flux from passing through them.
21. The domains of a permanent magnet remain aligned after the magnet is removed from the magnetizing field.
22. Magnetic materials are magnetized when put in the fields of other magnets.
23. Magnetomotive force can be increased by increasing the current or the turns in a coil.
24. When a material is saturated, increasing the mmf does not increase the flux.
25. Residual magnetism refers to the flux that remains in a temporary magnet after it is removed from a magnetic field.
26. Reluctance is the opposition to magnetic flux.
27. Magnetic materials have less reluctance than nonmagnetic materials.
28. A changing magnetic flux can induce a voltage in a conductor.
29. The ampere-turn is the base unit of mmf.
30. The base unit of magnetic flux (ϕ) is the weber (Wb).
31. The base unit of magnetic field strength (H) is the ampere-turn per meter (A · t/m).
32. Magnetic field strength is also known as field intensity and magnetizing force.
33. Flux density (B) is the flux per unit cross-sectional area of a magnetic material.
34. The tesla (T) is the base unit of flux density. One tesla is one weber per square meter.
35. Permeability refers to the ability of a material to pass or carry magnetic flux. It is equal to flux density divided by magnetic field strength.
36. Relative permeability is the ratio of the permeability of a material to the permeability of air.
37. The strength of an electromagnet is determined by the reluctance and permeability of the core and by the mmf.
38. Direct-current motors rotate because of attraction and repulsion between the magnetic fields of the field poles and the armature.
39. The plunger of a solenoid is made from a high-permeability material.
40. The pull of a solenoid depends on the mmf and on the reluctance of its magnetic circuit.
41. A relay is an electromagnetic switch.
42. The following formulas show some relationships between magnetic quantities:

   mmf = turns × current

   $$H = \frac{\text{mmf}}{\text{length}} \qquad B = \frac{\phi}{\text{area}} \qquad \mu = \frac{B}{H}$$

43. Common magnetic quantities and units are specified in Table 7-1.

TABLE 7-1	Magnetic Quantities and Units		
	Quantity	Unit	
Name	Symbol	Name	Symbol
Magneto-motive force	mmf	Ampere-turn	A · t
Flux	ϕ	Weber	Wb
Magnetic field strength	H	Ampere-turn per meter	A·t/m
Flux density	B	Tesla	T
Permeability	μ	Weber per ampere-turn-meter	Wb/A·t·m

Chapter Review Questions

For questions 7-1 to 7-10, determine whether each statement is true or false.

7-1. In a magnetic circuit flux leaves the north pole and enters the south pole.

7-2. Breaking a magnet into two pieces demagnetizes the magnet.

7-3. Unlike magnetic poles repel each other.

7-4. Both a north and a south pole is produced by all magnetic fields.

7-5. Lines of flux are continuous loops that follow the path of least opposition.

7-6. Decreasing space between turns of a coil will decrease the pole flux.

7-7. Permanent-magnet materials are usually used in motors and transformers.

7-8. An ac coil can be used either to magnetize or to demagnetize a permanent-magnet material.

7-9. Permanent-magnet materials have very little residual magnetism.

7-10. When the current in a conductor is flowing away from you, the direction of the flux will be counterclockwise.

For questions 7-11 to 7-21, supply the missing word or phrase in each statement.

7-11. Nonmagnetic materials do not have magnetic _____ .

7-12. Magnetic shields are made from _____ materials.

7-13. The _____ is the rotating part of a dc motor.

7-14. The _____ is the base unit of flux density.

7-15. The _____ is the base unit of flux.

7-16. The _____ is the base unit of mmf.

7-17. The _____ is the base unit of magnetizing force.

7-18. Inducing a voltage into a moving conductor is called _____ action.

7-19. Inducing a voltage into a conductor by a changing magnetic field is called _____ action.

7-20. The most magnetic element is _____ .

7-21. The invisible lines of force associated with a magnet are called _____ .

For questions 7-22 to 7-25, choose the letter that best completes each sentence.

7-22. If all other factors are equal, the material with the highest permeability will
a. Have the least reluctance
b. Saturate most easily
c. Have the most residual magnetism
d. Have the lowest flux density

7-23. The pull of a solenoid is increased by
a. Forcing more current through the coil
b. Using fewer turns of wire in the coil
c. Using a core material of lower permeability
d. Increasing the air gap around the plunger

7-24. A magnet is a device which
a. Attracts materials like paper and glass
b. Always has two poles
c. Produces a force field
d. Creates reluctance in a material

7-25. Doubling the distance between two magnetic poles
a. Doubles the force between the poles
b. Quadruples the force between the poles
c. Makes the force between the poles one-half as great
d. Makes the force between the poles one-fourth as great

Answer the following questions:

7-26. What is the mmf of a 300-t coil that draws 3 A of current if the coil is 3 in. long?

7-27. What is the flux density of a magnetic circuit with 0.1 base units of flux and cross-sectional dimensions of 0.02 by 0.03 m?

7-28. Determine the magnetic field strength of a circuit with an average core length of 0.4 m, a 300-turn coil, and a coil current of 0.6 A.

7-29. Calculate the permeability of a material that requires 150 base units of magnetizing force to create 0.3 T in the material.

7-30. A magnetic circuit has an average core length of 0.3 m and a coil current of 0.54 A. Determine the required number of turns in the coil to provide 360 base units of magnetic field strength.

Critical Thinking Questions

7-1. Why do the domains in some magnetic materials return to a random pattern when the material is removed from a magnetic field?

7-2. Determine the rate of flux change in a circuit in which 15 V is induced in a 30-turn coil.

7-3. How much current is required in an 85-t coil to produce a magnetic field strength (*H*) of 50 A·t/m in a core with an average length of 0.45 m?

7-4. The permeability of air is 1.26×10^{-6} Wb/(A·t·m). Determine the flux density in a core if its relative permeability is 5000 when the magnetizing force is 50 A·t/m.

7-5. If you want to magnetize many pieces of permanent magnetic material, would you use ac or dc to power the coil? Why?

7-6. Discuss some limitations of using an electromagnet to pick up, move, and release metals in a scrap-metal yard.

Answers to Tests

1. Magnetism is a force field that exists around certain materials and devices.
2. A magnetic field is the area of force extending around a magnet.
3. flux
4. north pole
5. north, south
6. quadruple
7. south
8. yes
9. no
10. no
11. perpendicular
12. F
13. counterclockwise
14. coil
15. above
16. decreases
17. F
18. iron
19. Silicon

20. magnetic domain
21. silicon
22. T
23. The flux disappears.
24. The steel becomes a permanent magnet, and the flux remains as its field.
25. magnetomotive force
26. saturated
27. F
28. F
29. residual magnetism
30. permanent-magnet material
31. because the reluctance of the magnetic material is much less than that of the air which it replaces
32. F
33. the amount and the rate of flux change
34. to make connections to the rotating con-

ductor and to provide dc output
35. A changing flux caused by one coil induces a voltage in another coil.
36. 80 A · t
37. magnetizing force, field intensity
38. ampere-turn per meter
39. weber
40. tesla (weber per square meter)
41. permeability
42. The dimensions of the circuit are unimportant when flux density is specified
43. Relative permeability is the ability of a material to pass flux compared with the ability of air to pass flux.

44. F
45. F
46. T
47. F
48. F
49. T
50. They periodically reverse the direction of current in the armature.
51. Because the plunger has a much higher permeability than air does. When the plunger is pulled into the coil, the reluctance of the circuit is greatly reduced.
52. current and voltage ratings for both the coil and the contacts as well as the switching arrangement of the contacts

Chapter 8

Alternating Current and Voltage

Chapter Objectives

This chapter will help you to:

1. *Differentiate* between the various forms of alternating and direct current.
2. *Explain and use* the relationship between time and frequency.
3. *Describe* four ways to express the magnitude of alternating current.
4. *Understand* how a sine wave can be generated.
5. *Understand* the difference between, and relationship of, mechanical and electrical degrees.
6. *Illustrate* how three-phase alternating current is produced.
7. *Explain* the characteristics and applications of delta- and wye-connected ac systems.

M ost electric energy is distributed in the form of alternating current and voltage. Our homes and our facto- ries are both powered by ac electricity. These, of course, are the major consumers of electric energy.

8-1 AC Terminology

Alternating current periodically changes the direction in which it is flowing. It also changes magnitude (amount), either continuously or periodically. With most types of alternating current, the magnitude is changing continuously.

Of course, if there is an *alternating current,* there must also be an *alternating voltage* and power. Although it is not proper English, we often refer to alternating voltage as ac voltage. When written or spoken in the abbreviated form, the term *ac voltage* looks and sounds all right. However, when spelled out as *alternating current voltage,* the term can be confusing. An ac voltage is a voltage which produces an alternating current when used to power a circuit. Likewise, *ac power* refers to power that is produced by alternating current and alternating voltage.

8-2 Waveforms

An *electrical waveform* is represented by a line on a graph, as shown in Fig. 8-1(*a*). The line

is produced by plotting points on a graph and then connecting the points. The points represent the value of some electrical quantity at different times. The magnitude and direction of the electrical quantity (voltage or current) are indicated on the vertical axis of the graph. Time is marked on the horizontal axis.

The waveform shown in Fig. 8-1(*a*) could represent the voltage across the resistor in Fig. 8-1(*b*). In this case the units on the vertical axis would be volts. The waveform shows that the voltage rapidly increases to its maximum value (4 V) when the circuit is energized. The voltage remains at its maximum value until the circuit is opened. When the circuit is opened, the voltage across the resistor suddenly drops to zero.

The waveform of Fig. 8-1(*a*) could just as well represent the current in Fig. 8-1(*b*). The only thing that would change would be the units on the vertical axis. For the current waveform, the units would be amperes.

Notice in Fig. 8-1(*a*) that the waveform is always above the zero reference line. This

Alternating current

Alternating voltage

AC power

Electrical waveform

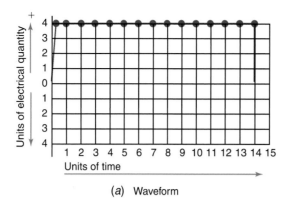

(a) Waveform

Square wave

Plot the voltage
at this point

**Sawtooth
waveform**

4 V

1 Ω

Reference point

(b) Circuit which produces the waveform

Fig. 8-1   Plotting a waveform. An electrical
quantity such as voltage is plotted
against time.

**Fluctuating
direct current**

**Pulsating direct
current**

Sine wave

means that the polarity of the voltage is always
positive with respect to some reference point.
Or, in the case of a current waveform, it means
that the direction of the current never reverses.
In other words, Fig. 8-1(a) is the waveform of
a pure (steady) direct current or voltage. The
magnitude of the voltage or current changes
only when the circuit is turned on or off.

Other common waveforms are shown in Fig.
8-2. The *fluctuating direct current* in Fig.
8-2(a) is the type of current produced in an
amplifying transistor. The pulsating waveform
of Fig. 8-2(b) represents the type of current (or
voltage) produced by a battery charger. Notice
that *pulsating direct current* periodically reduces to zero while fluctuating direct current
does not. Figure 8-2(c) shows one type of alternating current. Notice that the ac waveform
goes below the zero reference line. This means
that the polarity of the voltage reverses and
that the direction of the current changes.

8-3   Types of AC Waveforms

The most common type of ac waveform is the
sine wave, shown in Fig. 8-3(a). An alternating

current with this type of waveform is referred
to as *sinusoidal* alternating current. The alternating current (and voltage) supplied to homes
and factories is sinusoidal. A pure sine wave
has a very specific shape. Its shape can be precisely defined by mathematics. All the formulas we will be using to solve ac circuit problems are based on the sine wave. Therefore,
these formulas are appropriate only when
working with sinusoidal alternating current.

Figure 8-3(b) shows a *square wave*. This
type of ac waveform is used extensively in
computer circuits. With a square wave the magnitude of the current (or voltage) is not continuously varying. However, both amplitude and
direction do periodically change.

The *sawtooth waveform* of Fig. 8-3(c) is
used in television receivers, radar receivers,
and other electronic devices. Sawtooth voltages and currents are used in the circuits which
produce the picture on a television screen.

Wave shapes other than those shown in Fig.
8-3 are possible. In fact, alternating current can
be electronically produced in an almost infinite
variety of waveforms. Electronic music is created by producing and mixing together a wide
variety of waveforms.

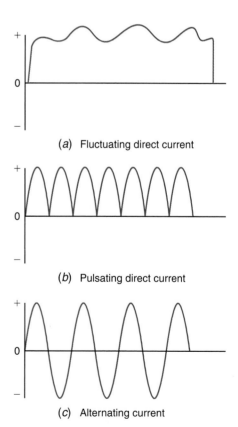

(a) Fluctuating direct current

(b) Pulsating direct current

(c) Alternating current

Fig. 8-2   Electrical waveforms.

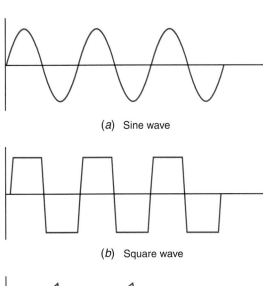

(a) Sine wave

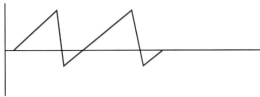

(b) Square wave

(c) Sawtooth wave

Fig. 8-3 Common types of ac waveforms.

■ TEST

Answer the following questions.

1. True or false. The vertical axis of a waveform graph is usually marked in units of time.

2. True or false. The horizontal axis of a waveform graph is usually marked in units of power.

3. True or false. A pulsating dc waveform periodically returns to the zero axis.

4. True or false. Fluctuating direct current periodically fluctuates above and below the zero reference line.

5. A _____ is produced when values of voltage and time are plotted on a graph.

6. Three common types of alternating current are _____ , _____ , and _____ .

8-4 Quantifying Alternating Current

Fully describing alternating current (or ac voltage) requires the use of a number of terms. Some of these terms have general meanings, but when used in electricity they have very specific meanings.

Cycle

The waveform in Fig. 8-4 shows four *cycles* of alternating current. A cycle is that part of a waveform which does not repeat or duplicate itself. Each cycle in Fig. 8-4 is a duplicate of every other cycle in the figure.

    The part of the cycle above the horizontal line in Fig. 8-4 is called the positive half-cycle. A half-cycle is also called an *alternation*. Therefore, the positive half-cycle could be called the positive alternation. The negative half-cycle, of course, is that part below the horizontal reference line.

Cycles

Alternation

Period

The time required to complete one cycle is the *period (T)* of a waveform. In Fig. 8-4 it takes

Period (T)

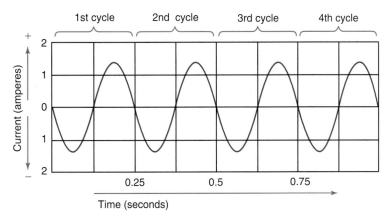

Fig. 8-4 Cycle, period, and frequency. The waveform has a period of 0.25 s and completes 4 cycles per second.

0.25 s to complete one cycle. Therefore the period T of that waveform is 0.25 s.

Frequency

Frequency (f)

The rate at which cycles are produced is called the *frequency (f)* of an ac current or ac voltage. Frequency, then, refers to how rapidly the current reverses or how often the voltage changes polarity.

Unit of Frequency—The Hertz

Hertz (Hz)

The base unit of frequency is the *hertz* (Hz). One hertz is equal to one cycle per second. The sine wave in Fig. 8-4 goes through four cycles in 1 s. Therefore, it has a frequency f of 4 Hz.

Commercial power in North America is distributed at a frequency of 60 Hz. It is sinusoidal alternating current. In many European countries a frequency of 50 Hz is used for electric power.

Electronic circuits use a wide range of frequencies. For example, an audio amplifier usually amplifies all frequencies between 20 Hz and 20 kHz. Frequencies between 540 and 1600 kHz are used for AM radio, and television uses frequencies up to 980 MHz.

Frequency and period are reciprocally related. That is:

$T = \dfrac{1}{f}$

$f = \dfrac{1}{T}$

$$T = \frac{1}{f} \quad \text{and} \quad f = \frac{1}{T}$$

 ... that 1 Hz is equal to one cycle per second. Therefore, the period is in seconds when the frequency is in hertz.

About ⬛ Electronics

Heinrich Hertz In 1887 German physicist Heinrich Hertz demonstrated the effect of electromagnetic radiation through space. In his honor, the hertz (Hz) is now the standard unit for the measurement of frequency (1 Hz equals 1 complete cycle per second).

Example 8-1

How much time is required to complete one cycle if the frequency is 60 Hz?

Given:	$f = 60$ Hz
Find:	T
Known:	$T = \dfrac{1}{f}$
Solution:	$T = \dfrac{1}{60 \text{ Hz}} = 0.0167$ s
Answer:	The time required is 0.0167 s, or 16.7 ms.

Example 8-2

What is the period of a 2-MHz sine wave?

Given:	$f = 2$ MHz
Find:	T
Known:	$T = \dfrac{1}{f}$
Solution:	$T = \dfrac{1}{2,000,000} = 0.0000005$ s
Answer:	The period is 0.0000005 s, or 0.5 μs.

If we had expressed the data in example 8-2 in powers of 10, the solution would be

$$T = \frac{1}{2 \times 10^{6}} = 0.5 \times 10^{-6} \text{ s}$$

Notice how the use of powers of 10 can simplify working with large and small numbers.

Example 8-3

What is the frequency of a waveform that requires 0.01 s to complete one cycle?

Given:	$T = 0.01$ s
Find:	f
Known:	$f = \dfrac{1}{T}$
Solution:	$f = \dfrac{1}{0.01} = 100$ Hz
Answer:	The frequency is 100 Hz.

Specifying Amplitude

The *amplitude* of an ac waveform can be specified in several different ways, as shown in Fig. 8-5. These ways of specifying amplitude are appropriate for either voltage or current.

For symmetrical waveforms the *peak* value (V_p) (also called the maximum value) of the negative and positive half-cycle are equal. Therefore, the *peak-to-peak* value ($V_{p\text{-}p}$) is twice as great as the peak value. For a sinusoidal ac voltage we can write

$$V_{p\text{-}p} = 2\,V_p$$

Many of the waveforms in electronic circuits are not symmetrical. Such waveforms can be specified as either peak or peak-to-peak values. When specifying a peak value, one must indicate whether it is the positive or the negative peak.

The *average* value (V_{av}) of a waveform is the arithmetical mean. Finding the arithmetical mean of a waveform is illustrated in Fig. 8-6(*a*). The instantaneous value of the waveform is determined at a number of equally spaced intervals along the horizontal axis. Then, all the instantaneous values are added together and divided by the total number of intervals. The closer the intervals, the more accurate the estimate of the average value of the waveform. Figure 8-6(*b*) provides a further idea of the meaning of the average value of a

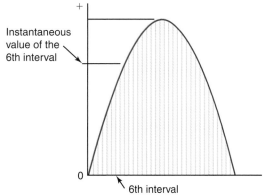

(*a*) Method of determining the average value

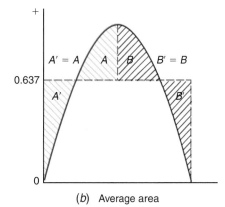

(*b*) Average area

Fig. 8-6 Average value of a waveform. The combined area of *A* and *B* is equal to the area of *A'* plus *B'*.

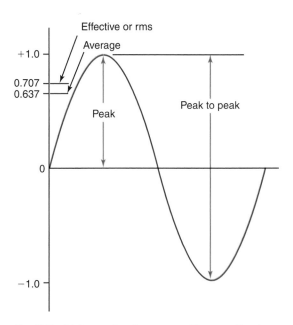

Fig. 8-5 Values of a sine wave. The amplitude of a sine wave can be specified in any one of four ways.

Amplitude

Peak value

Peak-to-peak value

$V_{p\text{-}p} = 2\,V_p$

Average value

$V_{av} = 0.637\,V_p$

$V_p = 1.57\,V_{av}$

Effective, or rms, value

waveform. The area of the waveform above the average value is equal to the "missing" areas below the average value. The area labeled *A* is equal to the area labeled *A'* in Fig. 8-6(*b*). Area *B* is equal to area *B'*.

For a sine wave, the relationship between peak value and average value is indicated in Fig. 8-5. For voltage, the relationship can be written as

$$V_{av} = 0.637\,V_p$$

Or, solving for V_p, the relationship is

$$V_p = 1.57\,V_{av}$$

The relationship between average and peak voltage is used when analyzing circuits which convert ac voltage to pulsating dc voltage.

The most common way of specifying the amount of alternating current is by stating its *effective*, or *rms*, value. The effective value of an alternating current is that value that produces the same heat in a resistive circuit as a direct current of the same value. Also, equal amounts of dc voltage and effective ac voltage

Example 8-4

What is the average voltage of a sinusoidal waveform if its peak-to-peak voltage is 300 V?

Given: $V_{\text{p-p}} = 300$ V

Find: V_{av}

Known: $V_{\text{p}} = \dfrac{V_{\text{p-p}}}{2}$

$V_{\text{av}} = 0.637\,V_{\text{p}}$

Solution: $V_{\text{p}} = \dfrac{300 \text{ V}}{2} = 150$ V

$V_{\text{av}} = 0.637 \times 150 \text{ V} = 95.6$ V

Answer: The average voltage is 95.6 V.

---

produce equal power across resistors of equal value. In Fig. 8-7(*a*), 20 V of effective ac voltage forces 2 A of effective alternating current through a 10-Ω resistor. This causes the 10-Ω resistor to dissipate 40 W:

$$P = IV$$
$$= 2 \text{ A} \times 20 \text{ V} = 40 \text{ W}$$

As shown in Fig. 8-7(*b*), 20 V of dc voltage across a 10-Ω resistor also dissipates 40 W.

Now that you know what *effective value* means, let us examine why it is also the *rms value*. The effective value of a waveform can be determined by a mathematical process

$V_{\text{rms}} = 0.707\,V_{\text{p}}$

$V_{\text{p}} = 1.414\,V_{\text{rms}}$

known as root mean square (rms). With this process the waveform is divided into many intervals, as was done in Fig. 8-6(*a*). The instantaneous value of each interval is squared. The mean (average) of the squared values is then determined. Finally, the square root of the mean is calculated. When many intervals are used, this process yields the effective value.

An example may help clarify how this mathematical process yields the effective value. Suppose a sine wave represents the current through a 1-Ω resistor. Because $P = I^2R$, and $R = 1\ \Omega$, the squared value at any interval represents the instantaneous power at that interval. Therefore, the mean of all of these squared values yields the mean power. Solving $P = I^2R$ for I, gives $I = \sqrt{P/R}$. And, since $R = 1\ \Omega$, the square root of P yields the effective value of I. Thus, we can use the terms *rms value* and *effective value* interchangeably.

For sinusoidal alternating current, the rms (effective) value and the peak value are related by the following formulas:

$$V_{\text{rms}} = 0.707\,V_{\text{p}} \quad \text{and} \quad V_{\text{p}} = 1.414\,V_{\text{rms}}$$

Example 8-5

What is the peak voltage of 120-V rms?

Given: $V_{\text{rms}} = 120$ V

Find: V_{p}

Known: $V_{\text{p}} = 1.414\,V_{\text{rms}}$

Solution: $V_{\text{p}} = 1.414 \times 120 \text{ V} = 169.7$ V

Answer: The peak voltage is 169.7 V.

Alternating currents and ac voltages are usually specified as rms values. It is common practice to assume that alternating current is in rms units unless a subscript (p-p, p, or av) is included. For example, the voltage at the outlets in your home may be specified as 120 V. This is the rms value of the ac voltage.

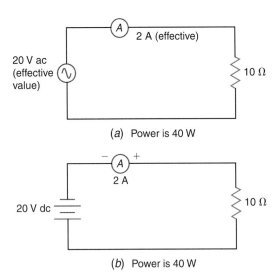

(*a*) Power is 40 W

(*b*) Power is 40 W

Fig. 8-7 Effective current and voltage. Direct current and effective alternating current produce the same amount of heat.

■ TEST_____

Answer the following questions.

7. The base unit of frequency is the _____ and its abbreviation is _____ .

8. One cycle per second is equal to 1 _____ .

174 ✦ **Chapter 8** Alternating Current and Voltage

9. The part of a waveform that does not duplicate any part of itself is called a _____ .
10. The segment of a waveform below the zero reference line is the negative _____ or the negative _____ .
11. How rapidly a waveform is produced is determined by the _____ of the waveform.
12. The _____ value of a waveform is also known as the effective value.
13. The symbol for frequency is _____ .
14. Determine the following for a 400-Hz, 28-V sine wave:
   a. Period
   b. V_{av}
   c. $V_{p\text{-}p}$
15. Determine the peak current when $I_{av} =$ 6 A.
16. Determine the rms current when $I_p = 8$ A.
17. Determine the frequency of a waveform that requires 0.005 s to complete one alternation.

8-5 The Sine Wave

YOU MAY **RECALL** . . . that the principles of *induced voltage* were introduced in Sec. 7-12.

To understand how a sine wave is produced, however, we must know more about induced voltages.

Induced Voltage and Current

When a conductor cuts across magnetic flux, a voltage is induced in the conductor. This voltage, of course, causes a current to flow if there is a complete circuit. Therefore, one can also say that a current is induced in the conductor.

The amount of induced voltage is a function of the amount of flux cut by the conductor per unit of time. The amount of *flux cut per unit of time* is in turn determined by:

Flux cut per unit of time

1. The speed of the conductor
2. The flux density
3. The angle at which the conductor crosses the magnetic flux

As shown in Fig. 8-8(*a*), no voltage (or current) is induced when the conductor moves parallel to the flux. This is because no lines of flux are crossed or cut. When the motion is 45° to the flux, as in Fig. 8-8(*b*), some voltage is induced because some flux is cut. For a given conductor speed, a conductor provides maximum induced current when the motion is 90° to the flux, as shown in Fig. 8-8(*c*).

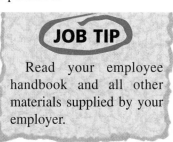

JOB TIP

Read your employee handbook and all other materials supplied by your employer.

YOU MAY **RECALL** . . . that the × and · in the center of the conductors in Figs. 8-8 and 8-9 indicate the direction of current, as explained in Sec. 7-3.

Induced voltage

The direction of an induced current (or the polarity of an induced voltage) is determined by two factors:

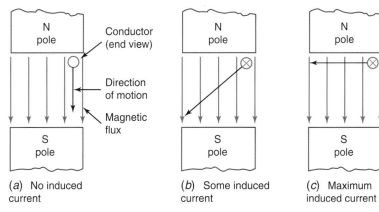

(*a*) No induced current

(*b*) Some induced current

(*c*) Maximum induced current

Fig. 8-8 Moving a conductor in a magnetic field. The amount of induced current depends on the angle at which the conductor cuts the flux.

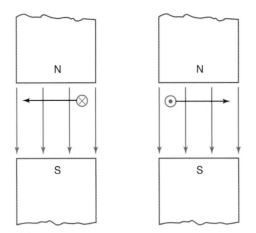

(a) Changing direction of conductor movement

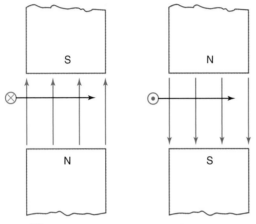

(b) Changing the polarity of the field

Fig. 8-9 Changing the direction of movement or the polarity of the field changes the direction of the induced current.

1. The direction in which the conductor is moving
2. The polarity of the magnetic field or the direction of the flux

Figure 8-9 shows the four possible combinations of conductor movement and polarity.

Left-hand rule   The direction of induced current can be determined by using the *left-hand rule,* which is

illustrated in Fig. 8-10. In this figure, the thumb indicates the direction of conductor movement. The index finger (pointing straight out) is aligned with the direction of the flux. The middle finger (bent at 90° to the palm) indicates the direction the current flows. Apply the left-hand rule to the diagrams in Fig. 8-9. Your middle finger should point into the page for those diagrams with an x in the conductor.

Producing the Sine Wave

A sine wave is produced when a conductor is rotated in a magnetic field. The conductor must rotate in a perfect circle and at a constant speed. The magnetic field has to have a uniform flux density. Production of a sine wave is illustrated in Fig. 8-11. At position 1, the conductor is moving parallel to the flux and no voltage is induced in the conductor. (The instantaneous direction of the conductor is indicated by the arrow.) As the conductor rotates from position 1 to position 2, it starts to cut the flux at a slight angle. Thus, a small voltage is induced in the conductor. The current produced by this voltage is flowing out of the page, as indicated by the dot in the conductor. As the conductor rotates on through positions 3 and 4, the angle at which the conductor cuts the flux increases. At position 4, the conductor is moving perpendicular to the flux. Maximum flux per unit of time is now being cut. Therefore, maximum voltage V_p is being produced, as shown in Fig.

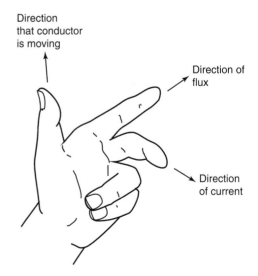

Fig. 8-10 Determining the direction of induced current using the left-hand rule.

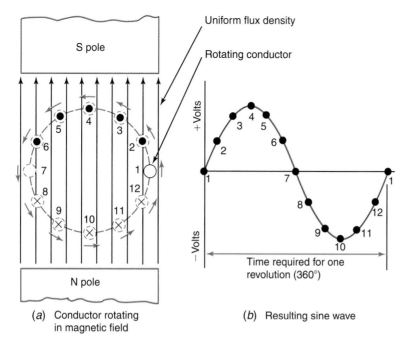

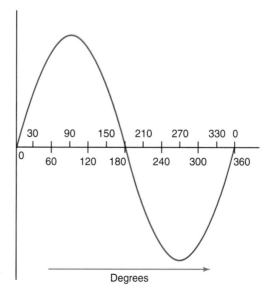

Fig. 8-11   Generating a sine wave by rotating a conductor through a magnetic field.

8-11(*b*). The positive peak of the sine wave is produced. As the conductor moves on to positions 5 and 6, less flux is cut. This is because the conductor's direction of movement again starts to parallel the flux. (For a given length of movement of the conductor, less flux is being cut.) By the time the conductor reaches position 7, no voltage is being induced. The first alternation (positive half-cycle) of the sine wave has been produced. Notice in Fig. 8-11(*a*) that the direction in which the conductor cuts the flux reverses as the conductor leaves position 7. Thus, the polarity of the induced voltage is reversed and the negative half of the cycle is started. At position 10, the conductor is producing its peak negative voltage. Finally, as the conductor returns to position 1, the induced voltage drops to zero. The first cycle is completed; a sine wave has been produced. Each additional revolution of the conductor produces another cycle of a sine wave.

Electrical and Mechanical Degrees

Notice in Fig. 8-11(*a*) that the conductor rotates 360° (one revolution) in producing one cycle of a sine wave. The horizontal axis of the waveform can therefore be marked off in degrees rather than units of time. This is done in Fig. 8-12 for each of the 12 positions of the conductor in Fig. 8-11. It can be seen from Fig.

8-12 that a sine wave reaches its peak values at 90° and 270°. Its value is zero at 0° (or 360°) and 180°. When degrees are used to mark off the horizontal axis of a waveform, they are referred to as *electrical degrees*. One alternation of a sine wave contains 180 electrical degrees; one cycle has 360 electrical degrees.

Electrical degrees

The advantage of marking the horizontal axis in degrees (rather than seconds) is that electrical degrees are independent of frequency. A cycle has 360° regardless of its frequency.

Fig. 8-12   Degrees of a waveform. One cycle has 360 electrical degrees.

However, the time required for a cycle (that is, the period) is totally dependent on the frequency. When the horizontal axis is marked off in seconds, the numbers are different for each frequency. Thus, you can see that the use of electrical degrees is very convenient and practical.

In Fig. 8-11 the number of electrical degrees is equal to the number of *mechanical degrees*. That is, the conductor rotates 360 mechanical degrees to produce the 360 electrical degrees of the sine wave. This one-to-one relationship of mechanical and electrical degrees is not essential. In fact, most ac generators produce more than one cycle per revolution.

Figure 8-13 shows the magnetic field produced by two pairs of magnetic poles. This field produces two cycles per revolution of the conductor. Every 45° of rotation (position 1 to position 2, for example) produces 90 electrical degrees. The conductor in Fig. 8-13 cuts flux under a south pole and a north pole for each 180° of rotation. This produces one cycle (360 electrical degrees) for 180 mechanical degrees.

Mechanical degrees

JOB TIP

Material safety data sheets are required in the workplace when hazardous materials are present. Read them.

8-6 AC Generator

In the preceding section we saw how a conductor rotating in a magnetic field produces an ac voltage. Now, let us see how this voltage can be connected to a load.

AC generator

The essential electric parts of an *ac generator* are shown in Fig. 8-14(*a*). In the drawing only one loop of wire (conductor) is shown. In a generator a coil consisting of many loops of wire would be used. This multilooped coil is called a *winding*. The winding is wound on a silicon-steel core called an *armature*. The armature winding is composed of many coils (armature coils).

Slip ring

Brush

Winding

Armature

As seen from Fig. 8-14(*a*), both the top and the bottom of the loop move through the magnetic field. The voltage induced in the top of the loop aids the voltage induced in the bottom of the loop. Both voltages force current in the same direction through the load. Figure 8-14(*b*) shows the top and bottom conductors of the loop and uses the symbols developed in the previous section. This sketch again shows that the voltages (and currents) induced into the

Field coils

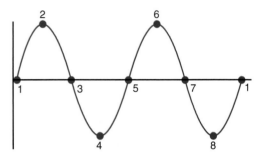

(*a*) Four-pole magnetic field

(*b*) 720 electrical degrees per revolution

Fig. 8-13 Electrical and mechanical degrees. The mechanical and electrical degrees of a generator need not be equal.

conductors of the rotating loop reinforce each other.

Notice in Fig. 8-14(*a*) that the bottom conductor always makes contact with *slip ring* 1. This ring always rotates on brush 1. Notice further that *brush* 1 is negative with respect to the other brush when the loop is in the position shown. It remains negative until the loop rotates another 90°. After 90° it becomes positive as the bottom loop starts to pass under the south pole. It remains positive for the next 180° while the conductor is moving under the south pole. Then it again becomes negative while the conductor moves across the face of the north pole. Thus, the polarity of the terminals of an ac generator reverses every 180 electrical degrees.

In Fig. 8-14 the magnetic field for the generator is created by a permanent magnet. In many generators the magnetic field is created by an electromagnet. The coils which create the electromagnet are called *field coils*. The

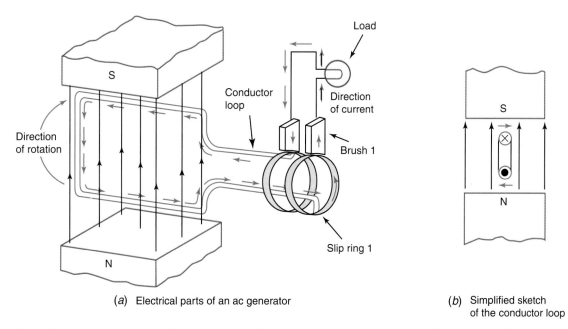

(a) Electrical parts of an ac generator

(b) Simplified sketch of the conductor loop

Fig. 8-14 An ac generator. The rotating conductor is connected to the load through the slip rings and brushes.

field coils are powered by a direct current so that the polarity of the field never changes. Also, in many large generators the conductors (loops) do not rotate. Instead, they are held stationary and the magnetic field is rotated around them. The end result is, of course, the same: voltage is induced in the conductors.

Generator Voltage

YOU MAY RECALL . . . that the voltage induced in a rotating conductor is determined by the flux density and the rotational speed.

In a generator, the rotating winding consists of multiturn coils (many loops). The voltage induced in each turn of a coil aids the voltage induced in every other turn. Thus, the output voltage of a generator is dependent on

1. The number of turns in the rotating (armature) coils
2. The speed at which the coils rotate
3. The flux density of the magnetic field

Generator Frequency

The speed at which the generator coils rotate influences the frequency as well as the output voltage of a generator. The frequency of the sine wave produced by a generator is determined by (1) the number of *pairs of magnetic poles* and (2) the rotational speed of the coils

(the revolutions per minute of the generator shaft).

A two-pole (one-pair) generator, like the one in Fig. 8-14, produces one cycle for each revolution. If this generator is rotating at 60 revolutions per minute (r/min), which is 1 r/s, the frequency is 1 Hz (one cycle per second). With a four-pole generator, 60 r/min produces a frequency of 2 Hz. Stated as a formula, the frequency of a generator's output is

$$\text{Frequency} = \frac{\text{r/min}}{60} \times \text{pairs of poles}$$

Using this formula, the frequency is in base units of hertz.

Example 8-6

What is the frequency of a six-pole generator rotating at 1200 r/min?

Given: Pairs of poles = 3
r/min = 1200

Find: f

Known: $f = \dfrac{\text{r/min}}{60} \times \text{pairs of poles}$

Solution: $f = \dfrac{1200}{60} \times 3 = 60 \text{ Hz}$

Answer: The frequency of the generator's output is 60 Hz.

Pairs of magnetic poles

8-7 Advantages of Alternating Current

It is easier to transform (change) alternating current from one voltage level to another than to transform direct current. That is why electric energy is distributed in the form of alternating current. In rural areas, power transmission lines can be operated at very high voltages (more than 400,000 V). When the power lines approach a city, the voltage can be reduced (by a transformer) to a few thousand volts. This lower voltage is then distributed in the city.

Another advantage of alternating current is that ac motors are less complex than dc motors. Alternating current motors can be built without brushes and commutators. This greatly reduces the amount of maintenance needed on an ac motor.

■ TEST

Answer the following questions.

18. True or false. When operated at the same speed, an eight-pole generator produces a higher frequency than a six-pole generator.
19. True or false. The field coils of an ac generator are excited with alternating current.
20. True or false. No voltage is induced when a conductor moves parallel to lines of flux.
21. True or false. A conductor moving at 30° to the lines of flux produces more voltage than one moving at 60°.
22. Two revolutions of a conductor rotating in a four-pole magnetic field produce _____ electrical degrees.
23. Voltage is induced in the _____ coils of an ac generator.
24. The brushes in an ac generator make contact with the _____ .
25. A six-pole generator rotating at 1000 r/min produces a frequency of _____ .
26. What determines the magnitude of the voltage induced in a conductor?
27. Determine the speed (r/min) of an 8-pole generator that produces a 400-Hz, 80-V sine wave.
28. List the factors which determine the polarity of an induced voltage.
29. List the necessary conditions for inducing a sine wave conductor.
30. List the factors which determine the output voltage of a generator.
31. List the factors which determine the frequency of a generator's output.

Did You Know?

The Power of the Wind
Wind power can be used as a source of electricity. About 1 percent of the electricity used in California is generated by wind. In some areas of the United States, wind-farm electricity costs as little as 4 cents per kilowatt-hour.

32. List two advantages of alternating over direct current.

8-8 Three-Phase Alternating Current

Electricity is produced at an electric power plant in the form of *three-phase alternating current*. *Phase* is often abbreviated with the symbol ϕ. Therefore, *three-phase* is sometimes written as 3-ϕ. In three-phase alternating current (Fig. 8-15), each phase is a sine wave. Each phase is displaced from the other two phases by *120 electrical degrees*. Phase 2 in Fig. 8-15 starts its positive alternation 120° after phase 1 but 120° before phase 3.

Figure 8-15 shows that the algebraic sum of the three phases at any instant is zero. For instance, at 60°, phase 1 is +8.66 V, phase 2 is −8.66 V, and phase 3 is zero; the sum of the three is zero. At 150°, phase 1 and 2 are both at +5 V while phase 3 is at −10 V. Again the algebraic sum is zero. At 260° the positive voltages of phases 2 and 3 are exactly canceled by the negative voltage of phase 1. Pick any instant of time you desire to compare the three phases. You will find that the sum of two of the

phases is always equal in magnitude to the remaining phase and opposite it in sign.

Generation of Three-Phase Alternating Current

Suppose three loops of wire (as in Fig. 8-16) are spaced 120° apart and rotating in a magnetic field. If each loop is electrically isolated from the others and has its own slip rings and brushes, then each produces a sine wave. Each sine wave is displaced 120 electrical degrees from the other two sine waves. In other words, we have a *three-phase generator*.

Figure 8-16 is drawn to represent the beginning (zero electrical degrees of phase 1) of the waveforms in Fig. 8-15. In Fig. 8-16 the colored end of a loop is considered to be the reference point for that phase. Note that each phase has its own reference point. Thus, phase 2 is now negative and will become more negative as the generator turns a few more degrees. Phase 1 is now zero and will become positive, and phase 3 is positive but will decrease in value. Notice how this agrees with Fig. 8-15. When phase 1 in Fig. 8-15 becomes positive, phase 3 becomes less positive and

Three-phase alternating current

120 electrical degrees

Three-phase generator

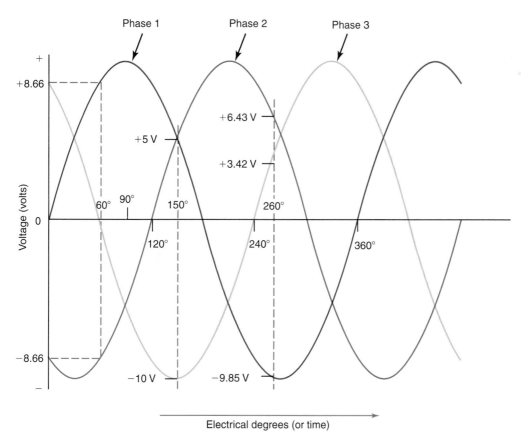

Fig. 8-15 Three-phase ac waveform. Each phase is separated by 120 electrical degrees.

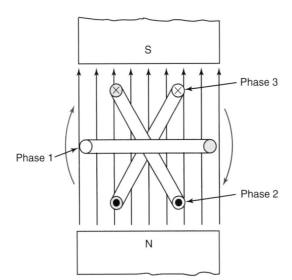

Reference end

Phase winding

Phase voltages

Fig. 8-16 Three-phase generator. The rotating conductors are mechanically locked together but electrically isolated.

Delta Connection

A delta connection is shown in Fig. 8-17. In this figure, each of the generator symbols represents one of the phases of a three-phase generator. The dot on one end of the generator symbol indicates the *reference end* of the *phase winding*. Notice that the reference ends of the generators are never connected together for the delta connection. The polarities and voltages indicated for the generators are momentary polarities and voltages. They are the ones that occur at 150° in Fig. 8-15. The following discussion assumes a balanced system. That is, all *phase voltages* are equal and all loads connected across them are equal.

The three phases are connected in one continuous loop for a delta connection. Yet no current flows in the loop until a load is applied to the generator. This is because the sum of the voltages of any two of the phases is equal in magnitude, but opposite in sign, to the other phase. For instance, the voltages of phases 1 and 2 are aiding each other. The sum of these two voltages is 10 V (Fig. 8-17), which exactly cancels the 10 V of phase 3.

Although the three phase voltages cancel each other within the continuous loop, each individual phase voltage is readily available. The full sinusoidal voltage produced by phase 1 of the generator is available between lines 1 and 3 in Fig. 8-17. Phase 2 is available between lines 1 and 2, and phase 3 exists between lines 2 and 3. Thus, the three phases can serve a load with a total of three lines.

phase 2 becomes more negative. After the generator in Fig. 8-16 has rotated 90°, phase 1 will be at peak positive voltage, phase 3 will be negative (and increasing in value), and phase 2 will be less negative.

The three phases of a generator can be connected so that the load can be carried on only three conductors. Thus, only three conductors are needed to carry three-phase power from a power plant to its place of use. The three phases of a generator can be connected in either a *delta connection* or a *wye* connection.

Delta connection

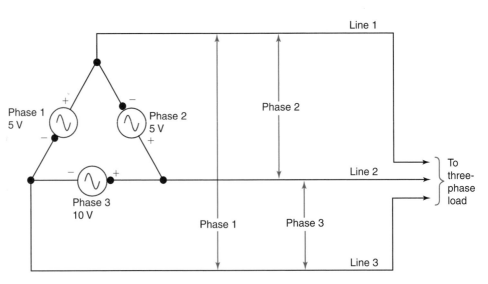

Fig. 8-17 Delta connection of generator. The line-to-line voltage is equal to the phase voltage.

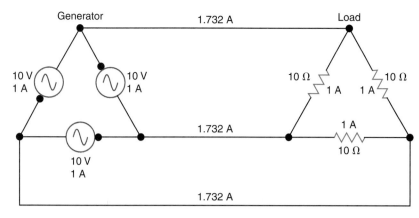

Fig. 8-18 Currents in a three-phase delta system. The line current is 1.732 times as great as the phase current.

Notice in Fig. 8-17 that the voltage between any two lines is exactly equal to one of the phase voltages. In other words, the *line voltages* and the phase voltages are the same. For a delta connection

$$V_{\text{line}} = V_{\text{phase}}$$

When a delta-connected generator is loaded (Fig. 8-18), currents flow in the loads, lines, and phase windings. However, the *line currents* and the *phase currents* are not equal. With equal loads connected to each phase, the line current is 1.732 times as great as the phase current. For a delta connection

$$I_{\text{line}} = 1.732 I_{\text{phase}}$$

The generator's phase currents and the load currents are equal. Therefore, the line currents are also 1.732 times as great as the load current. Line currents are not equal to phase currents because each line carries current from two phases. The currents from any two phases are 120° out of phase.

Notice in Fig. 8-18 that neither voltage polarity nor current direction is indicated. This is because it is an ac system in which polarity and direction reverse periodically. The current and voltage are assumed to be in rms values because they are not specified otherwise.

Remember the following points about a *balanced delta* connection:

1. No current flows in the phase windings until a load is connected.
2. The line voltage and the phase voltage are equal.
3. The line current is 1.732 times as great as the phase current or the load current.

Wye Connection

Figure 8-19 shows a *wye-connected* three-phase generator and load. Again in this dia-

Line voltages

Line currents

Phase currents

Balanced delta

Wye connection

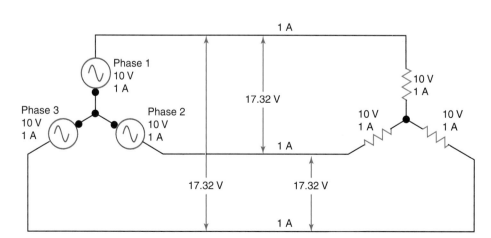

Fig. 8-19 Three-phase wye system. The line voltage is 1.732 times as great as the phase voltage.

gram, each generator symbol represents one phase of the three-phase generator. The voltages and currents are rms (effective) values. With the wye connection and balanced (equal) loads, all currents are the same. For a wye connection

$$I_{line} = I_{phase}$$

However, notice in Fig. 8-19 that the line voltages are 1.732 times as great as the phase voltages. For a wye connection.

$$V_{line} = 1.732\ V_{phase}$$

The *voltage* between any *two lines* is the result of two phase voltages. Since the two phase voltages are separated by 120 electrical degrees, their values cannot be simply added together.

Adding phase voltages to obtain line voltages is further complicated by the different reference points used for these voltages. Referring to Fig. 8-20 will help you understand reference points and addition of phase voltages. Figure 8-20(*b*) shows that phase 1 is +0.5 V at 30 electrical degrees. What does "+0.5 V" mean? Nothing, unless we have a reference point.

 . . . that voltage is a potential energy difference *between two points*.

The *reference point* for all three phase voltages is the center of the wye in Fig. 8-20(*a*). With this reference point established, the voltages read from the waveform graph in Fig. 8-20(*b*) have meaning. At 30° both phases 1 and 3 are 0.5 V positive with respect to the center of the wye. Phase 2 is 1 V negative with respect to the center of the wye. These voltages and polarities are marked on the diagram in Fig. 8-20(*a*). Notice that the *instantaneous voltages* of phases 1 and 2 are in series between lines 1 and 2. These two voltages (phases 1 and 2) are aiding each other at this instant, and so they can be added together. Together, they produce 1.5 V between lines 1 and 2. Referring back to Fig. 8-20(*b*), notice that phases 1 and 2 at 30° are of opposite polarities. When connected in a wye configuration, however, the two voltages aid each other. Also notice in Fig. 8-20(*a*) that phases 1 and 3 are series-opposing. Their momentary voltages cancel each other to provide a net voltage of zero between lines 1 and 3.

Reference points

Instantaneous voltage

Line-to-line voltage

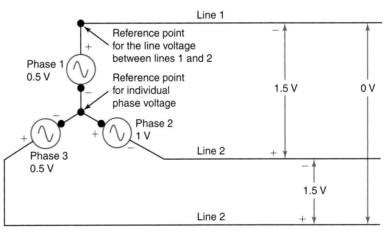

(*a*) Instantaneous (30°) phase and line voltage and polarity

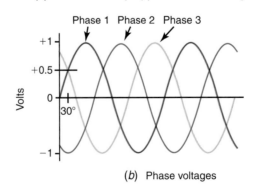

(*b*) Phase voltages

Fig. 8-20 Instantaneous voltages in a wye system.

In Fig. 8-20 we are looking only at instantaneous values taken at 30° on the three-phase waveform. Of course, these instantaneous values are different for each degree (or part of a degree) of the waveform. If all possible instantaneous values between two lines are plotted on a graph, the result is a sine wave that is 1.732 times as large as the sine wave of the phases which produced it. In plotting a waveform of this line voltage, a reference point is needed. The selection of this point is arbitrary. Let us select line 1 as the reference point. Then we can plot a waveform of the line voltage between lines 1 and 2. This waveform is shown in Fig. 8-21(a) superimposed on the waveforms of phases 1 and 2. Figure 8-21(b) lists some of the instantaneous values used to plot the waveform of the line voltage. Notice in Fig. 8-21 that the line voltage is displaced 30° from the phase-2 voltage. The line voltage lags the phase-2 voltage by 30°. That is, the negative peak of the line voltage occurs 30° after the negative peak of the phase voltage.

Suppose line 2 had been used as the reference point for the line voltage in Fig. 8-20. Then the polarity of the line voltage would have started out positive rather than negative. The line voltage waveform in Fig. 8-21 would be completely inverted. It would start at +0.866 V rather than −0.866 V. The *line voltage* would then lead the phase-1 voltage by 30°.

In summary, the major characteristics of the *balanced wye*-connected three-phase system are

1. The phase and line currents are equal.
2. The line voltage is 1.732 times greater than the phase voltage.
3. The line voltage and the phase voltage are separated by 30°.

Four-Wire Wye System

Figure 8-22 illustrates a *four-wire wye* system. In this figure, the generator symbol for each phase has been replaced by a coil symbol. The coil symbol represents one of the phases in a three-phase generator. This is the more traditional way of representing a three-phase generator.

The fourth wire of the four-wire system comes from the common center connection of the phase windings. (This common connection

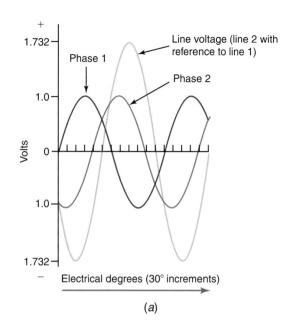

Electrical degrees	Instantaneous values (volts)		
	Phase 1	Phase 2	Line
0	0	−0.866	−0.866
30	+0.5	−1	−1.5
60	+0.866	−0.866	−1.732
90	+1	−0.5	−1.5
120	+0.866	0	−0.866
150	+0.5	+0.5	0
180	0	+0.866	+0.866
210	−0.5	+1	+1.5
240	−0.866	+0.866	+1.732

(b)

Fig. 8-21 Adding wye-connected phase voltages to obtain the line voltages.

is sometimes referred to as the *star point* or connection.) This fourth wire is often called the *neutral wire* because it is electrically connected to ground (earth). Thus, the fourth wire is neutral (has no voltage) *with respect to ground.*

With a balanced three-phase load, the neutral wire carries no current. However, it is needed when single-phase loading is required. In that case both single-phase and three-phase loads are connected to the wye system.

Inspection of Fig. 8-22 reveals the main feature of the four-wire wye system. This system can supply (with only four wires) all the voltage requirements of a small manufacturing facility. Single-phase, 120-V circuits are connected between the neutral and any line; single-phase, 208-V circuits are connected between any two of the three lines; and three-phase, 208-V devices are connected to the three

Phase and line currents

Line voltage

Balanced wye

Star point

Neutral wire

Four-wire wye

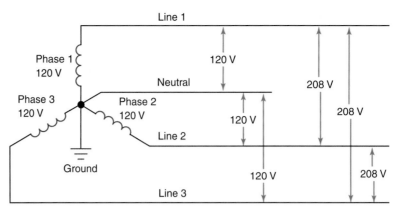

Fig. 8-22 Four-wire wye system. Four wires provide 120-V single-phase, 208-V single-phase, and 208-V three-phase.

Balanced three-phase load

lines. Common voltages in four-wire wye systems are 120/208 V and 277/480 V. The first value in each case is the phase, or single-phase, voltage. The second value in each case is the three-phase voltage. Notice that the three-phase voltage is 1.732 times the single-phase voltage.

Advantages of Three-Phase Systems

The three-phase system has advantages over the single-phase system. It

1. Makes more efficient use of copper
2. Provides a more constant load on the generator
3. Operates a motor that is less complex and provides more constant torque
4. Produces a smoother dc voltage and current when rectified

Rectification

Suppose a three-phase system and a single-phase system are delivering the same amount of power over the same length of lines. The single-phase system requires about 1.15 times as much copper for power lines as the three-phase system does.

The power required by a resistive load is equal to I^2R. With single-phase current, the power required from the generator follows the current variations. Thus, the load on the generator goes from zero to maximum power and back to zero with each alternation. With three-phase current at least two of the phases are providing current (and thus power) at any instant. The load on the three-phase generator never re-

duces to zero. The more uniform load provides for smoother mechanical operation of the generator.

Most ac motors can start only if the magnetic field behaves as if it were rotating. With single-phase alternating current, this rotating magnetic field is created through an auxiliary circuit in the motor. Once the motor has started, this auxiliary circuit is disconnected by a mechanically operated switch. In a three-phase motor, the rotating field is created by having each phase produce part of the magnetic field of each pole. No internal switch (or mechanical mechanism to operate it) is needed with the three-phase motor. Of course, the three-phase motor has more uniform torque because at least two phases are always producing magnetic fields.

Alternating current can be converted to direct current by a process known as *rectification*. The results of rectifying both single-phase alternating current and three-phase alternating current are illustrated in Fig. 8-23. Rectified single-phase alternating current produces pulsating direct current in which the current periodically reduces to zero. However, when three-phase alternating current is rectified, the direct current never reduces below 0.866 of its peak value. Obviously the three-phase system produces a dc output with much less fluctuation.

■ TEST

Answer the following questions.

33. The sine waves of a three-phase system are separated by _____ electrical degrees.

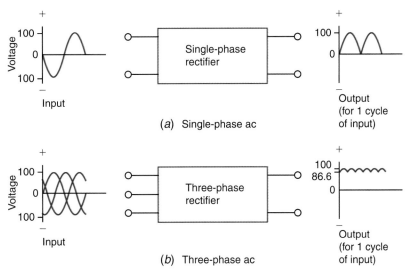

Fig. 8-23 Rectification of alternating current. When rectified, three-phase alternating current produces a smoother direct current than single-phase alternating current does.

34. The instantaneous voltage of phase 2 is _____ when phase 1 is −67 V and phase 3 is +18 V.

35. The windings of a three-phase generator are connected in either the _____ or _____ configuration.

36. Transmission of three-phase power requires _____ conductors.

37. For a _____ connection, the three-phase windings are connected in a continuous loop.

38. In the _____ connection, the phase current equals the line current.

39. In the _____ connection, the line current is _____ times as large as the phase current.

40. In the _____ connection, the three phases are connected together at the _____ point.

41. In the _____ connection, the line voltage is _____ times the phase voltage.

42. In a four-wire, three-phase system, the fourth wire is called the _____ wire.

43. Under what conditions does the fourth wire in a three-phase system carry current?

44. List the three voltages that are available from a four-wire, three-phase, 208-V system.

45. A three-phase generator produces a phase voltage of 277 V. What is the largest line voltage it can produce?

Summary

1. The great majority of electric energy is used in the form of alternating current and ac voltage. A sine wave is the most common waveform for alternating current.
2. Alternating current periodically reverses the direction in which it flows.
3. Alternating current periodically or continuously changes magnitude.
4. A dc waveform never crosses the zero reference line.
5. A cycle is that part of a periodic waveform which occurs without repeating itself. It is composed of two alternations: a negative alternation and a positive alternation.
6. The period (T) is the time required to complete one cycle.
7. The frequency (f) is the rate at which cycles are produced. Its base unit is the hertz, abbreviated Hz.
8. One hertz is equal to one cycle per second.
9. Power in North America is distributed at a frequency of 60 Hz.
10. Frequency and period are reciprocally related:

$$T = \frac{1}{f} \qquad f = \frac{1}{T}$$

11. Unless otherwise indicated, ac quantities are assumed to be in rms (effective) values.
12. Effective, or rms, values of alternating current produce the same heating effect as the same value of direct current.
13. A conductor rotating in a perfect circle at constant speed in a uniform magnetic field produces a sine wave.
14. There are 180 electrical degrees per alternation and 360 electrical degrees per cycle.
15. Generator voltage is determined by (1) speed of rotation, (2) number of turns in the coils, and (3) flux density.
16. Generator frequency is determined by (1) speed of rotation and (2) number of pairs of magnetic poles:

$$f = \frac{\text{r/min}}{60} \times \text{pairs of poles}$$

17. The magnitude of ac voltage is easier to change than the magnitude of dc voltage.
18. Electric power is usually transmitted from a power plant as three-phase alternating current.
19. The algebraic sum of the instantaneous phase values of three-phase voltage or current is always zero.
20. The voltages in a three-phase system are separated by 120 electrical degrees.
21. In a delta connection, the line and phase relationships are

$$V_{\text{line}} = V_{\text{phase}}$$
$$I_{\text{line}} = 1.732\, I_{\text{phase}}$$

22. In a wye connection, the line and phase relationships are

$$I_{\text{line}} = I_{\text{phase}}$$
$$V_{\text{line}} = 1.732\, V_{\text{phase}}$$

23. The fourth wire in a four-wire system is the neutral wire. When the loads on all three phases are equal (balanced), the fourth wire carries no current.
24. The advantages of three-phase alternating current over single-phase alternating current are (1) more efficient transfer of power, (2) more constant load on the generator, (3) more constant torque from motors, (4) less fluctuation when rectified to direct current.

Chapter Review Questions

For questions 8-1 to 8-16, determine whether each statement is true or false.

8-1. One half-cycle is also known as an alternation.

8-2. One alternation per second is equal to 2 Hz.

8-3. Fluctuating direct current never crosses the zero line of a waveform graph.

8-4. The average value of a sine wave is smaller than the rms value.

8-5. The mechanical degrees and the electrical degrees are equal for a four-pole generator.

8-6. Both frequency and output voltage are a function of the rotational speed of a generator.

8-7. DC motors require more maintenance than ac motors.

8-8. The algebraic sum of the phase voltages in a three-phase system is always zero at any given instant.

8-9. A three-phase system requires six conductors to transfer power between two locations.

8-10. The line voltages and phase voltages are equal in a wye-connected system.

8-11. The line currents and phase currents are equal in a wye-connected system.

8-12. When line currents and load currents are not equal in a balanced three-phase system, they are related by a factor of 1.732.

8-13. The line voltage in a wye-connected system is always displaced 30° from one of the phase voltages.

8-14. In a 208-V, four-wire system, 120 V is available between the neutral and any line.

8-15. In a delta connection, the reference ends of the phase windings of a generator are all connected together.

8-16. In some ac generators, the magnetic field is rotated.

For questions 8-17 to 8-23, supply the missing word or phrase in each statement.

8-17. The waveform used to distribute electric energy is the _____ .

8-18. The base unit of frequency is the _____, which is abbreviated _____ .

8-19. The time required to complete a cycle is called the _____, which is abbreviated _____ .

8-20. In a _____ four-wire, three-phase system, the _____ wire carries no current.

8-21. V_p = 20 V is meaningful for a _____ wave.

8-22. One hertz can be defined as a _____ .

8-23. The _____ wire is connected to the star point of a four-wire, three-phase system.

Answer the following questions:

8-24. Determine the frequency and the period of an electrical waveform that goes from zero to maximum positive and back to zero in 0.005 s.

8-25. Determine the period of a 200-Hz sine wave.

8-26. What is the rms voltage of a sine wave with a peak-to-peak value of 280 V?

8-27. Determine the average value of a 120-V electric outlet.

8-28. What determines the amount of voltage induced in a conductor moving through a magnetic field?

8-29. List three factors that determine the output voltage of a generator.

8-30. What is the frequency of an eight-pole generator turning at 900 r/min?

8-31. List four advantages of a three-phase system over a single-phase system.

8-32. What value of $I_{p\text{-}p}$ is equivalent to 4.24 A of direct current?

Critical Thinking Questions

8-1. Draw a schematic diagram that shows how to connect an automatic-reset circuit breaker, two cells, a relay, and a resistor so that square-wave current flows in the resistor. The symbol for a relay is shown below.

8-2. Using only the symbols presented in previous chapters, draw a schematic diagram for a circuit that will produce a pulsating direct current in a resistor. Also, draw the waveform for the voltage across the resistor.

8-3. How could you double the frequency of a generator and not change the output voltage?

8-4. The neutral wire in a four-wire wye system opens at the star point of the generator. How will this affect the 208-V loads? How will this affect the 120-V loads?

8-5. A phase coil of a delta-connected generator opens. How will this affect the operation of single-phase motors and three-phase delta-connected motors?

8-6. A phase coil of a wye-connected generator opens. How will this affect the operation of single-phase motors and three-phase wye-connected motors?

8-7. What is the magnitude relationship between the phase current of the source and the phase current of the load when a balanced wye load is connected to a delta source?

8-8. What is the magnitude relationship between the phase current of the source and the phase current of the load when a balanced delta load is connected to a wye source?

8-9. Why is frequency expressed in hertz when solving this practical working formula?

$$f = \frac{r/min}{60} \times \text{pairs of poles}$$

Answers to Tests

1. F
2. F
3. T
4. F
5. waveform
6. square wave, sine wave, sawtooth wave
7. hertz, Hz
8. hertz
9. cycle
10. alternation, half-cycle
11. frequency
12. Root mean square (rms)
13. f
14. 2.5 ms, 25.2 V_{av}, 79.2 V_{p-p}

15. 9.42 $A_{(peak)}$
16. 5.656 $A_{(peak)}$
17. 100 Hz
18. T
19. F
20. T
21. F
22. 1440
23. armature
24. slip rings
25. 50 Hz
26. the flux cut per unit of time
27. 6000 r/min
28. the polarity of the magnetic field and the direction of conductor movement
29. uniform magnetic

field, constant speed of rotation of the conductor, and rotation in a perfect circle
30. number of turns in the armature coils, strength of the poles, and speed at which the armature rotates
31. number of magnetic poles and revolutions per minute of the armature
32. It is easier to change voltage, and ac motors are easier to construct and maintain.

33. 120
34. +49 V
35. delta, wye
36. three
37. delta
38. wye
39. delta, 1.732
40. wye, star
41. wye, 1.732
42. neutral
43. when the phases are unequally loaded
44. 120-V, single-phase; 208-V, single-phase; 208-V, three-phase
45. 480 V

Chapter 9

Power in AC Circuits

Chapter Objectives

This chapter will help you to:

1. *Understand* phase relationships in ac circuits.
2. *Use* phasor diagrams to represent circuit currents and voltages.
3. *Apply* right-triangle relationships to electric circuits to determine phase angles.
4. *Use* trigonometric functions to determine resistive and reactive currents and voltages.
5. *Understand* the relationship between true power and apparent power.
6. *Explain* the importance of the power factor of an electric distribution system.

The most common form of electric power is the power produced by alternating current and ac voltage. In many ac circuits, the current and the voltage do not rise and fall in step with each other. Thus, we need to learn some new techniques to determine the power in ac circuits.

9-1 Power in Resistive AC circuits

When the load on an ac source contains only resistance, the current and voltage are *in phase*. *In phase* means that two waveforms arrive at their peak positive values, zero values, and peak negative values simultaneously. The current in Fig. 9-1 is in phase with the voltage. The *resistive load* in Fig. 9-1(*a*) might represent an electric heater, a clothes iron, or a stove. As with direct current, power is the product of the current and voltage. With alternating current, the values of current and voltage vary with time. At any instant, the power is equal to the current at that instant multiplied by the voltage at that instant. As shown in Fig. 9-1(*b*), the product of the instantaneous current and voltage is the instantaneous power. Plotting all the instantaneous powers produces a power waveform. However, notice that the power waveform is always positive for in-phase currents and voltages. A negative current multi-

plied by a negative voltage yields a positive power. This means that the resistive load is converting electric energy into heat energy during the complete cycle. For in-phase circuits, that is, resistive circuits, we can state

$$P = IV = I^2R = \frac{V^2}{R}$$

where I and V are the rms current and voltage.

In phase

Resistive load

Example 9-1

How much power is used by the resistive load in Fig. 9-1(*a*)?

Given: $V = 3.1$ V
            $I = 1.2$ A
Find: P
Known: $P = IV$
Solution: $P = 1.2$ A $\times$ 3.1 V $= 3.72$ W
Answer: The power used is 3.72 W.

Example 9-2

A 400-Ω resistor is connected to a 100-V ac source. How much power is used?

Given: $V = 100$ V
              $R = 400$ Ω
Find: P
Known: $P = V^2/R$
Solution: $P = \dfrac{V^2}{R} = \dfrac{100 \text{ V} \times 100 \text{ V}}{400 \text{ Ω}}$
              $= 25$ W

Answer: The resistor dissipates 25 W.

Notice that power calculations in resistive ac circuits are no different from those in dc circuits.

9-2 Power in Out-of-Phase Circuits

An ac circuit may contain a load that is not entirely composed of resistance. The load may also contain reactance. *Reactance* is the name given to the opposition to current caused by capacitors and inductors. How and why capacitors and inductors provide reactance is discussed in later chapters. Here, we will be studying only the effect of reactance (and thus capacitors and inductors) on power in ac circuits.

Although reactance opposes current in ac circuits, it is different from resistance. It causes current and voltage to be 90° out of phase with each other. That is, reactance causes a phase shift to occur. The reactance of a capacitor, which is called *capacitive reactance,* causes the current to lead the voltage. In Fig. 9-2(a) the voltage is zero at time zero. At this time, however, the current has already reached its maximum value. Thus, the current leads the voltage by 90°.

Inductors produce *inductive reactance.* The effect of inductive reactance is to make the current lag the voltage by 90°. In Fig. 9-2(b) the voltage has already reached its peak positive value when the current is only beginning to go positive.

Loads containing only capacitance or only inductance cause the current and voltage to be exactly 90° out of phase. In other words, such loads cause a 90° phase shift between current and voltage. As shown in Fig. 9-2(c) and (d), loads containing a combination of resistance and reactance cause a phase shift of less than 90°. The exact amount of phase shift is determined by the relative amounts of resistance and reactance.

A good example of a combination load is an electric motor. It contains both inductance and resistance. When a small ac motor has no load on its shaft, the phase shift is about 70°. When the motor is loaded, it acts more like a resis-

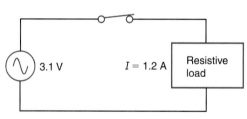

(a) Resistive ac circuit

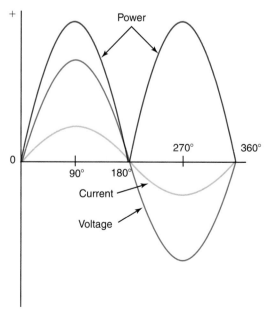

(b) Current, voltage, and power waveforms

Fig. 9-1 Power in a resistive ac circuit.

Reactance

Capacitive reactance

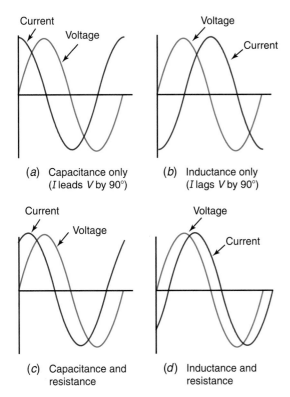

(a) Capacitance only
(I leads V by 90°)

(b) Inductance only
(I lags V by 90°)

(c) Capacitance and
resistance

(d) Inductance and
resistance

Fig. 9-2 Phase shift in ac circuits.

tance. When the motor is loaded, the phase shift reduces to about 30°. Of course, in a purely resistive circuit, the phase shift is 0°.

For circuits in which *phase shift* occurs, calculating power is no longer a simple matter of multiplying current and voltage. In Fig. 9-3, three examples of phase shift and the resulting power are shown. Notice that in some cases the current and voltage are of opposite polarities. This yields a negative power for that part of the cycle. What does it mean to have negative power? It means that the load is not taking power from the source during that period of time. In fact, it means that the load is returning power to the source for that part of the cycle. Therefore, the net power used by the load is the difference between the positive power and the negative power. Notice also in Fig. 9-3 that the greater the phase shift, the smaller the net power taken from the source. At 90° phase shift, no net power (thus no energy) is taken from the source.

Since capacitive and inductive reactances cause a 90° phase shift, we can conclude that reactance uses no power. Only resistance uses power. Therefore, in circuits with combination loads we can conclude several things (Fig. 9-4). First, the resistance part of the load uses

power. In the resistance part, the current and the voltage are in phase and the power is simply $P = IV$. Second, the reactance part of the load uses no power. The current and the voltage are a full 90° out of phase. Third, the phase relationship between the total voltage and current is determined by the relative amount of resistance and reactance.

We need a way in which to determine what part of the total current and voltage is associated with the resistance of the load. Then we can determine the net (or real) power used by any combination load with any amount of phase shift. Therefore, let us next develop a way of breaking a current or voltage into a resistive part and a reactive part. We will do this through

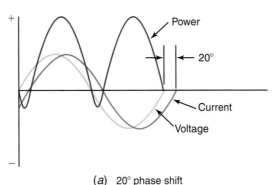

(a) 20° phase shift

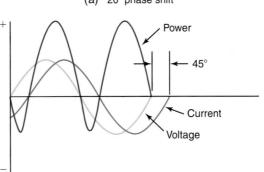

(b) 45° phase shift

(c) 90° phase shift

Fig. 9-3 Power in phase-shifted circuits. At 90° of phase shift, the power is zero.

Phase shift

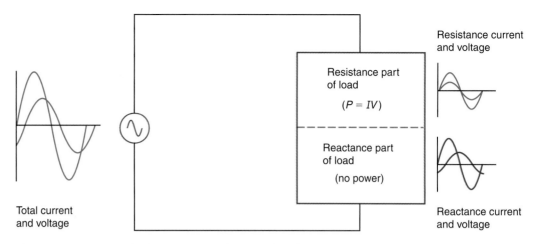

Fig. 9-4   Current and voltage in a combination load.

the use of *phasors* and *right triangle* relationships.

TEST

Answer the following questions.

1. _____ causes current and voltage to be 90° out of phase.
2. _____ causes no phase shift between the current and the voltage.
3. When the phase shift is between 0 and 90°, the circuit contains both _____ and _____.
4. The current in an ac motor will _____ the voltage.
5. A motor has _____ reactance.
6. _____ does not convert electric energy to heat energy.
7. $P = IV$ can be used in ac circuits when the load is purely _____.
8. How much power does a 400-Ω resistor use when it carries 25 mA of alternating current?
9. How much power is used by a 500-Ω resistor connected to a 100-V_{p-p} source?

Phasors

Drawing sinusoidal waveforms to show amplitude and phase shift is slow and tedious work. A much simpler technique of showing amplitude and phase shift is the use of phasors.

A *phasor* is a line whose *direction* represents the phase angle in electrical degrees and whose length represents the *magnitude* of the electrical quantity. You may have studied *vectors* in mathematics or physics classes. From the definition of a phasor, you can see that phasors and vectors are very similar. In electricity and electronics, the term *phasor* is used to emphasize the fact that the position of the line changes with time as the wave moves through its cycle from 0 to 360°. The phasor in Fig. 9-5 represents 10 A of current at 60°. By convention, 0 electrical degrees is on the right-hand side of the horizontal axis. Figure 9-6 shows the phasors of two voltages that are 45° out of phase. In this figure the two phasors have to use the same scale since they both represent voltage. For convenience, one phasor of a group is typically drawn on the 0° line. This phasor, then, becomes the reference to which the other phasors are related.

Remember that phasors are actually moving with time. By convention, phasors are assumed to rotate in a counterclockwise direction. By

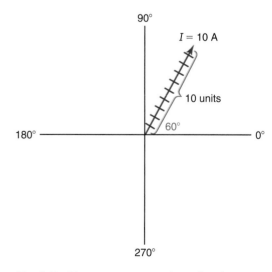

Fig. 9-5   Phasor representation of a sine wave.

Phasors

Direction

Magnitude

Vectors

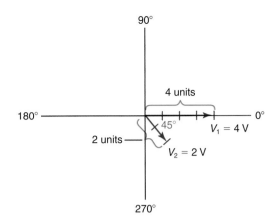

Fig. 9-6 Out-of-phase voltages. Amplitude and phase shift can be shown by phasors.

showing V_1 on the 0° line, we stopped its motion at that instant. Of course, the motion of V_2 is also stopped at the same instant. Thus, the phase relationship between V_1 and V_2 does not change. Since the motion of phasors is counterclockwise, V_2 in Fig. 9-6 is lagging behind V_1, which is the reference phasor. Since V_2 is out of phase with V_1 by 45°, we can conclude from Fig. 9-6 that V_2 lags V_1 by 45°.

A *voltage phasor* and a *current phasor* are shown in Fig. 9-7. Since current and voltage are different quantities, these two phasors can use different scales for their magnitude. The phasor diagram shows that the current leads the voltage by 45°. The power resulting from this current and voltage can be found by breaking the current phasor into two smaller phasors. One of the smaller phasors represents the current in the resistive portion of the load. It is

in phase with the voltage phasor. The other smaller phasor represents the current in the reactive portion of the load. It, of course, is 90° out of phase with the voltage. The two smaller phasors are separated by 90°.

The idea of breaking a phasor into two smaller phasors is illustrated in Fig. 9-8. In this figure, we are using vectors because the positions and magnitudes do not change with time. One vector of 500 newtons (N), approximately 112 lb, of force is exerted on the spring in the northeast direction. In the first case [Fig. 9-8(a)], the force is created by one person pulling in a northeast direction with a force of 500 N. In the second case [Fig. 9-8(b)], one person pulls north with 300 N of force and another person pulls east with 400 N. The net result is 500 N of force in the northeast direction. The

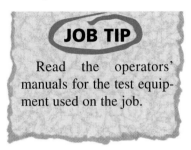

JOB TIP

Read the operators' manuals for the test equipment used on the job.

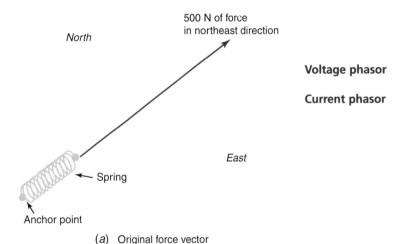

(a) Original force vector

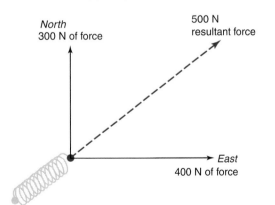

(b) Equivalent force vectors

Fig. 9-8 Equivalency of vectors. A single vector can be represented by two smaller vectors separated by 90°.

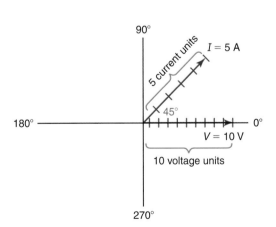

Fig. 9-7 Current and voltage phase relationship represented by phasors. *I* leads *V* by 45°.

effect on the spring is exactly the same in both cases.

The vectors in Fig. 9-8(*b*) represent two forces at a right angle (90°) to each other. However, they could equally well represent any other quantity that has magnitude and direction. Thus, they could be phasors representing resistive current and reactive current in an electric circuit.

The magnitude of resistive and reactive current can be determined graphically. The process is shown in Fig. 9-9. In this diagram, *V* is at the 0° reference line. The total current I_T is 5 A and is leading the voltage by 36.9°. The first step is to draw the original phasor I_T to scale at the correct angle. Then draw a vertical

line (line 1) from the tip of the I_T phasor through the horizontal axis. Line 1 must be parallel to the vertical axis. The intersection of line 1 and the horizontal axis locates the end of the resistive current phasor I_R. The end of the reactive current phasor I_X is found by drawing line 2 from the tip of I_T to the intersection of the vertical axis. Again, line 2 must be parallel to the horizontal axis. Finally, the magnitude of the resistive and reactive currents are measured with the scale used to measure I_T. The 4-A resistive current in Fig. 9-9 is the current that is in phase with the voltage. It is the current that can be used to calculate the power when 5 A of total current leads the total voltage by 36.9°.

▣ TEST

Answer the following questions.

10. True or false. A phasor represents the magnitude and the phase angle of an electrical quantity.
11. True or false. The same scale must be used for all voltage phasors in a phasor diagram.
12. True or false. Current and voltage phasors must use the same scale in a phasor diagram.
13. Draw a phasor for 10 V at 280°.
14. Draw a phasor for 8 A at 45°.
15. A current of 6 A lags a voltage of 10 V by 90°. Draw a phasor diagram of this current and voltage, using current as the reference phasor.
16. Draw a sketch that shows the resistive part of the current phasor in Fig. 9-5.

Functions of a Right Triangle

The currents determined by the *graphical method* illustrated in Fig. 9-9 are as accurate as the drawing itself. For most applications the graphical method is accurate enough. However, it is also a slow and tedious method. Let us develop a faster, easier way of determining these currents. This easier way is through the use of the properties of a right triangle. Notice in Fig. 9-9 that the reactive current phasor is exactly the same length as line 1. Also, notice that it is parallel to line 1. Therefore, line 1 can be replaced by the reactive current phasor. This has been done in

Graphical method

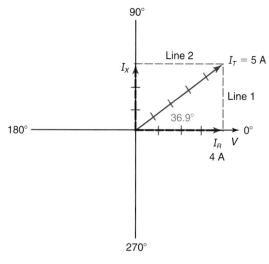

Fig. 9-9 Graphical analysis of a phasor.

Fig. 9-10. It can be clearly seen in Fig. 9-10 that the three current phasors form a right triangle. Now can you see why we need to know how to work with right triangles?

The right triangle of Fig. 9-10 is redrawn and relabeled in Fig. 9-11. The new labels are terms used to describe any right triangle. The symbol θ (which is the Greek letter *theta*) is used to indicate the angle that will be used to analyze the triangle. In the phasor diagram of Fig. 9-9, θ is the angle by which the total current leads the voltage. The *adjacent side* is always the shorter side that makes up angle θ. The side across the triangle from angle θ is always called the *opposite side*. The longest side of a right triangle is always the *hypotenuse*.

The three angles of any triangle always contain a total of 180°. This means that specifying θ for a right triangle also determines the other angle. Thus, the only possible difference between right triangles with the same angle θ is their size. This is illustrated in Fig. 9-12. Triangles *ABG*, *ACF*, and *ADE* all have the same angle θ. Only their sizes differ.

Notice in Fig. 9-12 that the sides and the hypotenuse increase proportionately. That is, when the hypotenuse is doubled in length, the opposite and adjacent sides must also be doubled. For example, all the sides of triangle *ACF* are twice as great as the corresponding sides of triangle *ABG*. The sides of triangle *ADE* are four times as large as the corresponding sides of triangle *ABG*.

Since the sides of a right triangle increase proportionately, the ratio of any two sides remains constant. In Fig. 9-12, side *BG* divided by side *AG* is exactly equal to side *CF* divided by side *AF*. The opposite side divided by the adjacent side is a constant for all right triangles

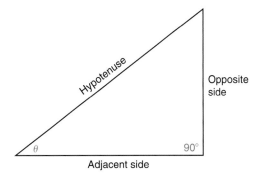

Theta

Fig. 9-11 Right triangle terminology.

that have the same angle θ. This same statement can be made for any other combination of sides (including the hypotenuse) of a right triangle.

The ratios of the sides of a right triangle are called the *trigonometric functions*. Three of the common ones used in electricity are shown in Fig. 9-12. The ratio of the *opposite side* to the adjacent side is called the *tangent* of angle θ and abbreviated tan θ. The ratio of the *adjacent side* to the hypotenuse is called cos θ, which is the abbreviation for the *cosine* of θ. The ratio of the opposite side to the *hypotenuse* is called sin θ, which stands for the *sine* of θ.

For each possible value of angle θ, there is a specific value for the sine, the tangent, and the cosine. A table of trigonometric functions for even angles between 0 and 90° is given in

Trigonometric functions

Opposite side

Tangent

Adjacent side

Cosine

Hypotenuse

Sine

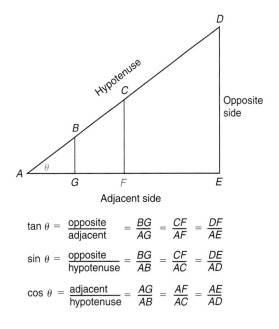

$$\tan \theta = \frac{\text{opposite}}{\text{adjacent}} = \frac{BG}{AG} = \frac{CF}{AF} = \frac{DF}{AE}$$

$$\sin \theta = \frac{\text{opposite}}{\text{hypotenuse}} = \frac{BG}{AB} = \frac{CF}{AC} = \frac{DE}{AD}$$

$$\cos \theta = \frac{\text{adjacent}}{\text{hypotenuse}} = \frac{AG}{AB} = \frac{AF}{AC} = \frac{AE}{AD}$$

Fig. 9-12 Trigonometric functions of a right triangle.

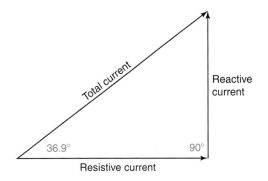

Fig. 9-10 Right triangle formed by current phasors.

Appendix F. Using this table (or a calculator) and the formulas in Fig. 9-12, we can easily solve right triangle problems. For example, if we know the angle and the length of the hypotenuse, we can solve for the length of the adjacent side.

Example 9-3

How long is the adjacent side of a right triangle if the hypotenuse is 7 cm and the angle θ is 40°?

Given: Hypotenuse = 7 cm
$\theta = 40°$

Find: Adjacent side

Known: $\cos \theta = \dfrac{\text{adjacent}}{\text{hypotenuse}}$

Solution: From the cosine formula we can solve for the adjacent side:

Adjacent = hypotenuse $\times \cos \theta$

From a table of trigonometric functions (Appendix F) we can find the value of cos 40°:

$$\cos 40° = 0.766$$

Then,

Adjacent = 7 cm $\times$ 0.766
= 5.362 cm

Answer: The adjacent side is 5.362 cm long.

In an electric circuit the hypotenuse might represent the total current, and the adjacent side the resistive current (Fig. 9-10). In that case,

$$\cos \theta = \frac{\text{resistive current}}{\text{total current}}$$

Let us now use the functions of a right triangle to solve an electrical problem.

P = IV cos θ

Example 9-4

The current in a circuit leads the voltage by 25°. The total current is 8.4 A. What is the value of the resistive current?

Given: $\theta = 25°$
$I_T = 8.4$ A

Find: Resistive current

Known: $\cos \theta = \dfrac{I_R}{I_T}$

Solution: From the cosine formula:

$$I_R = I_T \times \cos \theta$$

From Appendix F:

$$\cos 25° = 0.906$$

Then,

$$I_R = 8.4 \text{ A} \times 0.906$$
$$= 7.610 \text{ A}$$

Answer: The resistive current is 7.61 A.

Notice in the trigonometric table of Appendix F that the cosine varies from 1 to 0 as the angle changes from 0° to 90°. In other words, at 0° the adjacent side and the hypotenuse are the same length and cos θ is 1. At 90° the adjacent side reduces to 0 and cos θ is 0. Referring back to Fig. 9-10, we can see that there is no restrictive current when the angle is 90°. When θ is 0°, there is no reactive current. Notice, then, that cos θ tells what part, or percentage, of the total current is in phase with the voltage. That is, cos θ is the portion of the total current that is resistive. For example, when θ is 45°, cos θ is 0.707. Therefore, 0.707 of the total current is resistive current. That portion (0.707) of the current uses power.

Because cos θ determines the amount of resistive current, it is included in the formula for calculating power. The power formula is

$$P = IV \cos \theta$$

This is the general formula for power. It is the correct formula for all types of circuits. However, when a circuit contains only resistance, cos θ is 1. Therefore, cos θ can be dropped when working only with resistance.

Example 9-5

A 208-V motor draws 12 A. The current lags the voltage by 35°. What is the power input to the motor?

Given: $V = 208$ V
$I = 12$ A
$\theta = 35°$

Find: P

Known: $P = IV \cos\theta$

Solution: From Appendix F:

$$\cos 35° = 0.819$$

Then,

$$P = IV \cos\theta$$
$$= 12\text{ A} \times 208\text{ V} \times 0.819$$
$$= 2044\text{ W}$$

Answer: The power required by the motor is 2044 W.

So far in developing our power equation, we have always shown the current out of phase with the voltage, with the voltage on the 0° reference line. Thus, we broke the total current into two parts: the resistive and the reactive. It is also possible (and common) for the current to be shown on the 0° line and the voltage to be shown at some other angle. In that case, the total voltage would be separated into a resistive voltage and a reactive voltage. The phasor diagram for such a case is illustrated in Fig. 9-13. In this case, we still use cos θ. Only now we use it to determine the part of the total voltage that is resistive. Then we will know the resistive voltage and current, and from that we can find the power.

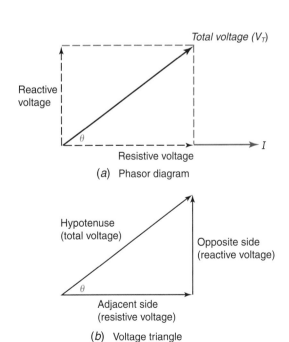

(a) Phasor diagram

(b) Voltage triangle

Fig. 9-13 Determining the resistive voltage. The total voltage phasor is represented by its two equivalent phasors.

Example 9-6

In an electric circuit the voltage leads the current by 73°. The voltage is 80 V, and the current is 4 A. What is the power used?

Given: $\theta = 73°$
$V = 80$ V
$I = 4$ A

Find: P

Known: $P = IV \cos\theta$

Solution: $\cos 73° = 0.292$
$P = 4\text{ A} \times 80\text{ V} \times 0.292$
$= 93.4\text{ W}$

Answer: The power used by the circuit is 93.4 W.

In the above example the voltage was leading the current by 73°. However, the power would be the same if the voltage were lagging the current.

■ TEST

Answer the following questions.

17. Draw a right triangle which represents the force vectors shown in Fig. 9-8.
18. Draw a right triangle that has a 30° angle θ. Label all three sides of the triangle and mark the degrees in each angle.
19. The symbol that represents the angle between two phasors is _____ .
20. The resistive current is represented by the _____ side of a right triangle.
21. The _____ side of a right triangle represents the reactive voltage.
22. The sin θ is equal to the _____ divided by the hypotenuse.
23. The resistive current divided by the total current yields the _____ of angle theta (θ).
24. How much resistive voltage is there when the total voltage is 80 V at 30°?
25. How much resistive current is there when 5 A of total current leads the source voltage by 70°?
26. Determine the power of a 120-V circuit in which the 7 A of current leads the voltage by 300°.
27. Determine the power used by a 440-V motor that draws 8 A and causes the current to lag the voltage by 35°.

9-3 True Power and Apparent Power

True power

So far we have been calculating the *true power* of ac circuits. That is, we have determined the *actual* power used by the circuit. We have done this by taking into account the phase shift between the total current and the total voltage. We can also determine true power in a circuit by measuring it with a *wattmeter*. A wattmeter is constructed so that it takes into account any phase difference between current and voltage.

$$\text{Cos } \theta = \frac{P}{P_{app}}$$

Wattmeter

Apparent power

Sometimes it is as important to know the apparent power in a circuit as it is to know the true power. The *apparent power* is the power that appears to be present when the voltage and the current in a circuit are measured separately. The apparent power, then, is the product of the voltage and the current regardless of the phase angle θ. In a circuit containing both resistance and reactance, the apparent power is always greater than the true power.

Apparent power is calculated by the formula

$$P_{app} = IV$$

Voltamperes

It has a base unit of *voltampere* (VA) to indicate that it is not true power, which has units of watts.

 ... that one watt of true power means that electric energy is converted into heat energy at a rate of 1 J/s.

One voltampere of apparent power converts *less than* 1 J/s. How much less depends on how far the voltage and current are out of phase.

Example 9-7

Determine the apparent power for the circuit specified in example 9-6.

Given: $V = 80$ V
             $I = 4$ A
Find: P_{app}
Known: $P_{app} = IV$
Solution: $P_{app} = 4\text{ A} \times 80\text{ V} = 320$ VA
Answer: The apparent power of the circuit is 320 VA.

Look closely at the formulas for apparent power and true power:

$$P = IV \cos \theta$$
$$P_{app} = IV$$

Notice that the only difference between the two is that the true power includes a cos θ term. Combining these two formulas yields

$$\cos \theta = \frac{P}{P_{app}}$$

This relationship makes it relatively easy to determine cos θ and thus the phase relationship between current and voltage. All we need to know is the current, the voltage, and the true power. All three of these are easily measured, as shown in Fig. 9-14. With the measured current and voltage, we can calculate the apparent power. Then, with the measured power and the calculated apparent power, we can calculate cos θ. After calculating the value of cos θ, we can use a calculator, or the table of trigonometric functions, to find θ.

Example 9-8

Refer to Fig. 9-14. Suppose the wattmeter reads 813 W, the voltmeter reads 220 V, and the ammeter reads 6 A. What is the phase angle (θ) between the current and the voltage?

Given: $P = 813$ W
             $V = 220$ V
             $I = 6$ A
Find: θ
Known: $P_{app} = IV$, $\cos \theta = \dfrac{P}{P_{app}}$
Solution: $P_{app} = 6\text{ A} \times 220\text{ V} = 1320$ VA
             $\cos \theta = \dfrac{813\text{ W}}{1320\text{ VA}} = 0.616$
             Using Appendix F, we find 0.616 in the cosine column. The angle whose cosine is 0.616 is 52°.
Answer: The current and voltage are 52° out of phase ($\theta = 52°$).

If the reactive part of the load in the above example were inductance, the current would lag the voltage by 52°. If it were capacitance,

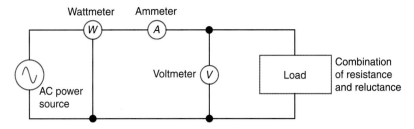

Fig. 9-14 Measuring power and apparent power. The apparent power is found by multiplying the current by the voltage. The true power is indicated by the wattmeter.

the current would lead the voltage by 52°. Phasor diagrams for both cases are drawn in Fig. 9-15.

■ TEST

Answer the following questions.

28. _____ power does not account for any phase difference between current and voltage.
29. Voltampere is the base unit of _____.
30. _____ power can never be greater than _____ power.
31. The measured power, current, and voltage in a circuit are 470 W, 3.6 A, and 240 V respectively. Determine θ, $\cos \theta$, and P_{app}.

32. Determine P_{app}, PF, and I_T for a circuit in which I_T leads V_T by 40°, $P = 780$ W, and $V_T = 240$ V.
33. True or False. A wattmeter measures apparent power.
34. True or False. A wattmeter compensates for phase shift when measuring power.
35. True or False. One voltampere converts one joule per second.

9-4   Power Factor

The ratio of the true power to the apparent power in a circuit is known as the *power factor* (PF). Since this ratio also yields cos θ, power factor is just another way of specifying cos θ. Mathematically we can write

Power factor (PF)

$$\text{PF} = \cos \theta = \frac{P}{P_{app}}$$

Power factor, therefore, is a way of indicating that portion of the total current and voltage that is producing power. When current and voltage are in phase, the power factor is 1; the true power is the same as the apparent power. When current and voltage are 90° out of phase, the power factor is 0. Thus, the power factor of a circuit can be any value between 0 and 1. Sometimes power factor is expressed as a percentage. Then the power factor can vary from 0 to 100 percent.

Power factor is very important to the utility company that sells electric energy. Its importance can be best shown by an example. Suppose a utility company is selling energy to two different manufacturers. The two manufacturers are located the same distance from the utility's distribution center (Fig.

JOB TIP

Periodic instrument calibration is required in many companies. Be aware of items such as calibration labels.

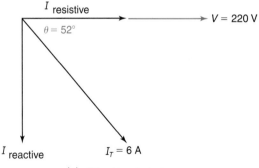

(a) If load has inductance

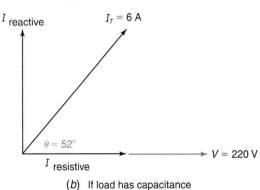

(b) If load has capacitance

Fig. 9-15 Phasor diagrams for example 9-7.

Example 9-9

An electric motor draws 18 A of current from a 240-V source. A wattmeter connected to the circuit indicates 3024 W. What is the power factor of the circuit?

Given: $I = 18$ A
$V = 240$ V
$P = 3024$ W

Find: Power factor

Known: $PF = \dfrac{P}{P_{app}}, \ P_{app} = IV$

Solution: $P_{app} = 18 \text{ A} \times 240 \text{ V}$
$= 4320$ VA
$PF = \dfrac{3024 \text{ W}}{4320 \text{ VA}} = 0.7$

Answer: The power factor is 0.7, or 70 percent.

JOB TIP

Interviewers will file a negative report about applicants whom they believe might ignore safety regulations and company policies.

9-16). Both manufacturers receive power at the same voltage (4700 V), and both require the same true power (1.5 MW). However, manufacturer A uses a lot of reactive loads (motors) and operates with a power factor of 60 percent. Manufacturer B uses mostly resistive loads (heaters) and operates with a power factor of 96 percent. In order to serve manufacturer A,

the utility company must provide the following number of voltamperes:

$$P_{app} = \frac{P}{PF} = \frac{1,500,000 \text{ W}}{0.6}$$
$$= 2,500,000 \text{ VA} = 2.5 \text{ MVA}$$

In order to serve this load, the utility company feeders (conductors) must be able to carry the following current:

$$I = \frac{P_{app}}{V} = \frac{2,500,000 \text{ VA}}{4700 \text{ V}}$$
$$= 532 \text{ A}$$

Manufacturer B uses the same true power as manufacturer A and thus pays the same amount to the utility company. Manufacturer B, however, requires the following apparent power:

$$P_{app} = \frac{P}{PF} = \frac{1,500,000}{0.96}$$
$$= 1,562,500 \text{ VA}$$

or about 1.56 MVA. This is almost 1 MVA less than required by manufacturer A. In addition, the current drawn by manufacturer B would be

$$I = \frac{P_{app}}{V} = \frac{1,560,000 \text{ VA}}{4700 \text{ V}}$$
$$= 332 \text{ A}$$

Again, manufacturer B draws less current than manufacturer A (332 A versus 532 A) to obtain *exactly the same true power.*

The utility company (as well as the customer) designs its transmission and distribution systems according to the *apparent power* and current it must deliver. Since the customer is billed for the *true power* used, the utility company encourages the use of high-power-factor systems. Power factor can be improved (or corrected) by inserting a reactance opposite the one causing the low power factor. Thus, a lagging power factor can be improved by inserting a leading-power-factor device, such as a capacitor, in the system.

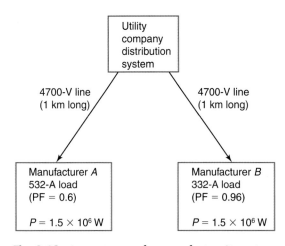

Fig. 9-16 Importance of power factor. It costs more to deliver a unit of power when the power factor is low.

■ TEST

Answer the following questions.

36. PF is numerically equal to _____ .
37. Power divided by apparent power equals
_____ .
38. PF can be no larger than _____ .
39. What is the power factor of a circuit in

which the current and the voltage are 70° out of phase?

40. The apparent power in a circuit is 360 VA. The power is 295 W. What is the power factor expressed as a percentage?

41. Two manufacturing plants use equal amounts of true power for the same length of time. Plant X operates with a power factor of 0.75, and plant Y operates with a power factor of 0.89. If both plants operate at the same voltage, which plant draws more current?

42. Which plant in question 41 is utilizing the distribution equipment more efficiently?

43. How can the PF be improved in an industrial plant that primarily uses motors?

Summary

1. In a pure resistance circuit, current and voltage are in phase.
2. When I and V are in phase, $P = IV$.
3. Reactance causes current and voltage to be 90° out of phase.
4. Circuits with inductive reactance cause current to lag voltage.
5. Circuits with capacitive reactance cause current to lead voltage.
6. A phasor represents the magnitude and direction of a quantity.
7. A single phasor can be broken into two smaller phasors which are at right angles to each other.
8. Only the resistive part of a current or a voltage can use power (convert electric energy).
9. The cosine of angle θ is the ratio of the adjacent side to the hypotenuse.
10. The cosine of θ can be the ratio of the resistive current to the total current.
11. The cosine of θ can be the ratio of the resistive voltage to the total voltage.
12. In any electric circuit, power can be calculated with the formula $P = IV \cos \theta$.
13. Apparent power does not take into account the phase relationship between current and voltage.
14. The unit for apparent power is the voltampere (VA).
15. The ratio of power to apparent power is equal to $\cos \theta$. It is also equal to the power factor.
16. Power factor varies from 0 to 1 or from 0 to 100 percent.
17. Energy used at a low power factor is more costly than the same energy used at a high power factor.

Chapter Review Questions

For questions 9-1 to 9-9, determine whether each statement is true or false.

9-1. Resistance causes electric energy to be converted into heat energy.
9-2. In a circuit containing only resistance, the current and the voltage are 90° out of phase.
9-3. All voltage phasors in a phasor diagram must be drawn to the same scale.
9-4. The sine of an angle is equal to the adjacent side divided by the hypotenuse.
9-5. A current that leads its voltage source by 30° uses more power than a current that lags its voltage source by 30°.
9-6. The apparent power and the true power are equal in a circuit containing both resistance and inductance.
9-7. The true power in a circuit is often greater than the apparent power.
9-8. Power factor and $\cos \theta$ have the same numerical value.
9-9. Industry pays more for a unit of energy used at a high power factor than one used at a low power factor.

For questions 9-10 to 9-22, supply the missing word or phrase in each statement.

9-10. Phasors are like vectors except they are _____ .
9-11. In a phasor diagram, the direction of a phasor represents _____ .
9-12. Most motors cause the current to _____ the voltage.
9-13. Capacitance causes current to _____ voltage by _____ electrical degrees.
9-14. _____ is a form of opposition to current which uses no power.
9-15. The base unit of apparent power is the _____ .
9-16. Any phasor can be represented by two other phasors that are separated by _____ degrees.
9-17. θ is the symbol for _____ .
9-18. Power factor can be determined by dividing the _____ power by the _____ power.
9-19. In some circuits, the total current can be divided into _____ current and _____ current.

9-20. A wattmeter measures _____ power.

9-21. Unless theta = 0°, _____ power is _____ than true power.

9-22. A current that leads the voltage by 340° also lags the voltage by _____ degrees.

Answer the following questions.

9-23. A pure capacitance load draws 6 A from an 80-V ac source. Determine the power, apparent power, and θ.

9-24. Determine the reactive current when the total current is 7 A at 40° and the total voltage is 240 V at 0°.

9-25. Determine the power, apparent power, and power factor for question 9-24.

9-26. Determine θ when the measured current, voltage, and power in a circuit are 5 A, 120 V, and 400 W, respectively.

9-27. Does a load with a PF of 0.6 or a PF of 0.9 make the most efficient use of the power distribution equipment? Why?

Critical Thinking Questions

9-1. Could the force of attraction (or repulsion) between two electromagnets be used to measure power in an ac circuit? How?

9-2. To optimize the power factor, should electric motors be loaded to rated value or lightly loaded? Why?

9-3. Draw a phasor diagram for the line voltages of a wye-connected three-phase generator that has a phase voltage of 100 V.

9-4. A 500-Ω resistor (R_1) in series with a 1000-Ω resistor (R_2) is connected to a 30-V ac source. Draw the phasor diagram for the circuit.

9-5. What is the efficiency of a motor that provides 5/8 hp of shaft power when it draws 4 A at 240 V and has a power factor of 0.72?

9-6. Does it matter whether we say "the current leads the voltage by 290 degrees" or "the current lags the voltage by 70 degrees"? Why?

9-7. When a resistance load and a reactance load are in series, which phasor (voltage or current) is used as the reference? Why?

9-8. When a resistance load and a reactance load are in parallel, which phasor (voltage or current) is used as the reference? Why?

9-9. When calculating power, does it matter whether the current or the voltage is the reference? Why?

Answers to Tests

1. reactance
2. resistance
3. reactance, resistance
4. lag
5. inductive
6. Reactance
7. resistive
8. 0.25 W
9. 2.5 W
10. T
11. T
12. F
13.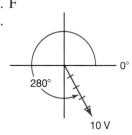

14.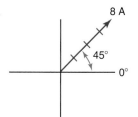

15.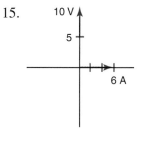

16.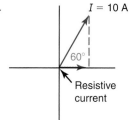

17.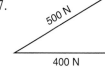

18.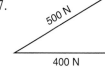

19. (θ) (theta)
20. adjacent
21. opposite
22. opposite
23. $\cos \theta$
24. 69 V
25. 1.7 A
26. 420 W
27. 2883 W
28. apparent

29. apparent power
30. true, apparent
31. $\theta = 57°$
    $\cos \theta = 0.544$
    $P_{app} = 864$ VA
32. $P_{app} = 1018.2$ VA
    PF = 76.6 percent
    $I_T = 4.24$ A
33. F
34. T

35. F
36. $\cos \theta$
37. $\cos \theta$ of PF
38. 1 or 100 percent
39. 0.342 or 34.2 percent
40. 81.9 percent
41. plant X
42. plant Y
43. Improve the PF by inserting a capacitor in the system.

Chapter 10

Capacitance

Chapter Objectives

This chapter will help you to:

1. *Explain* the construction of capacitors and the purpose of each part.
2. *Understand* how capacitors behave in ac and dc circuits.
3. *Understand* why a capacitor causes current to lead voltage by 90°.
4. *Determine* the values of reactance, voltage, and current in capacitive circuits.

5. *Write* the specifications needed to order capacitors.
6. *Test* capacitors for opens and shorts.
7. *Determine* the time required to charge a capacitor in a resistor-capacitor circuit.

> C *apacitors,* devices which possess capacitance, are almost as common as resistors in electric and electronic circuits. They are used in electric motors, automobile ignition systems, electronic photoflashes, and fluorescent lamp starters, to name just a few applications.

10-1 Terminology

The ability to store energy in the form of electric charge is called *capacitance.* The symbol used to represent capacitance is *C.* A device designed to possess capacitance is called a *capacitor.* In its simplest form a capacitor is nothing more than two conductors separated by an insulator. Figure 10-1 illustrates a capacitor consisting of two pieces of aluminum (a conductor) separated by a sheet of polyester plastic (an insulator). In capacitors the conductors are called *plates* and the insulator is called a *dielectric.*

10-2 Basic Capacitor Action

A capacitor stores energy when an electric charge is forced onto its plates by some other energy source, such as a battery. When a capacitor is connected to a battery, a current flows until the capacitor becomes charged (Fig. 10-2). Electrons from the negative terminal of the battery move through the connecting lead and pile up on one of the plates. At the same instant, electrons from the other plate move through the connecting lead to the positive terminal of the battery. The net result is that one plate of the capacitor ends up with an excess of electrons (negative charge). The other plate ends up with a deficiency of electrons (positive charge). These charges on the plates of the capacitor represent a voltage source just as the charges on the plates of a cell do. Note that in the process of charging the capacitor, no electrons move from one plate through the dielectric to the other plate.

Because of opposing voltages, the current stops once the capacitor is charged. The opposite charges on the plates of the capacitor create a new energy source. The energy stored in the capacitor produces a voltage equal to that of the battery (the original energy source). Since the capacitor's voltage is equal to, and in opposition to, the battery voltage, a state of equilibrium exists. No current can flow in either direction.

Capacitance (C)

Plates

Dielectric

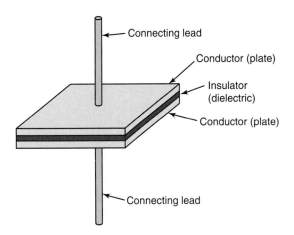

Fig. 10-1 Simple capacitor. The plates of a capacitor are always insulated from each other.

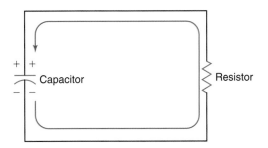

Fig. 10-3 A capacitor is an energy source. A capacitor forces current through a resistor until the capacitor is discharged.

Charged capacitor

A *charged capacitor* can be disconnected from the original energy source (the battery) and used as a new energy source. If a voltmeter is connected to the capacitor, it will register a voltage. If a resistor is connected to the capacitor (Fig. 10-3), current will flow through the resistor. However, a capacitor has limited use as a primary energy source for two reasons:

1. For its weight and size, the amount of energy it can store is small compared with what a battery can store.
2. The voltage available from the capacitor rapidly diminishes as energy is removed from the capacitor.

Although the amount of energy stored in a capacitor is small, a capacitor can deliver a shock. The shock can be very severe (even fatal) if the capacitor is large and charged to a high voltage. Treat charged capacitors as you would any other electric energy source.

When a capacitor is charged, an electric field is established between the two charged plates.

The lines of Fig. 10-4 represent an electric field in the same way that lines between magnetic poles represent a magnetic field. The electric field exerts its force on any charge within the field. A negatively charged particle, such as an electron, is forced toward the positive plate of the capacitor. The dielectric material between the plates of a capacitor is under stress because of the force of the electric field. One way to visualize this stress is illustrated in Fig. 10-5. In Fig. 10-5 the electric field modifies the orbital path of an electron by forcing it toward the positive plate and away from the negative plate. The modified orbital path of the electron raises the energy level of the electron. This is the way the capacitor stores energy.

10-3 Voltage Rating

A dielectric material, such as that illustrated in Fig. 10-5, can withstand only a certain amount of stress before electrons in the material break free of the parent atoms. These free electrons are then available as current carriers. Then the

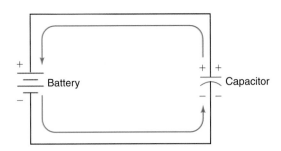

Fig. 10-2 Capacitor being charged. The current stops when the capacitor is fully charged.

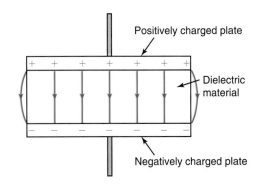

Fig. 10-4 Electric field of a charged capacitor. The field stresses the dielectric material.

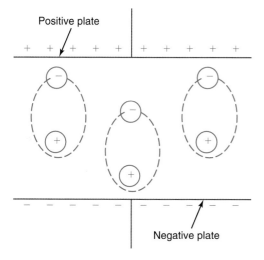

Positive plate

Negative plate

Fig. 10-5 Stress on the dielectric material. Energy is stored in the stressed dielectric material.

capacitor's negative plate can discharge through the dielectric to the positive plate. This burns a hole in the dielectric material and ruins the capacitor.

The strength of the electric field in a capacitor is related to the distance between the plates and to the voltage across the plates. Therefore, a specific capacitor can withstand being charged only to a given voltage before its dielectric breaks down. The maximum voltage to which a capacitor should be charged is determined by the manufacturer. It is listed as part of the specifications of the capacitor. The manufacturer's voltage rating is usually listed as the *direct current working voltage* (DCWV). When a capacitor is rated at 100 DCWV, it can be operated at 100 V for long periods of time with no deterioration of the dielectric. If the capacitor is subjected to 125 or 150 V dc, the dielectric may not immediately break down. However, its life expectancy will be greatly shortened.

◼ TEST

Answer the following questions.

1. A _____ stores energy in the form of electric charges.
2. The symbol, or abbreviation, for capacitance is _____.
3. A capacitor has two plates and a _____.
4. The _____ in a charged capacitor is stressed by the electric field.

5. True or false. A capacitor rapidly discharges once it is disconnected from the battery that charged it.
6. True or false. Electrons travel from one plate through the dielectric to the other plate when a capacitor is charged.
7. Where is the energy in a capacitor stored?
8. What abbreviation is used to specify the voltage rating of a capacitor?
9. Why do capacitors have voltage ratings?
10. What two factors determine the electric field strength in a capacitor?

10-4 Unit of Capacitance

The base unit of capacitance is the *farad*. The abbreviation for farad is F. One farad is that amount of capacitance which stores 1 C of charge when the capacitor is charged to 1 V. Thus, a farad is a coulomb per volt (C/V).

Farad (F)

YOU MAY **RECALL** . . . from Sec. 2-20 that one letter is often used to denote two different quantities. In that section it was pointed out that the letter "W" is used as a symbol for energy and as the abbreviation for the base unit for power, the watt.

Now, notice that the letter "C" is used as the symbol for capacitance and as the abbreviation for the base unit of charge, the coulomb.

Farad is the unit of capacitance (C), coulomb is the unit of charge (Q), and volt is the unit of voltage (V). Therefore, capacitance can be mathematically expressed as

Working voltage

$$C = \frac{Q}{V}$$

Example 10-1

What is the capacitance of a capacitor that requires 0.5 C to charge it to 25 V?

Given:	Charge (Q) = 0.5 C
	Voltage (V) = 25 V
Find:	Capacitance (C)
Known:	$C = \dfrac{Q}{V}$
Solution:	$C = \dfrac{0.5 \text{ C}}{25 \text{ V}} = 0.02 \text{ F}$
Answer:	The capacitance is 0.02 F.

$C = \dfrac{Q}{V}$

Michael Faraday The unit of measure for capacitance, the farad, was named for Michael Faraday, an English chemist and physicist who discovered the principle of induction.

Temperature coefficient

JOB TIP

Companies whose business is not directly related to electronics might hire electronics technicians.

Plate area

This relationship is very useful in understanding voltage distribution in series capacitor circuits.

The 0.02 F just calculated is a very high value of capacitance. In most circuits the capacitances used are much lower. In fact, they are usually so much lower that the base unit of farad is too high to conveniently express their values. The microfarad (μF), which is 1/1,000,000 or 1×10^{-6} farad, and the picofarad (pF), which is 1/1,000,000,000,000 or 1×10^{-12} farad, are more convenient units. Both are used extensively in specifying values of capacitors in electronic circuits. Often it is necessary to convert from one unit to another. For example, a manufacturer may specify a 1000-pF capacitor in a parts list. The technician repairing the equipment may not have a capacitor marked 1000 pF. However, an equivalent capacitor marked in microfarads may be available. The technician must convert picofarads into microfarads by multiplying picofarads by 10^{-6}. The net result is

$$1000 \text{ pF} = 1000 \times 10^{-6} \text{ } \mu\text{F} = 0.001 \text{ } \mu\text{F}$$

To convert from microfarads to picofarads, the reverse process is used. Divide the microfarads value by 10^{-6} to obtain the equivalent picofarad value. Since dividing by 10^{-6} is the same as multiplying by 10^6, the conversion proceeds as follows:

$$0.001 \text{ } \mu\text{F} = 0.001 \times 10^6 \text{ pF}$$
$$= 1000 \text{ pF}$$

which agrees with the results above.

10-5 Determining Capacitance

The capacitance of a capacitor is determined by four factors:

1. Area of the plates
2. Distance between the plates
3. Type of dielectric material
4. Temperature

The temperature of the capacitor is the least significant of four factors. It need not be considered for many general applications of capacitors. However, in more critical applications (such as oscillator circuits), the temperature characteristic of a capacitor is very important. The temperature characteristic of a capacitor is determined primarily by the type of dielectric material used in constructing the capacitor. Some dielectric materials cause an increase in capacitance as the temperature increases. This is referred to as a positive *temperature coefficient*. Other dielectric materials have negative temperature coefficients; that is, the capacitance increases as the temperature decreases. Still other dielectric materials have zero temperature coefficients. Their capacitance is independent of temperature. The temperature coefficient of a capacitor is specified by the manufacturer in *parts per million per degree Celsius*. Negative temperature coefficient is abbreviated N, positive temperature coefficient is P, and zero temperature coefficient is NPO. The reference temperature at which the capacitor is rated is 25°C. (Note that the letter "C" is used here as the abbreviation for Celsius. With only 26 letters in the alphabet and thousands of abbreviations and symbols, it is no wonder that some letters are used more than once.)

The capacitance of a capacitor is directly proportional to the area of its plates (or the area of its dielectric). All other factors remaining the same, doubling the plate area doubles the capacitance [Fig. 10-6(*a*)]. This is so because doubling the *plate area* also doubles the area of dielectric material. For a given voltage across the capacitor, the strength of the electric field is independent of the plate area. Thus, when the dielectric area is increased, the amount of energy stored in the dielectric is increased and the capacitance is also increased. (Remember, capacitance is defined as *the ability to store energy*.)

Other factors being equal, the amount of ca-

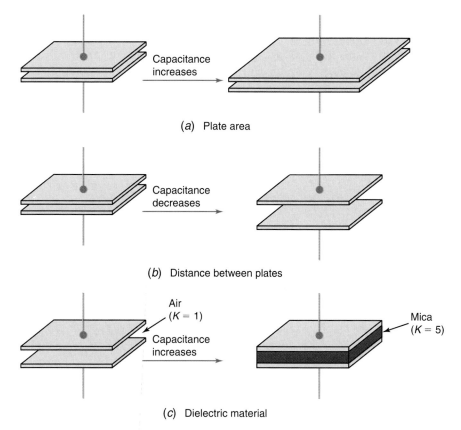

Fig. 10-6 Major factors affecting capacitance.

pacitance is inversely proportional to the distance between the plates [Fig. 10-6(b)]. As the *distance between the plates* is doubled, the amount of dielectric material also doubles. However, doubling the distance causes the strength of the electric field to decrease by a factor of 4. The net effect is that the amount of energy stored in the capacitor, for a given voltage applied to the capacitor, decreases. Thus, the capacitance decreases.

The capacitance of a capacitor is also controlled by the type of dielectric material used [Fig. 10-6(c)]. When subjected to the same electric field, some materials undergo greater molecular distortion than others. In general, those materials which undergo the greatest distortion store the most energy. The ability of a dielectric material to distort and store energy is indicated by its *dielectric constant K*. The dielectric constant of a material is a pure number (that is, it has no units). It compares the material's ability to distort and store energy when in an electric field with the ability of air to do the same. Since air is used as the reference, it has been given a *K* equal to 1. Mica, often used as a dielectric, has a dielectric constant ap-

proximately five times that of air. Therefore, for mica, *K* = 5 (approximately). Suppose all other factors (plate area, distance between plates, and temperature) are the same. Then a capacitor with a mica dielectric will have five times as much capacitance as one using air as its dielectric. Dielectric constants for materials commonly used for dielectrics range from 1 for air to more than 4000 for some types of ceramics.

Distance between the plates

◣ TEST

Answer the following questions.

11. The base unit for capacitance is the _____.

12. The abbreviation for the base unit of capacitance is _____.

13. A 4700-pF capacitor is equal to a _____-μF capacitor.

14. The pF rating of a 0.003-μF capacitor would be _____.

15. A farad is equal to a _____ divided by a _____.

16. List four factors that determine the capacitance of a capacitor.

Dielectric constant (*K*)

17. Which of the four factors in question 16 has the least effect on capacitance?
18. What happens to the capacitance of a capacitor when the plate area is doubled?
19. Suppose the distance between the plates of a capacitor is reduced by half. What happens to the capacitance?
20. What does the dielectric constant of a material indicate?
21. Does air have a smaller or larger dielectric constant than other common dielectric materials used in making capacitors?
22. How much charge is on a 6-μF capacitor that is charged 150 V?

(a) Disk ceramic

(b) Tuning (variable)

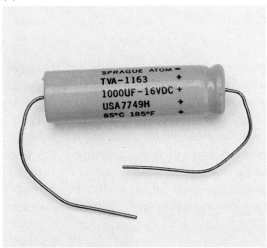

(c) Electrolytic (axial lead)

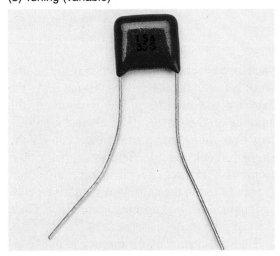

(d) Dipped mica

(e) Electrolytic (radial lead)

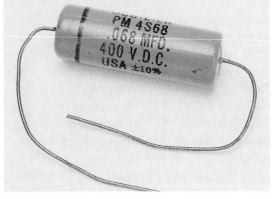

(f) Molded (film or paper)

Fig. 10-7 Types of capacitors. *(Parts a to f, Mark Steinmetz)*

23. How much capacitance is needed to store 0.5 C when the capacitor is charged to 30 V?

24. How much voltage is required to store 0.8 C of charge on a 2000-μF capacitor?

10-6 Types of Capacitors

Many different types and styles of capacitors (Fig. 10-7) are manufactured to satisfy the needs of the electronics industry. Capacitors may be named to indicate their dielectric material, their enclosure, the process used in their construction, or their intended use.

Electrolytic Capacitors

Electrolytic capacitors [Fig. 10-7(c) and (e)] provide more capacitance for their size and weight than any of the other types of capacitors. This is their primary advantage over other capacitors.

A common electrolytic capacitor consists of two aluminum plates separated by a layer of fine gauze or other absorbent material. The plates and separators are long, narrow strips. These strips are rolled up and inserted into an aluminum container (Fig. 10-8). One plate (the negative plate) is usually electrically connected to the aluminum can (container). Electric terminals from the plates are brought out one end

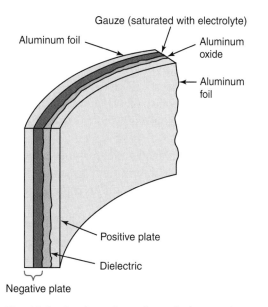

Fig. 10-9 Section of an electrolytic capacitor.

Electrolytic capacitor

of the container. Then the container is sealed.

One of the aluminum plates in Fig. 10-8 is oxidized. Since aluminum oxide is an insulator, it is used for the dielectric. The separators are saturated with a chemical solution (such as borax) which is called the *electrolyte*. The electrolyte, having relatively good conductivity, serves as part of one of the plates of the capacitor (Fig. 10-9). It is in direct contact with both the dielectric and the pure aluminum plate. The electrolyte is also necessary in forming and maintaining the oxide on one of the plates.

Electrolyte

The method used to produce electrolytic capacitors results in the plates being *polarized*. When using these capacitors, always keep the same voltage polarity as was used in manufacturing them. This polarity is always marked on the body of the capacitor. Again, the aluminum container (which may be encased in an insulating tube) is usually connected to the negative plate. It is never connected to the positive plate. Reverse polarity on an electrolytic capacitor causes excessively high current in the capacitor. It causes the capacitor to heat up and possibly to explode. Thus, the common electrolytic capacitor is limited to use in dc circuits.

Polarized

Special electrolytic capacitors are manufactured for use in ac circuits. These capacitors are usually listed in parts catalogs as *nonpolarized* or *ac* electrolytic capacitors. An ac electrolytic capacitor is really two electrolytic capacitors packaged in a single container (Fig.

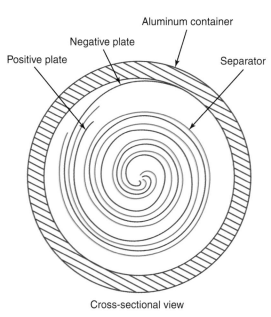

Fig. 10-8 Electrolytic capacitor. The container is electrically connected to the negative plate.

10-10). The two internal capacitors are in series, with their positive ends connected together. Regardless of the polarity on the leads of the ac electrolytic capacitor, one of the two internal capacitors will be correctly polarized. The correctly polarized capacitor will limit the current flowing through the reverse-polarized capacitor.

Many nonpolarized electrolytic capacitors are designed for intermittent use only. These capacitors overheat (and often rupture) if used for extended periods of time.

The oxide on the positive plate of an electrolytic capacitor (Fig. 10-9) is very thin (approximately 0.25 μm thick) and rough. Its thinness keeps the plates very close together. Its roughness effectively increases the plate area. These two factors make the capacitance of an electrolytic capacitor very high for its weight and volume.

Tantalum, a conductive metallic element, is also used for the plates of electrolytic capacitors. Tantalum electrolytic capacitors are smaller, more stable, and more reliable than aluminum electrolytic capacitors. An oxide dielectric is far from a perfect insulator; therefore, some electrons do find their way from one plate to the other. This produces *leakage current.* Tantalum electrolytic capacitors have less leakage current (higher dielectric resistance) than aluminum electrolytic capacitors do. However, tantalum electrolytic capacitors are more expensive than their aluminum counterparts.

Film and Paper Capacitors

Paper capacitors and *film capacitors* are constructed using a rolled-foil technique, illus-

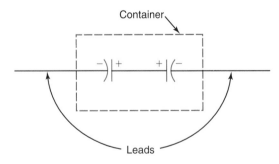

Fig. 10-10   Nonpolarized electrolytic capacitor. Regardless of lead polarity, one of the internal capacitors will be correctly polarized.

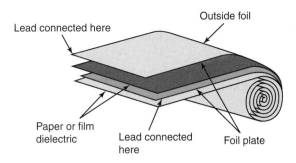

Fig. 10-11   Paper or film capacitor construction.

trated in Fig. 10-11. Notice in this figure that the two foil plates are offset so that one plate extends out each end. The leads for the capacitor are then connected to the ends of the completed roll. After the leads are attached, the complete assembly (plates, lead connections, and dielectric) is covered with a protective coating of insulation. Sometimes the insulation is molded around the assembly to produce a *molded capacitor.* Sometimes the capacitor is dipped in a plastic insulating material to produce a *dipped capacitor.* Sometimes it is placed in an insulated tube and the ends of the tube are then sealed. This type is sometimes referred to as a *tubular capacitor.*

Paper as a dielectric material is rapidly being replaced by plastic film. Plastic film yields a capacitor that is smaller, more stable, and has less leakage current. Such materials as polystyrene, polyester (Mylar), and polycarbonate are popular replacements for paper.

Many of the plastic-film capacitors use metallized plates. With these *metallized-film capacitors,* the metal plate is deposited directly onto the film. This keeps the distance between the plates as small as possible and produces a small, compact capacitor. Some metallized capacitors are self-healing. When a surge voltage arcs through the dielectric, it vaporizes the thin metallized plate around the charred area of the dielectric material. Therefore, the charred dielectric cannot short the plates together. Thus, the temporary short has been "healed."

Film and paper capacitors used in industrial applications can be as large as several hundred microfarads. They are often rated in terms of apparent power (VA) as well as DCWV. These high-value capacitors are usually enclosed in a metal container filled with a special insulation oil. However, in most electronic circuits, paper and film capacitors have a value of less than 1 μF.

Tantalum

Molded capacitor

Dipped capacitor

Tubular capacitor

Leakage current

Metallized-film capacitor

Paper capacitor

Film capacitor

Ceramic Capacitors

For low-value capacitors (less than 0.1 μF), ceramic is a popular dielectric material. The most common style of ceramic capacitor is the disc ceramic [Fig. 10-7(a)].

The structure of a typical ceramic capacitor is detailed in Fig. 10-12. As indicated in this figure, these capacitors have a simple, strong structure. This makes them a tough, reliable, general-purpose capacitor. They are used in a wide variety of applications where low values of capacitance are needed.

Mica Capacitors

Mica capacitors are limited to even lower values than ceramic capacitors. This is because mica has a lower dielectric constant than ceramic. However, it is easier to control production tolerances with mica dielectrics, and mica has good high-temperature characteristics.

It takes less space to provide a given value of capacitance with ceramic than with mica. Also, it is easier to construct a ceramic capacitor. Therefore, mica capacitors are not as common as ceramic capacitors.

Specific-Use Capacitors

So far we have classified capacitors by the type of dielectric materials used in their manufacture. Capacitors can also be classified by their function or by the way they are connected in a circuit. Some of the more common names in this classification are feed-through, filter, padder, trimmer, tuning, and stand-off.

Fig. 10-13 Trimmer capacitor.

Padders, trimmers, and tuning capacitors are all variable capacitors. They are used in circuits which tune radio and television sets to a particular station. An example of a miniature *trimmer* capacitor is pictured in Fig. 10-13.

One of the plates of a *feed-through capacitor* is the tubular metal case which encloses the capacitor (Fig. 10-14). The two leads extending from the ends of the capacitor are both connected to the other plate. In use, the capacitor protrudes through a hole in the electrical (metal) shield which separates two parts of an electronic system. The case (one plate) is electrically connected to the shield through which it protrudes. The case may be soldered as in Fig. 10-14, or it may be threaded and held in place with a nut. This arrangement allows dc current to feed through the grounded shield, while radio frequencies are routed to ground by the capacitor. The radio frequencies are bypassed from either lead.

Stand-off capacitors (Fig. 10-15) are also used to bypass high frequencies to the ground of an electronic system. These capacitors have

Ceramic capacitor

Mica capacitor

Trimmers

Feed-through capacitor

Stand-off capacitor

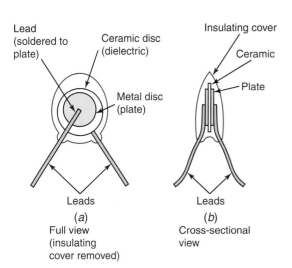

Fig. 10-12 Ceramic capacitor construction.

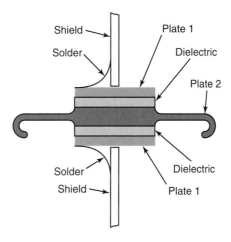

Fig. 10-14 Feed-through capacitor. The two leads connect to the same plate.

Potentially Fatal Capacity The energy stored in a charged, high-voltage capacitor can deliver a fatal shock.

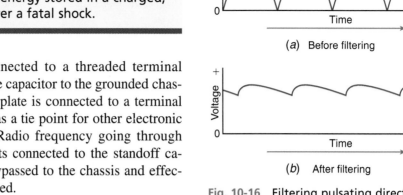

(*a*)   Before filtering

(*b*)   After filtering

Fig. 10-16   Filtering pulsating direct current.

one plate connected to a threaded terminal which bolts the capacitor to the grounded chassis. The other plate is connected to a terminal which serves as a tie point for other electronic components. Radio frequency going through the components connected to the standoff capacitors are bypassed to the chassis and effectively eliminated.

Filter capacitor

Filter capacitors are high-value capacitors used to filter (flatten) out pulsations in dc voltage. For example, a filter capacitor can change the pulsating direct current in Fig. 10-16(*a*) to that shown in Fig. 10-16(*b*). Many filter capacitors are electrolytic capacitors which are polarized. They must be used in circuits where the polarity of the voltage applied to them will never be reversed.

Most filter capacitors have a value somewhere between 1 and 5000 μF. The voltage rating may be anywhere from a few volts to thousands of volts.

Energy-Storage Capacitors

Energy storage

$W = 0.5CV^2$

Capacitors designed to release their stored energy (discharge) during a short time interval are sometimes called *energy-storage capacitors*. They may be either electrolytic or non-electrolytic capacitors. During the rapid discharge of an energy-storage capacitor, the terminal leads and plates must carry a very high current. The dielectric material, plates, and enclosure must be able to withstand the extreme stress created by such a discharge. Therefore, energy-storage capacitors often

Plate — Dielectric

— Plate

Chassis

Nut

Fig. 10-15   Stand-off capacitor.

have a peak current rating. This rating allows one to calculate the minimum resistance through which the fully charged capacitor can be discharged. Exceeding the peak current rating one time will not immediately destroy the capacitor. However, repeated discharging at excess current levels will cause premature failure in a capacitor.

In addition to the standard voltage and capacitance ratings, most energy-storage capacitors also have an energy rating. The energy rating specifies how many joules of energy the capacitor stores when charged to its rated voltage. The amount of energy a capacitor can store depends upon its capacitance and voltage rating. We can calculate the amount of stored energy with the formula

$$W = 0.5CV^2$$

The calculated energy is in joules when capacitance is in farads and voltage is in volts.

Example 10-2

An energy-storage capacitor is rated at 300 μF and 450 V. How much energy can it store?

Given:	$C = 300 \ \mu$F
	$V = 450$ V
Find:	W
Known:	$W = 0.5CV^2$
Solution:	$W = 0.5 \times (300 \times 10^{-6} \text{ F})$
	$\times 450 \text{ V} \times 450 \text{ V}$
	$= 30.4$ J
Answer:	The capacitor stores 30.4 J of energy.

The amount of energy in the above example is not very great. A 40-W lamp uses more energy every second than this capacitor stores. However, the capacitor can provide high power if it is discharged rapidly. For instance, if the capacitor in example 10-2 is discharged in 2 ms, it will produce the following power:

$$P = \frac{W}{t} = \frac{30.4 \text{ J}}{0.002 \text{ s}} = 15,200 \text{ W}$$

Large, high-voltage capacitors like those shown in Fig. 10-17 can be charged to many thousands of volts. They can store several thousand joules of energy. These capacitors are used in such industrial applications as capacitor-discharge welding. They can also be used for less demanding applications, such as filtering.

SMD Capacitors

Surface-mount capacitors, commonly called *chip capacitors,* are about the same size and shape as the chip resistors shown in Fig. 4-20 in Chap. 4. Depending on their specific characteristics, typical chip capacitors range from 0.050 to 0.180 in. in length, 0.040 to 0.169 in. in width, and 0.035 to 0.110 in. in thickness.

Ceramic chip capacitors are commonly available with capacitance values ranging from 1 pF to 0.27 μF and with voltage ratings of 25, 50, and 100 V. However, they are available with voltage ratings in the thousands of volts.

Solid-tantalum chip capacitors are also

Fig. 10-17 Energy-storage capacitors. *(Courtesy of Aerovax Industries, Inc.)*

available. These surface-mount devices are, of course, polarized. They typically have voltage ratings in the 6- to 25-DCWV range. Capacitances up to 100 μF are common.

■ TEST

Answer the following questions.

25. True or false. For a given value of capacitance, a metallized-film capacitor is smaller than a foil-and-film capacitor.
26. True or false. The electrolyte is the dielectric material in an electrolytic capacitor.
27. True or false. The negative plate of an electrolytic capacitor is often connected to the aluminum container.
28. True or false. The plates in an electrolytic capacitor are oxidized to increase their effective plate area.
29. True or false. Single electrolytic capacitors are polarized.
30. True or false. Mica capacitors are more common than ceramic capacitors are.
31. _____ capacitors are sometimes self-healing.
32. A trimmer capacitor is classified as a _____ capacitor.
33. A _____ capacitor has both leads connected to the same plate.
34. Electrolytic capacitors use either _____ or _____ for plate material.
35. The dielectric material in an electrolytic capacitor is a metal _____ .
36. A _____ chip capacitor is polarized.
37. What happens when an electrolytic capacitor has reverse-polarity voltage connected to it?
38. List four types of capacitors named after the dielectric material used in them.
39. What are the advantages of film capacitors over paper capacitors?
40. What name is given to a capacitor that is used to smooth out pulsating dc voltage?
41. Where are energy-storage capacitors used?
42. How much energy is stored in a 150-μF capacitor that is charged to 300 V?

SMD capacitor

Chip capacitor

10-7 Schematic Symbols

Capacitors can be broadly classified as either fixed or variable and as either polarized or nonpolarized. All variable capacitors are nonpolarized. *Schematic symbols for capacitors* are shown in Fig. 10-18. The curved line representing one of the plates in the symbols of Fig. 10-18 has no particular significance for some types of capacitors. For other types it indicates a specific lead of the capacitor.

For a polarized capacitor, the curved plate of the symbol indicates the negative terminal of the capacitor. With variable capacitors, the curved line identifies the plate which is electrically connected to the frame or adjusting mechanism. The curved line identifies the outside foil (Fig. 10-11) on a rolled-foil (film or paper) capacitor. On the capacitor itself, the lead connecting to the outside foil has to be identified. This is done with a color band around the end of the capacitor from which this lead protrudes.

JOB TIP

Many organizations look for employees who will have a customer focus.

10-8 Capacitors in DC Circuits

When the switch of a dc circuit containing only capacitance is closed, a *surge of current* charges the capacitor. Once this surge is over, no more current flows (Fig. 10-19). The capacitor, once charged, appears to the battery just like an open switch. For practical purposes it has infinite opposition.

The magnitude of the surge of current in Fig. 10-19 is controlled by the resistance in the circuit. In this circuit the resistance is composed of the lead resistance and the internal resistances of the battery and the capacitor. These resistances are usually very low; therefore, the surge current can be very high. However, this

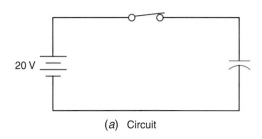

(a) Circuit

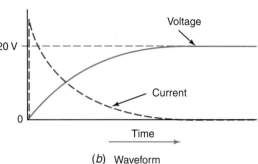

(b) Waveform

Fig. 10-19 Capacitor charging in dc circuit. The current reaches its maximum value almost the instant the switch is closed. The voltage developed across the capacitor increases more slowly.

surge current lasts only for an extremely short period of time.

Capacitor Time Constants

The time required for a capacitor to charge or discharge is determined by the amount of resistance and capacitance in the circuit. Substituting appropriate and equivalent units for resistance and capacitance, as has been done below, shows that multiplying resistance by capacitance yields time:

$$RC = \text{ohms} \times \text{farads} = \frac{\text{volts}}{\text{amperes}} \times \frac{\text{coulombs}}{\text{volts}}$$
$$= \frac{\text{coulombs}}{\text{amperes}} = \frac{\text{coulombs}}{\text{coulombs/second}}$$
$$= \text{seconds}$$

The time obtained by multiplying resistance by capacitance is called a *time constant* and is abbreviated T. Therefore, we can write

Time constant (T) = resistance (R) $\times$ capacitance (C)
$$T = RC$$

If R and C are in base units of ohms and farads, T is in base units of seconds.

When a capacitor is charging, the time constant represents the time required for the ca-

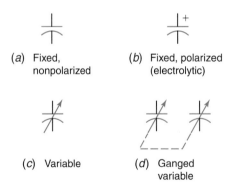

(a) Fixed,
nonpolarized

(b) Fixed, polarized
(electrolytic)

(c) Variable

(d) Ganged
variable

Fig. 10-18 Schematic symbols for capacitors.

pacitor to charge to 63.2 percent of the available voltage. When a charged capacitor is discharging, the time constant represents the time required for the capacitor to lose 63.2 percent of its available voltage. The available voltage when the capacitor is charging is the difference between the capacitor's voltage and the source voltage. When the capacitor is discharging, the available voltage is the voltage remaining across the capacitor. The curves in Fig. 10-20 demonstrate the meaning of a time constant. Figure 10-20(*a*) shows that a capacitor charges from 0 to 63.2 percent of the source voltage in one time constant. During the next time constant, it charges another 63.2 percent of the remaining available voltage. Thus, at the end of two time constants the capacitor is 86.5 percent charged [63.2 + (0.632 × 36.8) = 86.5]. The percentage of charge at the end of additional time constants is tabulated in Fig. 10-20(*a*).

The amount of time required to charge to 86.5 percent of the source voltage is strictly a function of resistance and capacitance. We can determine the exact time by calculating the value of a time constant. For example, the time constant for the resistance-capacitance circuit in Fig. 10-21 is

$$T = RC = 2 \times 10^6 \, \Omega \times 4 \times 10^{-6} \, \text{F}$$
$$= 8 \, \text{s}$$

The capacitor requires two time constants to charge to 86.5 percent of the source voltage. Therefore, it takes 16 s for the capacitor in Fig. 10-21 to charge to 86.5 percent of the battery voltage.

The exact amount of voltage on the capacitor after two time constants is dependent on the value of the source voltage. If the source voltage in Fig. 10-21 is 10 V, the voltage on the capacitor will be 8.65 V. If the source is 100 V, it will be 86.5 V.

The discharge of a capacitor is illustrated in Fig. 10-22. Notice the values of the resistance and the capacitance. They are the same as used in Fig. 10-21; therefore, the time constant is

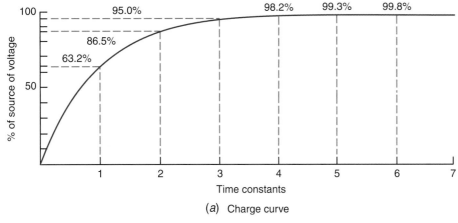

(*a*) Charge curve

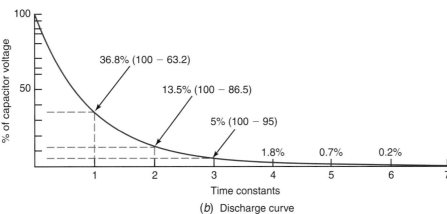

(*b*) Discharge curve

Fig. 10-20 Universal charge-discharge curves. After five or six time constants, the capacitor is considered to be fully charged or discharged.

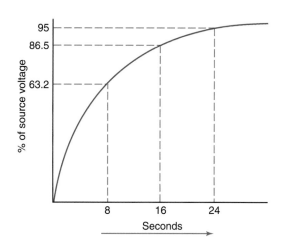

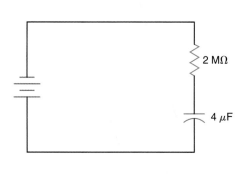

Fig. 10-21   Charging time for a capacitor. After 24 s (three time constants) the capacitor is 95 percent charged.

still 8 s. When the switch is closed in Fig. 10-22, the capacitor will be charged to 200 V. When the switch is opened, the capacitor will discharge through the resistor. After 8 s (one time constant) the capacitor will have discharged 63.2 percent of its voltage. The voltage across the capacitance will therefore decrease by 126.4 V. It will still have 73.6 V left. During the next time constant, it will discharge 63.2 percent of the remaining 73.6 V. Thus, it

will discharge 46.5 V (0.632 × 73.6 = 46.5). As shown in Fig. 10-22, there will be about 27 V left on the capacitor at the end of two time constants (16 s).

Inspect the curves in Fig. 10-20 again. Notice that a capacitor is essentially charged or discharged by the end of six time constants. For analyzing and designing circuits, many technicians consider a capacitor to be charged (or discharged) after only five time constants.

Resistance-capacitance circuits and their time constants are used extensively in electronic circuits. For example, they are used to control the frequency of many nonsinusoidal waveforms. When you adjust the vertical-hold control on a television receiver, you are adjusting a time constant.

Voltage Distribution

A dc source voltage divides among series capacitors in inverse proportion to their capacitance (Fig. 10-23). The voltmeter across the 1-μF capacitor indicates twice as much voltage as the one across the 2-μF capacitor. This happens because capacitors in series must receive the same quantity of charge. (There is only one path for the charging current.) The same amount of charge produces less voltage in a large capacitor than it does in a small capacitor. Remember that $C = Q/V$; therefore, $V = Q/C$. If Q is the same for both capacitors, then the higher the capacitance, the lower the voltage.

Sometimes capacitors are connected in series because the voltage rating of either capac-

Fig. 10-22   Discharging time for a capacitor.

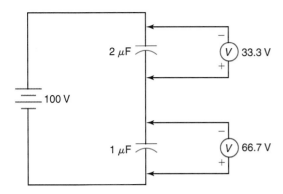

Fig. 10-23 The dc voltage distribution within series capacitors. The smallest capacitor develops the most voltage.

itor is lower than the source voltage. In this case, the capacitance of the two capacitors should be as close to equal as possible. Then the source voltage will divide equally between the two capacitors.

▮ TEST

Answer the following questions.

43. The curved line of the capacitor symbol represents the _____ plate of an electrolytic capacitor.
44. Technicians figure it takes _____ time constants to charge a capacitor.
45. The outside foil of a film capacitor is identified by a _____.
46. Define *time constant*.
47. What is the time constant of a 50-kΩ resistor and a 2.2-μF capacitor?
48. A dc source is connected to an 8-μF and a 5-μF capacitor connected in series. Which capacitor develops the larger voltage? Why?

10-9 Capacitors in AC Circuits

A capacitor controls the current in an ac circuit. It controls the current by storing energy, which produces a voltage in the capacitor. The voltage produced by the capacitor's stored energy is always in opposition to the source voltage.

Careful study of Fig. 10-24 shows exactly how a capacitor controls circuit current. The voltage waveform represents the source (generator) voltage applied across the capacitor. The current waveform represents the current

flowing through the ammeter. Assume the source voltage in Fig. 10-24(a) has just changed polarity and is increasing in the positive direction. Also assume the capacitor is completely discharged at the instant the source voltage changes polarity. Now, as the generator (source) voltage builds up, the capacitor *must charge* to keep up with the increasing generator voltage. The current in the circuit is

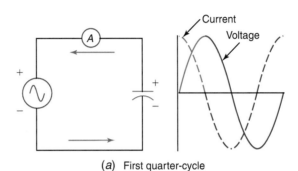

(a) First quarter-cycle

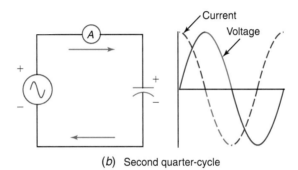

(b) Second quarter-cycle

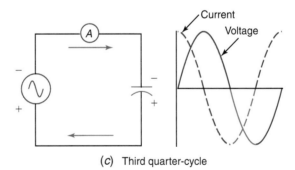

(c) Third quarter-cycle

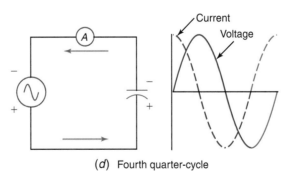

(d) Fourth quarter-cycle

Fig. 10-24 Energy transfer in a capacitor circuit. Every other quarter-cycle the capacitor becomes the energy source.

maximum at this time because the generator's voltage is changing most rapidly at this time. (Remember that the faster the capacitor's voltage must rise, the more charge per second it requires. And more charge per second means more current.) As the first quarter-cycle progresses in Fig. 10-24(a), the generator's voltage increases less rapidly. Consequently, the current in the circuit decreases. At the end of the first quarter-cycle there is an instant when the generator's voltage is stable (neither increasing nor decreasing). At that instant the circuit current is zero. The capacitor's voltage is equal to and opposite the source voltage.

Let us note four things from observing the action during this first quarter-cycle:

1. The generator is charging the capacitor. Thus electrons (current) leave the negative terminal of the generator and enter the positive terminal of the generator.
2. The generator is transferring energy to the capacitor as it charges the capacitor.
3. The capacitor allows just enough current to charge it to the value of the source voltage.
4. The current and voltage are 90° out of phase with each other. Specifically, the *current leads the voltage* by 90°.

Now let's look at the second quarter-cycle [Fig. 10-24(b)]. During this part of the cycle, the generator's voltage is decreasing. Since the capacitor is connected directly across the generator, its voltage must also decrease. For the capacitor to lose voltage, it must discharge. This means that electrons must leave its negative terminal and enter its positive terminal. To discharge, the capacitor has to force electrons through the generator against (in opposition to) the generator's voltage. Thus, the capacitor is now returning its stored energy to the generator. Figure 10-24(b) shows that the decrease in the generator's voltage causes an increase in

the current in the circuit in the opposite direction. The current is maximum when the generator voltage is zero. The current in this case, however, is a result of the capacitor's discharging. Again, let us summarize the major action during the second quarter-cycle:

1. The capacitor is discharging through the generator.
2. The capacitor is *transferring energy* back to the source (generator) as it discharges.
3. The capacitor furnishes just enough current to discharge it to the value of the source voltage.
4. The current still leads the voltage by 90°.

Inspection of the third quarter-cycle [Fig. 10-24(c)] shows that it is essentially the same as the first quarter-cycle. The only difference is that the voltage and current have reversed. The generator again takes control and furnishes the energy needed to charge the capacitor. The capacitor stores energy during this third quarter of the cycle.

The fourth quarter-cycle [Fig. 10-24(d)] is like the second quarter-cycle. The capacitor now discharges through the generator and returns its stored energy to the source. As in the previous three quarters of the cycle, the current and voltage remain 90° out of phase.

Capacitive Reactance

A capacitor's opposition to alternating current is known as *reactance*. The symbol for reactance is X. Since inductors also provide reactance, we use a subscript with the symbol X. Thus X_C is used to signify capacitive reactance. That is, X_C means the reactance of a capacitor. Reactance is like resistance in that it is an opposition that can control the current in a circuit. Therefore, the base unit of reactance is also the ohm (Ω). Reactance is unlike resistance in that it does not convert electric energy into heat energy. It is incorrect to interchange the terms *resistance* and *reactance* even though both are expressed in ohms.

Capacitive reactance is controlled by two factors: the frequency of the current (and voltage) and the amount of capacitance. Further, the reactance is inversely proportional to both these factors. If either frequency or capacitance is doubled, the reactance is halved. If either frequency or capacitance is halved, the reactance is doubled. Why reactance is inversely proportional to both frequency and capacitance can be explained by the example which follows.

Figure 10-25 shows a capacitor circuit in which frequency, capacitance, and voltage are indicated. We know that the current in the circuit is controlled by the capacitive reactance. If the reactance increases, the current decreases and vice versa. Therefore, if we determine what happens to the current in the circuit, we will know what happens to the reactance. First let us determine the circuit current for the frequency, voltage, and capacitance given in Fig. 10-25. Then we can change either frequency or capacitance and see what happens to the current. To determine the current, we first determine the amount of charge needed to charge the capacitor to the source voltage. This can be done with the basic formula for capacitance:

$$C = \frac{Q}{V}$$

Rearranging yields

$$Q = CV$$

In Fig. 10-25 then, the charge is

$$Q = 1\,F \times 1\,V = 1\,C$$

Next we determine the amount of time available to transfer the charge from the source to the capacitor. The time available is the time required for one-quarter of a cycle. The period of 1 Hz is 1 s. Thus the capacitor has 0.25 s in which to charge. Finally, we know that

$$\text{Current } (I) = \frac{\text{charge } (Q)}{\text{time } (t)}$$
$$= \frac{1\,C}{0.25\,s}$$
$$= 4\,C/s = 4\,A$$

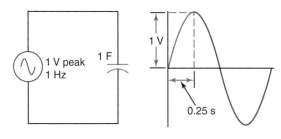

Fig. 10-25 Determining charge and average current. In 0.25 s the capacitor must receive 1 C of charge.

A coulomb per second is an ampere, and so the average current for the first quarter-cycle in Fig. 10-25 is 4 A. The remaining quarters of the cycle handle the same charge in the same amount of time. Therefore, the average current in the circuit of Fig. 10-25 is 4 A.

Now suppose we double the frequency to 2 Hz. The new period is 0.5 s, and the time for one quarter-cycle is 0.125 s. Since the voltage and the capacitance are the same, the charge is the same. The current now increases to

$$I = \frac{Q}{t}$$
$$= \frac{1\,C}{0.125\,s}$$
$$= 8\,C/s = 8\,A$$

Since doubling the frequency doubles the current, the reactance of the circuit must be halved.

Now suppose we leave the frequency at 1 Hz and double the capacitance to 2 F. The new charge is

$$Q = CV = 2\,F \times 1\,V = 2\,C$$

The time for one quarter-cycle at 1 Hz is still 0.25 s. Therefore, the new current is

$$I = \frac{Q}{t}$$
$$= \frac{2\,C}{0.25\,s}$$
$$= 8\,C/s = 8\,A$$

Again, the current doubles and so the reactance must be half as great as it was with 1 F.

The exact value of capacitive reactance can be calculated with the formula

$$X_C = \frac{1}{2\pi f C}$$

where the Greek letter "pi" (π) is a unitless constant equal to approximately 3.14. Using this value for pi, the formula becomes

$$X_C = \frac{1}{6.28\,f\,C} \qquad X_C = \frac{1}{6.28\,f\,C}$$

Capacitive reactance is in ohms when frequency is in hertz and capacitance is in farads.

Example 10-3

What is the reactance of a 0.01-μF capacitor at 400 Hz?

Given: $C = 0.01\ \mu$F
             $f = 400$ Hz

Find: X_C

Known: $X_C = \dfrac{1}{6.28\,f\,C}$

Solution: $X_C =$

$$\dfrac{1}{6.28 \times 400 \times 0.01 \times 10^{-6}}$$

$$= 39{,}800\ \Omega$$

Answer: The capacitive reactance therefore equals 39,800 Ω, or about 40 kΩ.

Notice that in the above example microfarads were converted to farads by multiplying by 10^{-6}.

From the above example we can quickly estimate the reactance of a 0.005-μF capacitor at 400 Hz. Since 0.005 μF is half as great as 0.01, the reactance will be twice as great. Thus, the 0.005-μF capacitor at 400 Hz has approximately 80 kΩ of capacitive reactance.

Reactance cannot be measured with an ohmmeter. It must be calculated using either the reactance formula or Ohm's law. Since reactance is a form of opposition, Ohm's law can be used by replacing R with X_C. Thus, we can use any of the following formulas:

$$X_C = \dfrac{V_C}{I_C} \qquad I_C = \dfrac{V_C}{X_C} \qquad V_C = I_C X_C$$

Example 10-4

How much current flows in the circuit shown in Fig. 10-26?

Given: $f = 2$ MHz
             $C = 500$ pF
             $V = 20$ V

Find: I

Known: $I = \dfrac{V}{X_C}$ and $X_C = \dfrac{1}{6.28\,f\,C}$

Solution: $X_C =$

$$\dfrac{1}{6.28 \times 2 \times 10^6 \times 500 \times 10^{-12}}$$

$$= 159\ \Omega$$

$$I = \dfrac{20\ \text{V}}{159\ \Omega} = 0.126\ \text{A}$$

$$= 126\ \text{mA}$$

Answer: The current in the circuit is 126 mA.

Quality

Capacitors (and inductors) are built to provide reactance so that current can be controlled without converting electric energy into heat energy. However, it is impossible to build capacitors and inductors (especially inductors) that do not also have some resistance. *Quality* is the term used to rate a capacitor (or inductor) on its ability to produce reactance with as little resistance as possible.

Quality (Q) is a ratio of reactance to resistance. Since it can be used to describe either capacitors or inductors, the formula for quality should include a subscript for the reactance. For capacitors, the formula is

$$Q = \dfrac{X_C}{R}$$

Since both X and R have units of ohms, Q has no units (it is a pure number). Because X (and also R to a lesser extent) changes with frequency, Q is frequency-dependent. That is, its value is always given for some specified frequency.

Power in Capacitor Circuits

Pure capacitance can use no energy. It merely stores energy for one quarter-cycle and returns it the next quarter-cycle. Since capacitance uses no energy, it cannot consume power either (power = energy/time). This idea is illustrated in Fig. 10-27. This figure shows power (or en-

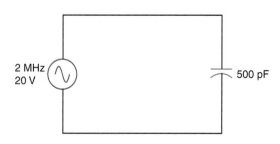

Fig. 10-26   Circuit for example 10-4.

ergy) taken from the source as positive power, and power (or energy) returned to the source as negative power. The negative power exactly cancels the positive power.

 . . . that you have seen the waveform of Fig. 10-27 before. It is the power waveform obtained in Chap. 9 for a current and voltage that are 90° out of phase.

Therefore, we can conclude two things about a circuit that contains only capacitance:

1. Current and voltage are 90° out of phase.
2. The circuit uses no net energy or power.

Notice that the term used is *capacitance,* not *capacitor. Capacitance* is used because capacitors do lose some energy.

Energy losses in capacitors can be classified as *resistance losses* and *dielectric losses.* The resistance loss is caused primarily by current flowing through the resistance of the capacitor leads and plates. There is also a very minute current flowing through the high resistance of the dielectric material. The dielectric resistance appears as a parallel resistance, whereas the plate and lead resistances appear as a series resistance (Fig. 10-28). Dielectric loss results from continually reversing the polarity of the electric field through the dielectric. Each time the molecular structure of the dielectric has to adjust to a new polarity, a little electric energy is converted into heat energy. This dielectric loss appears as a small resistance in series with the capacitor. The effects of all these resistances are often combined as the *equivalent series resistance* (ESR). A capacitor with a small ESR has a small energy loss.

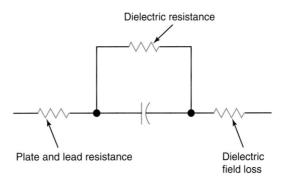

Fig. 10-28 Equivalent resistances in a capacitor. These resistances convert electric energy into heat energy.

The relative amount of energy lost in a capacitor can be indicated by any of three terms. The terms are *dissipation factor* (DF), *power factor* (PF), and *quality* (Q). All these terms give about the same information. *Power factor* is the ratio of power (P) to apparent power P_{app}. In a capacitor, power represents the energy per second lost, and apparent power represents the energy per second stored. If a capacitor has very little energy loss, its power factor is low (less than 0.01, or 1 percent). *Quality* (Q) is the ratio of reactance to resistance. If a capacitor has very little energy loss, R is low and Q is high (more than 100). *Dissipation factor* is the reciprocal of quality. In other words, dissipation factor is the ratio of resistance to reactance. Obviously, if Q is typically a large number, dissipation factor will be a small number or percentage. For all but the lowest-quality capacitors, power factor and dissipation factor are essentially equal to each other. In summary, remember that capacitors with small energy losses have a high Q and low dissipation and power factors.

With lower-quality electrolytic capacitors, energy losses can be considerable (1 to 10 percent of the stored energy). However, with high-quality nonelectrolytic capacitors the losses are so small that they can be ignored in most applications. In this text, we will assume capacitors have negligible energy loss. In terms of phase shift, this means we will treat nonelectrolytic capacitors as ideal capacitors. We will assume the current and voltage are a full 90° out of phase.

Dissipation factor (DF)

Power factor (PF)

Resistance loss

Dielectric loss

Quality (Q)

Equivalent series resistance (ESR)

JOB TIP

During the interview, avoid using too many short answers but do not elaborate in every case. Also ask the interviewer whether it is permissible to call a day or two later to ask for an assessment.

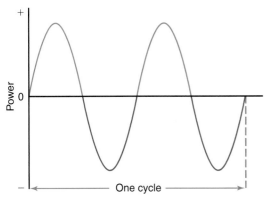

Fig. 10-27 Power waveform for capacitance. Capacitance uses no power.

Answer the following questions.

49. How does a capacitor control current in an ac circuit?
50. Why doesn't capacitance consume power?
51. What is the phase relationship between current and voltage in an ideal capacitor?
52. Define *quality.*
53. List three terms used to indicate relative energy loss in a capacitor.
54. The opposition of a capacitor to ac current is called _____ .
55. The symbol for a capacitor's opposition to ac current is _____ .
56. ESR is the abbreviation for _____ .
57. True or false. Capacitive reactance is inversely proportional to frequency.
58. True or false. Capacitive reactance is directly proportional to capacitance.
59. Calculate the reactance of a 2-μF capacitor at 150 Hz.
60. At what frequency will a 0.02-μF capacitor have a reactance of 400 Ω?
61. A 50-V, 1-kHz source is applied to a 0.1-μF capacitor. Determine the circuit current.

10-10 Capacitors in Series

When capacitors are connected in series, the total capacitance decreases. In fact, the total capacitance of series capacitors is always less than the capacitance of the smallest capacitor. The logic of this statement can be seen by referring to Fig. 10-29, which shows two capacitors connected in series. Assume these capacitors are made from flat metal plates separated by an insulator. Suppose the lead connecting the two capacitors is shortened until the plates of the capacitors touch each other. The result

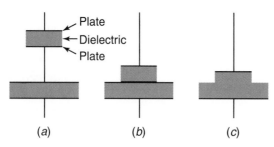

| (a) | (b) | (c) |

Fig. 10-29 Capacitors in series. The equivalent capacitance in (c) is less than the lower of the two capacitances in (a).

would be as indicated in Fig. 10-29(b), where the two plates combine to make one plate. The combined plates, isolated between two insulators, serve no useful purpose. If they are removed, as in Fig. 10-29(c), the result is a single equivalent capacitor. Look carefully at the plates of the equivalent capacitor. They are spaced farther apart than the plates of either of the original capacitors were. Also, the effective plate area is limited by the smaller plate. Therefore, the equivalent capacitor has less capacitance than either of the two series capacitors.

The total (or equivalent) capacitance of series capacitors can be calculated with any of these formulas:

General formula for series capacitors:

$$C_T = \frac{1}{\dfrac{1}{C_1} + \dfrac{1}{C_2} + \dfrac{1}{C_3} + \text{etc.}}$$

Two capacitors in series:

$$C_T = \frac{C_1 \times C_2}{C_1 + C_2}$$

n equal capacitors in series:

$$C_T = \frac{C}{n}$$

Notice that the general form of these formulas is the same as the form of the formulas used for parallel resistances. The formulas are indeed the same except that C is replaced by R.

Example 10-5

What is the total capacitance of a 0.001-μF capacitor in series with a 2000-pF capacitor?

Given: $C_1 = 0.001\ \mu\text{F}$
$C_2 = 2000\ \text{pF}$

Find: C_T

Known: $C_T = \dfrac{C_1 \times C_2}{C_1 + C_2}$

Solution: $2000\ \text{pF} = 0.002\ \mu\text{F}$

$$C_T = \frac{0.001\ \mu\text{F} \times 0.002\ \mu\text{F}}{0.001\ \mu\text{F} + 0.002\ \mu\text{F}}$$

$$= \frac{0.000002}{0.003}$$

$$C_T = 0.00067\ \mu\text{F}$$

Answer: The total series capacitance is 0.00067 μF.

Notice in the above example that the capacitances were converted into the same units (μF) before using the formula. This way the total (equivalent) capacitance comes out in the same unit as the two original capacitances.

The reactances of series capacitors are additive. The total reactance can be computed by the formula

$$X_{C_T} = X_{C_1} + X_{C_2} + X_{C_3}$$

Example 10-6

Determine the total reactance of the circuit shown in Fig. 10-30.

Given: $C_1 = 2\ \mu F$
$C_2 = 4\ \mu F$
$f = 100\ Hz$

Find: X_{C_T}

Known: $X_{C_T} = X_{C_1} + X_{C_2}$

$$X_C = \frac{1}{6.28\,f\,C}$$

Solution: $X_{C_1} = \dfrac{1}{6.28 \times 100 \times 2 \times 10^{-6}}$

$= 796\ \Omega$

$X_{C_2} = \dfrac{1}{6.28 \times 100 \times 4 \times 10^{-6}}$

$= 398\ \Omega$

$X_{C_T} = 796\ \Omega + 398\ \Omega$

$= 1194\ \Omega$

Answer: The total reactance of the circuit is 1194 Ω.

The answer to example 10-6 could also have been found by first determining the total capacitance and then finding the total reactance:

$$C_T = \frac{C_1 \times C_2}{C_1 + C_2} = \frac{2 \times 4}{2 + 4} = \frac{8}{6} = 1.33\ \mu F$$

$$X_{C_T} = \frac{1}{6.28\,f\,C_T}$$

$$= \frac{1}{6.28 \times 100 \times 1.33 \times 10^{-6}}$$

$$= 1194\ \Omega$$

Ohm's law is also applicable in capacitor circuits. Just replace R with X_C.

Example 10-7

Determine the circuit current and the voltage across each capacitor in Fig. 10-30.

Given: From the previous example:
$X_{C_1} = 796\ \Omega$, $X_{C_2} = 398\ \Omega$,
$X_{C_T} = 1194\ \Omega$.
From Fig. 10-30:
$V_T = 40\ V$

Find: I_T, V_{C_1}, V_{C_2}

Known: $I_T = \dfrac{V_T}{X_{C_T}}$

$I_T = I_{C_1} = I_{C_2}$

$V = I X_C$

Solution: $I_T = \dfrac{40\ V}{1194\ \Omega} = 0.0335\ A$

$= 33.5\ mA$

$V_{C_1} = 0.0335\ A \times 796\ \Omega$

$= 26.7\ V$

$V_{C_2} = 0.0335\ A \times 398\ \Omega$

$= 13.3\ V$

Answer: The total current is 33.5 mA. The voltage across C_1 is 26.7 V, and that across C_2 is 13.3 V.

As a check on the computations in the above example, we can apply Kirchhoff's voltage law:

$$V_T = V_{C_1} + V_{C_2} = 26.7\ V + 13.3\ V = 40\ V$$

Notice in the above example that the largest capacitor dropped the least amount of voltage. This is because capacitance and capacitive reactance are inversely proportional.

In electronic circuits, capacitors can be used as ac voltage dividers. The advantage of using

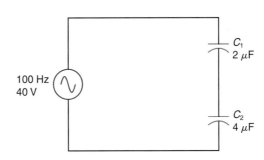

Fig. 10-30 Circuit for examples 10-6 and 10-7.

capacitors rather than resistors as ac voltage dividers is that capacitors use no power.

10-11 Capacitors in Parallel

The total capacitance of parallel capacitors is found by adding the individual capacitances:

$$C_T = C_1 + C_2 + C_3 + \text{etc.}$$

Figure 10-31 shows that connecting capacitors in parallel effectively increases the plate area. Suppose the two capacitors shown in cross section in Fig. 10-31(a) are moved together until the ends of their plates touch. Then the two connecting conductors can be removed. The result is an equivalent capacitor [Fig. 10-31(b)] with a larger plate area. A larger plate area, of course, means more capacitance.

The reactances of capacitors in parallel can be treated like resistances in parallel. The formulas are

General formula for parallel reactances:

$$X_{C_T} = \frac{1}{\dfrac{1}{X_{C_1}} + \dfrac{1}{X_{C_2}} + \dfrac{1}{X_{C_3}} + \text{etc.}}$$

Two reactances in parallel:

$$X_{C_T} = \frac{X_{C_1} \times X_{C_2}}{X_{C_1} + X_{C_2}}$$

n equal reactances in parallel:

$$X_{C_T} = \frac{X_C}{n}$$

The *total reactance* of parallel capacitors can also be obtained by either the reactance formula or Ohm's law. The formulas are

$$X_{C_T} = \frac{1}{6.28 f C_T} \quad \text{and} \quad X_{C_T} = \frac{V_T}{I_T}$$

Of course, total capacitance must be determined before the reactance formula can be applied.

Example 10-8

Refer to Fig. 10-32. Determine the total current, the current through C_1, the total capacitance, and the total reactance.

Given: $C_1 = 0.2\ \mu F$
              $C_2 = 0.3\ \mu F$
              $f = 1000\ Hz$
              $V = 50\ V$

Find: $I_T, I_{C_1}, C_T,$ and X_{C_T}

Known: $I_T = \dfrac{V_T}{X_C}$

             $I_{C_1} = \dfrac{V_{C_1}}{X_{C_1}}$

             $C_T = C_1 + C_2$

             $X_{C_T} = \dfrac{1}{6.28 f C_T}$

             $V_T = V_{C_1} = V_{C_2}$

Solution: $C_T = 0.2\ \mu F + 0.3\ \mu F = 0.5\ \mu F$

             $X_{C_T} =$
             $$\frac{1}{6.28 \times 1 \times 10^3 \times 0.5 \times 10^{-6}}$$
             $$= 318\ \Omega$$

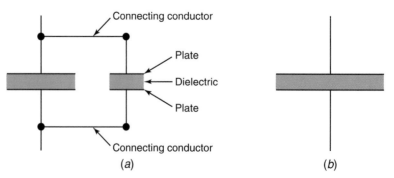

Fig. 10-31 Capacitors in parallel. The equivalent capacitance in (b) is equal to the sum of the capacitances in (a).

$$I_T = \frac{50 \text{ V}}{318 \text{ }\Omega} = 0.157 \text{ A}$$
$$= 157 \text{ mA}$$
$$X_{C_1} =$$
$$\frac{1}{6.28 \times 1 \times 10^3 \times 0.2 \times 10^{-6}}$$
$$= 796 \text{ }\Omega$$
$$I_{C_1} = \frac{50 \text{ V}}{796 \text{ }\Omega} = 0.0628 \text{ A}$$
$$= 62.8 \text{ mA}$$

Answer: The total current is 157 mA, and I_{C_1} is 62.8 mA. The total capacitance is 0.5 μF, and the total reactance is 318 Ω.

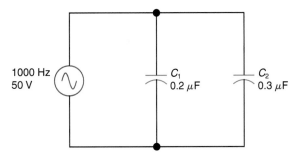

Fig. 10-32 Circuit for example 10-8.

In example 10-8, total reactance was determined without using a parallel-reactance formula or determining the value of X_{C_2}. Let us check the answers to this example by calculating X_{C_2} and then recalculating X_{C_T}:

$$X_{C_2} = \frac{1}{6.28 \times 1 \times 10^3 \times 0.3 \times 10^{-6}}$$
$$= 531 \text{ }\Omega$$
$$X_{C_T} = \frac{796 \times 531}{796 + 531}$$
$$= 318 \text{ }\Omega$$

Since the value obtained for X_{C_T} is the same in both cases, we have made no mathematical errors. If we wanted to know the current through C_2 in Fig. 10-32, we could use Kirchhoff's current law (we have already determined the value of I_T and I_{C_1}):

$$I_T = I_{C_1} + I_{C_2}$$
$$I_{C_2} = I_T - I_{C_1}$$
$$= 157 \text{ mA} - 62.8 \text{ mA}$$
$$= 94.2 \text{ mA}$$

We could also determine I_{C_2} by using Ohm's law because we know X_{C_2} and V_{C_2}:

$$I_{C_2} = \frac{V_{C_2}}{X_{C_2}} = \frac{50 \text{ V}}{531 \text{ }\Omega}$$
$$= 0.0942 \text{ A} = 94.2 \text{ mA}$$

■ TEST

Answer the following questions.

62. The equivalent capacitance of a 2-μF capacitor and a 1-μF capacitor is 3 μF.

Are the capacitors connected in series or in parallel?

63. What is the equivalent capacitance of a 2000-pF capacitor and a 0.002-μF capacitor connected in series?

64. What is the total reactance at 600 Hz of a 0.2-μF and a 0.4-μF capacitor connected in parallel?

65. If the two capacitors in question 64 are connected in series, what is the total reactance?

66. How much voltage is required to produce a current of 0.03 A in a capacitor that has 4500 Ω of reactance?

67. For the circuit of Fig. 10-33, find the following:
   a. X_{C_2}
   b. C_T
   c. V_{C_1}
   d. I_T

68. For the circuit of Fig. 10-34, find the following:
   a. X_{C_T}
   b. I_T
   c. C_T

69. In Fig. 10-34, which carries more current, C_1 or C_2?

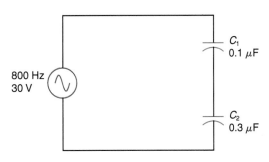

Fig. 10-33 Circuit for test question 67.

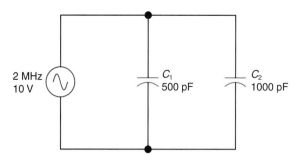

Fig. 10-34 Circuit for test questions 68 and 69.

10-12 Detecting Faulty Capacitors

An ohmmeter can be used to check the condition of a capacitor. When an ohmmeter is connected across a completely discharged capacitor, the capacitor will charge to a voltage equal to that of the battery or dc power supply used in the ohmmeter. How long it takes the capacitor to charge depends on two things: the size of the capacitor and the internal resistance of the ohmmeter. In other words, it depends on the time constant. When capacitors are being checked, the ohmmeter should be on a high range to provide a large time constant so you can "see" the capacitor charge.

With *high values of capacitance,* the ohmmeter pointer of the VOM will deflect toward 0 Ω and then return to a high reading. If the capacitor is nonelectrolytic, the final indication on the VOM should be as high as the meter can go. With the DMM, each new reading will be higher than the previous reading until the DMM indicates overrange. (After the capacitor has charged, the ohmmeter is just measuring the resistance of the dielectric material.) With low-quality electrolytic capacitors, the final reading may be as low as 100 kΩ. This is because the dielectric material is an oxide, which is not as good an insulator as plastics or ceramics.

With *low-value capacitors,* the charging current may not last long enough to cause a deflection on the ohmmeter. In this case, the meter just indicates a very high resistance. How large a capacitor must be to show a deflection depends on the type of meter. For a particular meter this can be determined by experiment.

So far we have dealt with ohmmeter action when a capacitor is not defective. How does it react to a defective capacitor? If the capacitor is open (a lead is not making contact with a plate), the meter won't deflect. If a capacitor is short-circuited, the pointer will never return to a high resistance reading. This means that current is flowing through the dielectric or that the plates are touching each other. If the ohmmeter indicates 0 Ω, we say the capacitor is *dead-shorted* or completely shorted. This usually means that the plates are making direct contact. If the dielectric material has become contaminated or deteriorated, the ohmmeter may indicate a few thousand ohms. We would say the capacitor is partially shorted, has a high-resistance short, or is leaky.

10-13 Undesired, or Stray, Capacitance

Any two conductive surfaces separated by insulation possess capacitance. Thus, every electric and electronic circuit has some undesired capacitance. For example, capacitance exists between

1. An insulated conductor and a metal chassis
2. Two conductors in an electric cable
3. The turns of a coil of wire
4. The input and output leads of a transistor

A complete list of places where *stray capacitance* is found would be almost endless.

Whether or not undesired capacitance affects a circuit depends on the amount of capacitance and the frequency of the circuit. For example, the capacitance between two parallel insulated conductors 5 cm long is very low. In dc and most ac power circuits, this amount of capacitance would have no effect. However, at the frequency used in a television tuner, even such a low capacitance would have a very pronounced effect. In fact, if the conductors were 6 cm rather than 5 cm long, the tuner might not function properly.

10-14 Capacitor Specifications

When ordering capacitors, one has to include the *specifications* needed to ensure receiving the desired capacitors. The minimum specifications for ordering capacitors for general use are

1. Capacitance
2. Tolerance
3. Direct current working voltage
4. Type (ceramic disc, polystyrene, variable, etc.)
5. Lead configuration (axial or radial)

Leaky capacitor

High values of capacitance

Stray capacitance

Low-value capacitors

Specifications

If the capacitors are for circuits in which the capacitance value is critical, then the temperature coefficient should also be specified. For extreme environmental conditions, the operating temperature range should be included in the specifications. When available space is limited, capacitor size and shape should also be specified.

10-15 Uses of Capacitors

Capacitors are used with other components to accomplish a wide variety of jobs. For example, capacitors are used

1. In power supply filters which smooth pulsating direct current into pure direct current
2. In oscillators which convert direct current into either sinusoidal or nonsinusoidal alternating current
3. In filters which separate low-frequency alternating current from high-frequency alternating current
4. For coupling amplifiers together (to separate alternating current from direct current)
5. For power-factor correction (their leading

current compensates for the lagging current of an inductive load)

The circuit in Fig. 10-35 illustrates how a capacitor works in a complete functional system. This circuit is called a *relaxation oscillator.* Its purpose is to produce a *sawtooth voltage* from a pure dc voltage. The neon lamp in the circuit fires (ionizes) at 80 V and deionizes at 70 V. When fired, its resistance is extremely low; when deionized, its resistance is high.

Figure 10-35(*a*) shows the current flow when the lamp is deionized; at this time the capacitor is charging. Since C_1 charges through R_1, the time constant is relatively long, as indicated in the waveform graph. Once the capacitor has charged to 80 V, the lamp fires [Fig. 10-35(*b*)]. The internal resistance of the lamp drops to a low value when it fires. This allows the capacitor to quickly discharge through the low resistance. When the capacitor has discharged to the lamp's critical voltage (about 70 V), the lamp deionizes. The capacitor then stops discharging and again starts to recharge.

The above charging and discharging process repeats itself endlessly as long as the battery

Relaxation oscillator

Sawtooth voltage

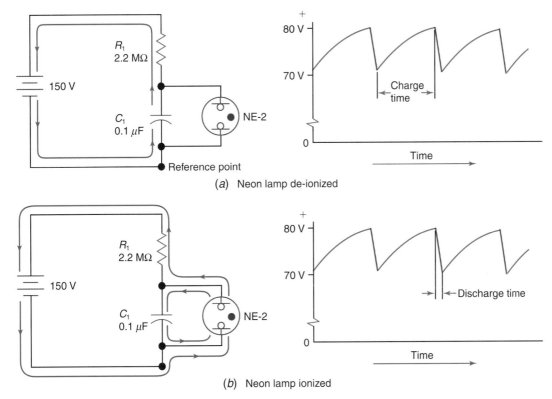

(*a*) Neon lamp de-ionized

(*b*) Neon lamp ionized

Fig. 10-35 Relaxation oscillator circuit. The capacitor discharges during the time that the lamp is ionized.

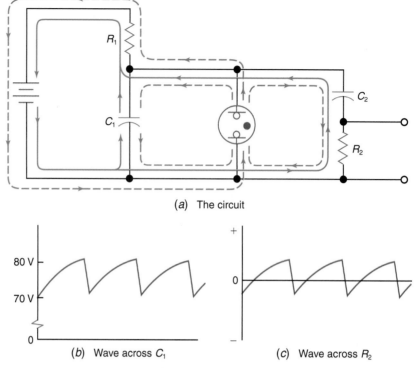

(a) The circuit

80 V

70 V

0

(b) Wave across C_1

+

0

−

(c) Wave across R_2

Fig. 10-36 Coupling capacitor. The coupling capacitor separates the alternating current from the direct current.

energy lasts. It produces a continuous sawtooth waveform with a 10-V peak-to-peak amplitude. Notice in Fig. 10-35 that the waveform is a fluctuating dc voltage. It varies from 70 to 80 V positive with respect to the reference point.

Suppose we wished to separate the ac sawtooth part of the waveform from the dc part (the +70 V). We can do this with another capacitor and resistor combination, such as C_2 and R_2 in Fig. 10-36. In this figure C_2 couples the sawtooth (ac) to R_2 while blocking the dc voltage. This is what is being referred to by the statement "A capacitor blocks direct current and passes alternating current." The broken lines in Fig. 10-36 show the paths of current flow when the lamp is fired. When fired, the lamp carries discharge current for both C_1 and C_2 as well as current for R_1. The solid line shows the paths when the lamp is deion-

ized. Notice that the current through R_2 is alternating current; it periodically reverses itself.

■ TEST_____

Answer the following questions.

70. How much resistance should an ohmmeter indicate across the terminals of a good ceramic capacitor?

71. When checking a capacitor with an ohmmeter, does no deflection of the pointer always indicate an open capacitor?

72. What is another name for undesired capacitance?

73. List the minimum specifications for ordering capacitors.

74. A coupling capacitor _____ direct current and _____ alternating current.

Summary

1. Capacitance is the ability to store electric energy.
2. The symbol for capacitance is C.
3. Capacitors are devices constructed to provide specific amounts of capacitance.
4. Capacitors have two plates and a dielectric.
5. A charged capacitor stores energy and creates a voltage between its plates.
6. A charged capacitor can provide energy for other parts of an electric circuit.
7. Current does not flow through the dielectric material.
8. The farad is the base unit of capacitance.
9. The abbreviation for farad is F.
10. Capacitance is determined by plate area, distance between plates, dielectric material, and temperature.
11. Capacitors are named after their dielectric material, enclosure, construction process, or intended use.
12. Electrolytic capacitors are usually polarized. They have maximum capacitance for their size and weight.
13. Typical paper and film capacitances are less than 1 μF.
14. Ceramic and mica capacitors are limited to low values.
15. Energy-storage capacitors can provide very high power levels.
16. The curved line on a capacitor symbol identifies a specific plate on variable, polarized, paper, and film capacitors.
17. Except for an initial charging surge, capacitors block direct current.
18. The smaller of two series capacitors develops the greater voltage.
19. Capacitors control current flow in an ac circuit.
20. Current leads voltage by 90° in an ideal capacitor.
21. Capacitance uses no power or energy.
22. Energy loss in a capacitor results from dielectric and resistance losses.
23. Dissipation factor, power factor, and quality all are ways of rating a capacitor's relative energy loss.
24. Capacitive reactance is the opposition of a capacitor to sinusoidal alternating current.
25. The symbol for capacitive reactance is X_C.
26. Capacitive reactance is inversely proportional to both frequency and capacitance.
27. The total capacitance of series capacitors is less than the smallest capacitance.
28. The total reactance of series capacitors is the sum of the individual reactances.
29. The total capacitance of parallel capacitors is the sum of the individual capacitances.
30. Reactances of parallel capacitors are added reciprocally.
31. Ohm's law and Kirchhoff's laws apply to capacitor circuits.
32. A time constant defines the rate at which a capacitor charges or discharges through a resistor.
33. The capacitance formulas are

$$C = \frac{Q}{V}$$
$$W = 0.5CV^2$$
$$X_C = \frac{1}{6.28\,f\,C}$$
$$T = RC$$

For series capacitors,

$$C_T = \frac{1}{\dfrac{1}{C_1} + \dfrac{1}{C_2} + \dfrac{1}{C_3} + \text{etc.}}$$

$$X_{C_T} = X_{C_1} + X_{C_2} + X_{C_3} + \text{etc.}$$

For parallel capacitors,

$$C_T = C_1 + C_2 + C_3 + \text{etc.}$$

$$X_{C_T} = \frac{1}{\dfrac{1}{X_{C_1}} + \dfrac{1}{X_{C_2}} + \dfrac{1}{X_{C_3}} + \text{etc.}}$$

For questions 10-1 to 10-16, determine whether each statement is true or false.

10-1. Metal oxide can be used for the dielectric in a capacitor.

10-2. In an ac circuit, a capacitor charges every other quarter-cycle.

10-3. For a given weight and size, a capacitor stores more energy than a battery.

10-4. The voltage rating of a foil capacitor is a dc voltage rating.

10-5. A farad can be defined as a coulomb per volt.

10-6. Electrons travel through the dielectric material when a capacitor is being charged.

10-7. Electrolytic capacitors are smaller and lighter than other types of capacitors that have the same capacitance and voltage rating.

10-8. Energy-storage capacitors rarely produce a peak current of more than 0.5 A while being discharged.

10-9. The color-band end of a paper or foil capacitor indicates the negative plate of the capacitor.

10-10. The negative end of a variable capacitor is indicated by the curved plate of the capacitor symbol.

10-11. Except for the charging current, a capacitor blocks direct current.

10-12. The lowest-value capacitor develops the highest voltage in either a series ac or series dc circuit.

10-13. The final reading on an ohmmeter connected to a good capacitor should never exceed 100 kΩ.

10-14. Some ceramic capacitors are self-healing.

10-15. The plates of a metallized-film capacitor are metal oxide.

10-16. The capacitor in a relaxation oscillator discharges through the neon lamp.

For questions 10-17 to 10-23, supply the missing word or phrase in each statement.

10-17. The base unit of capacitance is the _____.

10-18. The abbreviation for the base unit of capacitance is _____.

10-19. A 5600-pF capacitor has _____ μF of capacitance.

10-20. The symbol for capacitance is _____.

10-21. Capacitance causes current to _____ voltage by _____ degrees.

10-22. Reactance is _____ proportional to both _____ and _____.

10-23. After _____ time constants, a capacitor will be over 99 percent charged.

Answer the following questions.

10-24. What four factors determine the capacitance of a capacitor? Also indicate how the capacitance can be increased by changing each factor.

10-25. Where is the energy stored in a capacitor? In what form (chemical, heat, etc.) is it stored?

10-26. What two factors determine the strength of the electric field between the plates of a capacitor?

10-27. How are electrolytic capacitors used in ac circuits?

10-28. List the minimum specifications needed when ordering capacitors.

10-29. How much energy is stored in a 450-μF capacitor when it is charged to 250 V?

10-30. How much charge is stored on the capacitor in question 10-29?

10-31. What is the reactance of a 0.033-μF capacitor at 10 kHz

10-32. A 3-μF capacitor and a 2-μF capacitor are connected in parallel across a 50-V, 100-Hz source. Determine the
a. Total capacitance
b. Total reactance
c. Total current
d. Current through the 2-μF capacitor

10-33. If the capacitors in question 10-32 are connected in series, what is the value of
a. Total capacitance
b. Total reactance
c. Voltage across the 3-μF capacitor

10-34. What is the voltage across a capacitor after two time constants if the capacitor and a resistor are series-connected to a 200-V dc source?

10-35. What is the time constant of a 2200-pF capacitor in series with a 2.7-MΩ resistor and a 100-V dc source?

10-36. What is the Q of a 2-μF capacitor at 100 Hz if the ESR is 0.5 Ω?

Critical Thinking Questions

10-1. A series combination of a 2-μF capacitor (C_1) and a 3-μF capacitor (C_2) is connected in parallel with a 1.5-μF capacitor (C_3). Determine the total capacitance and the total reactance at 100 Hz.

10-2. What value of capacitance must be in series with a 2-μF capacitor to produce a total capacitance of 1.5 μF?

10-3. What percentage of the source voltage will develop across a 1-μF capacitor that is in series with a 0.68-μF capacitor? Show how you calculated your answer.

10-4. What percentage of the source current will flow through a 0.12-μF capacitor when it is paralleled by a 0.33-μF capacitor?

10-5. Does alternating current flow in any of the conductors in Fig. 10-35? Explain.

10-6. For the formula $W = 0.5\ CV^2$, prove that W will be in joules if C is in farads and V is in volts.

10-7. A capacitor stores 0.3 C of charge when it is charged to 500 V. How much power will it produce if it is discharged in 0.5 ms?

10-8. A 3-μF capacitor (C_1) and an 8-μF capacitor (C_2) are series-connected to a 40-V dc source. Determine the voltage on the 8-μF capacitor.

10-9. Which formula(s) would not be correct for series capacitor circuits if the capacitors were not assumed to be ideal? Why?

10-10. Using parallel circuit rules and Q, V, and C relationships, prove that $C_T = C_1 + C_2$ for parallel capacitors.

10-11. Using series circuit rules and Q, V, and C relationships, prove that

$$C_T = \frac{1}{\frac{1}{C_1} + \frac{1}{C_2}}$$

for series capacitors.

Answers to Tests

1. capacitor
2. C
3. dielectric
4. dielectric
5. F
6. F
7. in the dielectric material
8. DCWV
9. Because too much electric field strength between the plates can destroy the dielectric material.
10. the distance between the plates and the voltage to which the capacitor is charged
11. farad
12. F
13. 0.0047
14. 3000
15. coulomb, volt
16. plate area, distance between plates,
dielectric material, temperature
17. temperature
18. It doubles.
19. It doubles.
20. the material's ability to store energy when used as a dielectric
21. smaller
22. 0.9 mC
23. 16,667 μF
24. 400 V
25. T
26. F
27. T
28. F
29. T
30. F
31. Metallized
32. variable
33. feed-through
34. aluminum, tantalum
35. oxide
36. tantalum
37. It draws excessively
high current from the source and overheats. It may explode.
38. mica, ceramic, paper, film (plastics)
39. smaller, more stable, and higher dielectric resistance
40. filter capacitor
41. in applications that require very rapid capacitor discharge
42. 6.75 J
43. negative
44. five
45. color band
46. the time required to charge or discharge a capacitor by 63.2 percent of the available voltage
47. 0.11 s
48. The 5-μF capacitor, because $V = Q/C$
and both capacitors receive the same amount of Q.
49. by producing a voltage which opposes the source voltage
50. Because current and voltage are 90° out of phase.
51. Current leads voltage by 90°.
52. Quality is the ratio of reactance to resistance.
53. dissipation factor, power factor, and quality.
54. capacitive reactance
55. X_C
56. equivalent series resistance
57. T
58. F
59. 531 Ω
60. 19.9 kHz

61. 31.4 mA
62. parallel
63. 1000 pF or 0.001 μF
64. 442 Ω
65. 1990 Ω
66. 135 V
67. a. 663.5 Ω
    b. 0.075 μF

c. 22.5 V
d. 11.3 mA
68. a. 53.0 Ω
    b. 188.4 mA
    c. 1500 pF
69. C_2
70. nearly infinite
71. no

72. stray capacitance
73. capacitance, tolerance, type, lead arrangement, and voltage rating
74. blocks, passes

Chapter 11

Inductance

Chapter Objectives

This chapter will help you to:

1. *Understand* what inductance is and what causes inductance.
2. *Learn* the terminology associated with inductance and inductors.
3. *Identify* common types of inductors and write complete specifications for them.
4. *Understand* why inductance causes current to lag voltage by 90° and thus use no power.
5. *Learn and use* the relationship between inductance, frequency, and reactance.
6. *Solve* circuit values when inductors are connected in series or in parallel.
7. *Understand* what causes an inductor to have resistance and how this controls the quality of the inductor.

Many electric and electronic devices operate on the principle of inductance. Therefore, it is important that you develop a sound understanding of inductance and inductors.

11-1 Characteristics of Inductance

Inductance is the electrical property that opposes any change in the magnitude of current in a circuit. The letter "*L*" is the symbol used to represent inductance.

Devices that are used to provide the inductance in a circuit are called *inductors*. Inductors are also known as *chokes, reactors,* and *coils.* These three names are descriptive of the way inductance behaves in a circuit. Inductance, and thus an inductor, "chokes off" and restricts sudden changes in current. Inductance reacts against (resists) changes, either increases or decreases, in current. Inductors are usually coils of wire.

Inductance is the result of a voltage being induced in a conductor. The magnetic field that induces the voltage in the conductor is produced by the conductor itself.

YOU MAY RECALL . . . that in Chap. 7 we discussed how a magnetic field is formed around a current-carrying conductor.

When current begins to flow in a conductor, magnetic flux rings start to expand out from the conductor, as in Fig. 11-1(*a*). This expanding flux induces a small voltage in the conductor. The induced voltage has a polarity that opposes the increasing source voltage which is creating the increasing current. Thus, the inductance of the conductor opposes the rising current and tries to keep it constant. Of course, the inductance cannot completely stop the increase in current because the induced voltage is caused by the increasing flux. And the increasing flux depends on the increasing current. The inductance of the conductor, therefore, restricts only the rate at which the current can increase.

When the current in a conductor starts to decrease, as in Fig. 11-1(*b*), the flux starts to collapse. The collapsing flux reverses the polarity of the induced voltage from what it was when the flux was increasing. Thus, the voltage induced by a decreasing flux *aids* the source voltage and tends to keep the current from decreasing. Again, the inductance restricts the rate at which the current can change.

Inductance (*L*)

Inductors

Chokes

Reactors

Coils

The amount of voltage induced in a single conductor like that in Fig. 11-1 is very small. So small, in fact, that it has no practical significance in most low-frequency electric and electronic devices. However, at high frequencies, like those used in television systems, the inductance of a single conductor can be very significant.

The inductance of a conductor can be greatly increased by forming the conductor into a coil as in Fig. 11-2. Now the flux produced by one turn of the coil induces voltage not only in itself but in adjacent turns as well. The long, closed flux loops in Fig. 11-2 are the result of the magnetic fields of all three turns of the coil. They are stronger than the flux created by any one of the turns. Yet, they induce a voltage into each of the three turns. The inductance of the coil is much greater than the inductance of the straight length of conductor from which it was made.

Self-Inductance

Self-inductance

Mutual inductance

The inductance of an inductor is called *self-inductance*. It is given this name because the

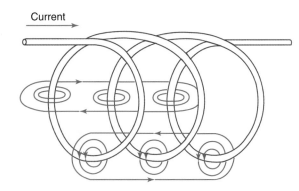

Fig. 11-2 Multiturn coil. Some of the flux created by one turn links to all the other turns.

inductor induces voltage in itself. That is, its own changing magnetic field induces voltage in its own turns of wire. In the case of a single straight conductor, its own field induces a voltage in it.

Mutual Inductance

When the magnetic flux from one conductor induces a voltage in another, electrically isolated conductor, it is called *mutual inductance*. With mutual inductance, circuits that are electrically separated can be magnetically coupled together. A transformer uses the principle of mutual inductance. Transformers are fully discussed in Chap. 12.

Lenz's Law and CEMF

The voltage induced in a conductor or coil by its own magnetic field is called a *counter electromotive force* (cemf).

Counter electromotive force (cemf)

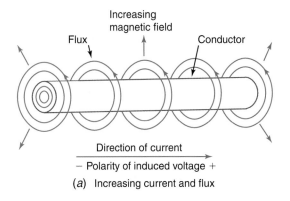

(a) Increasing current and flux

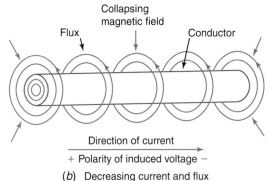

(b) Decreasing current and flux

Back electromotive force (bemf)

Fig. 11-1 Induced voltage in a conductor. The polarity of the induced voltage is dependent on whether the current is increasing or decreasing.

YOU MAY RECALL ... from Chap. 2 that electromotive force (emf) is another name for voltage.

Since the induced emf (voltage) is always opposing, or countering, the action of the source voltage, it is known as a cemf. Counter electromotive force is sometimes referred to as *back electromotive force* (bemf). The term *back* implies that the induced voltage is backward, or working against the effort of the source voltage.

The concept contained in Lenz's law is used to explain how inductance behaves. *Lenz's law* states that a cemf always has a polarity which opposes the force that created it. This idea is

238 ✦ **Chapter 11** Inductance

illustrated in Fig. 11-3(*a*), which shows, in schematic form, an inductor and an ac voltage source. When the voltage is increasing, as shown in the graph, the cemf opposes the source voltage. When the voltage is decreasing [Fig. 11-3(*b*)], the cemf aids the source voltage and tries to keep the current constant.

Energy Storage and Conversion

Another way to look at inductance is in terms of energy conversion and storage. When current flows through an inductor, the inductor builds up a magnetic field. In the process of building its magnetic field, the inductor converts electric energy into magnetic energy. When the current increases, more electric energy is converted into magnetic energy. The inductor's magnetic field now processes more energy than it had before the current increased. When current through an inductor decreases, its magnetic field decreases. Magnetic energy from the field is converted back into electric energy in the inductor. Thus, an inductor stores energy when its current increases and returns stored energy when its current decreases. *Inductance converts no electric energy into heat energy.* Only resistance is capable of converting electric energy into heat energy. Thus, if an inductor with absolutely no resistance could be constructed, its net use of energy would be zero. For two quarters of the ac cycle, it would take energy from the system. For the other two quarters of the cycle, it would return the same amount of energy to the system.

■ TEST

Answer the following questions.

1. The electric property that opposes changes in current is called _____.
2. The physical device that opposes changes in current can be called a(n) _____, _____, _____, or _____.
3. True or false. A straight wire possesses inductance.
4. True or false. A straight wire can be called an inductor.
5. True or false. Inductance converts electric energy to heat energy.
6. True or false. The cemf aids the source voltage when the current in an inductive circuit is increasing.
7. True or false. Transformers operate on the principle of self-inductance.
8. The symbol or abbreviation for inductance is _____.
9. _____ law can be used to find the polarity of the cemf in an inductor.
10. Another abbreviation for cemf is _____.
11. When current in an inductor is increasing, _____ energy is being converted to _____ energy.

11-2 Unit of Inductance— The Henry

The base unit of inductance is the *henry*. This unit, named in honor of an American scientist, is abbreviated H. The henry is defined in terms of the amount of cemf produced when the current through an inductor is changing amplitude. One henry of inductance develops 1 V of cemf when the current changes at a rate of 1 A/s. This definition of a henry is shown graphically in Fig. 11-4.

A wide range of inductances are used in electric and electronic circuits. Inductances in circuits of very high frequency are often less than 1 μH. For low-frequency circuits, inductors with more than 5 H of inductance are common.

Henry (H)

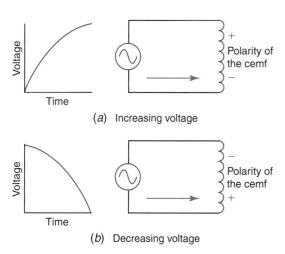

(*a*) Increasing voltage

(*b*) Decreasing voltage

Fig. 11-3 Polarity of counterelectromotive force (cemf).

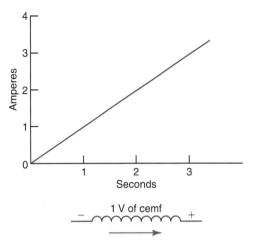

Air core

Iron core

1 V of cemf

Fig. 11-4 A 1-H inductor produces 1 V of cemf when the current changes at a rate of 1 A/s.

11-3 Factors Determining Inductance

The inductance of an inductor is primarily determined by four factors:

1. The type of core material
2. The number of turns of wire
3. The spacing between turns of wire
4. The diameter of the coil (or core)

The *core* of an inductor is the material that occupies the space enclosed by the turns of the inductor.

The amount of current in an *iron-core* inductor also influences its inductance. This is because the magnetic properties of the iron core change as the current changes.

About ◄═ Electronics

Joseph Henry American physicist Joseph Henry did extensive research on electromagnetism and discovered the principles that made the development of the telegraph possible. The fundamental unit for inductance, the henry, is named for him.

Ultimately, the amount of inductance is determined by the amount of cemf produced by a specified current change. Of course, the amount of cemf depends on how much flux interacts with the conductors of the coil.

If all other factors are equal, an iron-core inductor has more inductance than an *air-core* inductor. This is because the iron has a higher permeability, that is, it is able to carry more flux. With this higher permeability there is more flux change, and thus more cemf, for a given change in current.

Adding more turns to an inductor increases its inductance because each turn adds more magnetic field strength to the inductor. Increasing the magnetic field strength results in more flux to cut the turns of the inductor.

When the distance between the turns of wire in a coil is increased, the inductance of the coil decreases. Figure 11-5 illustrates why this is so. With widely spaced turns [Fig. 11-5(*a*)], many of the flux lines from adjacent turns do not link together. Those lines that do not link together produce a voltage only in the turn which produced them. As the turns come closer together [Fig. 11-5(*b*)], fewer lines of flux fail to link up.

When other factors are equal, the inductor with the largest-diameter core will have the most inductance. This is because all of the flux has to go through the core of an inductor. Thus a large-diameter core can handle more flux, at a specified flux density, than a small-diameter core can.

▚ TEST

Answer the following questions.

12. The base unit of inductance is the _____.

13. The abbreviation for the base unit of inductance is the _____.

14. In terms of the base units of voltage, current, and time, the base unit of inductance is equal to a _____.

15. Does the amount of inductance increase or decrease when more turns are added to an inductor?

16. List four ways to increase the inductance of an inductor.

11-4 Types of Inductors

One way of classifying inductors is by the type of material used for the core of the inductor.

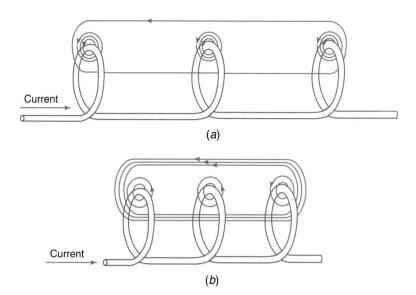

Fig. 11-5 Effect of turn spacing. (*a*) Widely spaced turns provide less inductance than (*b*) closely spaced turns do.

The core may be either a magnetic material or a nonmagnetic material. The symbols for inductors with these materials are shown in Fig. 11-6.

Inductors are also classified as either fixed or variable. Figure 11-7 shows two symbols used to indicate a *variable inductor*. The most common way of varying inductance is by adjusting the position of the core material. In Fig. 11-8 the position of the ferrite core material (called a *slug*) is adjustable within the coil form. Maximum inductance occurs when the slug is positioned directly in line with the coil of wire. Some variable inductors use a brass slug. Brass has more reluctance (opposition to flux) than air does. Therefore, the brass slug decreases inductance when it is centered in the coil.

Air-Core Inductors

An *air-core inductor*, used as part of a high-frequency circuit, is shown in Fig. 11-9. This inductor is self-supporting and requires no coil form. However, many inductors which are represented by the air-core symbol are wound on a coil form. The form may be either solid or hollow. These forms have about the same reluctance (opposition to magnetic flux) as air does. Therefore, the inductor is much like an air-core inductor; its core is nonmagnetic. These inductors may be wound on such core materials as ceramic or phenolic. They often look like the coil in Fig. 11-10(*a*). These inductors seldom have more than 5 mH of inductance.

Ferrite and Powdered-Iron Cores

The coil shown in Fig. 11-10(*a*) may also be an *iron-core inductor*. In this case the core material would be ferrite or powdered iron. The correct symbol would then be the iron-core symbol of Fig. 11-6(*a*). (On some schematic diagrams, the two solid lines in the iron-core symbol are replaced by two broken lines to

Variable inductor

Slug

Iron-core inductor

Air-core inductor

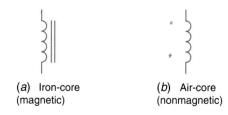

(*a*) Iron-core (magnetic) (*b*) Air-core (nonmagnetic)

Fig. 11-6 Fixed-value inductor symbols.

Fig. 11-7 Variable inductor symbols. Either symbol can be used for either magnetic or nonmagnetic cores.

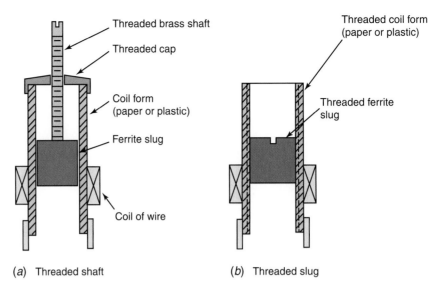

Threaded brass shaft

Threaded cap

Coil form
(paper or plastic)

Ferrite slug

Coil of wire

Threaded coil form
(paper or plastic)

Threaded ferrite
slug

(a) Threaded shaft

(b) Threaded slug

Fig. 11-8 Methods of adjusting inductance.

**Molded
inductors**

represent a ferrite or powdered iron core.) Most of the inductors of this type have less than 200 mH of inductance. They are used primarily at frequencies above the audio (sound) range. Fig. 11-10(b) shows another style of inductor wound on a ferrite core.

Toroid Cores

Toroid core

The cores of the inductors discussed so far have all been straight. The magnetic flux loops must extend through the air as well as through the core material. With a *toroid core,* the flux loops all exist within the core. Toroid cores (Fig. 11-11) are doughnut-shaped. Each turn of wire is threaded through the center of the core, as shown in Fig. 11-11. Inductors made with toroid cores are called toroidal inductors. The toroid core is usually made from powdered iron or ferrite. Toroidal inductors can have high inductance values for their size.

Molded Inductors

Some inductors look like resistors (Fig. 11-12). These inductors are enclosed in an insulating material to protect the inductor winding. Molded inductors can have cores of air, ferrite, or powdered iron.

Shielded Inductors

Inductors are often shielded to protect them from the influence of magnetic fields other

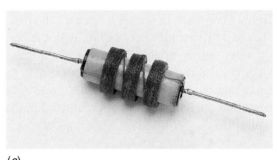

(a)

(b)

Fig. 11-10 Miniature high-frequency inductors.
(a) Either nonmagnetic or ferrite core.
(b) Ferrite core. *(Mark Steinmetz)*

Fig. 11-9 Air-core inductor.
(Mark Steinmetz)

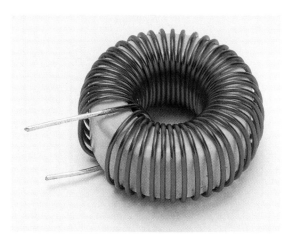

Fig. 11-11 Toroid core. All the flux produced by the coil is concentrated in low-reluctance core. *(Mark Steinmetz)*

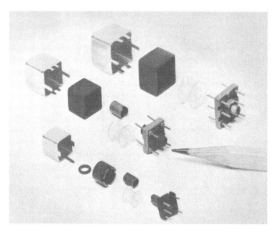

Fig. 11-13 Shielded inductors. *(Courtesy of Bell Industries, J. W. Miller Division)*

than their own. The shield is made from a magnetic material. Figure 11-13 shows an exploded view of the parts of three shielded, adjustable coil forms. The coil windings are not shown. They would be wound on the cylindrical tubes, or bobbins. The coil forms shown in Fig. 11-13 are the type used on printed circuit boards.

Some miniature chokes (inductors), like those in Fig. 11-12, are also shielded. Their shields are encased underneath the outside molding.

Laminated Iron Core

Nearly all the large inductors used at power frequencies (60 Hz, for example) use laminated iron cores. These inductors have inductances ranging from about 0.1 to 100 H.

The typical laminated core uses laminations like those in Fig. 11-14. From this illustration

it is easy to see why these laminations are called *E and I laminations*. The E laminations are stacked together to the desired thickness, as are the I laminations. The winding is put on the center leg (Fig. 11-15) of the E lamination. The I lamination is then positioned across the open end of the E lamination.

E and I laminations

As seen in Fig. 11-16, the E and I laminations form two parallel paths for flux. The center leg of the E lamination is twice as wide as either of the outside legs because it has to carry twice as much flux. For a given amount and rate of current change, the laminated-iron-core inductor creates more flux than other types of inductors. This changing flux, in turn, creates

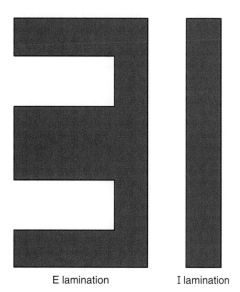

E lamination   I lamination

Fig. 11-14 E and I laminations. These laminations are stacked in various configurations to form cores for electromagnetic devices.

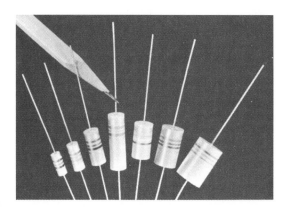

Fig. 11-12 Molded inductors. *(Courtesy of Bell Industries, J. W. Miller Division)*

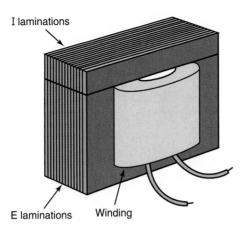

Fig. 11-15 Laminated-iron-core inductor. The coil fits over the center leg of the E laminations.

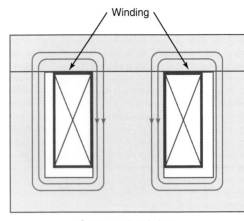

Cross-sectional view

Fig. 11-16 Flux paths in a laminated core.

cemf. This is why laminated-iron-core inductors can provide large amounts of inductance.

The inductance of an iron-core inductor is somewhat dependent on the amount of current flowing through it. The reason for this can be seen from Fig. 11-17, which illustrates the magnetic characteristics of the core material. The permeability of iron and iron alloys decreases as the magnetic field strength increases. The magnetic field of strength of an inductor is a function of the amount of current flowing in the winding. Refer to Fig. 11-17. Suppose the current through the inductor changes from point A to point B. The flux density would change from A' to B' and produce a certain amount of cemf. This amount of cemf would, of course, represent a certain amount of inductance. Now suppose the current in the inductor was greater and the current changed from C to

D in Fig. 11-17. Although C to D is the same amount of change as A to B, it produces a much smaller change (C' to D') in flux. Thus, the inductor has less cemf and inductance at the higher current. Points E and F in Fig. 11-17 show what happens when the core of an inductor is *saturated*. Since a change in current from E to F produces almost no change in flux, there is very little inductance. Except for special applications, inductors are never operated in the saturation region of the permeability curve.

Filter Chokes

Laminated-iron-core inductors are often referred to as *filter chokes*. These chokes are used in the filter circuits of power supplies in a wide variety of electrical and electronic

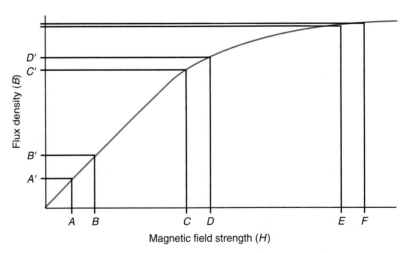

Fig. 11-17 Permeability curve. Permeability decreases as flux density and magnetic field strength increase.

equipment. The power supply is often the part of the equipment which converts alternating to direct current. The filter circuit, which includes the inductor, smooths out the fluctuating or pulsating direct current until it is nearly pure direct current.

There are two types of filter chokes: the *smoothing* choke and the *swinging* choke. The swinging choke is one in which the I and E laminates are butted together so that there is a minimum air gap between them. This makes the amount of inductance vary with the amount of current (Fig. 11-17). A typical swinging choke may be rated 20 H at 50 mA and 5 H at 200 mA.

The smoothing choke frequently has a small (0.1 mm) air gap between the I and E laminates. This makes the inductance less dependent on the amount of current because air does not saturate as easily as iron.

Radio-Frequency Chokes or Coils

Inductors which are used at higher frequencies are often called RF chokes or *RF coils*. Since radio was one of the early popular uses of high-frequency inductors, they became known as radio-frequency, or RF, coils. An RF coil or choke may have an air, powdered iron, or ferrite core. It may be either fixed or variable.

11-5   Ratings of Inductors

We have seen so far that one of the main ratings of an inductor is its inductance. Inductors are also rated for dc resistance, current, voltage, quality, and tolerance.

The dc resistance specifies the resistance of the wire in the winding of the inductor. This is the resistance between the terminals of the inductor that one would measure with an ohmmeter. Therefore this dc resistance is sometimes called the *ohmic resistance.*

The current rating of an inductor is important because it indicates how much current the inductor can continuously carry without overheating. With laminated-core inductors, the current rating also indicates the current level at which the inductance was measured. At lower current levels, the inductance is greater than the specified value.

The voltage rating indicates how much voltage the insulation on the inductor winding can

continuously withstand. Exceeding this voltage rating may not result in instantaneous breakdown of the insulation. However, it will shorten the life expectancy of the inductor's insulation. Voltage ratings are used mostly with laminated-core inductors. With these inductors, the core is often physically and electrically connected to the chassis of an electric device. However, the winding may be hundreds of volts positive or negative with respect to the chassis.

The *quality* of an inductor refers to the ratio of its reactance to its resistance. Generally it is desirable to have a high-quality inductor. All other factors being equal, the lower the dc resistance, the higher the quality of the inductor. Detailed information on quality is provided in Sec. 11-8.

Like all other components, inductors have manufacturer's tolerances. Precision inductors can be obtained with tolerances of less than ± 1 percent. However, they are expensive. Typical inductors used in mass-produced electric and electronic devices have tolerances of ± 10 percent or more.

11-6   Inductors in DC Circuits

The behavior of an inductor in a pure dc circuit is contrasted to that of a resistor in Fig. 11-18. With a resistor [Fig. 11-18(*a*)], the current jumps to its maximum value almost the instant the switch is closed. When the switch is opened, it drops back to zero just as fast. An inductor in a dc circuit [Fig. 11-18(*b*)] forces the current to rise more slowly. This is due to the inductor's cemf. The time required for the current to reach its maximum value is dependent on the amount of inductance and resistance. With inductors of typical quality, the time is much less than 1 s. Once the current reaches its peak value, the only opposition the inductor offers is its dc resistance. When the switch in Fig. 11-18(*b*) is opened, the cemf of the inductor prevents the current from instantaneously dropping to zero. It does this by ionizing the air between the switch contacts as the switch opens. As the energy stored in the inductor's magnetic field is used up, the switch contacts deionize and current stops.

When the switch in Fig. 11-18(*b*) is opened, the cemf of the inductor becomes much greater than the source voltage. The high voltage

Smoothing choke

Swinging choke

Quality

RF coils

Ohmic resistance

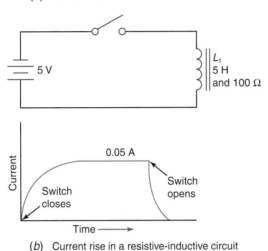

(a) Current rise in a resistive circuit

(b) Current rise in a resistive-inductive circuit

Fig. 11-18 Comparison of a resistive and a resistive-inductive dc circuit. The inductor opposes changes in current.

Inductive kick

(cemf) generated when an inductive circuit is opened is known as an *inductive kick*. It is the voltage which ionizes the air between the switch contacts and causes the contacts to arc and burn. The inductive kick of an inductor is very high because the current drops very rapidly when the switch is opened. The difference between the source voltage and the cemf is dropped across the ionized air between the switch contacts. Kirchhoff's voltage law still applies. That is, the voltage across the switch plus the inductor's cemf still equal the source voltage. Notice the polarities in Fig. 11-19. The inductor's cemf and the voltage across the switch are series-opposing. Thus, they both can be much greater than the battery voltage. The exact value of the cemf in Fig. 11-19 when

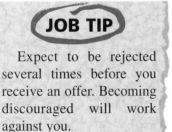

the switch opens depends upon two factors: the amount of inductance and the amount of current in the circuit before the switch is opened. The inductive kick of an inductor can be many thousands of volts. Inductive kick is the principle on which the ignition coil in an automobile operates.

The relationship between the current and voltage (cemf) in an inductor is illustrated in Fig. 11-20. The resistance R in this figure is very high relative to the ohmic resistance of the inductor. Therefore, the voltage across the inductor is almost zero once the current reaches its maximum value. Notice in Fig. 11-20 that the voltage across the inductor is maximum when the current through it is minimum. Also, the voltage is minimum when the current is maximum. Further, notice that the resistive current and voltage rise together.

■ TEST

Answer the following questions.

17. Draw the symbols for
    a. A fixed iron-core inductor
    b. A fixed air-core inductor
    c. A variable inductor
18. True or false. A slug-type core is used in an iron-core inductor.
19. True or false. When a brass core is centered inside a coil winding, the coil will have maximum inductance.
20. True or false. The air-core inductor symbol is used for all inductors that use non-magnetic core material.
21. True or false. A ferrite-core inductor would be represented by using the symbol for an iron-core inductor.
22. True or false. A 50-mH inductor would most likely have an air core.
23. True or false. To provide maximum inductance, a ferrite core should be centered in the coil winding.

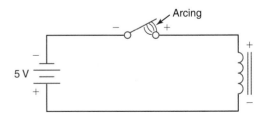

Fig. 11-19 Voltage polarities when an inductive circuit is opened.

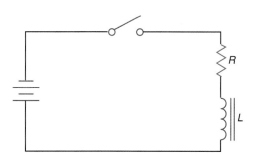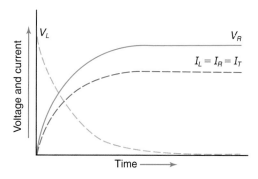

Fig. 11-20 Current and voltage relationships in an inductor. Maximum inductive voltage occurs before maximum current is reached.

24. What type of core would be used in a 5-H inductor?
25. What is the shape of a toroidal core?
26. Why is the center leg of an E lamination wider than the outside legs?
27. Why does an inductor have a current rating?
28. What is meant by the quality of an inductor?
29. What is an RF choke?
30. What type of core does a filter choke have?
31. What is meant by inductive kick?
32. An inductive circuit and a resistive circuit have equal currents. When the switch is opened in each circuit, which circuit produces more arcing across the switch contacts?
33. Does the cemf exceed the source voltage when an inductive dc circuit is opened?
34. Does an inductor's voltage (cemf) reach its maximum value before the current reaches its maximum value?

11-7 Ideal Inductors in AC Circuits

An *ideal inductor* is an inductor that has no resistance. It does not convert any electric energy into heat energy, and it has infinite quality. In the discussions that follow, we assume we have ideal inductors.

Inductive Reactance

Inductance, like capacitance, controls circuit current without using power. Therefore, the opposition of an inductor to alternating current is also called reactance X. To distinguish inductive reactance from capacitive reactance, we use the symbol X_L for *inductive reactance*.

Inductive reactance is the result of the cemf of the inductor. The inductor lets just enough ac flow to produce a cemf which is equal to (and opposite to) the source voltage. This idea is illustrated in Fig. 11-21. During each half-cycle of the source (an ac generator), the cemf of the inductor produces a matching half-cycle of sinusoidal voltage. At any instant of time the two voltages (source and cemf) are equal. The reason for this can be ascertained by referring to Fig. 11-22, which shows that the voltage leads the current by 90° in an inductive circuit. Notice from the figure that the current is changing direction at the instant the source voltage is at its peak value. When the current changes direction, two things happen: (1) the polarity of the mmf and the direction of the flux change, and (2) the flux changes from a collapsing flux to an expanding flux (or vice versa). Either of these happenings would change the polarity of the cemf; but when they occur simultaneously, the polarity cannot change. Also, notice from Fig. 11-22 that the rate of current change, and therefore the rate of flux change, is greatest as the current crosses the zero reference line. Thus, the cemf is greatest as the current is changing direction.

Inductive reactance (X_L)

Ideal inductor

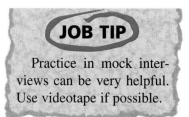

JOB TIP

Practice in mock interviews can be very helpful. Use videotape if possible.

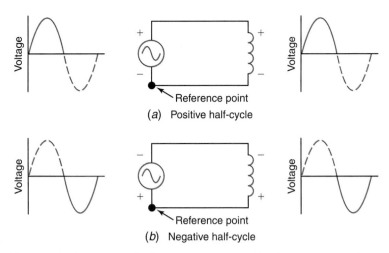

(a) Positive half-cycle

(b) Negative half-cycle

Fig. 11-21 Source and inductor ac voltages. The inductor's cemf opposes the source voltage.

The reactance of an inductor can be calculated with the following formula:

$$X_L = 2\pi f L = 6.28 f L$$

$X_L = 2\pi f L$

The inductive reactance is in ohms when the frequency is in hertz and the inductance is in henrys.

From the above formula it can be seen that inductive reactance is directly proportional to both frequency and inductance. Doubling either doubles the reactance. This direct proportional relationship makes sense when one recalls two things:

1. The higher the frequency, the more rapidly the current is changing. Thus more cemf and more reactance are produced.
2. The higher the inductance, the more flux change per unit of current change. Again, more cemf and reactance are produced.

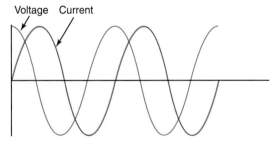

Voltage   Current

Fig. 11-22 Alternating current and voltage in an inductor. The voltage leads the current by 90°.

Example 11-1

What is the reactance of a 3-H inductor when the frequency is 120 Hz?

Given: Inductance $L = 3$ H
Frequency $f = 120$ Hz

Find: Inductive reactance (X_L)

Known: $X_L = 6.28 f L$

Solution: $X_L = 6.28 \times f \times L$
$= 6.28 \times 120$ Hz $\times 3$ H
$= 2261\ \Omega$

Answer: The 3-H inductor has 2261 Ω of reactance at 120 Hz.

Example 11-2

A 2.5-mH inductor is placed in a circuit where the frequency is 100 kHz. What is its inductive reactance?

Given: $L = 2.5$ mH
$f = 100$ kHz

Find: X_L

Known: $X_L = 6.28 f L$

Solution: 2.5 mH $= 0.0025$ H
100 kHz $= 100,000$ Hz
$X_L = 6.28 \times 100,000$ Hz
$\times 0.0025$ H
$= 1570\ \Omega$

Answer: The 2.5-mH inductor has 1570 Ω of reactance at 100 kHz.

When the inductor's current and voltage are known, its reactance can be calculated by using Ohm's law.

Example 11-3

The ac voltage measured across an inductor is 40 V. The current measured through the inductor is 10 mA. What is its reactance?

Given: $V_L = 40$ V
$I_L = 10$ mA

Find: X_L

Known: $X_L = \dfrac{V_L}{I_L}$

Solution: $X_L = \dfrac{40\ \text{V}}{0.01\ \text{A}} = 4000\ \Omega$

Answer: The inductor has 4000 Ω of reactance.

Phase Relationships of *I* and *V*

The sinusoidal cemf of the inductor in Fig. 11-21 is produced by a sinusoidal current through the inductor. This current wave, shown in Fig. 11-22, is 90° out of phase with the cemf and the source voltage. The current must be 90° out of phase because the cemf can be zero only when the current is not changing. The only instant when the current is not changing is when it is exactly at its peak value. That is, at the instant the current has just stopped rising and has

not yet begun to fall, it is effectively a constant value. This is the instant at which zero cemf occurs. It can be seen from Fig. 11-22 (and Fig. 11-20) that the *voltage leads the current* in an inductor circuit. More precisely, the voltage leads the current by exactly 90° in an ideal inductor.

Voltage leads the current

Power in an Inductor

The *ideal inductor* uses no power because its current and voltage are 90° out of phase. (Remember, $P = IV \cos \theta$, and $\cos 90° = 0$.) Thus, in a pure inductance (ideal inductor) both current and voltage are present but there is no net conversion of energy.

Ideal inductor

From an energy point of view, one can say that energy is transferred back and forth between the source and the inductor. During the quarter-cycle in which current is rising [Fig. 11-23(a)], energy is taken from the source (generator). The energy from the generator is converted into magnetic energy and stored in the inductor's field. During the next quarter-cycle [Fig. 11-23(b)], when the current is decreasing, the field of the inductor collapses. Its stored energy is converted back to electric energy and returned to the generator. As shown in Fig. 11-23(c), energy is again taken from the generator during the third quarter-cycle. This energy is again returned during the fourth quarter-cycle [Fig. 11-23(d)]. Notice in Fig. 11-23(b) and (c) that the polarity of the cemf remains the same.

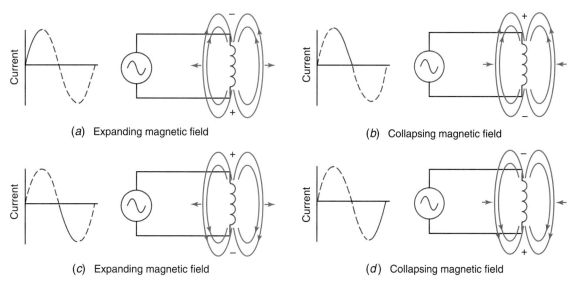

(a) Expanding magnetic field   (b) Collapsing magnetic field

(c) Expanding magnetic field   (d) Collapsing magnetic field

Fig. 11-23 Power in an inductor. Every other quarter-cycle (b and d) the inductor returns its stored energy to the source.

Yet the magnetic field changes from an expanding field to a collapsing field. As previously explained, this is because the polarity of the magnetic field also changes [between Fig. 11-23(b) and (c)] when the direction of the current reverses.

■ TEST

Answer the following questions.

35. Does an ideal inductor possess any resistance?
36. What causes inductive resistance?
37. The symbol for inductive reactance is _____.
38. The formula for calculating the reactance of an inductor is _____.
39. Doubling the frequency of an inductive circuit causes the reactance to _____.
40. Determine the reactance of the following inductances at the frequencies specified:
    a. 6 H at 60 Hz
    b. 150 mH at 10 kHz
    c. 30 μH at 250 MHz
41. What is the reactance of an inductor that drops 20 V ac when 0.5 A ac flows through it?
42. Inductance causes current to _____ voltage by _____ degrees.
43. The power consumed by an ideal inductor that draws 2 A from a 20-V ac source is _____.
44. How much inductance is needed to provide 3600 Ω of reactance to a 30-V, 400-Hz source?
45. What is the frequency when a 3-mH inductor produces 4200 Ω of reactance?

11-8 Real Inductors in AC Circuits

Real (nonideal) inductors use some power because all inductors possess resistance as well as reactance. The quality of real inductors is less than infinite.

. . . that quality is defined as reactance divided by resistance and that it is frequency-dependent. For an inductor, the formula for quality Q is

$$Q = \frac{X_L}{R}$$

Impedance (Z)

Example 11-4

What is the quality of a 10-mH coil (inductor) at 150 kHz if its resistance is 60 Ω?

Given: $L = 10$ mH
$f = 150$ kHz
$R = 60\ \Omega$

Find: Q

Known: $Q = \dfrac{X_L}{R}$ and $X_L = 6.28fL$

Solution: $X_L = 6.28 \times f \times L$
$= 6.28 \times 150{,}000$ Hz
$\times 0.01$ H
$= 9420\ \Omega$

$Q = \dfrac{9420\ \Omega}{60\ \Omega} = 157$

Answer: The Q of the coil is 157.

The quality of iron-core inductors used at low frequencies is often less than 10. With air-core inductors operating at high frequencies, the quality can be more than 200. Typical RF chokes have a quality ranging from 30 to 150. The higher the quality of the coil, the less power the inductor uses. Also, the higher the quality, the farther the current and voltage are out of phase. Current and voltage are 90° out of phase only when there is no resistance.

The combined opposition offered by resistance and reactance is called *impedance*. Since an inductor has both resistance and reactance, it offers impedance to an ac current. To be technically correct, we should specify the impedance of an inductor. However, the dominant form of opposition of an inductor is reactance. Therefore, we usually talk about its opposition in terms of reactance only. When the quality of the inductor is above 5, the difference between its reactance and its impedance is less than 2 percent. Thus, the use of reactance instead of impedance when calculating the current in an inductor is quite reasonable. It is especially reasonable when you consider that the inductance may be 20 percent above or below its rated value.

Power Losses in Inductors

It has already been emphasized that real inductors use power because of their resistance.

However, their resistance may actually be greater than the resistance measured by an ohmmeter. This higher-than-measured resistance is the result of the *skin effect*. The skin effect is caused by the tendency of electrons to travel close to the outer surface of a conductor (Fig. 11-24). The higher the frequency, the more pronounced the skin effect becomes. Because of the skin effect, the center of a conductor does not contribute to the current-carrying capacity of the conductor. Thus, the *effective resistance* of the conductor at high frequencies is greater than that measured by an ohmmeter.

The skin effect can be minimized by using *litz wire*, which is a multiple-conductor cable. Each conductor in litz wire has a very thin insulation on it. The conductors are very small in diameter (about 44 gage). These small, insulated conductors are twisted together to form a very small cable. When litz wire is used, the individual conductors are soldered together at the ends of the coil. This connects all the conductors in parallel to effectively make a single wire out of the multiple-conductor cable. For a given overall diameter, litz wire provides more surface area than a single-strand conductor. Because of this greater surface area, litz wire has a lower resistance at high frequencies.

Iron-core inductors have power losses in their core material as well as in their winding. Two actions cause the core to convert electric energy into heat energy. First, the magnetic field of the winding induces a voltage in the core material. The induced voltage causes a small current to flow in the core. This current produces heat in the core. The second action which causes power loss in the core is the periodic reversal of the magnetic field. Every time the polarity of the magnetic field reverses, it creates a small amount of heat in the core. Methods of minimizing these core losses are discussed in Chap. 12.

Skin effect

Effective resistance

■ TEST

Litz wire

Answer the following questions.

46. Does an iron-core inductor or an air-core inductor have a higher quality rating?
47. What is the quality of a 0.3-H inductor at 20 kHz if it has an effective resistance of 100 Ω?
48. What is impedance?
49. What causes the resistance of a conductor to be greater at high frequencies than at low frequencies?
50. What is litz wire?
51. What causes the core of a laminated-iron-core inductor to heat up?

11-9 Inductors in Parallel

Parallel inductors (with no mutual inductance) can be treated just like parallel resistors. The formulas used with resistors can be used with inductors by substituting L for R. The formulas are

Parallel inductors

General method:

$$L_T = \cfrac{1}{\cfrac{1}{L_1} + \cfrac{1}{L_2} + \cfrac{1}{L_3} + \text{etc.}}$$

Two inductors in parallel:

$$L_T = \frac{L_1 \times L_2}{L_1 + L_2}$$

n equal inductors in parallel:

$$L_T = \frac{L}{n}$$

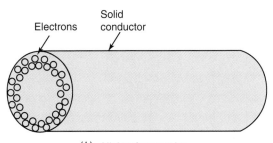

(a) DC and low frequencies

(b) Higher frequencies

Fig. 11-24 Skin effect. At high frequencies, the current concentrates near the surface of the conductor.

Example 11-5

What is the inductance of a 0.4-H inductor and a 600-mH inductor connected in parallel?

Given: $L_1 = 0.4$ H
$L_2 = 600$ mH $= 0.6$ H

Find: L_T

Known: $L_T = \dfrac{L_1 \times L_2}{L_1 + L_2}$

Solution: $L_T = \dfrac{0.4 \text{ H} \times 0.6 \text{ H}}{0.4 \text{ H} + 0.6 \text{ H}} = \dfrac{0.24}{1.0}$
$= 0.24$ H

Answer: The total (or equivalent) inductance is 0.24 H, or 240 mH.

Notice that the total inductance is less than the smallest of the parallel inductances. It always is with parallel inductors.

Total inductive reactance

The *total inductive reactance* of parallel inductors can be found by either of two methods. The first method is to find the total inductance as in example 11-5 and then find the total reactance by using the reactance formula. The second method is to determine the reactance of the individual inductors and then, using the parallel formula, combine the individual reactances to find the total reactance. The formulas for combining parallel inductive reactances have the same structure as those used for parallel resistances and parallel inductances:

General method:

$$X_L = \dfrac{1}{\dfrac{1}{X_{L_1}} + \dfrac{1}{X_{L_2}} + \dfrac{1}{X_{L_3}} + \text{etc.}}$$

Parallel inductive reactances

Two parallel inductive reactances:

$$X_L = \dfrac{X_{L_1} \times X_{L_2}}{X_{L_1} + X_{L_2}}$$

n equal inductive reactances in parallel:

$$X_L = \dfrac{X_L}{n}$$

Example 11-6

Using the first method method, find the total

inductive reactance for example 11-5 when the frequency is 20 kHz.

Given: $L_T = 0.24$ H
$f = 20$ kHz

Find: X_{L_T}

Known: $X_{L_T} = 6.28f\,L_T$

Solution: $X_{L_T} = 6.28f\,L_T$
$= 6.28 \times 20{,}000$ Hz
$\times 0.24$ H
$= 30{,}144\ \Omega$

Answer: The total inductive reactance is $30{,}144\ \Omega$, or $30.144\ \text{k}\Omega$.

Example 11-7

Using the second method, find the total inductive reactance for example 11-5 when the frequency is 20 kHz.

Given: $L_1 = 0.4$ H
$L_2 = 0.6$ H
$f = 20$ kHz

Find: X_{L_T}

Known: $X_L = 6.28f\,L$

$X_{L_T} = \dfrac{X_{L_1} \times X_{L_2}}{X_{L_1} + X_{L_2}}$

Solution: $X_{L_1} = 6.28 \times 20{,}000 \times 0.4$
$= 50{,}240\ \Omega$
$X_{L_2} = 6.28 \times 20{,}000 \times 0.6$
$= 75{,}360\ \Omega$
$X_{L_T} = \dfrac{50{,}240 \times 75{,}360}{50{,}240 + 75{,}360}$
$= 30{,}144\ \Omega$

Answer: The total inductive reactance is $30{,}144\ \Omega$.

As you might expect, the two methods used in examples 11-6 and 11-7 give exactly the same answer.

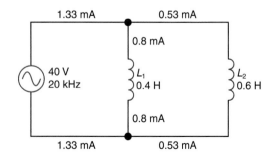

Fig. 11-25 Currents in parallel inductors. The smaller inductor carries the greater current.

In parallel-inductor circuits the total current splits up in inverse proportion to the inductance of the individual inductors. The lowest inductance carries the highest current (Fig. 11-25). The exact value of the current in each branch of the circuit can be found by using Ohm's law. Just replace the R in Ohm's law with X_L. The currents recorded in Fig. 11-25 were calculated by using the reactances in example 11-7:

$$I_{L_1} = \frac{V_{L_1}}{X_{L_1}} = \frac{40\text{ V}}{50,240\ \Omega} = 0.0008\text{ A} = 0.8\text{ mA}$$

$$I_{L_2} = \frac{V_{L_2}}{X_{L_2}} = \frac{40\text{ V}}{75,360\ \Omega} = 0.00053\text{ A} = 0.53\text{ mA}$$

$$I_T = \frac{V_T}{X_{L_T}} = \frac{40\text{ V}}{30,144\ \Omega} = 0.00133\text{ A} = 1.33\text{ mA}$$

Of course, I_T can also be found by using Kirchhoff's current law. For the circuit of Fig. 11-25, we have

$$I_T = I_{L_1} + I_{L_2}$$
$$= 0.8\text{ mA} + 0.53\text{ mA} = 1.33\text{ mA}$$

The current-divider formula used with parallel resistors and resistances can also be used with parallel inductors and reactances. If I_{L_1} in Fig. 11-25 was not known, it could be calculated with the current-divider formula in either of these two ways:

$$I_{L_1} = \frac{I_T \times X_{L_2}}{X_{L_1} \times X_{L_2}}$$
$$= \frac{1.33\text{ mA} \times 75.36\text{ k}\Omega}{50.24\text{ k}\Omega + 75.36\text{ k}\Omega} = 0.8\text{ mA}$$

$$I_{L_1} = \frac{I_T \times L_2}{L_1 + L_2} = \frac{1.33\text{ mA} \times 0.6\text{ H}}{0.4\text{ H} + 0.6\text{ H}} = 0.8\text{ mA}$$

11-10 Inductors in Series

Treat *series inductances* and *series reactances* the same way you treat series resistances. The formula for series inductances is

$$L_T = L_1 + L_2 + L_3 + \text{etc.}$$

For inductive reactance in series the formula is

$$X_{L_T} = X_{L_1} + X_{L_2} + X_{L_3} + \text{etc.}$$

Again the total reactance can also be determined by the reactance formula if the total inductance is known. That is,

Example 11-8

Using the reactance formula, find the total reactance at 60 Hz of a 3-H choke and a 5-H choke connected in series.

Given: $L_1 = 3$ H
             $L_2 = 5$ H
             $f = 60$ Hz

Find: X_{L_T}

Known: $X_{L_T} = 6.28fL_T$
            $L_T = L_1 + L_2$

Solution: $L_T = 3$ H $+ 5$ H $= 8$ H
            $X_{L_T} = 6.28 \times 60$ Hz $\times 8$ H
                 $= 3014\ \Omega$

Answer: The total inductive reactance is 3014 Ω.

Example 11-9

Find the total reactance for example 11-8 without first finding the total inductance.

Given: $L_1 = 3$ H
             $L_2 = 5$ H
             $f = 60$ Hz

Find: X_{L_T}

Known: $X_{L_T} = X_{L_1} + X_{L_2}$
            $X_{L_1} = 6.28fL_1$
            $X_{L_2} = 6.28fL_2$

Solution: $X_{L_1} = 6.28 \times 60$ Hz $\times 3$ H
                  $= 1130\ \Omega$
            $X_{L_2} = 6.28 \times 60$ Hz $\times 5$ H
                  $= 1884\ \Omega$
            $X_{L_T} = 1130\ \Omega + 1884\ \Omega$
                  $= 3014\ \Omega$

Answer: The total reactance is 3014 Ω.

$$X_{L_T} = 6.28f\,L_T$$

As shown in Fig. 11-26, the total voltage in a series inductor circuit splits in direct proportion to the individual inductances:

$$V_{L_1} = \frac{L_1}{L_T} \times V_T$$
$$= \frac{3\text{ H}}{8\text{ H}} \times 20\text{ V} = 7.5\text{ V}$$

$$V_{L_2} = \frac{5\text{ H}}{8\text{ H}} \times 20\text{ V} = 12.5\text{ V}$$

Series inductances

Series reactances

Fig. 11-26 Voltages in series inductors. The larger inductor drops the greater voltage.

The exact voltage values can also be found by using Ohm's law and the reactances of example 11-9. Again, Kirchhoff's laws apply. The total voltage equals the sum of the individual circuit voltages.

11-11 Time Constants for Inductors

$$T = \frac{L}{R}$$

YOU MAY RECALL ... that in Sec. 10-8 we looked at time constants for resistor-capacitor (R-C) combinations.

The concepts developed there can easily be extended to cover resistor-inductor (R-L) combinations. The only modification needed is to think in turns of current rather than voltage.

The time constant for an R-L circuit is defined as the time required for the current through the resistor-inductor to rise to 63.2 percent of its final value. For an increasing current, such as that shown in Fig. 11-27(a), the final value is the value determined by the resistance in the circuit and the voltage applied to the circuit.

For a circuit in which the current is decreasing, the time constant is defined as the time required for the inductor's current to be reduced by 63.2 percent of its starting value. As shown in Fig. 11-27(b), it is also the time it takes for the current to decay to 36.8 percent of its former value.

The time constant of an R-L circuit can be calculated using the formula

$$T = \frac{L}{R}$$

The time constant is in seconds when L is in henrys and R is in ohms. Why the time is in seconds is shown by substituting equivalent

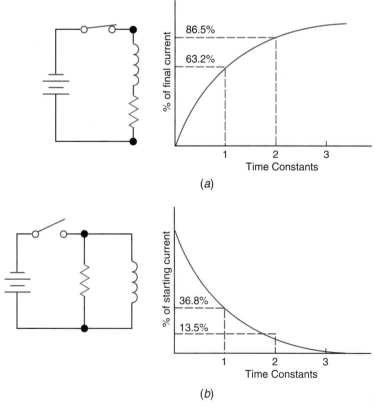

Fig. 11-27 *RL* time constants.

base units into the formula remembering that

$$\text{Henrys} = \frac{\text{volts}}{\text{amperes/seconds}}$$

and

$$\text{Ohms} = \frac{\text{volts}}{\text{amperes}}$$

Thus,

$$\text{Time constant} = \frac{\text{inductance}}{\text{resistance}} = \frac{\text{henrys}}{\text{ohms}}$$

$$= \frac{\dfrac{\text{volts}}{\text{amperes/seconds}}}{\dfrac{\text{volts}}{\text{amperes}}}$$

$$= \frac{\dfrac{\text{volts} \times \text{seconds}}{\text{amperes}}}{\dfrac{\text{volts}}{\text{amperes}}} = \text{seconds}$$

Example 11-10

Assume the inductor in Fig. 11-27 is an ideal inductor rated at 10 H. If the resistor is 50 Ω, what is the time constant of the circuit in Fig. 11-27?

Given: $L = 10$ H
           $R = 50$ Ω
Find: T
Known: $T = \dfrac{L}{R}$

Solution: $T = \dfrac{L}{R} = \dfrac{10 \text{ H}}{50 \text{ Ω}} = 0.2$ s

Answer: The time constant is 0.2 s.

It should be noted that T for practical values of R and L is usually in fractions of a second. Even if the resistor is removed from the circuit, the ohmic (dc) resistance of multihenry inductors keeps the time constant small.

11-12 Preventing Mutual Inductance

Mutual inductance can be reduced or prevented by the following methods:
1. Axis orientation
2. Physical separation
3. Shielding

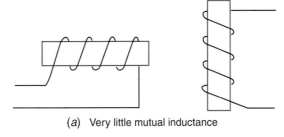

(a) Very little mutual inductance

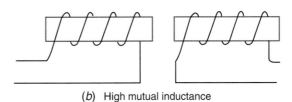

(b) High mutual inductance

Fig. 11-28 Axis orientation and mutual inductance.

Suppose the center axes of two coils are at 90° to each other, as shown in Fig. 11-28(a). Under this condition, very little of the flux from one coil cuts the other coil. We say that very little *coupling* occurs between the coils (inductors). When the axes of the coils are lined up and close together, as in Fig. 11-28(b), mutual inductance results.

Coupling

When inductors are physically separated, mutual inductance is reduced. The farther apart the inductors are, the less mutual inductance (coupling of flux) they have.

An inductor that is enclosed in a magnetic shield has very little mutual inductance with surrounding inductors. The flux from surrounding inductors passes through the low reluctance (high permeability) of the shield rather than through the inductor.

11-13 Undesired Inductance

As mentioned earlier, all conductors possess inductance. The inductance of a single wire, although low, is significant at very high frequencies. Often this inductance is undesirable because of its effect on the electric or electronic circuit. In high-frequency circuits, interconnecting leads are kept as short as possible to reduce inductance. Whenever possible, the inductance of the leads is used as part of the required inductance of the circuit.

Wire-wound and deposited-film resistors may have an appreciable amount of inductance. Their resistive elements are coils of conductive

Career and other information can be found at the Website for the Electronic Industries Association (EIA).

material wound on a nonmagnetic insulator form. In dc and low-frequency ac circuits, this undesired inductance has very little reactance and can be ignored. However, at higher frequencies the reactance becomes greater and the total opposition (impedance) of the resistor significantly exceeds its resistance. In many electronic circuits this is undesirable and unacceptable.

Noninductive wire-wound resistors

To minimize the above problem, special *noninductive wire-wound resistors* are produced. In these resistors, half the turns of wire are wound clockwise and half are wound counterclockwise. Thus the magnetic field of half of the turns cancels the field of the other half of the turns. These resistors are often used in high-power circuits when the load on the circuit must be independent of the frequency.

TEST

Answer the following questions.

52. A 0.3-H inductor and a 0.6-H inductor are connected in parallel. They are connected to a 15-V, 150-Hz source.

a. What is their total, or equivalent, inductance?
b. What is the total reactance?
c. What is the total current?
d. What is the current through the 0.6-H inductor?

53. Suppose the inductors in question 52 are now connected in series rather than parallel.

a. What is the total inductance?
b. What is the total reactance?
c. What is the total current?
d. What is the voltage across the 0.6-H inductor?

54. How can you minimize the mutual inductance between two coils?

55. How is a wire-wound noninductive resistor constructed?

56. Determine the time constant of a 30-H ideal inductor connected in series with a 60-Ω resistor and a 10-V dc source.

Summary

1. Inductance opposes changes in current.
2. Inductance results from induced voltage.
3. Inductors are devices which provide inductance. Chokes, coils, and reactors are other names for inductors.
4. The symbol for inductance is L.
5. The induced voltage in an inductor is known as counter electromotive force (cemf) or back electromotive force (bemf).
6. Lenz's law is concerned with the polarity of an induced voltage (cemf).
7. A cemf opposes the change that created it.
8. Inductors convert energy back and forth between the magnetic form and the electrical form.
9. The henry is the base unit of inductance. The abbreviation for henry is H.
10. Inductance is determined by (1) core material, (2) number of turns, (3) spacing of turns, and (4) diameter of turns.
11. Inductors are rated for inductance, dc resistance, current, voltage, quality, and tolerance.
12. The dc resistance of an inductor is also called ohmic resistance.
13. The quality Q of inductors ranges from less than 10 to more than 200:

$$Q = \frac{X_L}{R}$$

14. Current in a dc inductive circuit rises more slowly than in a dc resistive circuit.
15. In a dc inductive circuit, voltage (cemf) reaches its peak value before the current does.
16. Inductive kick causes arcing in switch contacts when an inductive circuit is opened.
17. Inductive reactance is the opposition of an inductor to alternating current.
18. The symbol for inductive reactance is X_L.

19. Inductive reactance is directly proportional to both frequency and inductance:

$$X_L = 6.28 f L$$

20. Ohm's law can be used in inductive circuits by replacing R with X_L.
21. In an inductive circuit, current lags voltage by 90°.
22. Ideal inductors use no power or energy.
23. Real inductors have resistance; therefore, they do use some power.
24. Impedance is the combined opposition of reactance and resistance.
25. The skin effect increases the effective resistance of a conductor at high frequencies.
26. Litz wire is multistrand wire designed to reduce the skin effect.
27. Core losses are caused by induced currents in the core and by periodic reversal of the magnetic field.
28. Inductors (and inductive reactances) in parallel behave like resistors in parallel. The same formulas are used except that R is replaced by L or X_L.
29. Inductors (and inductive reactances) in series behave like resistors in series.
30. The lowest series inductance drops the least voltage.
31. Mutual inductance can be reduced by axis orientation, physical separation, and shielding.
32. Time constant formula:

$$T = \frac{L}{R}$$

33. Undesired inductance occurs in conductors and resistors.

For questions 11-1 to 11-12, determine whether each statement is true or false.

11-1. A straight length of conductor has no inductance.

11-2. A 2-H inductor would most likely have a laminated iron core.

11-3. Magnetic shields for inductors are usually made from high-reluctance materials.

11-4. The reactance of an inductor can be measured with an ohmmeter.

11-5. The core material in a variable inductor is often called a toroid.

11-6. Maximum inductance in a variable inductor occurs when the brass slug is centered within the coil core.

11-7. The cemf exceeds the source voltage when an inductive circuit is opened.

11-8. The iron core of an inductor converts some electric energy into heat energy.

11-9. The quality of a coil is frequency-dependent.

11-10. A small induced current flows in the core of an iron-core inductor.

11-11. The lowest-value inductor drops the most voltage in a series-inductor circuit.

11-12. The lowest-value inductor draws the most current in a parallel-inductor circuit.

For questions 11-13 to 11-30, supply the missing word or phrase in each statement.

11-13. _____ wire is used to reduce the skin effect.

11-14. The base unit of inductance is the _____.

11-15. The abbreviation for the base unit of inductance is _____.

11-16. The base unit for inductive reactance is the _____.

11-17. The symbol for inductive reactance is _____.

11-18. When current in an inductor is increasing, the cemf _____ the source voltage.

11-19. Another name for cemf is _____.

11-20. Inductive reactance is _____ proportional to frequency.

11-21. When one coil induces a voltage in another coil, the process is called _____.

11-22. Voltage _____ current by _____ degrees in an ideal inductor.

11-23. Inductors are also known as _____, _____, and _____.

11-24. The center leg of an iron-core carries _____ as much flux as an outside leg does.

11-25. The polarity of the cemf can be determined by applying _____ law.

11-26. The electric quantity that opposes change in current is _____.

11-27. An inductor converts _____ energy to _____ energy while the current is increasing.

11-28. The resistance of the turns of wire in an inductor is called _____ or _____ resistance.

11-29. Arcing between the switch contacts when an inductive circuit is turned off is caused by _____.

11-30. The combined opposition of reactance and resistance is called _____.

Answer the following questions:

11-31. What are three techniques used to minimize or eliminate mutual inductance?

11-32. What four factors determine the inductance of an inductor? How can inductance be increased using each of these factors?

11-33. What electrical ratings are used to completely specify an iron-core inductor?

11-34. What is the time constant of a 500-mH ideal inductor connected in series with a 10-Ω resistor?

11-35. What are the reactance and current in a circuit which consists of a 300-mH inductor connected to a 20-V, 7.5-kHz source?

11-36. Determine the quality of a 70-mH inductor that has a resistance of 125-Ω at 35 kHz.

11-37. Determine the total inductance and the circuit current of a 4-H inductor and a 6-H inductor series connected to a 400-Hz, 80-V supply. Also determine the voltage across the 4-H inductor.

11-38. What is the inductance of a 5-mH inductor and a 7-mH inductor connected in parallel?

11-39. How much inductance is needed to limit the current from a 50-V, 200-Hz source to 25 mA?

11-40. Two inductors are connected in series to a 35-V source. Determine V_{L_1} if $L_1 = 0.36$ H and $L_2 = 0.54$ H.

11-41. Determine the value of L needed to produce a time constant of 0.003 s when $R = 20$ Ω and $V_T = 50$ V.

Critical Thinking Questions

11-1. Two variable inductors are identical except that one has a brass slug, and the other a ferrite slug. Which inductor would have the larger range of inductance?

11-2. Why aren't inductor shields made from nonferrous materials such as tin or plastic?

11-3. At a given current, would a swinging choke or a smoothing choke have the larger inductance if the choke had identical coils and the same size and shape of core? Why?

11-4. Why is the total inductance of two series inductors greater than the inductance of the larger inductor?

11-5. Why are parallel inductive reactances treated like parallel resistances when figuring the total reactance?

11-6. Why does periodic reversal of the magnetic field cause a magnetic core to produce heat?

11-7. What is the frequency of the source voltage when a 0.13-A current drops 20 V across a 0.04-H inductor?

11-8. Why can we use either the values of X_L or the values of L in the current-divider formula for parallel inductor circuits?

11-9. A 200-mH (L_1) and a 0.15H (L_2) inductor are connected in parallel and draw 46 mA from the power source. Determine I_{L_2}.

11-10. Would you expect the current and voltage to be in phase in a circuit containing impedance? Why?

Answers to Tests

1. inductance
2. choke, coil, reactor, inductor
3. T
4. F
5. F
6. F
7. F
8. L
9. Lenz's
10. bemf
11. electric, magnetic
12. henry
13. H
14. volt per ampere per second
15. increase
16. Use a core with higher permeability or less reluctance, increase the number of turns, put the turns closer together, increase the diameter of the turns.
17. a. See Fig. 11-6(a).

b. See Fig. 11-6(b).
c. See Fig. 11-7.
18. F
19. F
20. T
21. T
22. F
23. T
24. laminated-iron core
25. doughnut-shaped
26. Because it carries twice as much flux.
27. Because too much current will cause the winding to overheat.
28. the ratio of reactance to resistance
29. an inductor or coil designed to be used at radio frequencies
30. laminated-iron
31. the high cemf that occurs when an inductive circuit is opened

32. inductive
33. yes
34. yes
35. no
36. the cemf of the inductor
37. X_L
38. $X_L = 6.28fL$
39. double
40. a. 2262 Ω
    b. 9425 Ω
    c. 47124 Ω
41. 40 Ω
42. lag, 90°
43. zero
44. 1.43 H
45. 223 kHz
46. air-core inductor
47. 377
48. Impedance is the combined opposition of resistance and reactance.
49. skin effect
50. Litz wire is multistrand wire used to

reduce the skin effect.
51. induced currents and magnetic polarity reversal
52. a. 0.2 H
    b. 188.5 Ω
    c. 79.6 mA
    d. 26.5 mA
53. a. 0.9 H
    b. 848 Ω
    c. 17.7 mA
    d. 10 V
54. Orient the axes 90° to each other, separate the coils, and shield them.
55. Half the turns are wound in one direction and the other half are wound in the opposite direction.
56. 0.5 s

Chapter 12

Transformers

Chapter Objectives

This chapter will help you to:

1. *Draw* the correct symbol for each type of transformer.
2. *Understand and correctly use* transformer terminology.
3. *Understand* how a transformer can change voltage levels, match impedances, and provide electrical isolation.
4. *Select* a transformer with ratings that are appropriate for the job to be done.
5. *Connect* three-phase transformer windings in either a delta or a wye configuration.
6. *Connect* transformer windings in series and/or parallel to obtain the desired voltage and current capabilities.
7. *Calculate* transformer losses.

Transformers are multiple-winding inductors. They operate on the principle of mutual inductance. For a rela- tively simple device, they are extremely versatile. Without transformers, our present power distribution system could not exist.

12-1 Transformer Fundamentals

A transformer consists of two or more coils linked together by magnetic flux (Fig. 12-1). The changing flux from one coil (the *primary*) induces a voltage in the other coil (the *secondary*). In other words, the coils are coupled, or linked, together by mutual inductance.

Without *mutual inductance,* there would be no such thing as a transformer. The amount of mutual inductance, like the amount of self-inductance, is specified in henrys. There is 1 H of mutual induction when 1 V is induced in a coil by a current change of 1 A/s in another coil. Suppose the primary current in Fig. 12-1 changes at a rate of 1 A/s. Further, suppose the secondary voltage in Fig. 12-1 is 3 V. Then the transformer has 3 H of mutual inductance.

Symbols

The symbol for a transformer is basically two coils with their axes parallel to each

other. As illustrated in Fig. 12-2, the basic symbol can be modified in many ways. These modifications are needed to more fully describe the various types of transformers.

The dashed lines in Fig. 12-2(*c*) represent the metal enclosure which houses the windings (coils). When this enclosure is made of aluminum or copper or other non-magnetic materials, it is intended as a shield against electric fields rather than magnetic fields.

The symbols shown in Fig. 12-2 are not the only ones you may see for transformers. For example, iron-core transformers sometimes have shields. Thus, to indicate an iron-core transformer, the symbol of Fig. 12-2(*c*) would show lines indicating the iron core. Not all tapped secondaries are center-tapped. An off-center-tapped winding is represented by having more turns indicated on one side of the tap than on the other.

Primary

Secondary

Mutual inductance

Transformer symbols

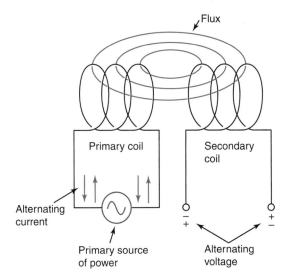

Fig. 12-1 Transformer action. The primary coil is linked to the secondary coil by magnetic flux.

Primary and Secondary Windings

A transformer is a device that transfers electric energy (or power) from a primary winding to a secondary winding. Except for the autotransformer, there is no electric connection between the primary winding and the secondary winding. The primary converts the electric energy into magnetic energy. The secondary converts the magnetic energy back into electric energy. The primary and the secondary winding are said to be electrically isolated from each other but magnetically connected or coupled to one another.

A transformer receives power from a power source, such as the 23-kV output of a power company generator or the 120-V outlet in your home. The transformer winding designed to receive power from the source is called the primary winding.

A load (Fig. 12-3) takes power from the secondary winding of a transformer. Therefore, the secondary winding becomes the power source for the load.

The secondary voltage need not be the same as the primary voltage. For example, the primary of the transformer in Fig. 12-3 is connected to a 120-V supply, and the secondary delivers 20 V to the load.

Transformers are two-way devices. The winding that the manufacturer calls the secondary can be used as a primary. Of course, a primary power source of the correct voltage

must be used. The winding designated as the primary then serves as a secondary. Of course, it provides power at the voltage for which it was originally designed. For example, the transformer in Fig. 12-3 can be used to provide a 120-V secondary source from a 20-V primary source, as shown in Fig. 12-4. As you can see, labeling one coil "primary" and another "secondary" is somewhat arbitrary. It is based on the intended use of the transformer, that is, the intended primary power source.

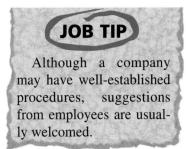

JOB TIP

Although a company may have well-established procedures, suggestions from employees are usually welcomed.

Coefficient of Coupling

The portion of the flux that links one coil to the other coil is referred to as the *coefficient of coupling*. The coefficient of coupling can range from 0 to 1. When all the flux is coupled, the coefficient of coupling is 1. Sometimes the coefficient of coupling is expressed as a percentage. Thus, 100 percent coupling means a coefficient of coupling of 1.

Coefficient of coupling

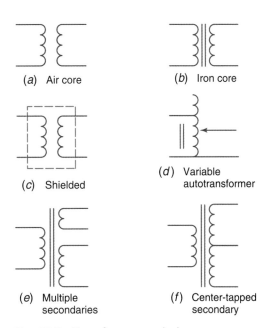

(a) Air core

(b) Iron core

(c) Shielded

(d) Variable autotransformer

(e) Multiple secondaries

(f) Center-tapped secondary

Fig. 12-2 Transformer symbols.

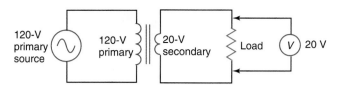

Fig. 12-3   Primary-secondary terminology. The primary
winding receives power from a power source.

YOU MAY
RECALL

. . . that Sec. 11-12 discussed ways of preventing mutual inductance. When no mutual inductance exists, the coefficient of coupling is 0.

Coupling in transformers with laminated-iron cores is very close to 100 percent. This is because all the flux is concentrated in the high-permeability core on which the coils are wound. The iron core provides a complete low-reluctance path for the flux loops. Therefore, essentially none of the flux leaks into the surrounding air.

On the other hand, air-core transformers can have very low coefficients of coupling. The paths through the air surrounding the inductors offer no more reluctance than the path through the cores. Therefore, much of the flux from one coil never links up with the other coil. This flux is called *leakage flux*. It just "leaks" off into surrounding air paths. The amount of coupling in an air-core transformer can be controlled by the spacing between the coils. The farther apart the coils are, the more flux leakage occurs and the lower the percentage of coupling. The percentage of coupling can also be controlled by the axis orientation of the coils. When the axes of the coils are perpendicular to each other, the coefficient of coupling is close to zero.

Leakage flux

Step up

Step down

■ TEST_____

Answer the following questions.

1. What is a transformer?
2. The two coils of a transformer are called the _____ and the _____.

3. What is the mutual inductance if a current change of 2 A/s in one coil induces 4 V in a second coil?
4. Draw the symbol for a shielded magnetic-core transformer.
5. What energy-conversion processes are involved in a transformer?
6. The _____ winding of a transformer receives power or energy from another source.
7. The _____ winding of a transformer provides power to the load.
8. Can the winding that the manufacturer of a transformer calls the secondary be used as a primary?
9. Define "coefficient of coupling."
10. Can the coefficient of coupling ever exceed 1?
11. Which have the lower coefficient of coupling, laminated-iron-core transformers or air-core transformers?
12. How can the coefficient of coupling in air-core transformers be varied?

Changing Voltage Values

A transformer can either *step up* or *step down* a voltage. If the primary voltage is greater than the secondary voltage, the transformer is stepping the voltage down. Thus, the transformer in Fig. 12-3 is a step-down transformer. If the voltage in the secondary exceeds the voltage in the primary (Fig. 12-4), the transformer is a step-up transformer. Some transformers with multiple secondaries have one or more step-up secondaries and one or more step-down secondaries.

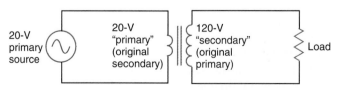

Fig. 12-4   Reversing the primary-secondary function.

Whether a secondary is a step-up or step-down winding is determined by the primary-to-secondary *turns ratio*. When the primary turns exceed the secondary turns *and the coupling is 100 percent,* the transformer steps down the voltage. In fact, with 100 percent coupling, the *turns ratio* and the *voltage ratio* are equal. Mathematically we can write

$$\frac{V_{pri}}{V_{sec}} = \frac{N_{pri}}{N_{sec}}$$

In this formula, N is the abbreviation for the number of turns. This formula can be rearranged to show that

$$\frac{N_{pri}}{V_{pri}} = \frac{N_{sec}}{V_{sec}}$$

In this new arrangement, the relationship between voltage and turns is very informative. It shows that the *turns-per-volt* ratio is the same for both the primary and the secondary. Furthermore, the turns-per-volt ratios of all secondary windings in a multiple-secondary transformer are equal. Thus, once you determine the turns-per-volt ratio of any winding, you know the ratios of all other windings.

Referring to Fig. 12-5 will help you understand why all windings have the same turns-per-volt ratio. Remember that the cemf of a coil is always equal to the ac voltage applied to the coil. Therefore, the cemf in the primary in Fig. 12-5 must be 1 V. Thus 1 V of cemf is created by the flux in the circuit. The flux, therefore, must induce 0.25 V in each turn of the primary. Notice that all the flux that produced cemf in the primary turns also goes through every secondary. Therefore, each turn in the secondary windings has 0.25 V induced in it. If it requires four turns in the primary for each volt, then four turns in a secondary will also develop 1 V.

The turns-per-volt ratio used for the windings of a laminated-iron-core transformer varies with the size of the transformer. Small transformers (less than 10-W rating) may have seven or eight turns per volt. Larger transformers (more than 500-W) may have less than one turn per volt.

The concept of turns per volt is useful in modifying the secondary voltage of a transformer.

Example 12-1

A transformer is to be rewound to provide a voltage of 14 V. The transformer's present secondary is rated at 6.3 V and contains 20 turns. How many turns will be required for the new 14-V winding?

Given: Present secondary winding delivers 6.3 V with 20 turns

Find: Number of turns necessary to provide 14 V

Known: $\dfrac{N_{pri}}{V_{pri}} = \dfrac{N_{sec}}{V_{sec}}$

Solution: $\dfrac{N_{sec}}{V_{sec}} = \dfrac{20}{6.3} = 3.175$ turns per volt

The transformer has a 3.175 turns-per-volt ratio. Since a 14-V secondary is required

N (new secondary)
= 3.175 turns per volt × 14 V
= 44.5 turns or 45 turns

Answer: A 14-V winding needs 45 turns.

Example 12-2

The designer of a transformer has calculated that the 120-V primary will have 2.6 turns per volt. How many turns will be required for a 400-V secondary?

Given: Turns per volt = 2.6
$V_{pri} = 120$ V
$V_{sec} = 400$ V

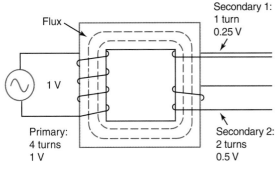

Fig. 12-5 Turns per volt. The turns-per-volt ratio is the same in all windings.

Find:	N_{sec}
Known:	$\dfrac{N_{pri}}{V_{pri}} = \dfrac{N_{sec}}{V_{sec}}$
Solution:	$\dfrac{N_{pri}}{V_{pri}} = 2.6$ turns per volt
	$N_{sec} = 400 \text{ V} \times 2.6$ turns per volt
	$= 1040$ turns
Answer:	The 400-V secondary will need 1040 turns.

Notice in example 12-2 that we did not need to calculate the number of turns in the primary. Also, we did not need to know the voltage of the primary. All we needed to know was the turns-per-volt ratio.

One of the chief uses of transformers is to change voltages from one value to another. An example will show why it is so necessary to be able to transform voltage levels. It is usually necessary to transmit electric power from the power plant where it is produced to the location where it is used. Often the distance between these two points is many hundreds of miles. Obviously it is desirable to use as small a conductor as possible in the power lines between the two points. The size of the conductor needed is directly dependent on the amount of current it must carry. So we need to keep the current as low as possible. Assuming a 100 percent power factor, power equals current times voltage ($P = IV$). Thus, the lower the current, the higher the voltage must be for a given amount of power. However, the generators at the power plant and the loads at the point of use have a limited operating voltage. The solution to this problem is illustrated in

Fig. 12-6. If the 10 MW from the generator were directly connected to the loads, the transmission lines would have to carry 500 A. With a step-up transformer at the power plant and a step-down transformer at a substation near the load, the power lines carry only 25 A. (This illustration assumes that the power factor is 1 and that the transformer uses no power.)

Figure 12-6 also illustrates another transformer relationship. When the voltage of a transformer is stepped up, the current is stepped down, and vice versa. The exact current ratio depends upon the power factor, the power consumed by the transformer, and the voltage ratio. For the ideal transformer (one with no losses), the primary power is approximately equal to the secondary power when the transformer has a full resistance load. Stating this as a formula, we have

$$P_{pri} \simeq P_{sec}$$

This leads to the formula for the current relationship:

$$\frac{V_{pri}}{V_{sec}} \simeq \frac{I_{sec}}{I_{pri}}$$

This points out that the current varies inversely to the voltage. In other words, a step-up voltage transformer is, in effect, a step-down current transformer.

◼ TEST

Answer the following questions.

13. True or false. With a step-down transformer, the secondary voltage is higher than the primary voltage.

14. True or false. The turns ratio equals the voltage ratio in a transformer with 100 percent coupling.

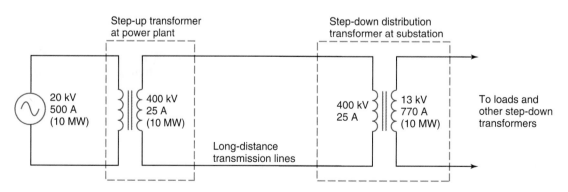

Fig. 12-6 Advantage of high-voltage power lines. The higher the voltage, the lower the current for a given amount of power.

15. True or false. With 100 percent coupling, the turns-per-volt ratio of the primary must be less than the turns-per-volt ratio of the secondary.
16. True or false. The turns-per-volt ratio of the primary is usually higher for a 2-kW transformer than for a 50-W transformer.
17. How many turns will be required for a 40-V secondary if the primary of a transformer has five turns per volt and is designed to operate from a 12-V, 60-Hz source?
18. What is one of the most common uses of a transformer?
19. Why do power systems transmit power at as high a voltage as possible?
20. When a transformer steps down voltage, does it step up or step down in current?

12-2 Efficiency of Transformers

The iron core and the copper coils of a transformer both convert some electric energy into heat energy. This, of course, is why a transformer heats up when in operation. The purpose of a transformer is not to provide heat but to transfer energy from the primary to the secondary. Therefore, any heat produced by the transformer represents inefficiency.

Since energy is equal to power times time, the efficiency of transformers is calculated in terms of power. The *efficiency* of a transformer (expressed as a percentage) is calculated by the following formula:

$$\text{Percent efficiency} = \frac{P_{sec}}{P_{pri}} \times 100$$

Example 12-3

What is the efficiency of a transformer that requires 1880 W of primary power to provide 1730 W of secondary power?

Given: $P_{pri} = 1880$ W
            $P_{sec} = 1730$ W
Find: Efficiency

Known: $\{\%\} \text{ eff.} = \dfrac{P_{sec}}{P_{pri}} \times 100$

Solution: $\{\%\} \text{ eff.} = \dfrac{1730 \text{ W}}{1880 \text{ W}} \times 100 = 92$

Answer: The transformer is 92 percent efficient.

In example 12-3, the 150-W difference between received power and delivered power is lost in the transformer. As indicated in Fig. 12-7, the power consumed by the transformer is referred to as a *power loss*. The power loss in a transformer is caused by

Power loss

1. Hysteresis loss
2. Eddy current loss
3. Copper (I^2R) loss

The first two of these losses occur in the transformer core material. The last occurs in the windings. All three convert electric energy to heat energy.

Hysteresis Loss

Hysteresis loss is caused by *residual magnetism,* that is, by the magnetism that remains in a material after the magnetizing force has been removed. The core of a transformer has to reverse its magnetic polarity every time the primary current reverses direction. Every time the magnetic polarity is reversed, the residual magnetism of the previous polarity has to be overcome. This produces heat. It requires energy from the primary to produce this heat. Hysteresis loss, then, refers to the energy required to reduce the residual magnetism to zero. This loss occurs once every half-cycle just before the core is remagnetized in the opposite direction.

Hysteresis loss

Residual magnetism

The *hysteresis loop* in Fig. 12-8 graphically illustrates hysteresis loss. The narrower the hysteresis loop, the lower the hysteresis loss. Therefore, the core material for transformers, and other magnetic devices that operate on alternating current, should have a narrow hysteresis loop. Laminated-iron cores are made from silicon steel. Silicon steel is an alloy which has a narrow hysteresis loop and still has high permeability.

Efficiency

Hysteresis loop

Hysteresis loss increases with an increase in the frequency of the primary current. This is one of the reasons that laminated-iron-core transformers are not used above the audio-frequency range.

Eddy Current Loss

The changing magnetic flux in the core of a transformer induces voltage into any conductors which surround it. Since the core is itself a conductor, the changing magnetic flux in-

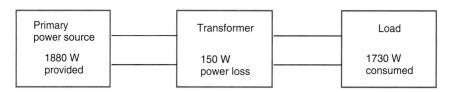

Fig. 12-7 Transformer efficiency. Power loss occurs because the transformer converts some electric energy into heat energy.

duces a voltage in the core as well as in the coil conductors. The voltage induced in the core causes current to circulate in the core. This current is called *eddy current*. The eddy current flowing through the resistance of the core produces heat.

The amount of heat due to eddy current is dependent on the values of both the eddy current and the induced voltage ($P = IV$). There is nothing we can do to reduce the value of the induced voltage. Therefore, we must reduce eddy current loss by reducing the value of the eddy current produced by the induced voltage. This can be done by increasing the resistance of the path through which the eddy current must flow. (Remember, $I = V/R$.) The resistance of the core, in the plane in which eddy current flows, is increased by laminating the core. Each *lamination* of the core is insulated with a thin layer of oxide [Fig. 12-9(a)]. The oxide has a much higher resistance than the rest of the silicon-steel lamination. Notice in Fig. 12-9(a) that the eddy current would have

Lamination

to flow through the oxide layers in order to circulate through the core. The equivalent circuit [Fig. 12-9(b)] for the core shows that the high resistance of the oxide on each lamination effectively reduces the flow of eddy current. Thus laminating the core reduces the eddy current and its associated heat loss.

The thinner the laminations, the more series resistance the core contains and the lower the eddy current will be. However, making the laminations thinner also increases the total amount of oxide in the core. The oxide has a lower permeability than the silicon steel does. Therefore, a core with thin laminations cannot carry as high a flux density as one with thicker laminations. Flux density, of course, controls the amount of magnetic energy a given size of

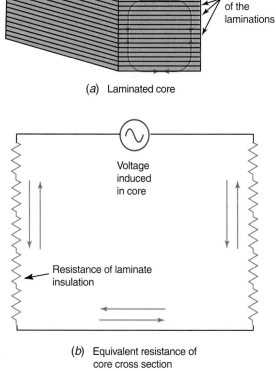

(*a*) Laminated core

(*b*) Equivalent resistance of core cross section

Fig. 12-9 Reducing eddy current loss.

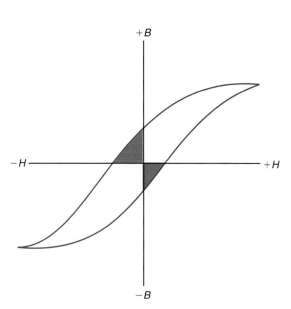

Fig. 12-8 Hysteresis loop. Magnetic energy is converted to heat energy in the shaded regions of the loop.

core can handle. This in turn controls the amount of power the transformer can handle. The transformer designer has to consider all these factors, and others, as the core is being designed.

Copper Loss

Copper loss refers to the power dissipated in the windings of a transformer. Since this loss can be calculated by $P = I^2R$, it is called the I^2R loss. The R in the formula is the ohmic, or dc, resistance of the turns in the winding.

Obviously, copper loss is minimized by using as large a conductor as possible in the windings. However, conductor size is limited by the area of the windows (openings) in the core into which the winding must fit.

Load and Efficiency

Maximum efficiency is obtained from a transformer when it is fully loaded. For small transformers (less than 10 W), maximum efficiency may be less than 70 percent. With transformers larger than 1000 W, it is more than 95 percent.

As the load is decreased, the efficiency of the transformer also decreases. This is because current flow in a transformer primary does not decrease in direct proportion to decreases in the load. The primary current still causes substantial core losses and copper losses even when the secondary is lightly loaded.

Data showing the changes in efficiency with changes in load are tabulated in Fig. 12-10. These data come from tests conducted on a transformer rated at 75 W.

Primary power (W)	Secondary power (W)	Efficiency %
7	3.8	54.3
11	7.3	66.4
18	12.8	71.1
53	45.5	85.8
71	62.0	87.3

Fig. 12-10 Efficiency versus load. The transformer is most efficient when operating at its rated power.

Answer the following questions.

21. What causes the inefficiency in a transformer?
22. Determine the efficiency of a transformer that requires 180 W to deliver 150 W.
23. How much power does a transformer require from its source if it is 93 percent efficient and delivers 750 W?
24. List the two causes of power loss in the core of a transformer.
25. I^2R loss occurs in the _____ of a transformer.
26. Hysteresis loss is caused by _____.
27. How often must residual magnetism be overcome in the core of a transformer?
28. What is a hysteresis loop?
29. What are core laminations made of? Why?
30. Is a narrow or a wide hysteresis loop desirable for the core of a transformer?
31. What is the relationship between hysteresis loss and frequency?
32. How are eddy current losses reduced?
33. For minimum eddy current loss, should core laminations be thick or thin?
34. What is I^2R loss? How can it be minimized?
35. Which has higher efficiency, a large transformer or a small transformer?
36. When does a transformer have maximum efficiency?

Copper loss

12-3 Loaded and Unloaded Transformers

It has been shown that a fully loaded transformer has higher efficiency than an unloaded transformer. Many other differences are also associated with the amount of load on a transformer. These will become apparent in the discussion which follows.

An unloaded transformer acts like a simple inductor. It is a highly inductive load on the primary power source to which it is connected. The current in the primary winding is nearly 90° out of phase with the voltage [Fig. 12-11(a)]. Only the copper and core losses keep the current from being a full 90° out of phase with the voltage. The current in the primary of an unloaded transformer is called the *energizing current*. It is the current needed to set up

Energizing current

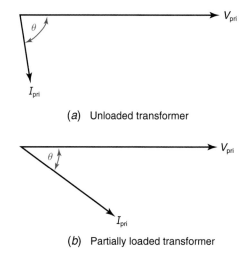

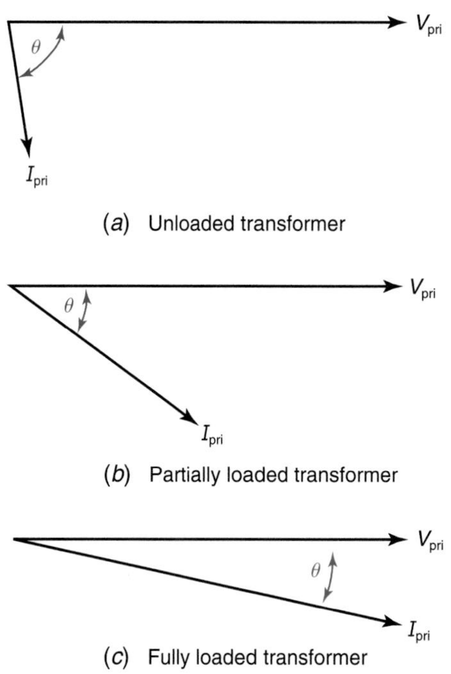

(a) Unloaded transformer

(b) Partially loaded transformer

(c) Fully loaded transformer

Fig. 12-11 Primary current and voltage phase relationships. When fully loaded, a transformer appears to be close to a resistive load.

Secondary magnetizing forces

is assumed to be resistive. Therefore, the transformer is furnishing power to the load.

Figure 12-12 presents experimental data collected by testing a 75-VA transformer. Look at the columns showing values of the secondary power and the angle θ. From them you can see that the transformer appears less inductive, and more resistive, as the load increases.

The last row of numbers in Fig. 12-12 indicates that the transformer is almost purely resistive at 72 VA. The apparent power and the true power of the primary are nearly equal; the cos θ (power factor) approaches 1. Thus, the primary is providing the power for the load connected to the secondary. In other words, the load on the secondary is reflected back to the primary. The primary, in turn, draws the required power from the primary power source.

The secondary load is reflected back to the primary by the interaction of the primary and secondary magnetizing forces. Until now we have not mentioned a *secondary magnetizing force*. But, as soon as the secondary is loaded, current starts to flow in the secondary coil. This current in the secondary coil creates a magnetizing force which tries to produce a flux in the core material. The polarity of the magnetizing force of the secondary always opposes the magnetizing force of the primary (Fig. 12-13). Obviously, the core cannot have flux flowing in opposite directions at the same time. Either the primary or the secondary magnetizing force must dominate. The primary magnetizing force always dominates. It must dominate because the flux created by the primary is what produces the secondary voltage and current. Whenever the secondary magnetizing force increases, the primary current increases to provide a greater primary magnetizing force.

the flux in the core. The energizing current can be rather high and still not use much power. This is because the current and voltage are so far out of phase. The amount of energizing current is determined primarily by the inductive reactance of the primary winding.

When a load is connected to the secondary of a transformer, both the primary current and the angle θ change. As indicated in Fig. 12-11(b) the (c), the amount of change depends upon how heavily the transformer is loaded. For this discussion, the load on the secondary

Measured values				Calculated values		
V_{pri} (volts)	I_{pri} (amperes)	P_{pri} (watts)	P_{sec} (watts)	Apparent P_{pri} (voltamperes)	Cos θ	θ (degrees)
120	0.125	3	0	15	0.20	78
120	0.135	7	3.8	16.2	0.432	64
120	0.15	11	7.3	18	0.611	52
120	0.19	18	12.8	22.8	0.789	38
120	0.46	53	45.5	55.2	0.960	16
120	0.60	71	62	72	0.986	9.6

Fig. 12-12 Primary current and angle theta (θ) versus load.

This is why the primary current increases whenever the secondary load current increases. Any primary current caused by a load on the secondary is an in-phase (resistive) current (Fig. 12-14). The total primary current (Fig. 12-14) is composed of this reflected resistive current and the energizing current (which is mostly inductive current). When the transformer is fully loaded, the resistive current (caused by the secondary load) dominates. Thus, the fully loaded transformer appears resistive to the primary power source.

◢ TEST

Answer the following questions.

37. Describe what happens to the following factors when a transformer's load is changed from no load to full load:
    a. Angle θ
    b. Cosine θ (power factor)
    c. Type of load the transformer presents to the primary source
38. The primary current drawn by an unloaded transformer is called the _____ current.
39. The primary current is mostly resistive when the transformer secondary is _____ .
40. True or false. The load on the secondary of a transformer is reflected back to the primary as an in-phase current.

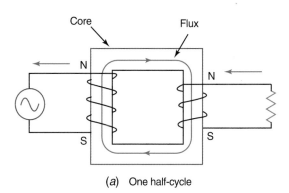

(a) One half-cycle

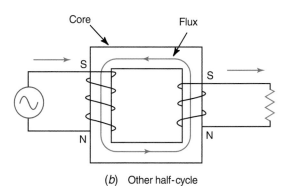

(b) Other half-cycle

Fig. 12-13 Primary and secondary magnetizing forces oppose each other. However, notice that the primary force dominates.

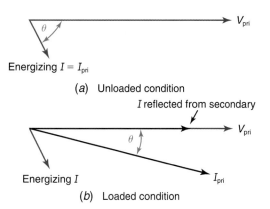

(a) Unloaded condition

(b) Loaded condition

Fig. 12-14 Reflected load current.

**I and E
laminations**

**Iron-core
transformers**

Toroidal cores

**Air-core
transformers**

41. True or false. the magnetizing force of the secondary current opposes the magnetizing force of the primary current.

12-4 Transformer Cores

Transformers can be broadly grouped by the type of core material they use. Like inductors, transformers can have either magnetic cores or air cores.

Iron-core transformers look like iron-core inductors. They use the same type of I and E laminations in their cores. However, the stacking of the I and E laminations in transformer cores is different from that in inductors.

YOU MAY RECALL . . . that with inductors, all the I laminates are stacked together and all the E laminates are stacked together. This leaves a small continuous air gap where the I and E stacks butt together. This small air gap in the inductor core aids in keeping the inductance more constant for different amounts of current. (Sometimes this air gap is increased by putting one or more layers of paper between the I and E joint.)

With transformers, the *I and E laminations* are rotated 180° every few layers (Fig. 12-15). This procedure breaks up the joint between the I and E laminations so that there is no continuous air gap. Thus flux leakage from the core is reduced to a minimum.

Laminated-iron-core transformers are used only at power and audio frequencies (frequencies up to 20 kHz). At frequencies above the audio range, their core losses become excessive.

Powdered iron and ferrite are also used as core material for magnetic-core transformers. When used in the audio range, the cores form a continuous path for the flux. *Toroidal cores* are often used at the higher audio frequencies. When used in the radio-frequency range, the core is often just a slug. Notice in Fig. 12-16 that the coils are physically separated from each other. When greater couplings is desired, the spacing between the coils is reduced. Frequently one coil is wound on top of the other to obtain maximum coupling.

Air-core transformers are used exclusively at high (radio) frequencies. Often they are made from wire that is heavy enough to allow the in-

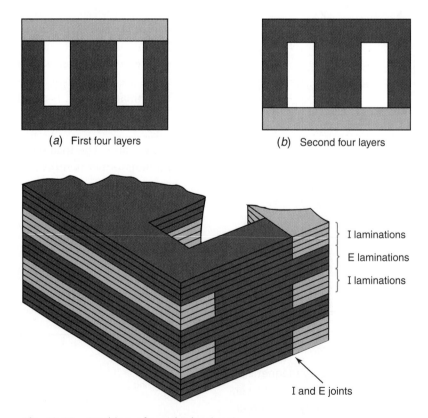

(*a*) First four layers   (*b*) Second four layers

I laminations
E laminations
I laminations

I and E joints

Fig. 12-15 Stacking of I and E laminations.

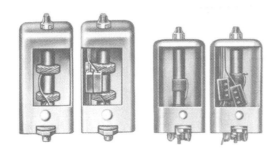

Fig. 12-16   Radio-frequency transformers. *(Courtesy of Bell Industries, J. W. Miller Division)*

dividual coils to be self-supporting. Some air-core transformers are made so that the coefficient of coupling is variable. This requires an arrangement that varies either the distance between the coils or the orientation of their axes.

12-5   Types of Transformers

Electric and electronic parts catalogs list many different classifications of transformers. Usually these catalogs classify a transformer according to the application for which it was designed. Some of the more common types and their applications are discussed below.

Power transformers are designed to operate at power-line frequencies and voltages (usually 60 Hz and from 115 to several thousand volts). The larger transformers are for power distribution and lighting. Smaller transformers are used for rectifier or control circuits in electronic systems. Rectifier transformers are used for providing low-voltage alternating current for rectification into direct current. These transformers are also used for providing low-voltage alternating current for control circuits (relays, solenoids, etc.). Therefore, they are also called *control transformers.*

Transformers designed to operate at frequencies up to 20 kHz are often referred to as *audio transformers.* They are further categorized as *input, output,* and *interstage* transformers. (These terms refer to audio amplifiers.) They are transformers to (1) receive the input to the amplifier, (2) deliver the output from the amplifier, or (3) process the audio signal in the amplifier.

Radio-frequency transformers perform functions similar to those of audio transformers, but at radio frequencies rather than audio frequencies. Radio-frequency transformers may be either air-core or magnetic-core. They are often enclosed in a metal container (Fig. 12-

16) which shields against electric fields. These transformers are used in such devices as radio and television receivers and transmitters.

Some electrical and electronic equipment (such as data processing equipment and computers) is very sensitive to voltage changes. Such equipment is often powered by *constant-voltage transformers* because regular power-line voltage may vary too much. A constant-voltage transformer provides a stable secondary voltage even when the primary voltage is very unstable. Typically the primary voltage can vary from 95 to 130 V without causing more than 1 percent variation in the secondary voltage.

Isolation transformers have equal primary and secondary voltages. Their purpose is to electrically isolate a piece of electrical equipment from the power distribution system. An important use of isolation transformers is illustrated in Fig. 12-17. Many pieces of electronic equipment use the metal chassis on which components are mounted as a *common conductor.* In Fig. 12-17, the currents for R_1, R_2, and L_1 flow through the chassis. Thus the chassis is part of the electric circuit. Technicians servicing this equipment can accidentally touch the chassis while power is being supplied to the equipment. If they do, they complete a circuit and receive a shock [Fig. 12-17(a)]. The shock can be fatal when the resistance through ground is low. Inserting an isolation transformer [Fig. 12-17(b)] breaks the circuit that includes the technician. Current can no longer flow from the ungrounded side of the power source through the chassis and the technician to ground. Yet the circuit containing R_1, R_2, and L_1 receives normal voltage, current, and power.

An *autotransformer* is somewhat different from the other transformers we have studied. Its primary is part of its secondary, and vice versa. With a step-up autotransformer [Fig. 12-18(a)], the secondary consists of the primary plus some additional turns. These additional turns are wound so that their induced voltage is series-aiding the cemf of the primary. A step-down autotransformer is shown in Fig. 12-18(b). Here the secondary is just a fraction of the primary. The cemf of that fraction of the primary provides the secondary voltage.

JOB TIP

Organizations change. Priorities come and go. It pays to be flexible and support new programs.

Constant-voltage transformers

Isolation transformers

Power transformers

Control transformers

Audio transformers

Autotransformers

Radio-frequency transformers

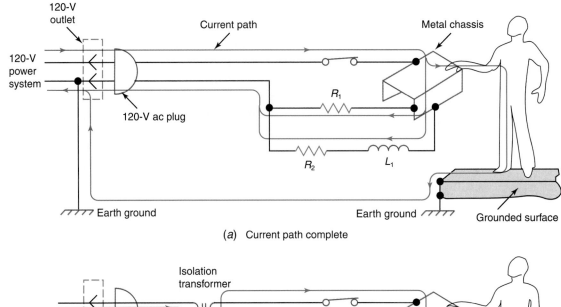

(a) Current path complete

(b) Current path isolated

Fig. 12-17 Importance of isolation transformer.

In most ways an autotransformer behaves like any other transformer. That is, a load on the secondary increases the primary current. Also, the turns-per-volt ratio of the primary and secondary is the same. The big difference between the autotransformer and other transformers is that the autotransformer does not provide electrical isolation.

Variable transformers that operate at power frequencies are usually autotransformers with

Variable transformers

an adjustable secondary. One secondary lead is connected to a carbon brush that can be adjusted up and down on the turns of the winding. The secondary voltage of most variable transformers can be adjusted from 0 to about 110 percent of the primary voltage.

■ TEST_____

Answer the following questions.

42. Why are I and E laminations rotated every few layers in transformers?
43. What types of transformers use laminated iron cores?
44. List several types of power transformers.
45. The _____ transformer provides a stable secondary voltage even though the primary voltage may vary between 95 and 130 V.
46. The _____ transformer has a secondary voltage that is equal to the primary voltage.

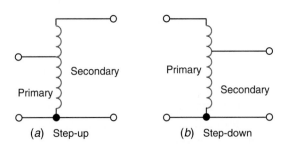

Fig. 12-18 Autotransformer. Part of the same coil is used for both the primary and the secondary.

47. The _____ transformer does not provide electrical isolation.

12-6 Impedance Matching

We have seen that transformers are used for changing voltage levels and for isolation. The third common use of transformers is impedance matching.

Before we see how a transformer can match impedances, let us review why impedance matching is desirable.

YOU MAY RECALL ... that in Chap. 5 it is shown that matching the resistance of a dc source with that of the load provides maximum transfer of power.

Impedance matching, in effect, does the same thing for ac circuits. When the internal impedance of a source is matched (equal) to the impedance of the load, maximum power is transferred from the source to the load. When the load impedance is either greater or less than the source impedance, the load power decreases. Although impedance includes both resistance and reactance, only resistance is used in the discussion which follows. Using only resistance does not change the concept; it just keeps the calculations simpler.

A transformer can make a load appear to the source to be either larger or smaller than its actual value. If the transformer is a step-up type, the load appears smaller. If it is a step-down transformer, the load appears larger. Figure 12-19 illustrates how a transformer changes the apparent value of a load. For purposes of calculation, the transformer in Fig. 12-19(a) is assumed to be an ideal transformer. That is, it has 100 percent efficiency, 100 percent coupling, and a 0° phase shift when loaded. If the transformer has no power losses, then its primary power equals its secondary power. To obtain 0.1 W of secondary power, the primary must provide 0.1 W. This means that the primary has to draw 0.01 A of current from the source. Notice this is the same value of current as the 1000-Ω resistor in Fig. 12-19(b) draws. Thus you can see that the transformer makes the 10-Ω resistor in the secondary appear as 1000 Ω to the source. The source's output voltage, current, and power are identical in the two circuits. The transformer has matched the impedance of a 1000-Ω source to a 10-Ω load.

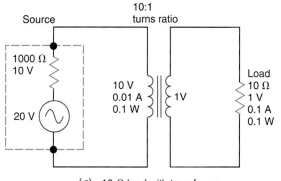

Impedance matching

(a) 10-Ω load with transformer

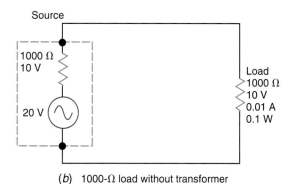

Fig. 12-19 Impedance-matching principle. The transformer makes the 10-Ω resistor appear to the source as a 1000-Ω resistor.

(b) 1000-Ω load without transformer

The 1000-Ω impedance that the primary seemed to have is called *reflected impedance*. In other words, the 10 Ω in the above example is reflected back to the primary as 1000 Ω. We could say that the primary impedance of the transformer is 1000 Ω and the secondary impedance is 10 Ω. Therefore, the primary-to-secondary *impedance ratio* is 100:1. Notice in Fig. 12-19(a) that the turns ratio (and also the voltage ratio) is 10:1. Thus the turns ratio squared is equal to the impedance ratio. Also, the voltage ratio squared is equal to the impedance ratio. If we use the symbol Z for impedance, these relationships can be written

Reflected impedance

Impedance ratio

$$\frac{Z_{pri}}{Z_{sec}} = \left(\frac{N_{pri}}{N_{sec}}\right)^2 \quad \text{or} \quad \frac{Z_{pri}}{Z_{sec}} = \left(\frac{V_{pri}}{V_{sec}}\right)^2$$

■ TEST_____

Answer the following questions.

48. What does "impedance matching" mean?
49. What is the significance of impedance matching?

Example 12-4

An 8-Ω load is to be matched to a source that has 8000 Ω of internal impedance. What must the turns ratio of the transformer be?

Given: $Z_{sec} = 8\ \Omega$
$Z_{pri} = 8000\ \Omega$

Find: $\dfrac{N_{pri}}{N_{sec}}$ (that is, the turns ratio)

Known: $\dfrac{Z_{pri}}{Z_{sec}} = \left(\dfrac{N_{pri}}{N_{sec}}\right)^2$

Solution: $\dfrac{Z_{pri}}{Z_{sec}} = \dfrac{8000\ \Omega}{8\ \Omega} = 1000$

$\left(\dfrac{N_{pri}}{N_{sec}}\right)^2 = 1000$

Turns ratio $= \dfrac{N_{pri}}{N_{sec}} = \sqrt{1000}$

$= 31.6$

Answer: The transformer must have a turns ratio of 31.6:1 (primary to secondary).

50. Does a load appear to the source as smaller or larger than its true value when a step-down transformer is used?
51. What does "reflected impedance" mean?
52. What is the impedance ratio of a transformer that steps down 240 V to 15 V?
53. If the transformer in question 52 is connected to a 10-Ω load, what is the value of the impedance reflected back to the source?

12-7 Transformer Ratings

To properly use a transformer, one must know its voltage and current ratings. Of course, from these ratings the power rating of the transformer can be calculated. Most transformers are also specified by their voltampere (apparent power) ratings.

Voltage Rating

Manufacturers always specify the voltage rating of the primary and secondary windings. Operating the primary above rated voltage usually causes the transformer to overheat. The additional stress placed on the transformer in-

sulation by the higher primary and secondary voltages can also be serious. Operating the primary below rated voltage does no harm, but this makes the secondary voltages lower than rated values.

The rated voltages of the secondaries are specified for full-load conditions with rated primary voltage. With no load, the secondary voltage is slightly higher than rated voltage (usually 5 to 10 percent higher).

There is considerable variation in the way in which manufacturers specify *center-tapped* secondaries. For example, the secondary in Fig. 12-20 may be specified in any of the following ways:

1. 40 V C.T.
2. 20 V-0-20 V
3. 20 V each side of center

Current Rating

Manufacturers usually specify current ratings for secondary windings only. As long as the secondary current rating is not exceeded, the primary current-carrying capacity cannot be exceeded.

Exceeding the current rating of a secondary causes its voltage to fall slightly below rated value. More serious than decreased voltage, however, is the increase in I^2R loss in the secondary. The increased I^2R loss causes the winding to overheat and eventually destroys the transformer.

Power Rating

Some manufacturers specify a power rating (in watts) for their transformers. This is understood to be the power the transformer can de-

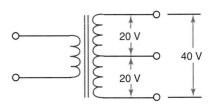

Fig. 12-20 Center-tapped secondary specifications. The secondary is classified as either 40 V center-tapped or 20 V each side of center.

liver *to a resistive load*. Thus the power rating is merely the product of the current rating and the voltage rating of the secondary ($P = IV$). For multiple-secondary transformers, the power rating is the sum of the powers available from the individual secondaries ($P_T = P_1 + P_2$ + etc.). The total power cannot be taken from a single secondary on a multiple-secondary transformer. The current rating of the individual secondaries must not be exceeded.

Voltampere Rating

The voltampere rating of a transformer is an apparent power rating. It is applicable to any type of load—resistive, reactive, or combination (impedance). The voltampere rating, like the power rating, is given for the total transformer instead of for individual secondaries. With a multiple-secondary transformer, the total voltampere rating cannot be taken from a single secondary.

A transformer can be loaded to its full voltampere rating and be delivering only a fraction of its power rating. Refer to Fig. 12-21 for an example. Here the load is a motor which has a power factor (cos θ) of 0.6. It is connected to a transformer which is rated at 750 VA and has a secondary voltage of 120 V. The motor draws 6.25 A. The transformer provides 750 VA (120 V $\times$ 6.25 A). However, the motor is drawing only 450 W ($P = IV$ cos θ = 6.25 A $\times$ 120 V $\times$ 0.6).

As you can see, manufacturers use many ways to rate transformers. You will never overload a transformer or exceed any of its ratings if you observe two rules.

1. Never apply more than the rated voltage to the primary.

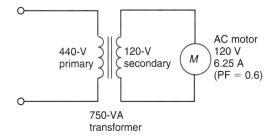

Fig. 12-21 Voltampere versus power rating. The limiting rating is the voltampere rating.

2. Never draw more than the rated current from any secondary.

■ TEST

Answer the following questions.

54. True or false. It is not possible to exceed the voltampere rating of a transformer without exceeding the power rating of the transformer.
55. True or false. The power rating of a transformer can be exceeded without exceeding either the primary voltage rating or the secondary current rating.
56. True or false. The voltampere rating of a transformer cannot be exceeded without exceeding either the primary voltage rating or the secondary current rating.
57. True or false. Secondary voltages are rated at full load current and rated primary voltage.

12-8 Series and Parallel Windings

Some transformers are made with more than one primary as well as more than one secondary. *Multiple* transformer *windings* can be connected in series or in parallel to change the voltage or current capabilities. Either primary or secondary windings (or both) can be connected in series or in parallel.

Multiple windings

Windings in Parallel

For parallel connections, windings should have identical ratings. They *must* have identical voltage ratings. Before windings are connected in parallel, their *phasing* must be correct. Phasing refers to the instantaneous polarity of the windings. *Correct phasing* for parallel connection is illustrated in Fig. 12-22. The polarities shown are instantaneous. Notice that the negative ends of the two secondary windings are connected together, as are the positive ends. With this phasing of the windings, no secondary current flows when there is no load. Neither winding can force current through the other because their voltages are equal and opposing. The output voltage is still 10 V, but the current capability has doubled from 3 to 6 A.

Some transformer manufacturers indicate the phasing of windings by terminal numbers

Correct phasing

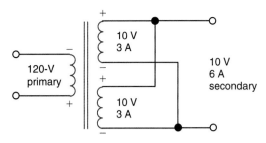

Fig. 12-22 Windings connected in parallel. The voltage remains the same, but the current capacity increases.

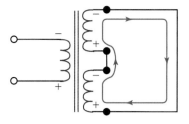

Fig. 12-24 Short circuit caused by incorrect phasing.

or color code. However, when phasing is not indicated, the method shown in Fig. 12-23 can be used to determine the correct phasing. If points 2 and 3 of Fig. 12-23 are of the same instantaneous polarity, the voltmeter across points 1 and 4 will indicate 0 V. This means that points 1 and 4 are also of the same polarity. If the two windings are to be operated in parallel, points 1 and 4 can be connected together. Points 2 and 3 form one terminal of the parallel winding, and points 1 and 4 form the other terminal.

> ## JOB TIP
>
> Newcomers do well to observe well-established and respected workers. Emulation of some of their characteristics can help with your success.

If, on the other hand, the meter in Fig. 12-23 indicates 20 V, points 2 and 3 have opposite polarities. Then, for a parallel connection, points 1 and 3 can be connected together, as can points 2 and 4.

Incorrect phasing (Fig. 12-24) causes a "dead" short on the secondaries of the transformer. The voltage of each winding aids the other winding in producing secondary current. The secondary current is limited only by the re-

sistance of the secondaries. Therefore, the current becomes very high. If not protected against overload, the transformer will soon burn out.

Some transformer primaries have two identical windings so that they can be powered from either of two voltages. When operated on the lower voltage, the primaries are connected in parallel. Proper phasing of the primary windings can also be determined using the method of Fig. 12-23. Just use one of the secondaries as a temporary primary and apply an appropriate voltage source to it. The voltage applied to the secondary can be any value equal to *or less than* its rated value.

Windings in Series

Transformer windings can be connected in series so that they either aid or oppose each other,

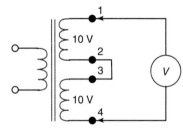

Fig. 12-23 Determining correct phasing for a parallel connection. If correctly phased, the meter will indicate 0 V.

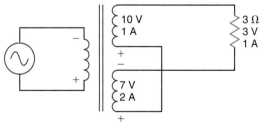

(a) Series-aiding connection

(b) Series-opposing connection

Fig. 12-25 Series-aiding and series-opposing winding connections.

as in Fig. 12-25. In this figure the polarity markings indicate the instantaneous polarities for one half-cycle. In the *series-aiding* configuration [Fig. 12-25(*a*)], the output voltage is the sum of the two secondary voltages. Notice, however, that the current capability is restricted to the lower rating of the two windings. This is because all the load current flows through both secondaries.

When connected in a *series-opposing* configuration [Fig. 12-25(*b*)], the two windings produce a voltage equal to the difference between the two voltages. We say the two voltages are *bucking* one another. Again note that the current is limited to the lower current rating of the two windings.

In summary, it is possible to obtain four voltages from two secondary windings:

1. The voltage of secondary 1
2. The voltage of secondary 2
3. The sum of the voltages of secondaries 1 and 2
4. The difference between the voltages of secondaries 1 and 2

Dual primary windings are connected in series for operation on the higher voltage. The connection must be series-aiding. Otherwise the magnetic field from one winding would cancel the field from the other. The primary current would be very high. The primary would burn out immediately unless it was properly protected against overloads. Phasing the primary windings for series-aiding connections can be done as described for parallel primary windings. The connection that yields the maximum voltmeter reading is the correct connection for series-aiding primaries. Power is applied to the two terminals where the voltmeter was connected.

12-9 Off-Center-Tapped Windings

Some transformers have an off-center-tapped primary and/or secondary. Often, the winding has two or more taps. The taps are usually near one end of the winding. Thus, changing taps produces relatively small changes in the winding's voltage rating.

The nominal voltage for homes and offices is 120 V. However, the actual voltage varies from area to area as well as according to the time of day. A transformer with primary taps for 115 V, 120 V, and 125 V can be connected to closely match the voltage typical of a given area.

Tapped secondary windings have various uses. For example, they can be used to adjust impedance-matching ratios in audio transformers. In power transformers, tapped secondaries are useful for adjusting the charging rate of a battery charger or the output of an arc welder.

■ TEST

Answer the following questions.

58. List two requirements for parallel-connected windings.
59. What does *phasing* refer to?
60. Two secondary windings are to be connected in series. One winding is rated for 11 V and 2 A. The other is rated for 8 V and 1.5 A.
    a. How much voltage and current are available if one instantaneous + is connected to the other instantaneous +?
    b. How much voltage and current are available if + is connected to −?
61. Are incorrectly phased parallel windings likely to damage a transformer?
62. Series-connected primary windings of a dual-voltage primary are connected series _____ .
63. True or False. Primary windings are never tapped for small voltage adjustments.
64. True or False. A tapped secondary winding is always center-tapped.

12-10 Three-Phase Transformers

YOU MAY **RECALL** . . . that three-phase circuits were discussed in Chap. 8. The ideas developed in Sec. 8-8 are essential in understanding three-phase transformers. Therefore, you should review that section before reading this section.

Three-phase voltages can be transformed either by a single three-phase transformer or by three single-phase transformers. The end results are the same: all three of the phase voltages are changed.

The structure of a three-phase transformer is illustrated in Fig. 12-26. The flux in the phase 1 leg is equal to the phase 2 flux plus the phase 3 flux. Phase 2 flux equals phase 3 flux plus

Series-aiding

Series-opposing

Bucking

Three-phase transformer

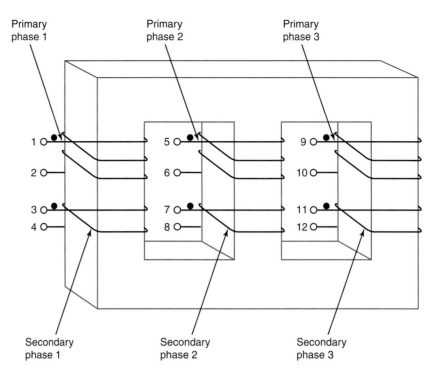

Primary
phase 1

Primary
phase 2

Primary
phase 3

Secondary
phase 1

Secondary
phase 2

Secondary
phase 3

Fig. 12-26 Three-phase transformer. One phase is wound on each leg of the transformer.

phase 1 flux, etc. This is because the flux, like the current, of each phase is displaced by 120°.

Figure 12-27 graphically presents the idea of how the flux splits up in a three-phase transformer core. In this figure, the flux of each phase is assumed to be in step with the current in the phase. (This assumption implies that the core has no hysteresis loss.) At instant *A* [Fig. 12-27(*a*)], the phase 1 current is zero. Therefore, the phase 1 flux is also zero. In Fig. 12-27(*b*) (at instant *B*), currents in phases 1 and 3 are both positive. This produces flux in the direction shown in both the phase 1 leg and the phase 3 leg of the core. At this same instant (instant *B*), phase 2 current is negative and of twice the value of either phase 1 current or phase 3 current. Thus, the phase 2 leg of the core has twice as much flux as either phase 1 or phase 3. Also, the direction of the flux in phase 2 is opposite to that of the flux in the other two phases. Close inspection of Fig. 12-27(*c*) and (*d*) shows how the flux continues to shift around in the core.

The primary and secondary windings of a three-phase transformer may be either *wye-connected* or *delta-connected*. The secondary does not have to have the same configuration (wye or delta) as the primary. Figure 12-28

shows four possible ways to connect a three-phase transformer. The dots on one end of each winding indicate the beginning of each winding. Refer back to Fig. 12-26. Notice that all windings are wound in the same direction (counterclockwise) when you start at the dotted end of the winding. Identifying the start of the windings is necessary before they can be properly phased.

The diagrams in Fig. 12-28 show one way of connecting the windings to obtain correct phasing. With a wye connection, correct phasing can also be obtained by connecting all the dotted ends to the star point. In the delta connection all three windings can be reversed; just be sure that two dotted ends are not connected together. On transformers, the dotted (start) end of a winding is identified by the manufacturer. The identification may be made in several ways. It may be a colored strip wrapped around the simulation of the start lead. It may be a number on a diagram mounted on the transformer.

Incorrect phasing of the primary, in either the wye or the delta configuration, causes excessively high primary current. If not protected against overload, the incorrectly phased primary can be destroyed by the excess current.

Wye-connected

Delta-connected

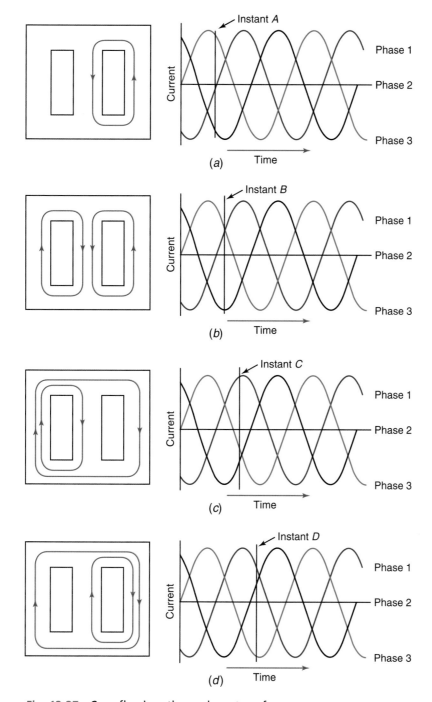

Fig. 12-27 Core flux in a three-phase transformer core.

Improper phasing of a delta-connected secondary also causes excessive, destructive current. A quick check for correct phasing of a delta-connected secondary is shown in Fig. 12-29. If the voltmeter indicates 0 V, the windings are properly phased. The ends of the windings to which the meter is connected can be connected together to complete the delta. If the meter indicates a high voltage (twice the phase voltage), incorrect phasing exists. Reverse the lead connections on one winding at a time until the meter indicates 0 V. With some wye-delta-connected transformers, the meter in Fig. 12-29 may not indicate 0 V when properly phased. In these cases the meter will indicate the lowest reading when the phase is correct.

A wye-connected secondary provides equal line voltages when correctly phased. If the line voltages are unequal, reverse one winding at a time until the line voltages are balanced.

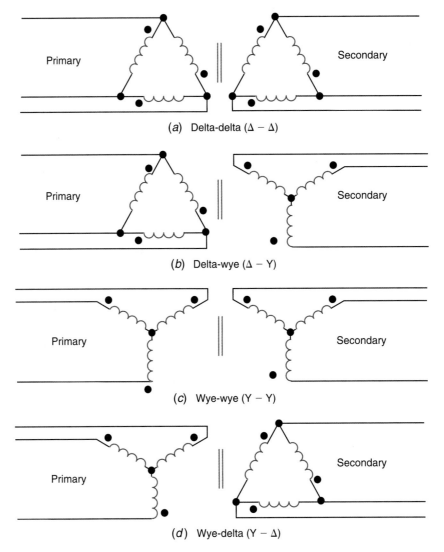

(a) Delta-delta (Δ − Δ)

(b) Delta-wye (Δ − Y)

(c) Wye-wye (Y − Y)

(d) Wye-delta (Y − Δ)

Fig. 12-28 Delta and wye connections.

. . . that the relationships between phase and line voltages and phase and line currents were explained in Sec. 8-8. These relationships apply to three-phase transformers as well as to three-phase generators.

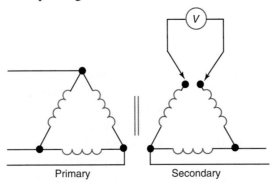

Fig. 12-29 Checking delta phasing. When properly phased, the meter will indicate 0 V.

■ TEST_____

Answer the following questions.

65. Three-phase transformer windings can be connected in a _____ or a _____ configuration.
66. When all the finish ends of the primary windings are connected together, the primary is connected in a _____ configuration.
67. What is the correct phasing for a delta connection?
68. Must the primary and secondary of a three-phase transformer be connected in the same configuration?
69. What happens when a delta winding is incorrectly phased?
70. Refer to Fig. 12-26. Which terminals should be connected together to form a wye-connected primary?

Summary

1. Transformers operate on the principle of mutual inductance.
2. Mutual inductance is measured in henrys.
3. Magnetic flux links, or couples, the two coils of a transformer.
4. Primaries receive power. Secondaries deliver power.
5. Primaries and secondaries, except on autotransformers, are electrically isolated.
6. Primaries and secondaries are reversible.
7. The coefficient of coupling specifies what portion of the primary flux links with the secondary.
8. The coefficient of coupling ranges from 0 to 1. It is greatest (almost 1) with iron-core transformers.
9. Flux leakage refers to primary flux that does not couple to the secondary.
10. Turns ratio and voltage ratio are the same when the coupling is 100 percent.
11. The turns-per-volt ratio is the same in all windings of a transformer.
12. One of the major uses of transformers is to step up and step down voltages in a power transmission and distribution system.
13. When voltage is stepped up, current is stepped down.
14. When voltage is stepped down, current is stepped up.
15. Transformer losses occur in both the core and the coils.
16. % eff. $= \dfrac{P_{\text{sec}}}{P_{\text{pri}}} \times 100$
17. Transformer loss consists of hysteresis, eddy current, and copper loss.
18. Copper loss is called I^2R loss.
19. Core loss consists of hysteresis and eddy current loss.
20. Hysteresis loss results from residual magnetism.
21. Hysteresis loss increases with increased frequency.
22. Eddy currents are currents induced in the core by primary flux.
23. Eddy currents are reduced by using laminations

and oxidizing or coating the surface of the lamination with an insulating material.
24. An unloaded transformer behaves just like an inductor. The energizing current and supply voltage are nearly 90° out of phase. In the unloaded condition, the input power is dissipated in the form of copper and core losses.
25. Energizing current is the current drawn by the primary when the transformer is unloaded. Its magnitude is controlled by the reactance of the primary.
26. A fully loaded transformer appears to be almost entirely resistive to the source.
27. Power factor (cos θ) approaches a value of 1 with a fully loaded transformer.
28. As secondary current increases, so does primary current.
29. Laminated-core transformers are used at power and audio frequencies.
30. Powdered iron and ferrite cores are used in the audio-frequency and lower radio-frequency ranges.
31. Air-core transformers are used only in the radio-frequency range. Their coupling can be controlled by the spacing and axis orientation of coils.
32. Constant voltage transformers provide a stable secondary voltage.
33. Isolation transformers have equal primary and secondary voltages.
34. Isolation transformers help protect the service technician from receiving a shock through the chassis of electrical equipment.
35. Autotransformers use a common primary-secondary winding. They are often used as variable transformers at power frequencies.
36. Matched impedances provide maximum power transfer.
37. Transformers may have voltage, current, power, and voltampere ratings.
38. Transformer power ratings apply to resistive loads only.
39. Power and voltampere ratings refer to the total of all secondaries.

40. Connecting windings in parallel increases the available current but does not change the voltage rating.

41. Connecting windings in series either increases or decreases the available voltage but does not increase the current rating.

42. Parallel windings must be properly phased and have identical voltages.

43. Three single-phase transformers can be used to transform three-phase voltages.

44. Three-phase transformer windings can be connected in either delta or wye configurations.

45. The primary and secondary of a three-phase transformer need not be connected in the same configuration.

Chapter Review Questions

For questions 12-1 to 12-13, determine whether each statement is true or false.

12-1. The same symbol is used for the air-core transformer and the powdered-iron-core transformer.

12-2. The winding of a transformer designated as the secondary can be used as a primary.

12-3. The secondary of a transformer converts electric energy to magnetic energy.

12-4. In an autotransformer, part or all of a coil is used as both the primary and the secondary.

12-5. The coefficient of coupling of an air-core transformer is greater than that of an iron-core transformer.

12-6. When the coupling in a transformer is 100 percent, the turns ratio is numerically equal to the voltage ratio.

12-7. In a step-up transformer, the turns-per-volt ratio in the primary is smaller than that in the secondary.

12-8. Large transformers usually have a higher turns-per-volt ratio than small transformers do.

12-9. A transformer is most efficient when the secondary is not loaded.

12-10. Power factor approaches 1 when a transformer is fully loaded.

12-11. The primary current of a transformer increases whenever the secondary current increases.

12-12. The primary and secondary of a three-phase transformer must be connected in the same configuration.

12-13. Three single-phase transformers can be used to transform a three-phase voltage.

For questions 12-14 to 12-35, supply the missing word or phrase in each statement.

12-14. Transformers operate on the principle of _____.

12-15. The unit of mutual inductance is the _____.

12-16. The _____ winding of a transformer receives power while the _____ winding delivers power.

12-17. The portion of the flux of one coil that links to another coil is specified by the _____.

12-18. Primary flux that does not couple to the secondary is known as _____ flux.

12-19. When voltage is stepped down, current is stepped _____.

12-20. Power loss occurs in the _____ and the _____ of a transformer.

12-21. I^2R loss is also called _____ loss.

12-22. In addition to I^2R loss, a transformer has _____ loss and _____ loss.

12-23. When a transformer is fully loaded, the primary current is mostly _____.

12-24. When a transformer has no load, the primary current is called the _____ current.

12-25. Constant voltage transformers provide a stable _____ voltage.

12-26. The _____ transformer has equal primary and secondary voltage.

12-27. The _____ is a transformer that does not provide electrical isolation.

12-28. Matched impedances will provide _____.

12-29. Transformer power ratings apply to _____ loads.

12-30. Connecting secondary windings in parallel increases the _____, but does not change the _____ rating of the combined windings.

12-31. Two unequal secondary windings can provide _____ different voltages.

12-32. Three-phase windings can be connected in either the _____ or the _____ configuration.

12-33. Transformer cores are laminated to minimize _____ loss.

12-34. Transformer cores are made of silicon steel to minimize _____ loss.

12-35. A step-up transformer causes the secondary

load resistance to appear to be _____ than its true value.

Answer the following questions.

12-36. List three common uses of transformers.

12-37. What is the maximum voltage and current available from a series-connected 14-V, 2-A secondary and a 7-V, 3-A secondary?

12-38. How many turns are required for a 40-V secondary if the primary of the transformer has five turns per volt?

12-39. Determine the percent efficiency of a transformer that requires 400 W of primary power when a 60-V secondary is connected to a 12-Ω resistive load.

12-40. A 1500-W transformer has one secondary rated at 230 V. Could this transformer power a 230-V motor which draws 7 A at a PF of 0.75? Why?

12-41. A 10-Ω load is connected to the 6-V secondary of a transformer with a 120-V primary. What value does the load appear to be to the source?

Critical Thinking Questions

12-1. The primary of a transformer has 5 turns per volt. What is the coefficient of coupling if a 40-turn secondary produces 6 V?

12-2. In the discussion of copper loss, it was stated that the size of wire in the primary winding was limited by the windows in the core. What would be the consequences of increasing the size of the window openings?

12-3. Why is it understood that the load must be pure resistance if a transformer is to be operated at its rated power output?

12-4. A transformer with 5 turns per volt has a 120-V primary, an 8-V secondary, and a 6-V secondary. If the 6-V secondary is connected in series with the primary and the series combination connected to a 100-V source, how much voltage will the 8-V secondary provide?

12-5. What might cause the voltage between the open terminals in the properly phased delta-connected secondary to be greater than 0 V?

12-6. Why doesn't an improperly phased wye-connected secondary produce destructive currents in the transformer?

12-7. A 120-V motor draws 6 A at a power factor (PF) of 0.75 from a transformer which is

90 percent efficient. The transformer draws 3 A from a 230-V source. What is the PF of the transformer primary?

12-8. How does increasing the number of primary turns affect the operation of an unloaded transformer?

12-9. Would a 10-Ω resistive load or a 10-Ω reactive load produce the larger shift of theta in the primary of a transformer? Why?

12-10. For each case below, determine the effect on system efficiency of adding an impedance-matching transformer (assume an ideal transformer). Also, determine the power of the load with and without the matching transformer.
   a. A 20-V source with 4 Ω of internal resistance driving a 16-Ω resistive load.
   b. A 15-V source with 10 Ω of internal resistance driving a 5-Ω resistive load.

12-11. Draw a schematic diagram showing how a 2-pole, 3-position rotary switch could be connected to the 4-V and 6-V secondaries of a transformer to provide 4, 6, or 10 V at the output terminals. The output voltage is to increase from 4 to 10 V as the switch is rotated clockwise.

Answers to Tests

1. two inductors coupled together by mutual inductance, or, a device that transfers power between electrically isolated circuits.
2. primary, secondary
3. 2 H
4.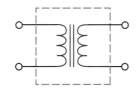
5. The primary converts electric energy to magnetic energy. The secondary converts magnetic energy to electric energy.
6. primary
7. secondary
8. yes
9. It is a number that indicates the portion of the primary flux that links the secondary coil.
10. no
11. air-core transformers

12. by varying the distance between primary and secondary or by changing the axis orientation between the primary and secondary coils
13. F
14. T
15. F
16. F
17. 200
18. changing voltage levels
19. to keep current low and thus wire size small
20. steps up
21. heat losses in the core and the windings
22. 83.3 percent
23. 806.5 W
24. hysteresis loss and eddy current loss
25. windings
26. residual magnetism
27. twice each cycle
28. a curve on a graph produced by plotting magnetizing force against flux density

29. silicon steel, because it has high permeability and low hysteresis loss
30. narrow
31. As frequency increases, so does hysteresis loss.
32. by using core laminations that have oxidized or insulation-coated surfaces
33. thin
34. Power loss in the transformer windings. I^2R loss is minimized by using large-diameter conductors in the coils.
35. large
36. when it is fully loaded
37. a. decreases from nearly 90° to nearly 0°
    b. increases from nearly 0 to nearly 1
    c. changes from inductive to resistive
38. energizing

39. fully loaded
40. T
41. T
42. to prevent a continuous air gap and thereby reduce flux leakage
43. power and audio-frequency transformers
44. distribution, lighting, rectifier, and control
45. constant voltage
46. isolation
47. auto
48. Impedance matching means making the load impedance equal to the source impedance.
49. It provides maximum power transfer from source to load.
50. larger
51. It means the impedance of the secondary load is reflected back to the source through the transformer.
52. 256:1 (primary to secondary)

53. 2560 Ω
54. F
55. F
56. T
57. T
58. equal voltages and correct phasing
59. the momentary polarity of transformer windings
60. a. 3 V, 1.5 A
    b. 19 V, 1.5 A
61. yes
62. Aiding
63. F
64. F
65. delta, wye
66. wye
67. The start end of one winding must be connected to the finish end of the next winding.
68. no
69. The winding has excessive current flowing in it.
70. 1, 5, and 9; or 2, 6, and 10

R, C, and *L* Circuits

Chapter Objectives

This chapter will help you to:

1. *Understand* the relationship between resistance, reactance, and impedance.
2. *Develop* your skills in using phasors to represent electrical qualities.
3. *Add* voltages, currents, or oppositions that are 90° out of phase.
4. *Calculate* impedance and phase angle for *RC*, *RL*, and *RCL* circuits when the components are connected in either series or parallel.
5. *Calculate* the resonant frequency, the quality, and the bandwidth of *LC* circuits.
6. *Understand* the purpose and operation of basic filter circuits.
7. *Identify and classify* simple filter circuits.

Previous chapters have dealt with resistance, inductance, and capacitance used individually in ac circuits. This chapter deals with the results of combining two or more of these quantities in a single circuit.

13-1 Impedance

The combined opposition to current due to resistance and reactance is called *impedance.* The symbol for impedance is *Z*. Like resistance and reactance, impedance has base units of ohms.

In a circuit that contains resistance and only one type of reactance, the current and voltage cannot be in phase. Neither can they be a full 90° out of phase. If the circuit contains resistance and capacitance, the current leads the voltage. If it contains resistance and inductance, the current lags the voltage. When the circuit contains all three (*R*, *C*, and *L*), the phase relationship depends upon the relative sizes of *L* and *C*.

Circuits containing resistance and only one type of reactance have more apparent power than true power. Their power factor is less than 1 (less than 100 percent).

Examples of common electric and electronic loads that possess impedance are motors, speakers, and earphones. These particular loads possess both inductive reactance and resistance.

■ TEST_____ **Impedance**

Answer the following questions.

1. True or false. In a circuit containing impedance, the current and voltage will be 90° out of phase.
2. Impedance is a combination of _____ and _____.
3. The symbol for impedance is _____.
4. The base unit for impedance is the _____.

13-2 Adding Phasors

Suppose we wish to add the two phasors Y_1 and Y_2 in Fig. 13-1(*a*). We can add them either graphically or mathematically. First we will do it graphically, and then we will develop the mathematical method.

The *graphical addition* of the phasors in **Graphical addition**

Pythagorean theorem

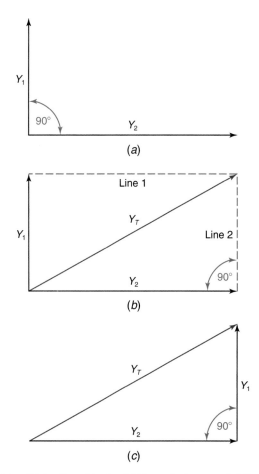

(a)

(b)

(c)

Fig. 13-1 Graphical addition of phasors. Adding Y_1 and Y_2 in (a) yields the resultant phasor (Y_T) in (b) and (c).

Fig. 13-1(a) is illustrated in Fig. 13-1(b). The process involves the construction of a rectangle. First, line 1 is drawn from the tip of the Y_1 phasor parallel to the Y_2 phasor. Then, line 2 is drawn from the tip of the Y_2 phasor parallel to the Y_1 phasor. The point at which lines 1 and 2 cross is the tip of the resultant Y_T phasor.

The three phasors of Fig. 13-1(b) can be rearranged into the configuration shown in Fig. 13-1(c). This rearrangement merely replaces line 2 of Fig. 13-1(b) with phasor Y_1. This is permissible because line 2 and phasor Y_1 are parallel and of the same length.

The rearrangement of phasors in Fig. 13-1(c) produces a right triangle. This right triangle is redrawn in Fig. 13-2 with the sides labeled. You are already familiar with the trigonometric functions listed in this figure. They are listed here merely as a review and easy reference for this chapter. Also listed in Fig. 13-2 is the formula for the relationship of the sides of a right triangle. This relationship,

known as the *Pythagorean theorem*, applies to all right triangles regardless of the size of the triangle or of the angle θ. Thus, knowing the length of any two sides of a right triangle, you can calculate the length of the third side using the formula

$$(\text{Hypotenuse})^2 = (\text{opposite side})^2 + (\text{adjacent side})^2$$

Example 13-1

Refer to Fig. 13-2. Suppose Y_1 represents 6 A and Y_2 represents 8 A. What is the value of the total current Y_T?

Given: $I_1 = 6$ A
$I_2 = 8$ A
Find: I_T
Known: $Y_T = \sqrt{Y_1^2 + Y_2^2}$
Solution: $I_T = \sqrt{I_1^2 + I_2^2}$
$= \sqrt{6^2 + 8^2}$
$= \sqrt{36 + 64}$
$= \sqrt{100} = 10$ A
Answer: The total current is 10 A.

Figure 13-3 illustrates how to add two phasors (Y_1 and Y_2) when the phasors are 180° apart. The resultant phasor (Y_X) is equal in length to the difference between the two pha-

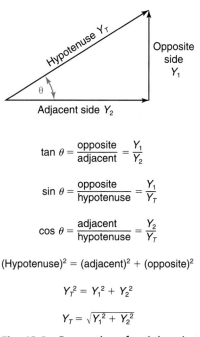

$$\tan \theta = \frac{\text{opposite}}{\text{adjacent}} = \frac{Y_1}{Y_2}$$

$$\sin \theta = \frac{\text{opposite}}{\text{hypotenuse}} = \frac{Y_1}{Y_T}$$

$$\cos \theta = \frac{\text{adjacent}}{\text{hypotenuse}} = \frac{Y_2}{Y_T}$$

$$(\text{Hypotenuse})^2 = (\text{adjacent})^2 + (\text{opposite})^2$$

$$Y_T^2 = Y_1^2 + Y_2^2$$

$$Y_T = \sqrt{Y_1^2 + Y_2^2}$$

Fig. 13-2 Properties of a right triangle.

286 **Chapter 13** *R, C, and L Circuits*

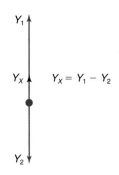

Fig. 13-3 Adding phasors separated by 180°.

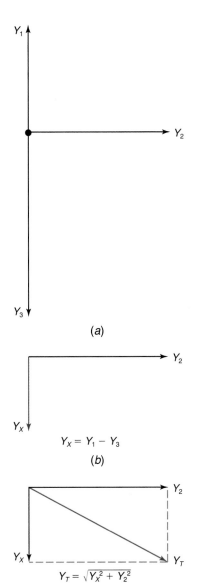

$Y_X = Y_1 - Y_2$

Y_2

(a)

Y_2

Y_X

$Y_X = Y_1 - Y_3$

(b)

Y_2

Y_X $\quad\quad\quad\quad Y_T$

$Y_T = \sqrt{Y_X^2 + Y_2^2}$

(c)

Fig. 13-4 Adding three phasors. Adding Y_1 and Y_3 in (a) results in two phasors in (b). Right triangle properties are used in (c) to find Y_T.

sors being added. The direction of the resultant phasor is the same as the direction of the longest phasor. As an example, assume Y_1 is 14 Ω and Y_2 is 10 Ω. Then Y_X would be 4 Ω in the direction of Y_1.

Suppose three phasors [Fig. 13-4(a)], all displaced 90° from each other, are to be added. The first step is to add the two phasors that are 180° apart. Adding Y_1 and Y_3 reduces the problem to the two phasors shown in Fig. 13-4(b). The second, and last, step involves adding the two phasors of Fig. 13-4(b). This is accomplished by using the Pythagorean theorem. The result [Fig. 13-4(c)] is a single phasor Y_T which is equivalent to the original three phasors of Fig. 13-4(a). The formula given in Fig. 13-4(c) can be expanded so that all three phasors can be added in one step. It can be expanded by substituting $Y_1 - Y_3$ for Y_X. The result is the general formula

$$Y_T = \sqrt{(Y_1 - Y_3)^2 + Y_2^2}$$

This formula can be used to combine any three phasors that are displaced 90° from each other. The three phasors may represent voltage, current, or impedance (and the components of impedance: resistance and reactance). The only requirement is that all the phasors have the same units, that is, volts, amperes, or ohms.

JOB TIP

Some companies invest heavily in research and development, and others do not. The nature of such investments often influences the types of positions open to electronic technicians.

Example 13-2

Find the total current represented by the three current phasors in Fig. 13-5.

Given: $I_1 = 5$ A
$I_2 = 4$ A
$I_3 = 8$ A

Find: I_T

Known: $I_T = \sqrt{(I_1 - I_3)^2 + I_2^2}$

Solution: $I_T = \sqrt{(5 - 8)^2 + 4^2}$
$= \sqrt{(-3)^2 + 4^2} = \sqrt{9 + 16}$
$= \sqrt{25} = 5$ A

Answer: The three phasors represent a total current of 5 A.

Now that we know how to add phasors, we can return to problems that involve impedance. That is, we can now calculate current, voltage,

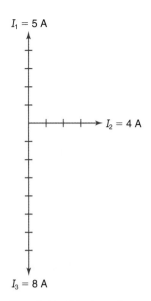

$I_1 = 5\ A$

$I_2 = 4\ A$

$I_3 = 8\ A$

Fig. 13-5 Diagram for example 13-2.

and impedance in circuits containing both reactance and resistance.

■ **TEST**_____

Answer the following questions

5. Refer to Fig. 13-1(*a*). If $Y_1 = 7$ m and $Y_2 = 9$ m, what is the length of Y_T?
6. What is the resultant force in Fig. 13-4(*a*) when $Y_1 = 45$ N, $Y_2 = 25$ N, and $Y_3 = 30$ N?
7. The _____ theorem shows the relationship between the sides of a right triangle.

13-3 Solving *RC* Circuits

Unless stated otherwise, we will assume that the components we use are ideal devices. That is, the resistors contain only resistance, and the capacitors contain only capacitance. Actual resistors usually have small amounts of inductance and capacitance, and actual capacitors have small amounts of resistance and inductance.

Series *RC* Circuits

Series *RC* circuit

Figure 13-6 shows a *series RC circuit* and the phasor diagrams that represent its voltages and oppositions. In a series circuit there is only one path for current; the capacitive current, the resistive current, and the total current are the

same current. Therefore, the current phasor is often used as the reference phasor in series circuits [Fig. 13-6(*b*)].

In any circuit, we know that the resistive voltage and the resistive current are in phase. This means that the resistive voltage phasor [V_R in Fig. 13-6(*b*)] is in the same direction as I_T. We also know that the current through a capacitor leads the voltage across the capacitor by 90°. Another way of looking at this is to say that the voltage lags the current by 90°.

YOU MAY **RECALL** . . . from Sec. 9-2 that phasors are assumed to rotate in a counterclockwise direction. Thus, the capacitive voltage phasor [V_C in Fig. 13-6(*b*)] is plotted 90° behind the current phasor.

(*a*) The circuit

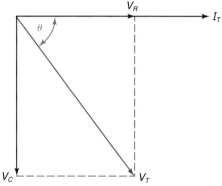

(*b*) Voltage phasors

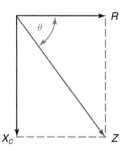

(*c*) *R*, *X_C*, and *Z* phasors

Fig. 13-6 Series *RC* circuit. The voltages and the oppositions are shown as phasors for this circuit.

Now that we have V_R and V_C, they can be added together to obtain V_T. Phasor V_T establishes angle θ. Angle θ shows how much the source current I_T leads the source voltage V_T. As you can see in Fig. 13-6, V_C is greater than V_R. This makes angle θ greater than $45°$.

Also from Fig. 13-6(b) we can conclude that X_C must be greater than R. We make this conclusion because of the following facts:

1. $I_R = I_C$
2. Ohm's law applies individually to both the resistor and the capacitor.
3. Ohm's law states that the voltage and the resistance or reactance are directly proportional if the current is held constant.

Thus, if V_C is greater than V_R, then X_C must also be greater than R.

Figure 13-6(c) shows that the resistance and capacitive reactance are also $90°$ out of phase. This is to be expected. The resistor and capacitor currents are in phase, but their voltages are out of phase. Therefore, X_C and R must also be $90°$ out of phase.

The phasors of Fig. 13-6(b) and (c) can be redrawn into triangles, as shown in Fig. 13-7. From Fig. 13-7 it can be seen that the Pythagorean theorem can be used to calculate V_T and Z:

$$V_T = \sqrt{V_R^2 + V_C^2} \quad \text{and} \quad Z = \sqrt{R^2 + X_C^2}$$

The ratio of R over X_C [in Fig. 13-6(c)] must be equal to the ratio of V_R over V_C [in Fig. 13-6(b)]. This statement is easily seen by looking at Ohm's law:

$$I_T = \frac{V_R}{R} \quad \text{and} \quad I_T = \frac{V_C}{X_C}$$

Therefore,

$$\frac{V_R}{R} = \frac{V_C}{X_C}$$

Transposing terms:

$$\frac{V_R}{V_C} = \frac{R}{X_C}$$

This means that angle θ can be figured equally well with voltages as with oppositions. In other words,

$$\cos\theta = \frac{\text{adjacent}}{\text{hypotenuse}} = \frac{V_R}{V_T} = \frac{R}{Z}$$

Now that we know the relationship between impedance, resistance, and reactance in series RC circuits, let us use it to solve an electrical problem.

Example 13-3

Find the impedance of the circuit in Fig. 13-8.

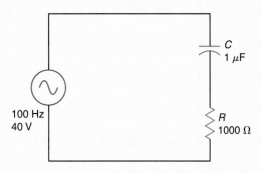

$$V_T = \sqrt{V_R^2 + V_C^2}$$

$$Z = \sqrt{R^2 + X_C^2}$$

Fig. 13-8   Circuit for examples 13-3, 13-4, and 13-5.

Given: $C = 1\ \mu F$
            $R = 1000\ \Omega$
            $f = 100\ Hz$

Find: Z
Known: $Z = \sqrt{R^2 + X_C^2}$

   $$X_C = \frac{1}{6.28\,f\,C}$$

Solution: $$X_C = \frac{1}{6.28 \times 100 \times 1 \times 10^{-6}}$$

   $$= 1592\ \Omega$$
   $$Z = \sqrt{1000^2 + 1592^2}$$
   $$= 1880\ \Omega$$

Answer: The circuit impedance is $1880\ \Omega$.

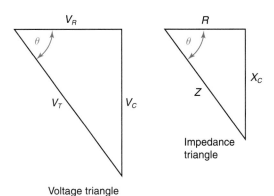

Fig. 13-7   Voltage and impedance triangles. Angle θ is the same in both triangles.

We can check the reasonableness of our answer very easily. Look again at Fig. 13-7 and notice that Z must be greater than either R or X_C. Also notice that Z must be less than the arithmetic sum of R and X_C. Since 1880 is greater than 1592 but less than 2592, it seems to be a reasonable answer.

Impedance can be used in Ohm's law just like resistance and reactance. For instance, the impedance and source voltage can be used to find the total current. Then the current can be used to calculate individual voltages.

Example 13-4

Find the current I_T, the resistive voltage V_R, and the capacitive voltage V_C for Fig. 13-8.

Given: $V_T = 40$ V
             $R = 1000\ \Omega$
             $X_C = 1592\ \Omega$
                   (from example 13-3)
             $Z = 1880\ \Omega$
                   (from example 13-3)

Find: I_T, V_C, and V_R

Known: Ohm's law

Solution: $I_T = \dfrac{40\ \text{V}}{1880\ \Omega}$

                   $= 0.0213$ A
                   $= 21.3$ mA

             $V_R = 0.0213\ \text{A} \times 1000\ \Omega$
                   $= 21.3$ V

             $V_C = 0.0213\ \text{A} \times 1592\ \Omega$
                   $= 33.9$ V

Answer: The circuit current is 21.3 mA, the resistive voltage is 21.3 V, and the capacitive voltage is 33.9 V.

$P_T = P_R$

We can verify our answer by checking whether the phasor sum of V_R and V_C equals the source voltage V_T:

$$V_T = \sqrt{21.3^2 + 33.9^2} = 40\ \text{V}$$

Since this calculated value of V_T agrees with the value given, we have not made any errors in our calculation.

Now we can continue our study of the circuit in Fig. 13-8 by determining angle θ and power.

Example 13-5

For the circuit of Fig. 13-8, find angle θ and the total true power.

Given: $R = 1000\ \Omega$
             $V_T = 40$ V
             $Z = 1880\ \Omega$
             $I_T = 21.3$ mA

Find: θ and P

Known: $\cos \theta = \dfrac{R}{Z}$, power formulas

Solution: $\cos \theta = \dfrac{1000\ \Omega}{1880\ \Omega} = 0.532$

             $\theta = 58°$ (from Appendix F)
             $P_T = IV \cos \theta$
                  $= 0.0213 \times 40 \times 0.532$
                  $= 0.453$ W

Answer: The current leads the voltage by 58°, and the power in the circuit is 0.453 W.

In the above example we could have found the total power without first calculating θ. Recall that only the resistance in a circuit can use power. Therefore, the power used by the resistor (P_R) must equal the total power. Thus,

$$P_T = P_R = I_R \times V_R = 0.0213\ \text{A} \times 21.3\ \text{V}$$
$$= 0.454\ \text{W}$$

This answer agrees (within round-off error) with the answer in example 13-5. Remember that $P_T = P_R$ for all circuits.

Also, in example 13-5 we could have found $\cos \theta$ using voltage ratios. This would give us

$$\cos \theta = \frac{V_R}{V_T} = \frac{21.3\ \text{V}}{40\ \text{V}} = 0.533$$

Again, the answer agrees (within round-off error) with that obtained in example 13-5.

Suppose the frequency of the source in Fig. 13-8 were decreased. What would happen to angle θ, the impedance, and the voltage distribution? The answer to this question is illustrated in Fig. 13-9(a) and (b). In Fig. 13-9(a) the phasor diagrams for the circuit operating at 100 Hz are shown. Figure 13-9(b) shows what happens when the frequency is decreased to 50 Hz. When the frequency decreases, the reactance of the capacitor increases. This causes an increase in the impedance and a corresponding decrease in current. Less current means less voltage across

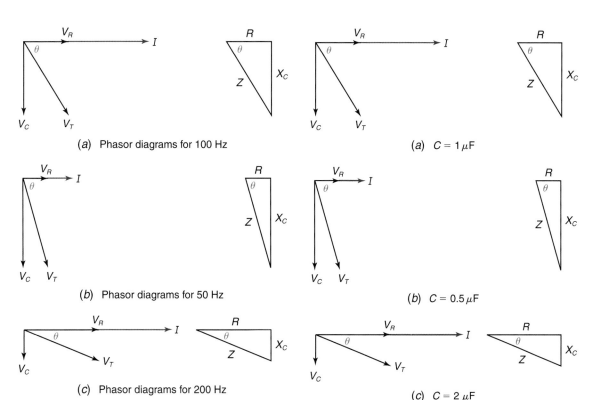

(a) Phasor diagrams for 100 Hz

(b) Phasor diagrams for 50 Hz

(c) Phasor diagrams for 200 Hz

Fig. 13-9 Effects of *f* on *Z, V,* and *θ.*

(a) C = 1 µF

(b) C = 0.5 µF

(c) C = 2 µF

Fig. 13-10 Effects of *C* on *Z, V,* and *θ.*

the resistor. However, the capacitive voltage increases because its reactance increased proportionately more than the current decreased. The net result is that the circuit is more capacitive than before. Angle *θ* has increased and the power in the circuit has decreased.

When the frequency *increases,* the results [Fig. 13-9(*c*)] are just the opposite: the circuit becomes more resistive, *θ* decreases, and the power increases.

Figure 13-10 illustrates what happens when the capacitor in Fig. 13-8 is changed. The diagrams in Fig. 13-10(*a*) show the situation with the original capacitor (1 µF). When the capacitance is decreased [Fig. 13-10(*b*)], the effects are the same as when the frequency is decreased. This is because capacitive reactance is inversely proportional to both capacitance and frequency. Decreasing capacitance causes the capacitive reactance to increase; that is, *θ* becomes larger. Increasing the capacitance, of course, decreases the capacitive reactance and causes the circuit to be more resistive [Fig. 13-10(*c*)].

An example of how an *RC* circuit is used is shown in Fig. 13-11. Amplifier 1 replaces the generator in Fig. 13-8. In other words, amplifier 1 is the source which provides the voltage across the *RC* circuit. Whatever voltage devel-

ops across *R* becomes the input voltage to the second amplifier. It is desirable to get as much as possible of the voltage from amplifier 1 to the input of amplifier 2. Therefore, the reactance of the capacitor must be low relative to the resistance of the resistor. The resistance

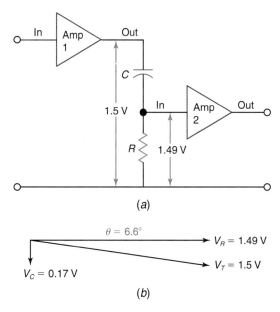

(a)

(b)

Fig. 13-11 Use of an *RC* circuit. The capacitor couples the amplifiers together with minimum voltage loss or phase shift.

should be about 10 times as high as the reactance. This makes the resistive voltage [Fig. 13-11(b)] nearly equal to the source voltage. Also, Fig. 13-11(b) shows that very little phase shift occurs when R is high relative to X_C. This means that the voltage input to amplifier 2 is nearly in phase with the output of amplifier 1.

$$I_T = \sqrt{I_C^2 + I_R^2}$$

Parallel RC Circuits

In a parallel circuit [Fig. 13-12(a)], the voltage is the same across all components. Obviously V_R, V_C, and V_T have to be in phase. Therefore, the voltage phasor is used as the *reference phasor* [Fig. 13-12(b)] in solving *parallel RC circuits*. The capacitive current still must lead the capacitive voltage by 90°. Thus, the I_C phasor is drawn 90° ahead of the voltage phasor [Fig. 13-12(b)]. As is always the case, the resistive current is in phase with the resistive voltage. The I_R phasor has to be in line with the V_T phasor. Adding phasors I_C and I_R yields the I_T phasor. Since the circuit has capacitance, the source current I_T leads the source voltage V_T.

The impedance of the circuit shown in Fig. 13-12(a) can be found by using Ohm's law with V_T and I_T:

$$Z = \frac{V_T}{I_T}$$

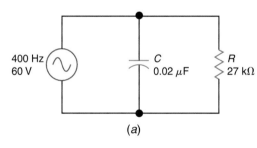

(a)

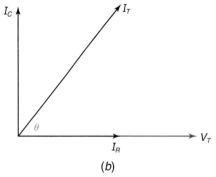

(b)

Fig. 13-12 Parallel *RC* circuit. The voltage phasor is the reference (0°) phasor. Notice the current phasors' relationships.

$$Z = \frac{X_C \times R}{\sqrt{X_C^2 + R^2}}$$

Of course, one must calculate I_T before Z can be determined. The value of I_T can be calculated by using the Pythagorean theorem. Just express the sides of the triangle in terms of currents. The formula is

$$I_T = \sqrt{I_C^2 + I_R^2}$$

Find the impedance of the circuit in Fig. 13-12.

Given: $V_T = 60$ V
$\quad\quad\quad f = 400$ Hz
$\quad\quad\quad C = 0.02\ \mu$F
$\quad\quad\quad R = 27$ kΩ

Find: Z

Known: Ohm's law

$$X_C = \frac{1}{6.28\,f\,C}$$

$$I_T = \sqrt{I_C^2 + I_R^2}$$

Solution: $X_C =$

$$\frac{1}{6.28 \times 400 \times 0.02 \times 10^{-6}}$$
$$= 19{,}900\ \Omega = 19.9\ \text{k}\Omega$$

$$I_C = \frac{60\ \text{V}}{19{,}900\ \Omega} = 3.0\ \text{mA}$$

$$I_R = \frac{60\ \text{V}}{27{,}000\ \Omega} = 0.00222\ \text{A}$$
$$= 2.22\ \text{mA}$$

$$I_T = \sqrt{0.003^2 + 0.00222^2}$$
$$= 0.00373\ \text{A} = 3.73\ \text{mA}$$

$$Z = \frac{60\ \text{V}}{0.00373\ \text{A}} = 16{,}100\ \Omega$$
$$= 16.1\ \text{k}\Omega$$

Answer: The impedance of the circuit is 16.1 kΩ.

Notice in this example that the impedance is less than the capacitive reactance or the resistance. This is because the reactance and the resistance are in parallel as well as 90° out of phase.

The impedance of a parallel *RC* circuit can also be calculated from the resistance and reactance. The formula is

$$Z = \frac{X_C \times R}{\sqrt{X_C^2 + R^2}}$$

This formula yields the same results as the Ohm's law method. Therefore, we can use it to check the answer arrived at in example 13-6:

$$Z = \frac{19,900 \times 27,000}{\sqrt{19,900^2 + 27,000^2}}$$
$$= 16,000 \ \Omega = 16.0 \ k\Omega$$

Since the two methods provide the same answer (within round-off error), we can assume no calculation errors were made.

The effects of changing f or C in a parallel RC circuit are illustrated in Fig. 13-13. Notice that the effect on angle θ is the exact opposite of the effect in a series RC circuit. Increasing C or f causes a decrease in reactance, which results in less impedance, more current, and a larger angle θ. Of course, decreasing C or f has the opposite effect.

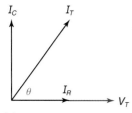

(a) Original phasor diagram

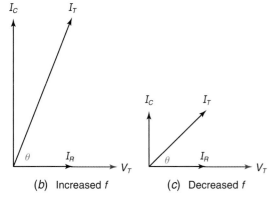

(b) Increased f (c) Decreased f

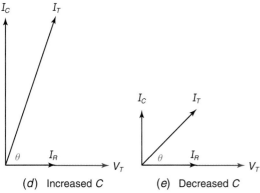

(d) Increased C (e) Decreased C

Fig. 13-13 Effects of f and C on I and θ.

The value of $\cos \theta$ can be calculated from either the currents or the powers. The formulas are

$$\cos \theta = \frac{P}{P_{app}} \quad \text{and} \quad \cos \theta = \frac{I_R}{I_T}$$

$$\cos \theta = \frac{I_R}{I_T}$$

These are also the power-factor formulas. Once $\cos \theta$ is determined, angle θ can be determined by using a calculator or a table of trigonometric functions. We can find the power factor of the circuit in example 13-6 using the values for I_R and I_T:

$$\cos \theta = PF = \frac{I_R}{I_T} = \frac{0.00222}{0.00373}$$
$$= 0.595, \text{ or } 59.5 \text{ percent}$$

Since the total current is leading the voltage, the power factor of the circuit is 59.5 percent, leading.

◼ TEST_____

Answer the following questions.

8. The current _____ the voltage in a parallel RC circuit.
9. True or false. In a series RC circuit, V_C is larger than V_T.
10. In a series RC circuit, the reference phasor is the _____ phasor.
11. True or false. If I_R is greater than I_C in a parallel RC circuit, θ will be greater than 45°.

12. In a series RC circuit, I_T leads V_T by 20°. What is the angle between R and Z for this circuit?

13. What is the phase relationship between resistance and reactance?

14. Write two formulas for finding the impedance of a series RC circuit.

15. Write the formula for combining the resistive and reactive voltages in a series RC circuit.

16. Write three formulas for finding $\cos\theta$ in a series RC circuit.

17. Which of the formulas in question 16 can also be used in a parallel RC circuit?

18. What other formula besides the one given in question 17 can be used to find $\cos\theta$ in a parallel RC circuit?

19. If the frequency of a parallel RC circuit decreases what happens to
    a. The impedance
    b. The resistive current
    c. The power
    d. $\cos\theta$

20. If the resistance of a series RC circuit increases, what happens to
    a. The impedance
    b. The resistive voltage
    c. The current
    d. $\cos\theta$

21. Refer to Fig. 13-12. Change the value of C to 0.03 μF and calculate the following:
    a. Z
    b. P
    c. θ
    d. I_C

22. Refer to Fig. 13-8. Change C to 1.5 μF and calculate the following:
    a. I_T
    b. V_R
    c. P
    d. Z

$$Z = \sqrt{R^2 + X_L^2}$$

$$V_T = \sqrt{V_R^2 + V_L^2}$$

13-4 Solving RL Circuits

Although RL circuits are not as common as RC circuits in electronics, they are important. One must know something of their characteristics to fully understand such devices as ac motors and high-frequency amplifiers.

Series RL Circuits

The series RL circuit [Fig. 13-14(a)] causes the source voltage to lead the source current. This is shown in Fig. 13-14(b), where V_T leads I_T (the reference phasor). The same general formulas used for series RC circuits can also be used for series RL circuits; merely replace the C with L and the X_C with X_L. The important formulas for series RL circuits are

$$Z = \sqrt{R^2 + X_L^2}$$
$$V_T = \sqrt{V_R^2 + V_L^2}$$
$$\cos\theta = \frac{V_R}{V_T} = \frac{R}{Z} = \frac{P}{P_{\text{app}}}$$

Example 13-7

Determine the impedance, the voltage across the inductance, and the power factor for the circuit in Fig. 13-14.

Given: $\quad V_T = 50$ V
$\quad\quad\quad\quad f = 50$ kHz
$\quad\quad\quad\quad L = 10$ mH
$\quad\quad\quad\quad R = 2700\ \Omega$

Find: $\quad\quad Z, V_L,$ and PF

Known: $\quad Z = \sqrt{R^2 + X_L^2}$
$\quad\quad\quad\quad X_L = 6.28fL$
$\quad\quad\quad\quad \text{PF} = \cos\theta = \dfrac{R}{Z}$
$\quad\quad\quad\quad$ Ohm's law

Solution: $\quad X_L =$
$\quad\quad 6.28 \times 50 \times 10^3 \times 10 \times 10^{-3}$
$\quad\quad\quad = 3140\ \Omega$
$\quad\quad Z = \sqrt{2700^2 + 3140^2}$
$\quad\quad\quad = 4141\ \Omega$
$\quad\quad I_T = \dfrac{50\ \text{V}}{4141\ \Omega}$
$\quad\quad\quad = 0.012$ A $= 12$ mA
$\quad\quad V_L = 0.012$ A $\times 3140\ \Omega$
$\quad\quad\quad = 37.7$ V
$\quad\quad \text{PF} = \dfrac{2700}{4141} = 0.652$

Answer: The impedance of the circuit is 4141 Ω, the voltage across the inductor is 37.7 V, and the power factor is 0.652.

Series RL and series RC circuits respond to changes in frequency in exactly opposite fashions. This is because inductive reactance is directly proportional to frequency while capacitive reactance is inversely proportional to frequency. Increasing the frequency of a series

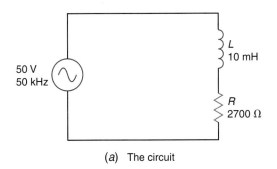

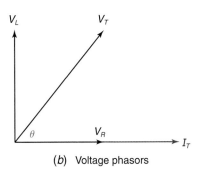

(a) The circuit

(b) Voltage phasors

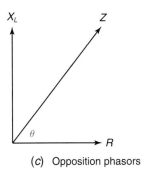

(c) Opposition phasors

Fig. 13-14 Series *RL* circuit.

RL circuit causes the following to occur: X_L increases, Z increases, I_T decreases, V_R decreases, V_L increases, and angle θ increases. Inductive reactance is also directly proportional to inductance. Therefore, increasing the inductance has the same effect as increasing the frequency.

Power and angle θ in series *RL* and series *RC* circuits are determined in exactly the same way. The difference in the two types of circuits is that the current lags the voltage in an *RL* circuit while the current leads the voltage in an *RC* circuit (compare Figs. 13-6 and 13-14).

Many small, single-phase *ac motors* are started by the phase shift between two *RL* circuits. These motors have two separate windings in them. The run winding uses large-diameter wire and therefore has only a small amount of resistance. The run winding can be represented by a small resistor in series with an ideal inductor [Fig. 13-15(*a*)]. Since the run

winding has a low resistance and a high reactance, its current lags far behind the voltage [Fig. 13-15(*b*)]. The start winding is wound with small-diameter wire; it also has fewer turns of wire than the run winding. Therefore, the start winding has more resistance and less inductance than the run winding. This causes the start-winding current to lag the voltage less than the run-winding current does. Thus, the start-winding current is out of phase with the run-winding current [Fig. 13-15(*b*)]. These out-of-phase currents produce the special magnetic field needed to start the motor. Once the motor has started, a switch [Fig. 13-15(*a*)] disconnects the start winding from the circuit. The motor continues to run on the run winding alone. Chapter 14 provides the details of how such motors operate.

JOB TIP

Technicians who work with engineers and scientists often make creative contributions to designs and help solve difficult problems.

Parallel *RL* Circuits

A parallel *RL* circuit and its phasor diagram are shown in Fig. 13-16. Parallel *RL* circuits can be treated like parallel *RC* circuits. There are only two differences between the two types of circuits:

Parallel *RL* circuit

1. The current in the *RL* circuit lags the voltage, whereas the current in the *RC* circuit leads the voltage. [Compare Figs. 13-12(*b*) and 13-16(*b*).]

2. The two circuits respond in opposite ways to changes in frequency and inductance or capacitance. When either *f* or *L* is increased in the parallel *RL* circuit, the impedance increases. This results in the inductive current and the total current decreasing and angle θ decreasing.

The following formulas are the major ones needed to work with parallel *RL* circuits:

$$I_T = \sqrt{I_L^2 + I_R^2}$$
$$Z = \frac{V_T}{I_T} = \frac{X_L \times R}{\sqrt{X_L^2 + R^2}}$$
$$\cos \theta = \frac{P}{P_{app}} = \frac{I_R}{I_T}$$

AC motors

$$I_T = \sqrt{I_L^2 + I_R^2}$$

$$Z = \frac{V_T}{I_T}$$
$$= \frac{X_L \times R}{\sqrt{X_L^2 + R^2}}$$

Notice that these formulas are the same as the formulas used for parallel *RC* circuits except that *L* is substituted for *C*.

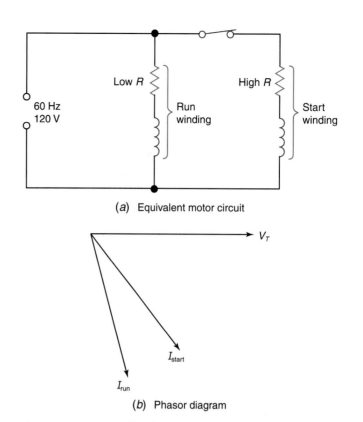

(a) Equivalent motor circuit

(b) Phasor diagram

Fig. 13-15 An ac split-phase motor circuit. (a) Equivalent circuit of a split-phase motor under starting conditions. (b) Phasor diagram showing the current in the run winding out of phase with the current in the start winding.

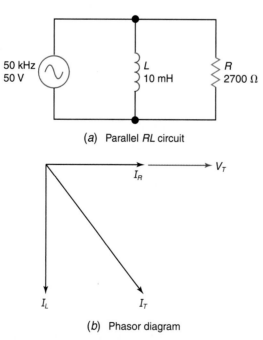

(a) Parallel RL circuit

(b) Phasor diagram

Fig. 13-16 Parallel RL circuits. The total current lags the total voltage.

Example 13-8

For the circuit of Fig. 13-16, calculate the total current, the impedance, and the power.

Given: $V_T = 50$
$f = 50$ kHz
$L = 10$ mH
$R = 2700 \ \Omega$

Find: I_T, Z, and P

Known: Ohm's law
$X_L = 6.28fL$
$I_T = \sqrt{I_L^2 + I_R^2}$
$P_T = P_R = I_R^2 R$

Solution: $I_R = \dfrac{50 \text{ V}}{2700 \ \Omega} = 0.0185$ A

$= 18.5$ mA

$X_L = 6.28 \times 50 \times 10^3$
$\times 10 \times 10^{-3}$

$= 3140 \ \Omega$

$$I_L = \frac{50 \text{ V}}{3140 \ \Omega} = 0.0159 \text{ A}$$

$$= 15.9 \text{ mA}$$

$$I_T = \sqrt{0.0185^2 + 0.0159^2}$$

$$= 0.0244 \text{ A} = 24.4 \text{ mA}$$

$$Z = \frac{50 \text{ V}}{0.0244 \text{ A}} = 2049 \ \Omega$$

$$P = 0.0185^2 \times 2700$$

$$= 924 \text{ mW}$$

Answer: The current is 24.4 mA, the impedance is 2049 Ω, and the power is 924 mW.

■ TEST

Answer the following questions.

23. In *RL* circuits, does the source current lead or lag the source voltage?

24. Suppose the inductance of a series *RL* circuit is reduced. What happens to I_T, V_L, and Z?

25. When the frequency of a parallel *RL* circuit is increased, what happens to Z and angle θ?

26. A 5-mH inductor and a 1500-Ω resistor are connected in series to an ac source. What is the impedance of this circuit at 30 kHz?

27. Determine the current and the power for the circuit in question 26 when the source voltage is 50 V.

28. Change *R* in Fig. 13-16 to 2.2 kΩ and solve for I_T and Z.

13-5 Solving *RCL* Circuits

The techniques used to solve *RC* and *RL* circuits can be combined to solve *RCL* circuits such as in Fig. 13-17(*a*). However, *RCL* circuits have some unusual characteristics. For instance, a reactive voltage or current can be higher than the source voltage or current.

Series *RCL* Circuits

Refer to the phasors in Fig. 13-17(*b*). Notice that the inductive and the capacitive voltage are 180° out of phase with each other. Adding these two voltage phasors results in V_X, which is the net reactive voltage. V_X is then added to V_R to determine the total, or source, voltage. The reactance and resistance phasors are added

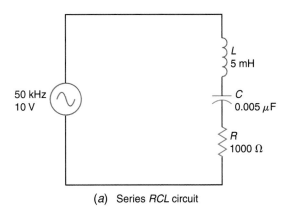

(a) Series *RCL* circuit

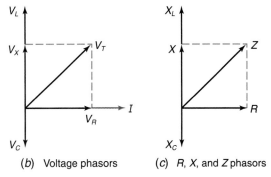

(b) Voltage phasors   (c) *R*, *X*, and *Z* phasors

Fig. 13-17 Series *RCL* circuit.

in the same sequence to determine the impedance.

In Fig. 13-17(*b*) you can see that V_L is greater than V_T. At first glance this may look like an exception to Kirchhoff's voltage law; however, it is not. At any instant the sum of the instantaneous voltages of the resistor, the capacitor, and the inductor equals the source voltage. In other words, the three voltages are never of the same polarity at the same instant. This idea is illustrated in Fig. 13-18. Notice the voltage drops across each component (*L*, *C*, and *R*) in Fig. 13-18(*a*). These voltage drops are a result of the series current passing through each component. The voltage across *L*, however, is 180° out of phase with the voltage across *C*. This means that the two voltages are directly opposing one another. The result is the *net reactive voltage* shown in Fig. 13-18(*b*) across the combination of *L* and *C*.

Notice in Fig. 13-18(*a*) that both the inductive voltage and the capacitive voltage are greater than the source voltage. If the individual voltages exceed the source voltage, then the individual reactances exceed the im-

Net reactive voltage

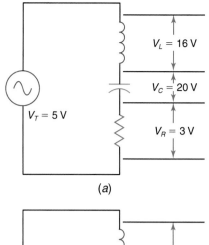

(a)

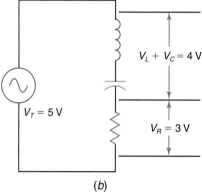

(b)

Fig. 13-18 Voltmeter readings in a series *RCL* circuit. Notice that both the inductive and the capacitive voltage in (a) exceed the source voltage.

pedance. This is common in series *RCL* circuits.

Impedance and total voltage in a series circuit can be calculated with the aid of the following formulas:

$$Z = \sqrt{(X_L - X_C)^2 + R^2}$$

$$Z = \frac{V_T}{I_T}$$

$$V_T = \sqrt{(V_L - V_C)^2 + V_R^2}$$

In these formulas, $V_L - V_C$ is the same as V_X in Fig. 13-17(*b*), and $X_L - X_C$ is the same as *X*.

The power factor (cos θ) and angle θ of the series *RCL* circuits can be found by using the same formulas used for series *RL* and *RC* circuits.

Example 13-9

Find the impedance and the voltage across the resistance for the circuit in Fig. 13-17(*a*).

Given: $V_T = 10$ V
               $f = 50$ kHz
               $L = 5$ mH
               $C = 0.005$ μF
               $R = 1000$ Ω

Find: Z and V_R
Known: Ohm's law
$$Z = \sqrt{(X_L - X_C)^2 + R^2}$$
$$X_L = 6.28 f L$$
$$X_C = \frac{1}{6.28\,fC}$$

Solution:
$$X_L = 6.28 \times 50 \times 10^3 \times 5 \times 10^{-3}$$
$$= 1570\ \Omega$$
$$X_C = \frac{1}{6.28 \times 50 \times 10^3 \times 0.005 \times 10^{-6}}$$
$$= 637\ \Omega$$
$$Z = \sqrt{(1570 - 637)^2 + 1000^2}$$
$$= 1368\ \Omega$$
$$I_T = \frac{10\ \text{V}}{1368\ \Omega} = 0.0073\ \text{A} = 7.3\ \text{mA}$$
$$V_R = 0.0073\ \text{A} \times 1000\ \Omega = 7.3\ \text{V}$$

Answer: The impedance is 1368 Ω, and the voltage across the resistance is 7.3 V.

Notice from example 13-9 that X_L is greater than X_C. This causes the circuit to be induc-

 Did You Know?

Sharks with Powerful Memories   The ocean floor is magnetized north to south except around underwater mountains, where lava flow makes magnetic spokes. Hammerhead sharks use the spokes as roads to relocate feeding grounds; they produce a current and then sense the magnetic differentials.

tive. The circuit produces a lagging power factor.

Suppose the frequency in Fig. 13-17(a) were reduced to 25 kHz. This would cause X_L to be 50 percent of its former value, and X_C twice its former value (785 Ω and 1274 Ω, respectively). Now the circuit would be capacitive; I_T would lead V_T, and the power factor would be leading.

Parallel RCL Circuits

Figure 13-19 shows a *parallel RCL circuit* and its current phasor diagram. From the phasor diagram it is obvious that a branch current (I_C) can exceed the total current. (This is because the inductive current and the capacitive currents are 180° out of phase.) In fact, both the capacitive and inductive currents can exceed the total current. For example, if L in Fig. 13-19(a) were halved, the inductive current would double. Both I_X and I_T would decrease; then I_L and I_C would exceed I_T.

The formulas for working with parallel RCL circuits are

$$I_T = \sqrt{(I_L - I_C)^2 + I_R^2}$$
$$\cos \theta = \frac{I_R}{I_T} = \frac{P}{P_{app}}$$

Example 13-10

Calculate the values of I_T and Z for the circuit in Fig. 13-19. (Note: X_L and X_C have the same values as in example 13-9.)

Given: $V_T = 10$ V
$R = 1000$ Ω
$X_L = 1570$ Ω
$X_C = 637$ Ω

Known: Ohm's law
$$I_T = \sqrt{(I_C - I_L)^2 + I_R^2}$$

Solution:

$$I_C = \frac{10 \text{ V}}{637 \text{ Ω}} = 0.0157 \text{ A} = 15.7 \text{ mA}$$

$$I_L = \frac{10 \text{ V}}{1570 \text{ Ω}} = 0.0064 \text{ A} = 6.4 \text{ mA}$$

$$I_R = \frac{10 \text{ V}}{1000 \text{ Ω}} = 0.0010 \text{ A} = 10 \text{ mA}$$

$$I_T = \sqrt{(0.0064 - 0.0157)^2 + 0.01^2}$$
$$= 0.0137 \text{ A} = 13.7 \text{ mA}$$

$$Z = \frac{10 \text{ V}}{0.0137 \text{ A}} = 730 \text{ Ω}$$

Answer: The impedance is 730 Ω, and the current is 13.7 mA.

Parallel *RCL* circuit

The impedance of example 13-10 can also be calculated directly from the values of R, C, and L (Appendix B).

The parallel *RCL* circuit of Fig. 13-19 and the series *RCL* circuit of Fig. 13-17 have the same component values. They also have the same source voltage and frequency. As expected, the parallel circuit has less impedance and draws more current than the series circuit does. Inspection of the phasor diagrams for the two circuits shows another difference between series and parallel *RCL* circuits: one of them is inductive and the other is capacitive. Notice in Fig. 13-17(b) that the source current lags the source voltage. Thus, this series *RCL* circuit is inductive. In Fig. 13-19(b), the source current leads the source voltage. Therefore, this parallel circuit (with the same components as the series circuit) is capacitive. In Figs. 13-17 and 13-19, X_L is greater than X_C. If X_C were greater

$I_T = \sqrt{(I_L - I_C)^2 + I_R^2}$

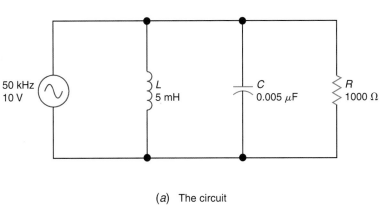

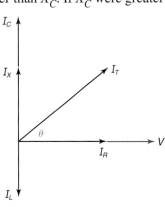

(a) The circuit

(b) Current phasors

Fig. 13-19   Parallel *RCL* circuit.

than X_L, then the series circuit would be capacitive and the parallel circuit would be inductive (Fig. 13-20).

■ TEST_____

Answer the following questions.

29. True or false. The reactance in a series *RCL* circuit can exceed the impedance.
30. True or false. The resistance in a series *RCL* circuit can exceed the impedance.
31. True or false. The reactive current in a parallel *RCL* circuit cannot exceed the total current.
32. Write two formulas for finding the impedance of a series *RCL* circuit.
33. Write the formula for adding the branch currents in a parallel *RCL* circuit.
34. Two *RCL* circuits, one series and the other parallel, are connected to the same frequency. If X_L is greater than X_C, which circuit
    a. Is inductive
    b. Is capacitive
    c. Draws the higher source current
35. A 9-mH inductor, a 0.005 μF capacitor, and a 2000-Ω resistor are connected in series to a 30-V, 40-kHz source. Find Z, I_T, and V_L.
36. Assume the components of question 35 are connected in parallel. Determine the impedance, the angle θ, and the current through the capacitor.
37. Is the PF in question 36 leading or lagging?

13-6 Resonance

Circuits in which $X_L = X_C$ are called *resonant circuits*. They can be either series or parallel circuits or either *RCL* circuits or *LC* circuits. Most often, resonant circuits are *LC* circuits; the only resistance in these circuits is that of the inductor and capacitor.

Resonant Frequency

For a given value of L and C, there is only one frequency at which X_L equals X_C. This frequency, called the *resonant frequency,* can be calculated with the following formula:

$$f_r = \frac{1}{6.28\sqrt{LC}}$$

Resonant circuits

Resonant frequency

$$f_r = \frac{1}{6.28\sqrt{LC}}$$

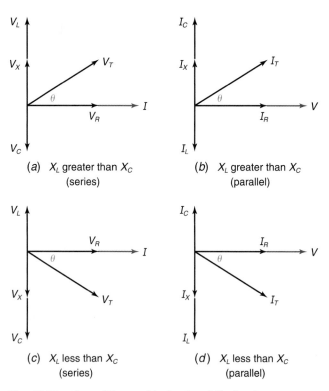

(a) X_L greater than X_C (series)

(b) X_L greater than X_C (parallel)

(c) X_L less than X_C (series)

(d) X_L less than X_C (parallel)

Fig. 13-20 Capacitive and inductive *RCL* circuits. Diagrams (*b*) and (*c*) are for capacitive circuits.

In this formula, the resonant frequency f_r is in hertz if inductance is in henrys and capacitance is in farads. The resonant frequency formula is derived from the two reactance formulas. By definition, the resonant frequency is that frequency at which $X_L = X_C$. Therefore,

$$6.28f\,L = \frac{1}{6.28f\,C}$$

Solving this equation for f yields the formula for f_r given above.

Figure 13-21 further explains the concept of resonant frequency. This figure is a plot of reactance against frequency. The reactance X_{L_1} of a specific value of inductance L_1, when plotted against frequency, produces a straight line. The reactance X_{C_1} of a specific value of capacitance C_1 produces a curved line. The resonant frequency of $L_1 C_1$ is the frequency at which the X_{L_1} line and the X_{C_1} line cross.

We have seen that $L_1 C_1$ can be resonant at only one frequency. However, any number of other values of L and C can also be resonant at the same frequency as $L_1 C_1$. Suppose a higher inductance and a lower capacitance, such as L_2 and C_2 of Fig. 13-21, are used. They are resonant at the same frequency as $L_1 C_1$. One could also use a lower value of L and a higher value of C to resonant at the same frequency $L_1 C_1$ and $L_2 C_2$.

Notice two other points in Fig. 13-21. If the inductance increases to L_2 and the capacitance remains at C_1, the resonant frequency drops to point A. On the other hand, if the inductance

remains at L_1 and the capacitance decreases (that is, X_C increases) to C_2, then the resonant frequency increases to point B.

The relationship between change in L or C and the resultant change in f_r is also evident from the resonant frequency formula. From the formula, it can be seen that increasing either L or C decreases f_r. Conversely, decreasing either L or C increases f_r.

What is the resonant frequency of a 10-mH inductor and a 0.005-μF capacitor;

Given: $L = 10$ mH
$C = 0.005\ \mu$F

Find: f_r

Known: $f_r = \dfrac{1}{6.28\sqrt{LC}}$

Solution:

$$f_r = \frac{1}{6.28\sqrt{10 \times 10^{-3} \times 0.005 \times 10^{-6}}}$$

$$= 22{,}500 \text{ Hz} = 22.5 \text{ kHz}$$

Answer: The resonant frequency is 22.5 kHz.

The capacitor and the inductor in the above example could be connected either in series or in parallel. The resonant frequency is the same in either case.

Parallel Resonant Circuits

The phasor diagram in Fig. 13-22 reveals the major characteristics of the *parallel resonant*

Parallel resonant circuit

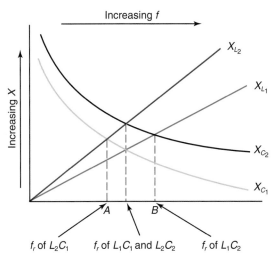

Fig. 13-21 Plot of X_L and X_C versus f. At resonance, $X_L = X_C$.

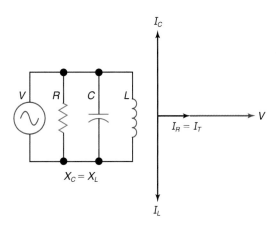

Fig. 13-22 Parallel resonant *RCL* circuit. Notice that $I_C = I_L$.

circuit. In this circuit, the inductive current exactly cancels the capacitive current. Therefore, the total current and the resistive current are the same. This means that the source current and voltage are in phase, angle θ is zero, and the power factor (cos θ) is 1. Thus, the parallel resonant circuit is purely resistive.

The only current drawn from the power source in Fig. 13-22 is the current that flows through the resistor. Yet, the capacitive and inductive branches have large currents flowing in them. This is because the inductor and the capacitor are just transferring energy back and forth between themselves. This idea is graphically represented in Fig. 13-23, which shows the capacitor and inductor removed from the rest of the circuit. Suppose the inductor and capacitor parts of the circuit are removed at the instant the capacitor is fully charged [Fig. 13-23(a)]. The capacitor then becomes the energy source and starts discharging through the inductor. This causes a magnetic field to build up in the inductor and store the energy being transferred from the capacitor. Once the capacitor has discharged, the inductor's field starts to collapse, as in Fig. 13-23(b). Now the inductor is the source. It provides the energy to recharge the capacitor in a reverse polarity, as shown in Fig. 13-23(b). After the inductor's field has collapsed and recharged the capacitor, the capacitor again takes over as the source. It discharges [Fig. 13-23(c)] and again transfers energy to the inductor's magnetic field. The cycle is completed, as shown in Fig. 13-23(d), when the inductor's collapsing field recharges the capacitor. Then the cycle starts over again with the conditions shown in Fig. 13-23(a).

The parallel *LC* circuit discussed above (and shown in Fig. 13-23) is often called a *tank circuit.* "Tank circuit" is a very descriptive name because the circuit stores energy the way a tank stores liquid. A tank circuit produces a sine wave, as shown in Fig. 13-24, as the capacitor charges and discharges repeatedly. If both the capacitor and inductor were ideal components (had no resistance), the tank circuit would produce a sine wave forever. Once the capacitor was given an initial charge and connected across the inductor, the cycling would continue indefinitely. However, all capacitors and inductors have some resistance. Therefore, some energy is converted to heat each time the capacitor charges and discharges. Thus, a real tank circuit produces a *damped waveform* like that shown in Fig. 13-24(b). How many cycles it takes for the waveform to completely dampen out depends on the quality of the circuit.

Tank circuit

Damped waveform

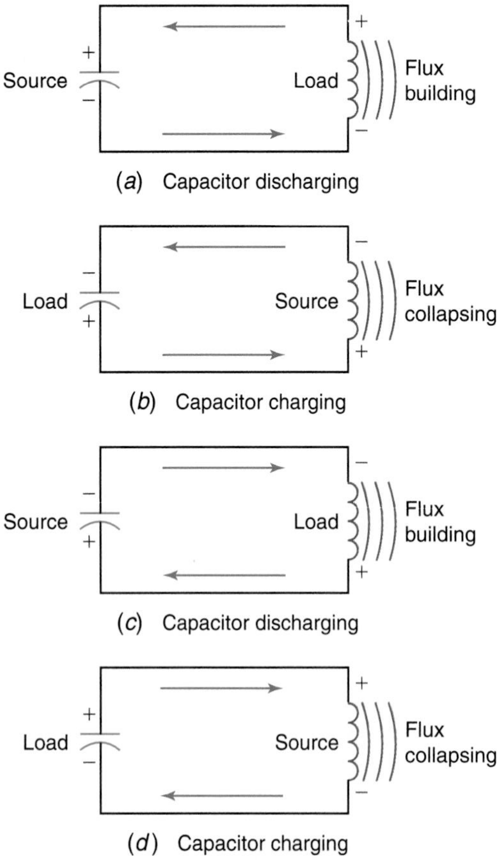

(a) Capacitor discharging

(b) Capacitor charging

(c) Capacitor discharging

(d) Capacitor charging

Fig. 13-23 Resonant tank circuit.

YOU MAY RECALL . . . that Q is the ratio X_L/R. Thus, for a given value of L and f, the quantity Q is inversely proportional to R. Therefore, a low Q circuit will have a large resistance which will convert the tank-circuit energy in very few cycles. The frequency of the tank circuit waveform is determined by the values of L and C. It can be calculated with the resonant frequency formula.

Suppose the *LC* circuit of Fig. 13-23 is connected to an ac source [Fig. 13-25(a)]. Further, let the source frequency be equal to the *LC* circuit's resonant frequency. If L and C were ideal components, no current would be required from the source. If no source current flowed, then

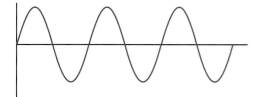

(*a*)   Ideal tank circuit output waveform

(*b*)   Practical tank circuit output waveform

Fig. 13-24   Tank circuit waveforms.

the impedance of the circuit would be infinite. When real components are used, a small source current does flow. This small current, together with the source voltage, furnishes the power

used by the very low resistance of the inductor and capacitor. In the phasor diagrams of Fig. 13-25, this low current is labeled I_R since it is caused by the resistance of L and C. With high-quality components, the source current is very low and the impedance is still very high.

Suppose the frequency of the source is below the resonant frequency of the LC circuit, as in Fig. 13-25(*b*). Then X_L is less than X_C, and I_L exceeds I_C. The difference between I_L and I_C becomes part of the current required from the source. The circuit is inductive, and its impedance has decreased from the value it had at its resonant frequency. Its impedance could be calculated by dividing the source voltage by the source current.

Figure 13-25(*c*) shows what happens when the applied frequency is greater than the resonant frequency. The circuit is capacitive, and again its impedance has decreased from the resonant frequency value.

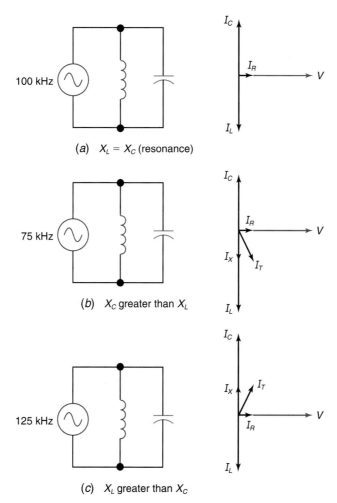

(*a*) $X_L = X_C$ (resonance)

(*b*) X_C greater than X_L

(*c*) X_L greater than X_C

Fig. 13-25   Parallel *LC* circuit. The impedance (voltage divided by total current) is greatest at resonance.

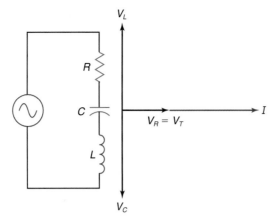

Fig. 13-26   Series resonant *RCL* circuit.

In summary, a parallel resonant *LC* circuit has the following major characteristics:

1. Maximum impedance

2. Minimum source current

3. Phase angle almost 0°

4. High circulating inductive and capacitive currents

Series Resonant Circuits

In the series *RCL* circuit in Fig. 13-26, the frequency of the source voltage is at the resonant frequency for *L* and *C*. This means $X_L = X_C$ and $V_L = V_C$. Since V_L and V_C are 180° apart, they cancel each other. The net voltage drop is that across the resistance. The source voltage and current are in phase, and the circuit has a power factor of 1. Since X_L and X_C cancel each other, the current in the circuit is limited only by the resistance.

Suppose the resistor in Fig. 13-26 is shorted out to produce an *LC* circuit like the one in Fig. 13-27(*a*). If *L* and *C* were ideal components, X_L would cancel X_C and this series resonant circuit would have zero impedance. It would be a short circuit across the source. The current would be limited only by the internal resistance of the source. The voltages across the inductor and capacitor would be extremely high. However, real inductors and capacitors have some resistance. This resistance prevents

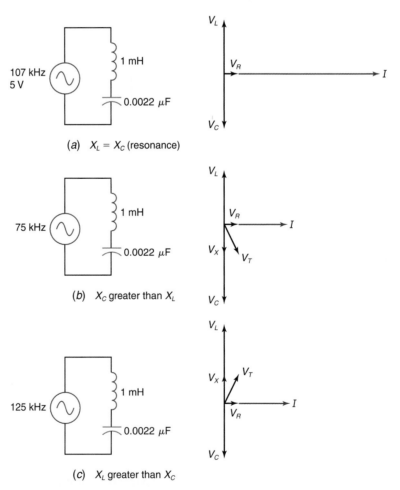

(*a*) $X_L = X_C$ (resonance)

(*b*) X_C greater than X_L

(*c*) X_L greater than X_C

Fig. 13-27   Series *LC* circuit. The impedance approaches zero at resonance.

the series resonant circuit from being a dead short. It drops an in-phase voltage (V_R in Fig. 13-27). This voltage and the line current represent the power used by the inductor and capacitor. The great majority of this power is used by the inductor.

The voltages across the capacitor and the inductor are many times greater than the source voltage in a series resonant circuit. Why this is so can be best explained by a practical example. The resistance of the typical inductor and capacitor in Fig. 13-27(a) is about 20 Ω. Therefore, the current in the circuit is

$$I_T = \frac{V_T}{Z} = \frac{5\text{ V}}{20\text{ Ω}} = 0.25\text{ A}$$

This 0.25 A is also the current through the inductor and the capacitor. The reactance of the inductor (which, in a resonant circuit, is the same as the reactance of the capacitor) is

$$X_L = 6.28f\,L$$
$$= 6.28\,(107 \times 10^3 \text{ Hz})(1 \times 10^{-3} \text{ H})$$
$$= 672\text{ Ω}$$

The voltage across the inductor can now be calculated:

$$V_L = I_L \times X_L = 0.25\text{ A} \times 672\text{ Ω} = 168\text{ V}$$

Thus, the voltage across the inductor (and across the capacitor) is more than 33 times (168 V versus 5 V) higher than the source voltage.

Figure 13-27(b) and (c) shows the results of operating a series LC circuit below and above its resonant frequency. When operating below its resonant frequency, the series circuit is capacitive because V_C is greater than V_L. Again, notice that parallel and series circuits are opposites in many ways. The parallel LC circuit operating below its resonant frequency [Fig. 13-25(b)] is inductive.

The major characteristics of the series resonant LC circuit can be summarized as follows:

1. Minimum impedance

2. Maximum source current

3. Phase angle almost 0°

4. High inductive and capacitive voltages

Notice that the only characteristic shared by the series and parallel resonant circuits is the almost 0° phase angle.

Response Curves, Bandwidth, and Selectivity

So far, only phasor diagrams have been used to show current, voltage, and impedance of LC circuits at several frequencies. A more complete picture of the behavior of LC circuits can be shown with the *response curves* of Fig. 13-28. These curves are made by either calculating or measuring the current, voltage, or impedance at numerous frequencies above and below resonance. The values are then plotted on a graph.

Compare the curves in Fig. 13-28(a) and (b). They clearly show the major differences between series and parallel LC circuits.

The *bandwidth* (BW) of an LC circuit is expressed in hertz. It is the range of frequencies to which the circuit provides 70.7 percent or more of its maximum response. Suppose the parallel LC circuit of Fig. 13-28(a) is resonant at 500 kHz and has a resonant impedance of 100 kΩ. Further, suppose its impedance drops to 70.7 kΩ at 495 kHz [f_{low} in Fig. 13-28(a)] and at 505 kHz [f_{high} in Fig. 13-28(a)]. Then the circuit's bandwidth is 10 kHz (505 to 495). The bandwidth of a series circuit can be determined from its current-versus-frequency curve [Fig. 13-28(b)]. The edges of the bandwidth (f_{low} and f_{high}) are those frequencies at which the circuit current is 70.7 percent of the current at resonance.

The bandwidth of a circuit determines the *selectivity* of the circuit. Selectivity refers to the ability of a circuit to select one frequency (the desired frequency) out of a group of frequencies. For example, the antenna of a radio receiver receives a signal from all the local radio stations. Each station's signal is at a different frequency. It is the job of the resonant circuits in the receiver to select the frequency of one station and reject the frequencies of all the other stations. (When you tune a receiver, you are changing capacitor or inductor values to adjust the resonant frequency of the receiver circuits.) If the bandwidths of the receiver's circuits are too wide, two or more stations will be heard simultaneously. In this case we would say that the receiver has poor selectivity.

Response curves

Bandwidth (BW)

Selectivity

Quality of Resonant Circuits

The term *quality* is also used to rate resonant circuits. It is the ratio of the reactance of the inductor or capacitor to the equivalent series resistance of both components. Recall that the

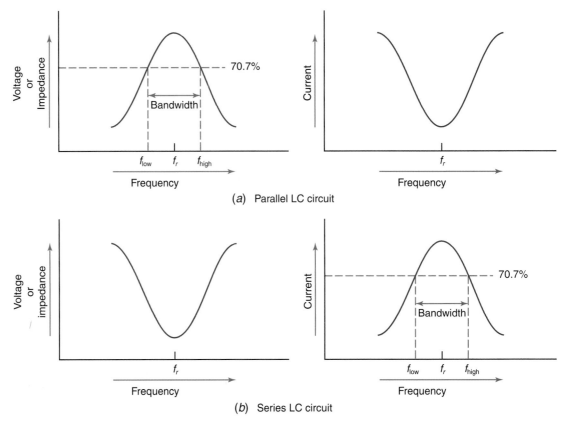

Fig. 13-28 Response curves of resonant circuits.

inductor possesses nearly all the resistance of a resonant circuit. Therefore, the quality of the inductor is, for all practical purposes, also the quality of the resonant circuit.

The quality of the resonant circuit is important for two reasons:

1. The quality determines the minimum Z of the series resonant circuit and the maximum Z of the parallel resonant circuit. These ideas are illustrated in Fig. 13-29. For the series circuit,

$$Z = \frac{V_R}{I}$$

and for the parallel circuit,

$$Z = \frac{V}{I_R}$$

From Fig. 13-29(*a*) you can see that the high-quality circuit yields the lowest Z for the series circuit. The high-quality parallel circuit yields the highest Z [Fig. 13-29(*b*)].

Quality determines bandwidth

2. The quality also determines the bandwidth of the resonant circuit.

The higher the quality, the narrower the bandwidth. A narrow bandwidth (high quality) provides a larger change in impedance for a given change in frequency than a wide bandwidth does. Figure 13-30 shows why a high-quality circuit provides a greater change in impedance than a low-quality circuit does. The phasor diagrams in Fig. 13-30(*a*) are for two parallel *LC* circuits operating at the same resonant frequency. The low-quality circuit has twice as much resistive current as the high-quality circuit. Since $Z = V/I$, the low-quality circuit also has only half as much impedance as the high-quality circuit. Figure 13-30(*b*) shows the results of increasing the source frequency the same amount for each circuit. Notice that the impedance of the high-quality circuit dropped from 100 to 35 kΩ; a change of approximately 65 percent. For the same change in frequency, the impedance of the low-quality circuit only changed about 40 percent (50 to 30 kΩ).

The result of the greater impedance change for the high-quality circuit of Fig. 13-30 is shown in Fig. 13-31. This figure shows that the high-quality circuit yields a narrower bandwidth for either series or parallel *LC* circuits. From Fig. 13-31 you can see that a high-quality circuit is more selective than a low-quality circuit.

Resonant frequency, quality, and bandwidth

are all interrelated. Their relationship can be expressed by the formula

$$BW = \frac{f_r}{Q}$$

Although this formula is actually only an approximation, it is accurate enough for predicting circuit behavior. If the circuit quality is 10 or more, the error will be less than 1 percent.

Example 13-12

What is the quality of a circuit which is resonant at 100 kHz and has a bandwidth of 4 kHz?

Given: $f_r = 100$ kHz
BW = 4 kHz

Find: Q

Known: $BW = \dfrac{f_r}{Q}$

Solution: $Q = \dfrac{f_r}{BW} = \dfrac{100 \text{ kHz}}{4 \text{ kHz}} = 25$

Answer: The quality of the circuit is 25.

$$BW = \frac{f_r}{Q}$$

The quality of a series resonant circuit can also be found from the following formula:

$$Q = \frac{V_L}{V_T}$$

In this formula the voltages are the voltages at resonance. You can understand where this formula comes from by referring to Fig. 13-26 or 13-27(a). At resonance, the resistive voltage and the source voltage are equal. Also, the resistance and the reactance have the same current flowing through them. Therefore, the ratio

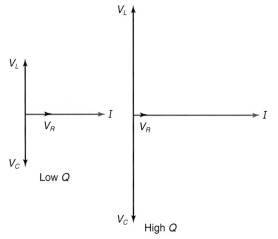

(a) Series resonant circuit

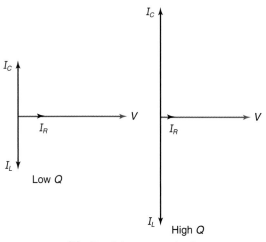

(b) Parallel resonant circuit

Fig. 13-29 Effects of Q on the resonant Z.

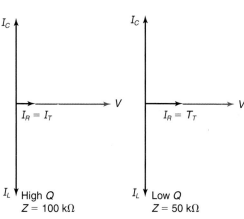

(a) At resonance

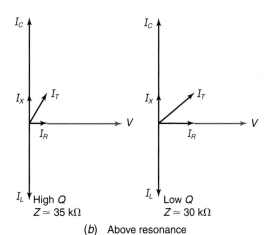

(b) Above resonance

Fig. 13-30 Effects of Q on the nonresonant Z.

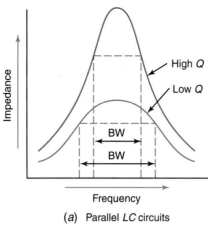

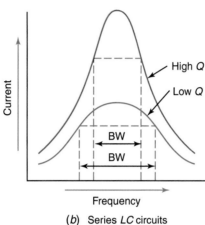

Fig. 13-31 Response curves for low- and high-Q circuits.

(a) Parallel LC circuits

(b) Series LC circuits

increased by _____ the value of either L or C.

44. A parallel LC circuit is often called a _____ circuit.

45. True or false. There is only one combination of L and C for each resonant frequency.

46. True or false. In a parallel resonant circuit, $X_L = Z$.

47. True or false. In any resonant circuit, $X_L = X_C$.

48. True or false. In a series resonant RCL circuit, $R = Z$.

49. True or false. High selectivity requires a wide bandwidth.

50. Determine the resonant frequency of a 5-mH inductor and a 0.03-μF capacitor.

51. What is the bandwidth of the circuit in question 50 if the resistance of the components is 5 Ω?

52. What is the bandwidth of a series LC circuit which allows 20 mA at its resonant frequency of 120 kHz and 14.14 mA at 118 kHz?

53. What is the quality of the circuit in question 52?

Filtering

of reactance to resistance has to be equal to the ratio of reactive voltage to resistive voltage.

The preceding formula, when rearranged to $V_L = QV_T$, shows that the inductive voltage is quality times as large as the source voltage. Since $V_L = V_C$, the same can be said for the capacitive voltage.

Low-pass filters

◼ TEST_____

Answer the following questions.

38. At resonance, θ is approximately _____.

39. Maximum opposition is provided by a _____ resonant circuit.

40. A parallel resonant circuit operating above resonance is _____.

41. A series resonant circuit operating below resonance is _____.

42. For a given resonant frequency, bandwidth is controlled by the _____ of the components.

43. The resonant frequency of a circuit can be

13-7 Filters

One of the major uses of RC, RL, LC, and RCL circuits is for filtering. *Filtering* refers to separating one group of frequencies from another group of frequencies. There are four general classes of filters: band-pass, band-reject, high-pass, and low-pass.

Low-Pass Filters

Low-pass filters offer very little opposition to low-frequency signals [Fig. 13-32(a)]. For low frequencies, most of the input signal appears at the output terminals. As the signal frequency increases, the filter provides more opposition. The filter drops more of the signal voltage and leaves less signal available at the output. As the graph of the output voltage illustrates [Fig. 13-32(a)], essentially none of the high-frequency signals appears at the output terminal. Although low-pass filters can be constructed with either RL or RC circuits, RC circuits are more common.

Inspection of Fig. 13-32(b) shows why the low-pass filter behaves as described in the para-

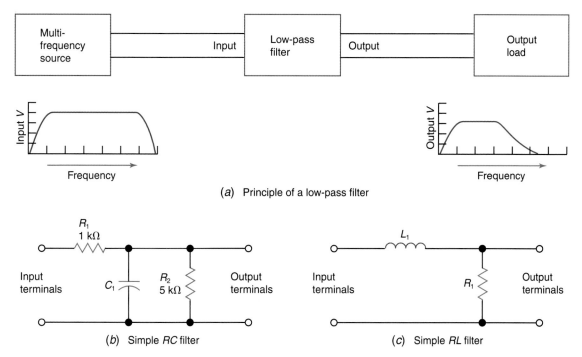

(a) Principle of a low-pass filter

(b) Simple RC filter

(c) Simple RL filter

Fig. 13-32 Low-pass filters.

graph above. At low frequencies, the reactance of C_1 is very high relative to R_2. The impedance of R_2C_1 is essentially 5 kΩ. Thus, five-sixths of the input voltage appears at the output terminals. At very high frequencies, the reactance of C_1 is very low relative to R_1. Now the impedance of R_2C_1 is very low (essentially equal to X_{C_1}). Therefore, nearly all the input signal drops across R_1.

The reactance of L_1 in Fig. 13-32(c) is low (relative to R_1) at low frequencies. Thus, low frequencies pass through the filter. However, at high frequencies, the inductive reactance of L_1 is high and most of the signal drops across L_1.

Very little of the high-frequency signal appears at the output terminals.

High-Pass Filters

Several simple high-pass filters are shown in Fig. 13-33. With the RC filter of Fig. 13-33(a), the capacitor drops nearly all the low-frequency signals. At high frequencies, the capacitive reactance is very low. Therefore, R_1 (which provides the output) drops nearly all the high-frequency voltage. In other words the circuit passes high-frequency signals through it and blocks low-frequency signals.

High-pass filters

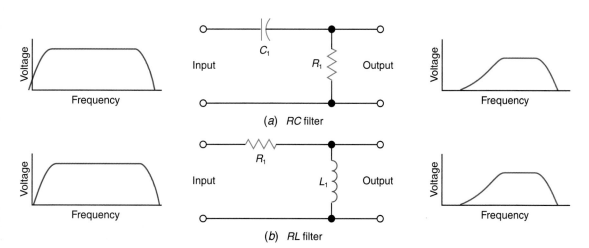

(a) RC filter

(b) RL filter

Fig. 13-33 High-pass filters.

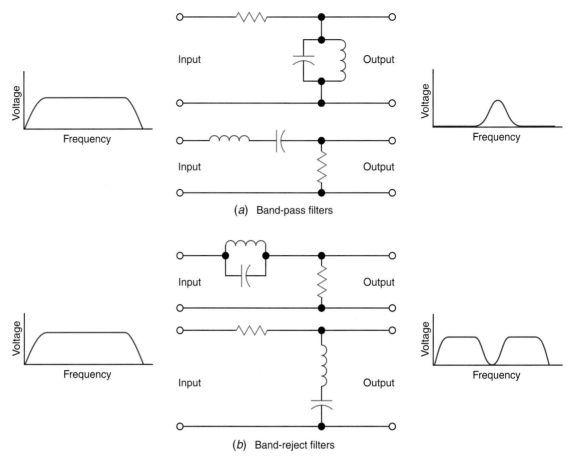

Fig. 13-34   Band-pass and band-reject filters.

YOU MAY **RECALL** . . . that $X_L = 6.28fL$. Therefore, L_1 in Fig. 13-33(b) drops most of the high-frequency voltage but practically none of the low-frequency voltage. Thus, a series RL circuit can serve as a high-pass filter.

Band-Pass and Band-Reject Filters

Band-pass filters

Band-reject filters

You are already familiar with band-pass and band-reject filters. They are just series LC and parallel LC circuits. Frequencies within their response curve range are either passed or rejected (Fig. 13-34).

Notice in Fig. 13-34 that a parallel LC circuit can serve as either a band-pass or a band-reject filter. The same thing can be said for the series LC circuit. The determining factor is where the input and output terminals are connected to the circuits.

▧ TEST

Answer the following questions.

54. List four classifications of filter circuits.
55. Which of the types listed in question 54 use resonant circuits?
56. If you wanted to eliminate low-frequency noise from a stereo record, what type of filter would you use?
57. Could the filter in question 56 be either an RL or an RC filter?

Summary

1. Impedance is a combination of resistance and reactance.
2. Impedance causes phase shift.
3. Impedance may cause either a leading or a lagging current.
4. The symbol for impedance is Z.
5. The ohm is the base unit of impedance.
6. In RC circuits. the current leads the voltage.
7. For series RC circuits. the following generalizations can be made:
   a. The total impedance is higher than either R or X_C.
   b. The arithmetic sum of R and X_C is greater than Z.
   c. Decreasing either f or C causes Z to increase, I to decrease, P to decrease, and θ to increase.
8. The formulas that are applicable to all circuits involving combinations of R, C, and L are

$$Z = \frac{V_T}{I_T} \qquad f_r = \frac{1}{6.28\sqrt{LC}}$$
$$\cos\theta = \frac{P}{P_{\text{app}}} = \frac{P}{I_T V_T}$$

9. The formula that is applicable to all series circuits using combinations of R, C, and L is

$$\cos\theta = \frac{R}{Z} = \frac{V_R}{V_T}$$

10. The formula that applies to all parallel circuits using combinations of R, C, and L is

$$\cos\theta = \frac{I_R}{I_T}$$

11. The following are characteristics of parallel RC circuits:
    a. The total impedance is less than R or X_C.
    b. Decreasing either f or C causes Z to increase, I to decrease, and θ to decrease. The power remains the same.
12. The current phasor is the reference phasor for series circuits.
13. The voltage phasor is the reference phasor for parallel circuits.

14. Resistance and reactance are 90° out of phase.
15. In all RL circuits, the current lags the voltage.
16. A series RL circuit has more Z than either R or X_L.
17. A series RL circuit has more V_T than either V_R or V_L.
18. A parallel RL circuit has less Z than either R or X_L.
19. A parallel RL circuit has more I_T than either I_R or I_L.
20. In a series RCL circuit, Z may be either less or more than X_L or X_C.
21. In a resonant RCL circuit, Z is equal to R.
22. In RCL circuits, reactive voltages and currents can be greater than the total current or voltage.
23. A series RCL circuit having X_L greater than X_C is inductive.
24. A parallel RCL circuit having X_L greater than X_C is capacitive.
25. The formulas used only for series RCL circuits are

$$Z = \sqrt{(X_L - X_C)^2 + R^2}$$
$$V_T = \sqrt{(V_L - V_C)^2 + V_R^2}$$

26. The formula used only for parallel RCL circuits is

$$I_T = \sqrt{(I_L - I_C)^2 + I_R^2}$$

27. Resonance occurs when $X_L = X_C$.
28. Increasing either L or C decreases the resonant frequency of an LC circuit.
29. A given L and C can produce only one resonant frequency.
30. For practical purposes, the quality of a resonant LC circuit is determined by the quality of the inductor.
31. All resonant circuits have a power factor ($\cos\theta$) of 1.
32. The impedance of a series resonant circuit approaches zero.
33. The impedance of a parallel resonant circuit approaches infinity.
34. A parallel LC circuit operating above its resonant frequency is capacitive.
35. A series LC circuit operating above its resonant frequency is inductive.

36. The higher the quality of a circuit, the narrower its bandwidth.
37. The narrower the bandwidth of a circuit, the more selective the circuit is.
38. The bandwidth of a circuit is equal to f_r/Q.
39. Either RC or RL circuits can be used for high-pass and low-pass filters.

Chapter Review Questions

For questions 13-1 to 13-13, determine whether each statement is true or false.

13-1. The Pythagorean theorem is used to determine the angles in a right triangle.

13-2. If the capacitive reactance in a series RC circuit is greater than the resistance, angle θ is less than 45°.

13-3. Resistance is 90° out of phase with reactance.

13-4. Reducing the inductance in a series RL circuit causes the total power to increase.

13-5. In a series RL circuit, the inductive voltage can be greater than the source voltage.

13-6. The resistance in a series RCL circuit can exceed the impedance.

13-7. In a parallel RCL circuit operating above its resonant frequency, the source current leads the source voltage.

13-8. There is only one value of inductance that resonates with a 0.001-μF capacitor at 10,000 Hz.

13-9. A high-quality circuit is more selective than a low-quality circuit.

13-10. The resonant frequency of a tank circuit can be increased by increasing the capacitance.

13-11. At resonance, the phase angle of a circuit is approximately 90°.

13-12. A series LC circuit can be used in a band-reject filter.

13-13. Either a series or a parallel LC circuit can be used in a band-pass filter.

For questions 13-14 to 13-27, supply the missing word or phrase in each statement.

13-14. Impedance is a combination of _____ and _____.

13-15. The base unit of impedance is the _____.

13-16. The bandwidth of a circuit is determined by the _____ and the _____ of the circuit.

13-17. The symbol for impedance is _____.

13-18. In any RC circuit, the current _____ the voltage.

13-19. In any RL circuit, the current _____ the voltage.

13-20. Decreasing the frequency of a series RL circuit causes θ to _____, P to _____, Z to _____, and V_R to _____.

13-21. Increasing the frequency of a parallel RC circuit causes Z to _____, θ to _____, I_C to _____, and P to _____.

13-22. The _____ phasor is the reference phasor for parallel circuits.

13-23. In a resonant RCL circuit, resistance and _____ have the same value.

13-24. The quality of a resonant LC circuit is primarily determined by the _____.

13-25. At resonance, the PF of a circuit is _____.

13-26. The impedance of a series resonant LC circuit approaches _____.

13-27. A series LC circuit operating above resonance will be _____.

Answer the following questions.

13-28. Calculate the bandwidth of a circuit which is resonant at 430 kHz and has a quality rating of 80.

13-29. List four types (classifications) of filter circuits.

13-30. Which of the filter circuits in question 13-29 use resonant circuits?

13-31. An 1800-Ω resistor and a 0.68-μF capacitor are connected in series to a 40-V, 200-Hz source. Calculate the following:
  a. Impedance
  b. Voltage across the resistor
  c. Power
  d. $\cos \theta$

13-32. Assume the resistor and capacitor in the circuit of question 13-31 are connected in parallel instead of in series. Calculate the following:
  a. Impedance
  b. Total current
  c. Power
  d. θ

13-33. What is the impedance of a 1500-Ω resistor connected in series with a 4-mH inductor and a 40-kHz source?

13-34. Determine the resonant frequency and the bandwidth of a 6-mH inductor and a 1200-pF capacitor connected in series if the quality of the circuit is 56.

13-35. A 0.001-μF capacitor, a 6-mH inductor, and a

1200-Ω resistor are connected in series to a
30-V, 60-kHz source. Determine the following:
a. Impedance
b. Total current
c. Voltage across the capacitor
d. Resonant frequency

13-36. Assume the components in question 13-35 are connected in parallel instead of in series.
Calculate the following:
a. Impedance
b. Total current
c. Inductive current
d. θ

Critical Thinking Questions

13-1. Two amplifiers are coupled together by an *RC* circuit. What factors will determine the amount of phase shift from the output of one amplifier to the input of the other amplifier? How would you control these factors to minimize the phase shift?

13-2. A series *RC* circuit has a 10-V, 200-Hz source and a 0.68-μF capacitor. How much resistance is required to cause the current to lead the voltage by 25°?

13-3. A parallel *RL* circuit has a 15-V, 500-Hz source and a 1200-Ω resistor. How much inductance is required to cause a 65° phase shift?

13-4. A series *RCL* circuit has an inductance of 0.1 H, a resistance of 1000 Ω, and a 20-V, 500-Hz source. Determine the capacitance required to produce a leading power factor of 0.82.

13-5. A series *RL* circuit has a 25-V, 250-Hz source

and a 1000-Ω resistor. What is the inductance if the power is 95.4 mW?

13-6. Determine the value of capacitance needed to resonate with a 10-mH inductor at 450 kHz.

13-7. Determine the quality of a series *LC* circuit when $L = 0.1$ mH, $C = 0.001$ μF, and the equivalent series resistance of *L* and *C* is 20Ω.

13-8. Refer to Fig. 13-4. Determine the value of Y_1 when $Y_3 = 18$ m, $Y_2 = 9$ m, and $Y_T = 12$ m at an angle between 270° and 360°.

13-9. Repeat question 13-8 when the angle of Y_T is between 0° and 90°.

13-10. Why can't we calculate *Z* in a parallel *RC* circuit using $Z = (R \times X_C)/(R + X_C)$, which is the form of the formula used for parallel resistors?

13-11. Why are *RC* filters used more often than *RL* filters at audio frequencies?

Answers to Tests

1. F
2. reactance, resistance
3. Z
4. ohm
5. 11.4 m
6. 29.15 N
7. Pythagorean
8. leads
9. F
10. current
11. F
12. 20°
13. They are separated by 90°.
14. $Z = \sqrt{R^2 + X_C^2}$ and $Z = \dfrac{V_T}{I_T}$

15. $V_T = \sqrt{V_R^2 + V_C^2}$
16. $\cos \theta = \dfrac{R}{Z} = \dfrac{V_R}{V_T}$
$= \dfrac{P}{P_{app}}$
17. $\cos \theta = \dfrac{P}{P_{app}}$
18. $\cos \theta = \dfrac{I_R}{I_T}$
19. a. increases
    b. stays the same
    c. stays the same
    d. increases
20. a. increases
    b. increases
    c. decreases

d. increases
21. a. 11.9 kΩ
    b. 0.133 W
    c. 63.8°
    d. 4.52 mA
22. a. 27.44 mA
    b. 27.44 V
    c. 753 mW
    d. 1458 Ω
23. lag
24. I_T increases, V_L decreases, and *Z* decreases.
25. *Z* increases and θ decreases.
26. 1771.5 Ω
27. 28.2 mA, 1.2 W

28. 27.7 mA, 1803 Ω
29. T
30. F
31. F
32. $Z = \dfrac{V_T}{I_T}$
$Z = \sqrt{(X_L - X_C)^2 + R^2}$
33. $I_T =$
$\sqrt{(I_L - I_C)^2 + I_R^2}$
34. a. series circuit
    b. parallel circuit
    c. parallel circuit
35. 2480 Ω, 12.1 mA, 27.4 V
36. 1047 Ω 58.4°, 37.7 mA

37. leading
38. zero
39. parallel
40. capacitive
41. capacitive
42. quality

43. decreasing
44. tank
45. F
46. F
47. T
48. T

49. F
50. 12,995 Hz
51. 159 Hz
52. 4 kHz
53. 30
54. band-pass,

band-reject, high-pass, and low-pass
55. band-pass, and band-reject
56. high-pass
57. yes

Chapter 14

Electric Motors

Chapter Objectives

This chapter will help you to:

1. *Classify* motors by type of power source, intended use, and special characteristics.
2. *Understand* motor ratings and use them in selecting an appropriate motor for a specific application.
3. *Identify* the parts of various squirrel-cage induction motors.
4. *Appreciate* the versatility of ac and dc motors.

5. *Understand and enumerate* various ways of creating the rotating magnetic field needed to start most induction motors.
6. *Understand* the special characteristics of each type of single-phase motor.
7. *Understand* the special characteristics of the major types of dc motors.

The electric motor is one of the more common electric devices in use today. It is used extensively in such diverse systems as household appliances, automobiles, computers, printers, and automatic cameras.

14-1 Motor Classifications

Electric motors can be broadly classified by the type of power source needed to operate them. The three major categories of motors, as shown in Fig. 14-1, are the *dc motor,* the *ac motor,* and the *universal motor.* The universal motor is designed to operate from either an ac or a dc power source.

Fig. 14-1 also lists many of the types of motors available within the three major categories of motors. Some of the types of motors listed in Fig. 14-1 can be further subdivided. For example, there are several kinds of dc stepper motors and numerous kinds of polyphase (especially three-phase) ac motors.

Motors can also be classified by their intended use or special characteristics. Some examples are gearmotors, synchronous motors, multispeed motors, and torque motors.

A *gearmotor* is a motor that has a gear train built into the motor housing to reduce the output shaft speed and increase the shaft torque.

The motor in a gearmotor may be almost any type of ac, dc, or universal motor depending on the intended application.

Synchronous motors are motors in which shaft rotation is in exact synchronization with the frequency of the power source. These motors are either single-phase or polyphase (usually three-phase) ac motors.

Multispeed motors are motors with two or more fixed speeds. Multispeed operation is often obtained by having either a tapped motor winding or a separate motor winding for each speed. This is in contrast to *variable-speed motors,* in which the speed is continuously variable and is typically controlled by varying the voltage and/or frequency of the power source.

Torque motors are designed to provide maximum, or near maximum, torque when the motor is stalled and still be able to operate for an extended period of time in the stalled (locked-rotor) mode without overheating. Either ac, dc, or universal motors can be designed to operate as torque motors.

Synchronous motors

DC motors

AC motors

Universal motors

Multispeed motors

Variable-speed motors

Torque motors

Gearmotors

Direct Current	Alternating Current	Universal
Series	Split-phase	Noncompensated
Shunt	Capacitor-start	Compensated
Compound	Permanent-split capacitor	
Permanent-magnet	Two-value-capacitor	
Brushless	Shaded-pole	
Stepper	Reluctance	
	Hysteresis	
	Repulsion	
	Repulsion-start	
	Repulsion-induction	
	Inductor	
	Consequent pole	
	Polyphase	

Fig. 14-1 Types of motors.

Motors can be grouped into one of three broad power ratings: *integral-horsepower* (ihp), *fractional-horsepower* (fhp), and *sub-fractional-horsepower* (sfhp). Motors rated at less than $\frac{1}{20}$ hp are classified as sfhp motors, and the power is usually expressed in milli-horsepower (mhp) rather than in fractional horsepower. Fractional-horsepower motors include those motors rated from $\frac{1}{20}$ to 1 hp. Any motor rated above 1 hp is an ihp motor.

Integral-horsepower

Fractional-horsepower

Subfractional-horsepower

■ TEST_____

Answer the following questions.

1. Based on power-source requirements, what are the three categories of electric motors?
2. True or false. The speed of a variable-speed motor is usually changed by selecting a different winding in the motor.
3. True or false. Synchronous motors can be classified as dc motors.
4. True or false. A $\frac{1}{10}$-hp motor would be classified as a fractional-horsepower motor.

Voltage rating

Motor nameplate

14-2 Motor Ratings

The parameters for which electric motors are commonly rated include voltage, current, power, speed, temperature, frequency, torque, duty cycle, service factor, and efficiency. Standards for these parameters have been established by the *National Electrical Manufacturers Association* (NEMA). The nameplate on a motor usually contains the information needed to determine whether or not a motor is operating within its specified ratings.

National Electrical Manufacturers Association (NEMA)

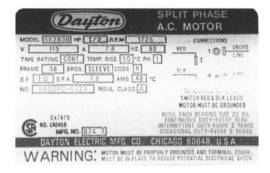

Fig. 14-2 Motor nameplate. The nameplate provides much useful information about the motor to which it is attached.

Voltage Rating

Each motor is designed to operate at a specified voltage. This is the *voltage rating* given on a *motor nameplate* such as the one shown in Fig. 14-2. As a general rule, motors are usually designed so that they will meet all other ratings as long as the supply voltage is within ±10 percent of the rated voltage. Therefore, a 115-V motor can be operated at full load on a nominal 120-V supply and still stay below its maximum temperature rating providing that the supply voltage does not exceed 126.5 V for extended periods of time [126.5 V = 115 V + (0.10 × 115 V)].

Example 14-1

Determine the minimum and maximum voltage that could be applied to a typical 230-V ac motor.

Given: 230-V rating and ±10 percent tolerance

Find: V_{min} and V_{max}

Known: $V_{min} = V_{rated} - (0.1 \times V_{rated})$
$V_{max} = V_{rated} + (0.1 \times V_{rated})$

Solution: $V_{min} = 230 \text{ V} - (0.1 \times 230 \text{ V})$
$= 207 \text{ V}$
$V_{max} = 230 \text{ V} + (0.1 \times 230 \text{ V})$
$= 253 \text{ V}$

Answer: The minimum and maximum voltage are 207 V and 253 V, respectively.

Example 14-2

If the motor in example 14-1 is fully loaded, should it be operated on a nominal 220-V supply line where the voltage may have long-term variations of ±10 percent? Why?

Given: $V_{min(motor)} = 207 \text{ V}$

Find: $V_{min(supply)}$

Known: $V_{min(supply)} = V_{supply} - (0.1 \times V_{supply})$

Solution: $V_{min(supply)} = 220 \text{ V} - (0.1 \times 220 \text{ V})$
$= 198 \text{ V}$

Answer: No, because $V_{min(supply)}$ is less than $V_{min(motor)}$.

Notice in example 14-2 that we had to check only the minimum voltages because the supply voltage was smaller than the rated voltage of the motor. Had the nominal supply voltage been greater than the motor's rated voltage, we would have checked only the maximum voltages.

Another factor to consider when determining if a given motor should be operated on a given supply voltage is the voltage drop caused by the lines (wires) connecting the motor to the supply. As shown in Fig. 14-3(*b*), the resistance of each connecting line appears in series with the motor. Therefore, if we know how much current the motor draws from the source and how much resistance the lines have, we can determine how much of the source voltage is applied to the motor. The resistance of the lines can be determined from the copper wire table in Appendix C when the gage and length of the lines are specified.

Example 14-3

Refer to Fig. 14-3. Assume the minimum secondary voltage is 112 V and the maximum current required by a 120-V motor is 12 A. Will a 12-gage cable supply adequate voltage to the 120-V motor?

Given: $V_{min(supply)} = 112 \text{ V}$
$V_{rated} = 120 \text{ V}$
$I_{max} = 12 \text{ A}$
Cable length = 100 ft
From Appendix C, 12-gage wire has 1.588 Ω per 1000 ft.

Find: $V_{min(motor)}$
R_{line} (resistance of the line)
V_{drop} (voltage drop of the line)
V_{motor} (voltage at the motor terminals)

Known:
$V_{min(motor)} = V_{rated} - (0.1 \times V_{rated})$
$R_{line} = (\Omega \text{ per } 1000 \text{ ft}/1000) \times \text{line length}$
$V_{drop} = R_{line} \times I_{max}$
$V_{motor} = V_{min(supply)} - V_{drop}$

Solution:
$V_{min(motor)} = 120 \text{ V} - (0.1 \times 120 \text{ V})$
$= 108 \text{ V}$
$R_{line} = [(1.588 \ \Omega/1000 \text{ ft})/1000] \times 200 \text{ ft} = 0.318 \ \Omega$
$V_{drop} = 0.318 \ \Omega \times 12 \text{ A} = 3.82 \text{ V}$
$V_{motor} = 112 \text{ V} - 3.82 \text{ V}$
$= 108.2 \text{ V}$

Answer: 12-gage cable will be adequate because V_{motor} is greater than $V_{min(motor)}$.

A typical 230-V ac motor can operate on a 220-V, 230-V, or 240-V system. If a 230-V supply is not available, it would be better to operate the motor on a 240-V rather than a 220-V supply. The higher voltage is preferred because, with respect to operation at the rated voltage, a higher voltage will

1. Provide greater starting or locked-rotor torque

2. Require less current at rated load

3. Run cooler at rated load

4. Be more efficient at rated load

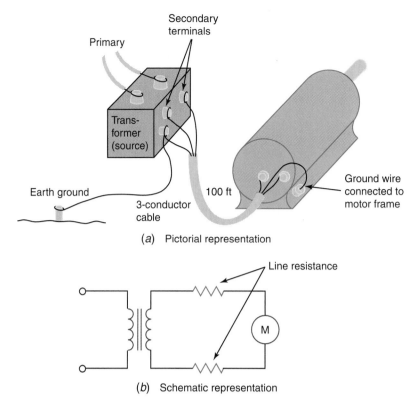

(a) Pictorial representation

(b) Schematic representation

Fig. 14-3 Voltage drop caused by the resistance of the supply lines.

Application of basic electrical principles will explain why higher voltage causes these effects.

For a given motor, the locked-rotor torque is dependent on the strength of the opposing or attracting magnetic fields, which in turn is dependent on the current in the coils of the motor. The current in the coils is a function of the applied voltage, coil resistance, and coil reactance. Since resistance and reactance are independent of voltage, the coil current is directly proportional to the voltage applied to the motor.

The horsepower taken from the shaft of a motor must come from the electric power source connected to the motor. For a given phase angle, the power taken from the source is directly proportional to the voltage and the current. Therefore, at full load, less current is required at a higher voltage.

If a higher voltage causes less current in the motor coils at rated power, then the copper loss (I^2R loss) will also be less at a higher supply voltage. Less copper loss means the motor will run cooler and its efficiency will be improved.

There is one disadvantage to operating a motor at higher-than-rated voltage. At a higher voltage, a motor tends to produce slightly more noise.

JOB TIP

Some companies recruit electronic technicians for positions called *quality technician* or *calibration technician*.

Running current

■ TEST_____

Answer the following questions.

5. What is NEMA the abbreviation for?
6. List five parameters on which motors are rated.
7. What is the typical tolerance for the rated voltage of a motor?
8. Determine the minimum and maximum voltage at which a 240-V motor should be operated.
9. Would it be better to operate a 115-V motor at 110 V or 120 V? Why?
10. What is the disadvantage of running a motor on a voltage 8 percent greater than rated voltage?
11. A 230-V, 18-A motor is to be operated 25 ft from a 220-V ±5 percent supply. What is the minimum gage wire that can be used to avoid excess voltage drop? At 350 cmil/A can this gage wire carry 18 A?

Current Ratings

The current rating given on the motor nameplate in Fig. 14-2 is the *running current*. The running current is the current the motor draws when the motor is loaded to rated horsepower and is operated at rated voltage, frequency, and temperature. For subfractional- and fractional-horsepower motors, this is usually the only current specified on the nameplate.

The *locked-rotor,* or *starting, current* of a motor can be many times greater than the running current. For example, a $\frac{1}{3}$-hp motor rated at 6 A may draw 40 A of *locked-rotor current.* This is the current the motor will draw the instant it is started. The large starting current greatly increases the instantaneous voltage drop in the supply line.

The ratio of starting current to running current can be extremely high in integral-horsepower motors. In fact, it is so high for many motors that the motors must be started on reduced voltage, or use series current-limiting resistors, to start the motor. As the motor builds up speed, the voltage is increased to the rated value or the series resistors are removed. The nameplate on an ihp motor includes the starting kilovolt-ampere (kVA) rating for the motor as well as the running current.

Power, Temperature, and Service Factor

The horsepower rating of a motor specifies the amount of power available at the motor shaft at the rated speed. Like many other electric devices, a motor is capable of delivering more than its rated power. However, unless the motor has a service factor (SF) greater than 1.0, a motor should not be operated with a load greater than its rated horsepower. To do so causes the motor's operating temperature to exceed its design limits.

Motors are designed to operate in an environment with a specified *maximum ambient temperature.* The standard ambient temperature used in most motor designs is 40° C. If a motor is operated within its other design limitations, that is, frequency, horsepower, and voltage, is operating temperature will also be within design limits when the ambient temperature is no more than 40° C.

The maximum temperature at which a motor should operate is dependent on the type of insulating material used on the motor windings. Motor insulating materials are grouped into the four classes shown in Fig. 14-4. Each

Class of Insulation	Maximum Operating Temperature (°C)
A	105
B	130
F	155
H	180

Fig. 14-4 Temperature ratings of different classes of insulation.

of the four classes, as shown in Fig. 14-4, has a maximum operating temperature. The difference between the maximum operating temperature and the rated ambient temperature is the allowable *temperature rise* for a motor. Thus, a motor with class B insulation has a permissible temperature rise of $130° - 40°$ C $= 90°$ C, while a class F insulation allows a 115° C temperature rise.

Some motors are designed and rated so that they can provide more than their rated horsepower under specified conditions. In other words, they have conservative ratings which provide a small safety margin to handle unexpected increases in power required from a motor. The magnitude of this safety margin is specified by the *service factor* (SF) rating of the motor. The service factor of a motor allows one to determine the absolute maximum horsepower that a motor can provide on a continuous basis. The service factor rating is a number that can be multiplied by the horsepower rating to determine the maximum continuous horsepower the motor can provide. The additional horsepower made possible by the SF should be used only when it is ensured that frequency and voltage will be at rated values and ambient temperature will not exceed the rated value.

Starting current

Locked-rotor current

Temperature rise

Service factor (SF)

Maximum ambient temperature

About ⬛⬛⬛ Electronics

Selling Power   Almost half of U.S.-generated power goes on the wholesale block. A distributor in one state may purchase electricity in other states; sources can change hourly depending on the prices. High costs are avoided by buying from unregulated suppliers.

Example 14-4

A certain mechanical system normally requires 0.7 hp, but under adverse conditions may require 0.9 hp. Can a ¾-hp motor with a service factor of 1.25 be used to power the system?

Given: $P_{rated} = 0.75$ hp
                  SF = 1.25
                  $P_{load} = 0.9$ hp

Find: P_{max} (maximum hp available from motor)

Known: $P_{max} = P_{rated} \times$ SF

Solution: $P_{max} = 0.75$ hp $\times 1.25$
                  $= 0.9375$ hp

Answer: A ¾-hp motor with SF = 1.25 will suffice because P_{max} is greater than P_{load}.

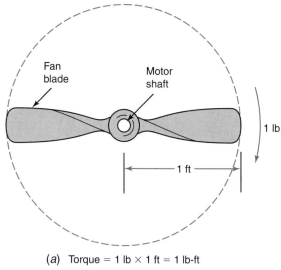

(a) Torque = 1 lb × 1 ft = 1 lb-ft

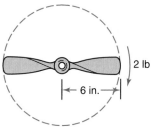

(b) Torque = 2 lb × 0.5 ft = 1 lb-ft

Fig. 14-5 Torque is the product of the distance from the point of rotation and the magnitude of the force.

The speed listed on the nameplate of a motor is the motor speed at the rated horsepower. Under light-load conditions, motor speed is **Rated speed** greater than the *rated speed*. When the load is greater than the rated horsepower, the speed will be less than the nameplate speed.

For ac motors, the frequency of the source voltage is also specified. In North America the **Starting torque** specified frequency is 60 Hz for motors intended for use on commercially available power. In many European countries, power is distributed at 50 Hz, so motors designed for use in these countries are, of course, designed **Frequency tolerance** for 50 Hz. Most motors have a *frequency tolerance* of ±5 percent from the design frequency providing the supply voltage is at rated values. Since the frequency of commercially distributed power is closely regulated, frequency tolerance is usually of concern only where power is locally generated.

Stall torque That which produces rotation, or tends to produce rotation, is known as *torque*. **Torque (T)** Common units for specifying torque are ounce-inches (oz-in.), pound-inches (lb-in.), pound-feet (lb-ft), gram-centimeters (g-cm), and newton-meter (Nm). As illustrated in Fig. 14-5, torque is a product of the force and the distance from the center of rotation at which **Running torque** the force exists. The torque in Fig. 14-5(*a*) and (*b*) is the same because the force doubles when the distance is halved.

The *starting torque* (locked-rotor torque) of a motor is an important factor in applications involving high-inertia loads. Many types of motors, such as repulsion and capacitor-start motors, have starting torques that are much greater than their torques while running. With other types of motors, such as shaded-pole and reluctance motors, the starting torque is much less than the torque while running.

The torque of a motor can be specified for different running conditions. When a motor is to be used in an application that produces short-term, heavy loads, the *stall torque* (breakdown or pull-out torque) of the motor is important. Stall torque is the torque produced just before the motor stalls out or starts to stall out and reengages its starting circuit. Stall torque is considerably higher than the torque at the rated horsepower and speed.

The torque at the rated horsepower and speed, sometimes called the *running torque* (full-load torque), can be calculated from the nameplate data. The relationship between speed, power, and torque is given by the formula

$$P = \frac{TS}{5252}$$

where P is the power in horsepower, T is the torque in pound-feet, and S is the speed in revolutions per minute (r/min or rpm). This formula is often written as

$$hp = \frac{\text{lb-ft} \times \text{r/min}}{5252}$$

Example 14-5

What is the running torque of a 2-hp, 1740-r/min motor?

Given: $P = 2$ hp, and $S = 1740$ r/min

Find: T (torque)

Known: $P = \dfrac{TS}{5252}$

therefore,

$T = \dfrac{5252P}{S}$

Solution: $T = \dfrac{5252 \times 2 \text{ hp}}{1740 \text{ r/min}} = 6$ lb-ft

Answer: The running torque is 6 lb-ft.

In some applications, such as an electric door opener, a motor operates only intermittently. Thus, some motors are rated for intermittent duty rather than continuous duty. The *duty cycle* of a motor is a ratio of the on time to the on time plus the off time. This ratio is often expressed as a percentage. For example, a motor that is on for 2 min and off for 5 min would have a duty cycle of $2/(2 + 5) = 0.29$, or 29 percent. Intermittent-duty motors are designed to operate for a specified time at rated horsepower followed by an off cycle of sufficient time to allow the motor to cool back to a specified temperature.

The efficiency of a motor is also an important characteristic of a motor. Large integral-horsepower motors are designed to be very efficient (around 95 percent), but small fractional-horsepower motors may be quite inefficient (30 to 40 percent). When a small motor is only operated intermittently, it is often cheaper to pay for the inefficiency of the motor over its useful life than it is to make the motor more efficient.

The efficiency of a given motor is a function of the load on the motor. A motor is designed to provide *maximum efficiency* at the rated horsepower. At half of the rated load, the efficiency typically decreases by about 10 percent. Of course, at no load, the efficiency is zero.

Maximum efficiency

One reason the efficiency increases with load is that the power factor also increases with load. A motor with no load on it is very inductive because the motor winding are coils of wire wound on a silicon-steel core. Power is only used for the copper and core losses. Because the ratio R/X is very low, the PF is also very low. When the motor is loaded, the PF increases as the motor becomes less inductive. With a higher PF, a given amount of current delivers more power without increasing the copper loss. Thus, the efficiency increases.

$P = \dfrac{TS}{5252}$

Example 14-6

Determine the efficiency of a motor that delivers 1.7 hp while drawing 12 A from a 240-V supply at a PF of 0.7.

Given: $P = 1.7$ hp
$I = 12$ A
$V = 240$ V
PF $= 0.7$

Find: % eff.

Known: % eff. $= \dfrac{P_{out}}{P_{in}} \times 100$

$P = IV \cos \theta$
$\cos \theta = $ PF
1 hp $= 746$ W

Duty cycle

Solution: $P_{in} = 12$ A $\times 240$ V $\times 0.7$
$= 2016$ W

$P_{out} = 1.7$ hp $\times \dfrac{746 \text{ W}}{\text{hp}}$
$= 1268.2$ W

% eff. $= \dfrac{1268.2 \text{ W}}{2016 \text{ W}} \times 100$
$= 62.9\%$

Answer: The efficiency of the motor is 62.9%.

■ TEST

Answer the following questions.

12. List two current ratings for a motor.

13. Which of the current ratings listed in question 12 is the larger value?

14. True or false. Some large motors are started on reduced-voltage or limited-current supplies.

15. True or false. The horsepower rating on the motor nameplate is the starting horsepower.

16. What is the standard ambient temperature used in rating motors?

17. What determines the maximum temperature at which a given motor should operate?

18. Define "allowable temperature rise."

19. What is the service factor of a motor?

20. It is desired to drive a 0.8-hp load with a ¾-hp motor. What would be the minimum service factor for this motor?

21. A 2.5-hp motor has a SF of 1.15. Under ideal conditions, how much continuous horsepower can it deliver?

22. True or false. The rated speed of a motor is its no-load speed.

23. True or false. The tolerance of the frequency rating of a motor is usually ±1 percent.

24. A 9-in. arm is attached to the shaft of a motor. When power is applied to the motor, the end of the arm exerts 1.3 lb of force. What is the locked-rotor torque of the motor?

25. What is the horsepower rating of a motor which produces 6.9 lb-ft of torque at its rated speed of 1140 r/min?

26. Define "stall torque."

27. Determine the efficiency of a 2-hp motor which requires 2 kW to operate at the rated horsepower.

28. True or false. For maximum efficiency, a motor should be operated at about 70 percent of its rated horsepower.

29. True or false. The PF of a motor is highest at its rated horsepower.

14-3 Motor Enclosures

The NEMA has developed standards for the dimensions of the various sizes of motor enclosures. The size (height, length, shaft diameter, etc.) is indicated by a *frame number.* There is considerable variation in the frame number for a given-horsepower motor. This variation is the result of using different insulating and core

Open motors

Totally enclosed motors

Dripproof motor

Splashproof motors

Frame number

Lintfree motors

Guarded motors

Fig. 14-6 The shaft end of a dripproof open motor.

materials. As better materials have been developed, a given frame has been able to house a motor with a larger horsepower rating. For example, a 56 frame has been for many years a common size for fhp motors; but it is now possible to get multihorsepower motors in this size frame. In general, with all other factors being equal, the larger the frame number, the larger the horsepower rating. For example, a 2-hp motor is commonly in a 145-T frame, while a 10-hp motor is in a 213-T frame. The 145-T frame is approximately 7 in. in diameter and 8 in. long excluding the shaft. Comparable dimensions for the 213-T frame are 10 and 11 in.

Depending on the style of enclosure, motors can be classified as either *open motors* or *totally enclosed motors.* There are many different enclosure types within each of these broad categories.

Open motors (see Fig. 14-6) have enclosures with ventilating openings which allow surrounding air to be forced through the enclosure to cool the motor windings. Various types of open motors put restrictions on the ventilating openings. These restrictions are associated with the type of environment in which the motor is intended to operate. For example, a *dripproof motor* is an open motor in which the openings are designed so that particles or drops striking the motor enclosure at an angle no greater than 15° from the vertical will not impair the operation of the motor. *Splashproof motors* are designed so that matter splashing within specified angles will not harm the motor. *Lintfree motors* have smooth, streamlined openings so that lint in the air will not build up and clog the openings. *Guarded motors* have

Fig. 14-7   A totally enclosed, fan-cooled motor.

the openings screened or grilled so that objects greater than a specified size and shape cannot enter the closure and make contact with electric or moving parts of the motor. Technical specifications of these and other types of open motors are published by the NEMA.

Totally enclosed motors (see Fig. 14-7) are motors which do not allow free exchange between surrounding air and air within the enclosures. Although they have no openings, they are not sealed or airtight. *Totally enclosed nonventilated* (TENV) *motors* have no external fans to force air over the external surface of the enclosure. *Totally enclosed fan-cooled* (TEFC) *motors* have an external fan, attached to the rotor shaft, which forces air circulation around the motor. An *explosionproof motor* is totally enclosed and designed so that an explosion of a gas inside the motor will not cause a like gas around the motor to explode also. Other

types of totally enclosed motors include *waterproof motors* which can be "hosed down" and *dust-ignitionproof motors* which prevent significant amounts of explosive dust from entering the enclosure and which will not ignite explosive dust around or on the enclosure.

Many totally enclosed motors have fins on the exterior of the enclosure to aid in transferring heat from the motor windings to the external air. Larger totally enclosed motors may use circulating water or air as well as heat exchangers to aid in cooling the motor.

■ TEST

Answer the following questions.

30. How is the physical size of a motor specified?
31. What does the abbreviation "TEFC" stand for?
32. List two categories used to classify motor enclosures.
33. True or false. Explosionproof motors are designed to prevent gas and vapor from exploding within the motor enclosure.

14-4 Squirrel-Cage Induction Motors

As shown in Fig. 14-8, many ac motors are classified as induction motors. An *induction*

Waterproof motors

Dust-ignitionproof motors

Totally enclosed nonventilated motors

Totally enclosed fan-cooled motors

Explosionproof motors

Induction motors

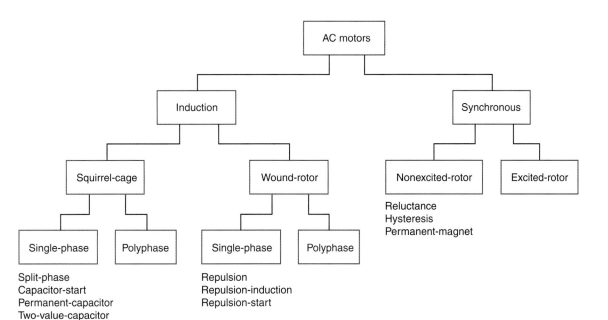

Fig. 14-8   Classification of ac motors by electrical characteristics.

motor is a motor that has no electric connections between the power sources and the rotor, yet the rotor has conductors which carry current. The current in the rotor conductors is an induced current. It is induced by the magnetic field created by the *stator windings*.

Figure 14-9 shows a stator, complete with windings, from an fhp induction motor. A single silicon-steel lamination from a stator core with 36 slots to hold the coils of the stator windings is shown in Fig. 14-10. Laminations like this one are stacked together to form the stator core shown in Fig. 14-10. Figure 14-10 also shows how a winding is insulated from the core by *slot insulators* and *slot wedges*. Slot insulators are installed in the core slots before the winding coils are inserted. Then the slot wedges, or slot keepers, are inserted over the winding coils to insulate and hold them in place. Notice that the core slots are shaped so that they provide a *core tooth* to hold the slot keeper in place. Some stators are insulated by a thin epoxy coating rather than by slot insulators.

All parts of the stator are bonded together by an insulating varnish which is applied to the finished stator by dipping it in a varnish vat.

A *squirrel cage rotor* is shown in Fig. 14-11. The rotor is called a squirrel-cage rotor because the configuration of the conductors imbedded in the silicon-steel rotor core resembles the rotary cage used to provide exercise

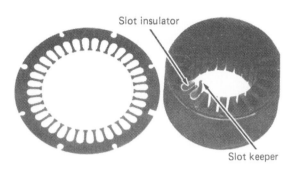

Fig. 14-10 Stator lamination and core.

for squirrels kept in captivity. The core of the rotor is a stack of disc laminations which have slots around the circumference. These disc laminations are pressed onto the rotor shaft. The rotor conductors occupy the slots in the core. The conductors are shorted together at each end by conducting rings. Careful inspection of Fig. 14-11 shows the *shorting rings* at each end of the *rotor core* and the conducting bars connected between the rings. If the iron core were removed, the conducting bars and rings would look like the illustration in Fig. 14-12. Notice that the rotor bars are *skewed* (not parallel to the shaft). Skewing the *rotor bars* allows the bars to enter the magnetic flux of one tooth of the stator before leaving the flux of the adjacent tooth. This provides a more constant torque and reduces vibration in the motor. No separate insulating material is needed to insulate the conducting bars and rings from the rotor core because the voltage induced into these single conductors is low and the conductor resistance is extremely low compared with the resistance of the core. Core resistance is high because each lamination is oxidized to insulate it from adjacent laminations. For many rotors, the bars and rings are cast into the core after it is assembled.

Fig. 14-11 Squirrel-cage rotor used in an induction motor.

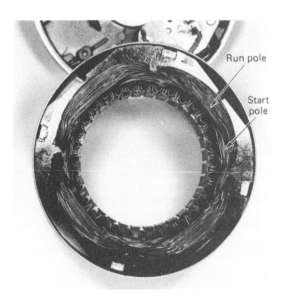

Fig. 14-9 Stator with a four-pole running winding and a four-pole starting winding. The start poles overlap, and are on top of, the run poles.

Stator windings

Slot insulators

Slot wedges

Shorting rings

Rotor core

Core tooth

Skewed

Rotor bars

Squirrel-cage rotor

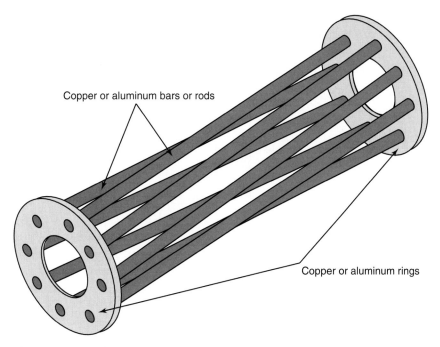

Copper or aluminum bars or rods

Copper or aluminum rings

Fig. 14-12 Conductor arrangement in a squirrel-cage rotor.

TEST

Answer the following questions.

34. What is the distinguishing characteristic of an induction motor?
35. What causes current in the squirrel-cage rotor?
36. True or false. Slot insulators are used in a squirrel-cage rotor to insulate the rotor bars from the rotor core.
37. True or false. The conductor bars in a squirrel-cage rotor are parallel to the rotor shaft.

Single-Phase Motors

In order to start a single-phase induction motor, some method must be found to produce two sources of magnetomotive force that are separated in both space and time. This creates a rotating magnetic field which develops a *rotational torque* on the rotor.

Why the field created by a single magnetomotive force does not produce a rotational torque to turn the rotor is illustrated in Fig. 14-13. This figure shows that when the stator coils are excited by single-phase alternating current, the pulsating magnetic field they produce induces a voltage into the rotor bars. Since the rotor is stationary, the induced voltage is caused by *transformer action* between stator coils and the shorted rotor bars. The shorted rotor bars act like shorted turns of a transformer secondary. Notice

that no voltage is induced into the top and bottom rotor bars in Fig. 14-13 because their axis is perpendicular to the axis of the stator coils. Also notice that maximum voltage is induced into the left and right rotor bars because their axis is parallel to that of the stator coils. The induced voltage causes a current to flow in the rotor bars and produces a magnetic field that is perfectly aligned with the stator field. Thus, while the rotor core is strongly attracted to the stator core, there is no rotational torque to turn the rotor.

A number of techniques are used to create a rotating magnetic field. We will look at these techniques as we discuss the various types of single-phase induction motors.

Split-Phase motor

The *split-phase internal-resistance* motor, commonly called the split-phase motor, is the most prevalent type of fhp single-phase ac motor. Its name is descriptive of the technique it uses to create a rotating magnetic field. The rotating field is created by two out-of-phase currents which are obtained by "splitting" the single-phase ac source to provide two out-of-phase currents.

 YOU MAY RECALL . . . that Sec. 13-4 shows how the phase shift caused by series *RL* circuits can produce two out-of-phase currents from a single-phase source.

Rotational torque

Split-phase internal-resistance motors

Transformer action

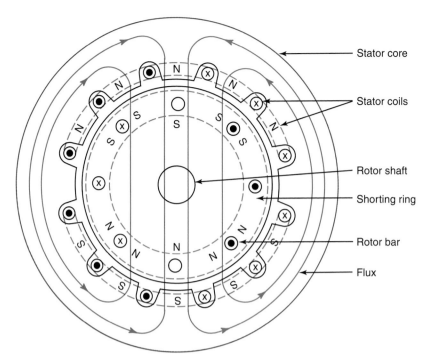

Fig. 14-13 Stator and rotor arrangement that does not provide any starting torque.

A split-phase motor has two windings in the stator as illustrated in Fig. 14-14. Notice in Fig. 14-14(*a*) that all slots hold a coil for both the *run winding* and the *start winding*. In many motors, some slots will hold a coil only for the run winding or for the start winding.

Both windings in Fig. 14-14(*a*) have two *poles*. A pole is a group of coils of wire all wound in the same direction so that the magnetic field of each coil is aiding all other coils in the pole. In all motors except the *consequent*

Run winding

Start winding

Poles

Consequent pole motor

pole motor, adjacent poles in a winding have opposite magnetic polarities. Figure 14-13 shows that the upper pole produces a north magnetic pole on the stator teeth while the bottom pole is producing a south magnetic pole. To produce opposite magnetic polarities, the coils in one pole are wound in the opposite direction from the coils in an adjacent pole. Of course, the polarities of the poles reverse each half-cycle when the currents through the coils reverse.

❓ Did You Know?

Power lines or Pollution Recent studies by the United States and Finland have shown that living near power lines may increase the occurrence of leukemia in children. However, scientists are not sure whether this increase is due to high levels of pollution occurring in these power line areas. Big power lines tend to run along busy streets in older neighborhoods, so the increase could be related to conditions in older homes or to pollution from traffic.

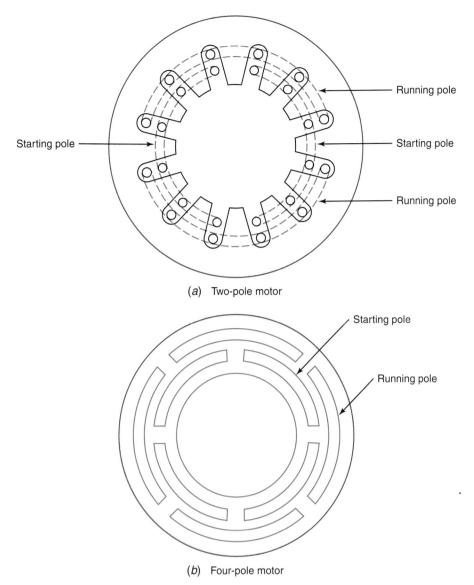

(a) Two-pole motor

(b) Four-pole motor

Fig. 14-14 Stators with run and start windings.

Figure 14-14(b) illustrates a simplified method of showing the windings in a four-pole stator. In this illustration, individual coils are not shown; instead, a complete pole is indicated by a circular segment. The four poles of the run winding are located farthest from the center of the drawing. The wires which connect the four poles in series are omitted, as are the wires that connect to the power source.

Notice in Fig. 14-14 that the centers of the poles for the start winding are placed equidistant from the centers of adjacent run poles. Thus, the start poles are displaced 90 mechanical degrees from the run poles for a two-pole motor and 45 mechanical degrees for a four-pole motor.

When out-of-phase currents flow in the run and start windings, a rotating magnetic field is developed. Figure 14-15 shows how the magnetic field rotates 90 mechanical degrees while the start-winding current and the run-winding current change 90°. In Fig. 14-15(a), at time t_1, the current in the run winding is zero, so the center of the magnetic field poles is in the center of the start-winding poles. As detailed in Fig. 14-15(b), 45 electrical degrees later, the center of the field is midway between the centers of the start and run poles because each pole is contributing equally to the magnetic field. In Fig. 14-15(c),

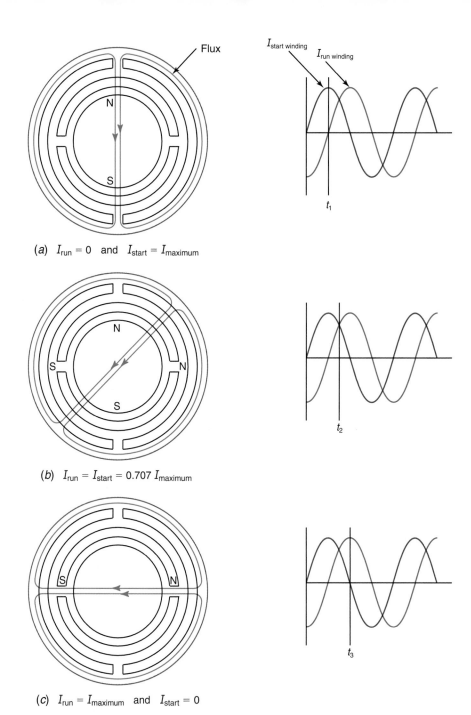

(a) $I_{run} = 0$ and $I_{start} = I_{maximum}$

(b) $I_{run} = I_{start} = 0.707 I_{maximum}$

(c) $I_{run} = I_{maximum}$ and $I_{start} = 0$

Fig. 14-15 Rotating magnetic field caused by out-of-phase currents in the run and start windings.

only the run winding is producing flux, so the field is now centered in the run poles. If you visualize the conditions 45° after time t_3 in Fig. 14-15(c), you can see that the start current is reversed, so the magnetic polarity of the start poles is reversed from that shown in Fig. 14-15(b). This results in the magnetic field being rotated another 45° in a clockwise direction.

The discussion based on Fig. 14-15 assumes that the run and start currents are 90° out of

phase and that the run and start windings are identical, so the contribution of each winding to the magnetic field is also equal. However, neither of these assumptions has to be met to cause a rotating field. As long as the two currents are neither in phase nor 180° out of phase, the field will rotate. However, inspection of Fig. 14-16 shows that currents that are unequal in magnitude and only a few degrees out of phase produce great fluctuation in the strength

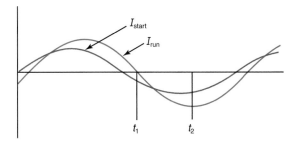

Fig. 14-16 Currents that are nearly in phase produce large pulsations in the rotating magnetic field.

of the rotating field. At time t_1 in Fig. 14-16, the field is very weak because the run current is zero and the start current is very small. At time t_2, both currents are close to their maximum values, so the magnetic field is very strong compared with this strength at time t_1. The current relationships illustrated in Fig. 14-16 would provide only a very weak starting torque in a motor. When the currents are equal and 90° out of phase (Fig. 14-15), the pulsations in the strength of the rotating field are minimized because one current is maximum when the other current is zero. Current rela-

tionships like those in Fig. 14-15 provide a very strong starting torque in a motor.

How the out-of-phase currents are obtained in a split-phase motor is illustrated in Fig. 14-17. The electrical diagram for this motor is given in Fig. 14-17(a). Notice that the four-pole run winding is in parallel with the four-pole start winding while the *start switch* is closed. The equivalent circuit for the stator is shown in Fig. 14-17(b), in which each winding is represented by its equivalent resistance and ideal inductance. By controlling Q, that is, the X_L/R ratio, in each parallel branch, we can control the phase shift of the branch currents with respect to the source voltage. In the split-phase motor, the diameter of the run-winding wire is typically three or four times larger than that of the start winding. Also, the number of turns in the coils of the run winding is typically two or three times greater than in the start winding. This difference in the windings causes the run winding to have less resistance and more inductance than the start winding. The higher Q of the run winding results in the run current lagging behind the start current by 20° to 30°, as illustrated in Fig. 14-17(c). These out-of-

Start switch

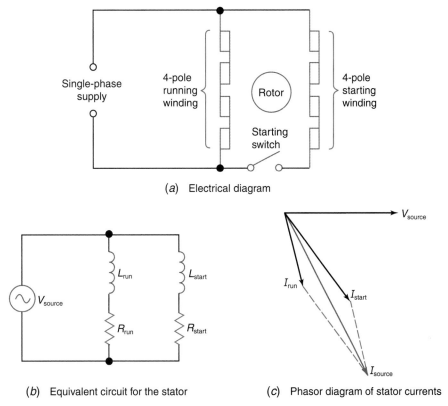

(a) Electrical diagram

(b) Equivalent circuit for the stator

(c) Phasor diagram of stator currents

Fig. 14-17 Phase relationship of the start current and run current in a split-phase motor.

phase currents provide the *rotating magnetic field* needed to start a single-phase motor.

The rotating magnetic field produced by the stator cuts the rotor bars and induces a voltage, by *generator action,* into the rotor bars. Thus, the rotor has two voltages induced into it, one by generator action and one by transformer action. However, as shown in Fig. 14-13, the transformer-action voltage does not create a flux which is out of alignment with the stator flux. Thus, for our analysis, the transformer-action voltage can be ignored. Maximum generator-action voltage is induced in the bars directly under the moving stator flux when the flux is at its maximum value. As shown in Fig. 14-18(*a*), this voltage causes a current in the rotor bars, and the resulting current creates a field which is 90° to the stator field. Of course, two flux paths cannot exist 90° to each other, so the *resultant flux* in the rotor is as shown in Fig. 14-18(*b*). The shifting of the flux in the rotor by the cross fields also shifts the magnetic poles on the rotor so that they are no longer aligned with the poles in the stator. Thus, as illustrated in Fig. 14-18(*c*), a rotational torque is developed in the motor. Notice that this torque causes the rotor to turn in the same direction as the flux is rotating.

Once the rotor builds up speed, the stator field no longer has to rotate for generator-action voltage to be induced in the rotor. The start windings are removed from the motor circuit by opening the start switch when the motor obtains about 70 to 80 percent of its operating speed. The moving rotor bars cut the stationary stator flux as this flux sinusoidally increases and decreases. The rotor poles are still shifted out of alignment with the stator poles, and the resulting rotational torque keeps the rotor turning as long as the load does not exceed the stall torque of the motor.

Once a motor is running, the rotor flux interacts with the stator windings. This interaction increases the effective inductance of the run winding and helps to reduce the current required by the motor. This, along with removal of the start-winding current, is why the run current is only a fraction of the start current in a split-phase motor.

The starting (locked-rotor) torque of fhp split-phase motors ranges from about 130 to 200 percent of full-load running torque. A speed-versus-torque curve for a typical split-

phase motor is shown in Fig. 14-19. The motor's locked-rotor torque is 1.5 times as great as its full-load running torque. Notice from this curve that the torque increases as the motor increases speed up to about 60 percent of synchronous speed. Above 60 percent of synchronous speed, torque starts to decrease. The discontinuity in the curve is the result of the

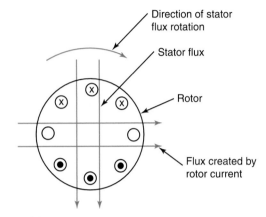

(*a*) Flux created by generator-action voltage

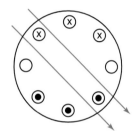

(*b*) Resultant flux in rotor

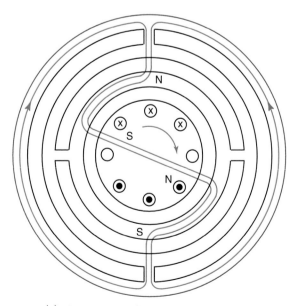

(*c*) Instantaneous flux path in rotor and stator

Fig. 14-18 Development of rotational torque in a motor.

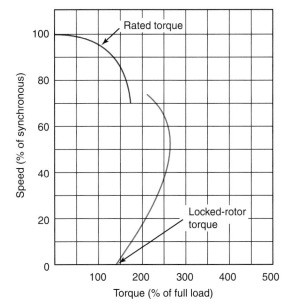

Fig. 14-19 Speed-versus-torque curve for a typical split-phase motor.

start-winding's being switched in and out of the motor circuit.

The synchronous speed of an induction motor, split-phase or any other type, is determined by the number of poles in the run winding and the frequency of the supply voltage. Figure 14-15 shows that for a two-pole motor, the field rotates 45 mechanical degrees for each 45 electrical degrees. Since the rotor can turn only as fast as the field rotates, the synchronous speed for the two-pole motor is equal to the speed of the flux rotation, which is one revolution per cycle. For a 60-Hz supply, the synchronous speed is calculated as follows:

$$\text{r/min} = 60 \text{ cycles/s} \times 60 \text{ s/min}$$
$$\times 1 \text{ r/cycle} = 3600 \text{ r/min}$$

With a four-pole motor, the field rotates only 0.5 mechanical degrees for each electrical degree. Thus, the four-pole motor operating on a 60-Hz supply has a synchronous speed of 1800 r/min. The general formula for determining the synchronous speed of an induction motor is

$$\text{Synchronous speed} = \frac{120 f}{\text{no. of poles}}$$

where f is the frequency in hertz and synchronous speed is in revolutions per minute.

Example 14-7

Determine the synchronous speed of a six-pole motor operating from a 220-V, 50-Hz source.

Given: No. of poles = 6
                f = 50 Hz
                V = 220 V

Find: Synchronous speed

Known:

$$\text{Synchronous speed} = \frac{120 f}{\text{no. of poles}}$$

Solution:

$$\text{Synchronous speed} = \frac{120 \times 50}{6} = 1000 \text{ r/min}$$

Answer: The synchronous speed of the six-pole, 50-Hz motor is 1000 r/min.

Notice that the synchronous speed of a motor is independent of the source voltage.

Figure 14-19 shows that the split-phase motor runs at less than the synchronous speed when it is loaded. This is true of all induction motors. At the rated full-load torque, most induction motors operate at 90 and 98 percent of the synchronous speed. The difference between the synchronous speed and the rated speed of a motor is called *slip*. The slip of a motor is usually expressed as a percentage. The formula for calculating the percent slip is

Percent slip =
$$\frac{\text{synchronous speed} - \text{rated speed}}{\text{synchronous speed}} \times 100$$

Slip

Did You Know?

Ford Isn't the Only Henry Known for His Motor   One of the earliest electric motors was designed and built by Joseph Henry.

Example 14-8

Determine the percent slip of a four-pole, 60-Hz, split-phase motor with a rated speed of 1725 r/min.

Given: No. of poles = 4
$f = 60$ Hz
Rated speed = 1725 r/min

Find: % slip

Known: % slip =

$$\frac{\text{synchronous speed} - \text{rated speed}}{\text{synchronous speed}} \times 100$$

$$\text{Synchronous speed} = \frac{120 f}{\text{no. of poles}}$$

Solution:

$$\text{Synchronous speed} = \frac{120 \times 60}{4} = 1800 \text{ r/min}$$

% slip =

$$\frac{1800 \text{ r/min} - 1725 \text{ r/min}}{1800 \text{ r/min}} \times 100 = 4.2\%$$

Answer: The slip is 4.2 percent.

Direction of rotation

Centrifugal switch actuator

A disassembled split-phase motor is pictured in Fig. 14-20 to show how the start switch is operated. When assembled, the spring-loaded bobbin part of the *centrifugal switch actuator* on the rotor shaft contacts the start switch and closes the switch. When power is applied to the motor, the bobbin remains in contact with the switch as the rotor starts to turn. When the rotor is at about 75 percent of the operating speed, the weights on the centrifugal actuator produce enough force to overcome the spring tension that holds the bobbin part of the actuator against the start switch. At this time the bobbin snaps back away from the switch and the switch opens and disconnects the start winding from the power source.

Solenoid-operated switch

The start winding can also be disconnected by a relay or a *solenoid-operated switch.* Many of the solenoid-operated switches rely on gravity to open the switch contacts, so they must be operated in the correct physical position. The diagram for a motor using a relay starting switch is drawn in Fig. 14-21. The coil of the relay is in series with the run winding. When the motor is turned on, its run-winding current is very high, so the relay energizes and connects the start winding to the power source. When a relay is energized, the air gap between the armature and the core is eliminated, so it takes much less current to hold the armature down than it takes to pull it down. Thus, as the motor starts to increase its speed and the run-winding current starts to decrease, the relay remains closed and the start winding remains active. By the time the motor speed reaches about 75 percent of the rated speed, the run-winding current will have reduced to a value that is insufficient to hold the relay closed; the relay will open, and the start winding will be disconnected.

The *direction of rotation* of a split-phase motor can be changed by reversing the lead connections of *either* the run winding or the start winding. Reversing the leads on one winding causes the winding current to shift 180°. How this 180° shift reverses the rotation can be seen by referring back to Fig. 14-15. Notice in this figure that the flux rotates clockwise because the start current is leading the run current. Visualize inverting either waveform in

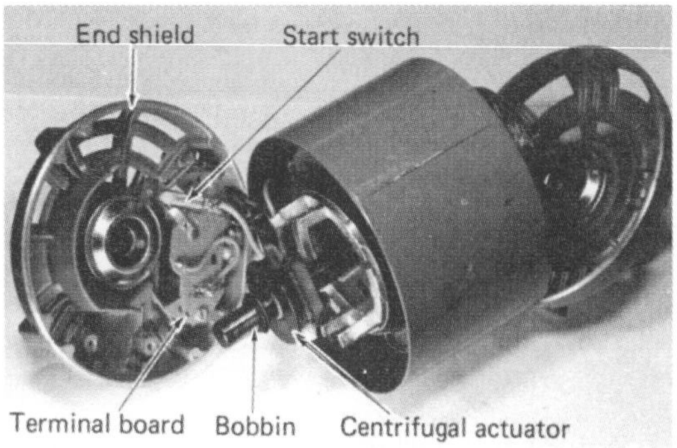

Fig. 14-20 Parts of a split-phase induction motor.

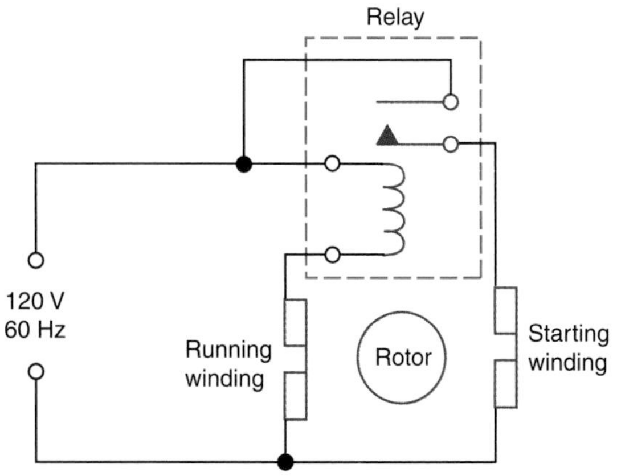

Fig. 14-21 A relay can be used to control power to the start winding in a split-phase motor.

this figure and you will see that the run current will lead the start current, and the flux will then rotate counterclockwise.

■ TEST_____

Answer the following questions.

38. True or false. A rotating magnetic field in the stator is needed to develop locked-rotor torque.
39. True or false. The voltage induced into a stationary rotor by transformer action produces a rotational torque on the rotor.
40. How are two out-of-phase currents developed in the internal-resistance split-phase motor?
41. Which winding in a split-phase motor
    a. Has the larger resistance
    b. Has the larger number of turns
    c. Uses the smaller-diameter wire
    d. Causes the larger phase shift of its current
42. Define the term "pole" as it is used in discussing motors.
43. Do adjacent poles in a motor stator have the same or opposite magnetic polarities?
44. In a six-pole motor, there are _____ mechanical degrees between adjacent start poles and _____ mechanical degrees between the center of a start pole and the center of the nearest run pole.
45. In a six-pole stator, the field rotates _____ mechanical degrees for every 90 electrical degrees.
46. For maximum starting torque, the start and run currents should be _____ degrees out of phase.
47. True or false. In an internal-resistance, split-phase motor, the start and run currents are 90° out of phase.
48. True or false. The locked-rotor torque of a split-phase motor is usually less than its full-load torque.
49. True or false. The coil of a relay used in a motor is connected in series with the start winding.
50. Determine the rated speed of a six-pole, 60-Hz motor if its slip is 4.16 percent.
51. Determine the synchronous speed of an eight-pole, 240-V, 60-Hz motor.
52. Determine the slip of a two-pole, 120-V, 60-Hz motor with a full-load speed of 3450 r/min.

53. The start switch usually opens when the motor obtains between _____ and _____ percent of the rated speed.
54. List two techniques used to connect and disconnect the start winding in a motor.
55. How can one reverse the direction of rotation in a split-phase motor?

Capacitor-Start Motor

Another popular type of single-phase motor is the *capacitor-start motor*. This motor is also a type of split-phase motor, but it does not rely on internal resistance to obtain out-of-phase currents in its run and start windings. Instead, it uses a capacitor in series with the start winding as diagrammed in Fig. 14-22(*a*). The capacitance value is selected so that its reactance is slightly greater than the reactance of the start winding under locked-rotor conditions. As shown by the phasor diagram in Fig. 14-22(*b*), this makes the start-winding circuit capacitive and causes the start current to lead the source voltage. Thus, with the capacitor-start motor, the two winding currents can be a full 90° out of phase.

Capacitor-start motor

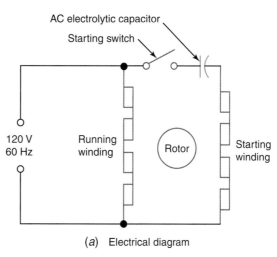

(*a*) Electrical diagram

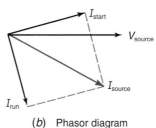

(*b*) Phasor diagram

Fig. 14-22 Four-pole capacitor-start induction motor.

The start winding in the capacitor-start motor uses only slightly smaller wire than the run winding, and it has as many, or sometimes more, turns as the run winding. This makes the start-winding field about the same strength as the run-winding field.

Because the start and run currents are 90° out of phase and the fields created by these currents are about equal, the capacitor-start motor has a much stronger starting torque than the split-phase motor. The starting torque of the capacitor-start motor ranges from about 300 to 450 percent of its full-load torque.

Not only does the capacitor-start motor have more starting torque, but it requires less source current to develop a given amount of starting torque than a split-phase motor does. One of the reasons for this can be ascertained by comparing Figs. 14-17(c) and 14-22(b). Notice that the phasor sum of two currents which are approximately 25° out of phase is much greater than that of two comparable currents that are approximately 90° out of phase. Another reason, as mentioned previously, is that equal currents that are 90° out of phase produce optimum starting torque.

Capacitor-start motors are available in standard fhp sizes. They are also available in ihp sizes up to about 7 hp.

The capacitor used in a capacitor-start motor is an ac electrolytic type. It is not designed for continuous duty. Its typical duty cycle is about 20 starts per hour. *Motor-start capacitors* are usually in the 50- to 600-μF range. A typical one is pictured in Fig. 14-23.

Except for the difference already mentioned, the capacitor-start motor and the split-phase motor are constructed in the same way. For the same size motors, the two types have the same characteristics after the start switch opens. The

procedure for reversing the direction of rotation is also the same, that is, reversal of the leads to either winding.

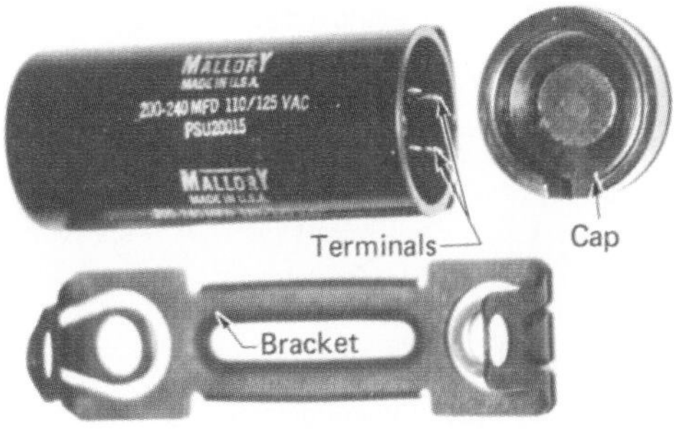

■ TEST

Answer the following questions.

56. In a capacitor-start motor, the run and start currents are approximately _____ degrees out of phase.
57. In terms of percentage of full-load torque, what is the range of the starting torque of capacitor-start motors?
58. What type of capacitor is used in the start circuit of a capacitor-start motor?
59. What are the advantages of the capacitor-start motor over the split-phase motor?

Permanent-Split Capacitor Motor

The *permanent-split capacitor motor* has two windings in its stator with a capacitor in series with one of the windings [see Fig. 14-24(a)]. However, it has no start switch; both windings and the capacitor are left in the circuit at all times. The rotor for this motor looks like the rotor for other split-phase motors. These motors are available in the fractional- and subfractional-horsepower sizes.

Capacitors for permanent-split capacitor motors must be rated for continuous duty. Usually an *oil-filled film capacitor* is used. The capacitance value is selected to provide a 90° phase shift between the winding currents when the motor is producing between 80 and 100 percent of the rated power [see Fig. 14-24(b)]. At this speed, the effective inductance and inductive reactance of the winding are large, so the capacitive reactance must also be large to make one of the winding circuits appear capacitive. Thus, compared with the value of a start capacitor, this capacitor is relatively small.

The permanent-split motor has a higher power factor, requires less line current, and is more efficient than other single-phase motors. The source current, as seen in Fig. 14-24(b), is small because it is the phasor sum of the two 90° winding currents. At the rated speed, the source current is shifted by the capacitive current in the winding containing the capacitor; this increases the power factor. With smaller line currents, the I^2R losses are decreased and the efficiency is increased.

Fig. 14-23 Typical motor-starting capacitor and mounting bracket.

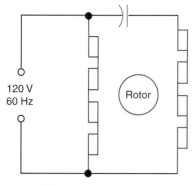

(a) Electrical diagram

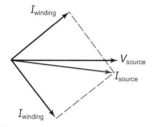

(b) Phasor diagram at rated speed

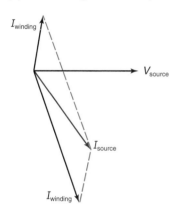

(c) Phasor diagram for locked-rotor

Fig. 14-24 Diagrams for a permanent-split capaci-
tor motor.

Figure 14-24(c) shows that the low value of capacitance and the reduced effective inductance of the windings causes the phase difference between the winding currents to be much greater than 90° under locked-rotor conditions. This results in a very weak starting torque. The starting torque of the permanent-split capacitor is typically 50 to 80 percent of the full-load torque.

Permanent-split capacitor motors are often used in variable-speed applications. By controlling the voltage applied to one or both windings, one can control the speed of this motor over the upper 50 percent of its rated speed. As the applied voltage is decreased, the slip in-

creases and the voltage is decreased, the slip increases and the speed decreases.

The direction of rotation of the permanent-split capacitor motor can be changed by either of two methods. First, the leads of either winding can be reversed as was done with the split-phase and capacitor-start motor. Second, as shown in Fig. 14-25, the capacitor can be switched from one winding to the other winding. This latter method has the advantage that it is easy to do while the motor is running. If the switch in Fig. 14-25 is thrown while power is applied, the motor will rapidly decelerate until rotation stops and then immediately accelerate in the opposite direction of rotation. This method of reversal works best when the two windings are identical. With unbalanced windings, the torques are different for the different directions of rotation.

Another desirable feature of the permanent-slip motor is that it is quieter than the split-phase and capacitor-start motors. This is the result of having both windings continuously energized to provide a uniform rotating field in the stator.

Two-Value Capacitor Motor

An electrical diagram for a *two-value capacitor motor* is drawn in Fig. 14-26. The start switch, which mechanically looks and operates like the start switch in a split-phase motor, connects a large value ac electrolytic capacitor in parallel with the *run capacitor* for starting the

**Two-value
capacitor motor**

Run capacitor

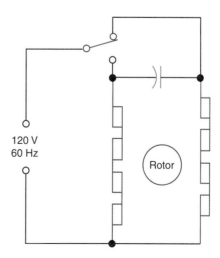

Fig. 14-25 Method of reversing
rotation in a permanent-
split capacitor motor.

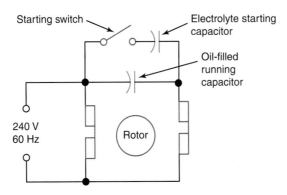

Fig. 14-26 Circuit diagram of a two-pole, two-value-capacitor motor.

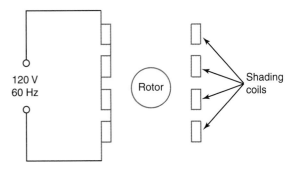

Fig. 14-27 Electrical diagram of a four-pole, shaded-pole motor.

Salient-pole stator

motor. Thus, the motor has the high-starting-torque and low-starting-current characteristic of a capacitor-start motor. When the motor reaches about 75 percent of its rated speed, the start switch opens and the motor runs as a permanent-split capacitor motor with its desirable characteristics of high efficiency, low source current, high power factor, and quiet operation.

Shading coil

Two-value capacitor motors are also called capacitor-start, capacitor-run motors. They are available in fhp and ihp motors up to about 20 hp.

■ TEST

Answer the following questions.

60. What is the major limitation of a permanent-split capacitor motor?
61. Which type of motor is easily reversed while power is applied to the motor?
62. List five advantages of the permanent-split capacitor motor over the capacitor-start motor.
63. What function does the start switch serve in the two-value capacitor motor?
64. What type of capacitor is used in a permanent-split capacitor motor?
65. True or false. The locked-rotor winding currents in a permanent-split capacitor motor are less than 90° out of phase.

Shaded-Pole Motor

Shaded-pole motor

A circuit diagram for a *shaded-pole motor* is shown in Fig. 14-27. Notice that this motor has no start switch and that only one winding is connected to the power source.

Distributed stator winding

Instead of a *distributed stator winding* like those used in the previously discussed motors,

this motor uses a *salient-pole stator* like that pictured in Fig. 14-28. With the salient-pole winding, the winding for a pole is a single coil rather than a group of coils. In some sfhp motors, a single coil wrapped around the core iron may provide both poles for a two-pole motor (see Fig. 14-29).

All shaded-pole motors have a *shading coil* wrapped around part of each salient pole. The shading coils are visible in the pictures in Figs. 14-28 and 14-29. The shading coil is often just a shorted single turn of heavy copper wire, as seen in Fig. 14-30, where the main winding has been removed.

The operation of the shaded-pole motor can be understood by applying basic principles of induced voltages and currents.

Fig. 14-28 Stator for a shaded-pole motor in which there is a coil for each salient pole. One coil has been removed to show the heavy single-turn copper shading coil.

Fig. 14-29 Stator of a two-pole shaded-pole motor. The motor uses one coil for two salient poles.

YOU MAY RECALL
. . . from Lenz's law that induced voltages and currents always oppose the action that produced them.

Thus, when an increasing flux from the main coil induces a voltage in the shading coil, the resulting current in the shading coil creates a flux that opposes and cancels part of the main flux in the shaded part of the pole. When the flux caused by the main coil starts to decrease, the shading-coil current reverses and creates a flux which aids, and adds to, the flux in the shaded part of the pole. The net result is that the strong part of the magnetic pole shifts, or rotates, from the unshaded to the shaded part of the pole. This rotating flux is sufficient to induce, by generator action, a current in the squirrel-cage rotor and start the motor. Once started, the shaded-pole motor operates like other induction motors except that the shading coils continue to produce I^2R losses.

Shaded-pole motors are cheap to construct and easy to maintain. However, they are very inefficient, have a low PF, and have a very weak starting torque. Therefore, they are only used for subfractional and fractional horsepower to about $\frac{1}{4}$ hp.

Shaded-pole motors like those we have been discussing cannot be electrically reversed. However, some small motors can be disassembled and the stator turned end-for-end to produce reverse rotation.

Reluctance-Start Motor

Occasionally a motor has a stator with salient poles shaped like the one shown in Fig. 14-31. The left side of this pole has more reluctance than the right side because of the wider air gap. More importantly, the flux in the left side obtains its maximum value before the flux in the right side. This is because air causes no hysteresis; that is, in air the flux is in phase with the magnetomotive force. In silicon steel, the flux lags behind the magnetomotive force. Since flux in the left part of the pole travels through more air, it leads the flux in the right part of the pole. Therefore, the field in Fig. 14-31 travels from left to right and the rotor turns clockwise. Of course,

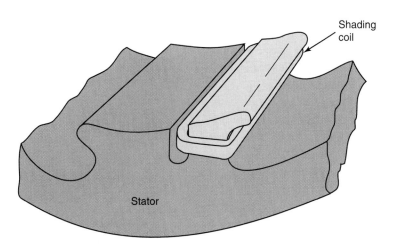

Fig. 14-30 Shading coil on a salient pole.

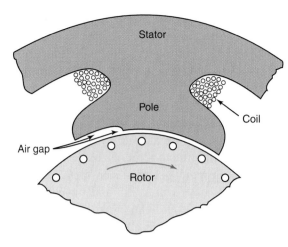

Fig. 14-31 Salient pole in a reluctance-start motor.

Reluctance-start motor

Synchronous motors

Reluctance motor

Dual-voltage motor

Dual-speed motor

the starting torque of the *reluctance-start motor* is very weak.

Some motors use a combination of shading coil and reluctance to provide increased starting torque. One side of the pole has the extra air gap and the other side has the shading coil. A stator for such a motor is pictured in Fig. 14-32.

Dual-Speed and Dual-Voltage Motors

Many motors are designed to operate on either of two source voltages which have a 1:2 ratio. For example, a two-pole, *dual-voltage motor* may be run on either 120 V or 240 V. When the motor is operated on 120 V, the coils for one pole are connected in parallel with the coils for the other pole. For 240-V operation, the coils of the two poles are series-connected. Thus, each coil receives the same voltage and current regardless of which source voltage is used.

A common way to make a *dual-speed motor* is to wind it with two run windings and one start winding. For example, a stator may have a four-pole run winding, a six-pole run winding, and a four-pole start winding. Regardless of the speed (1725 or 1150 r/min) at which the motor is to operate, it is always started as a four-pole motor. Then if 1150-r/min operation is desired, the power is switched from the four-pole run winding to the six-pole run winding

by an extra set of contacts on the start-switch mechanism.

■ TEST

Answer the following questions.

66. A shaded-pole motor usually uses a _____ pole stator winding.
67. List the main electrical characteristics of the shaded-pole motor.
68. Describe the shading coil in a shaded-pole motor.
69. What causes flux rotation in a reluctance-start motor?
70. Does flux shift toward or away from the shaded part of a pole?
71. True or false. Dual-voltage motors have two sets of run windings.
72. True or false. Dual-speed motors usually start on the low-speed windings.

14-5 Synchronous Motors

As an example of the many types of *synchronous motors* listed in Fig. 14-8, we will look at a *reluctance motor.* The reluctance motor is a type of induction motor which can use any of the methods discussed in Sec. 14-4 to develop a rotating stator field to get the rotor in motion. The squirrel-cage rotor in the reluctance motor is modified so that the iron core has salient poles. Figure 14-33 shows a lamination from the rotor of a typical four-pole reluctance motor.

The reluctance motor starts, like any other induction motor, by having currents induced into the rotor bars by the rotating stator flux. As the rotor gets up to about 90 to 95 percent of the synchronous speed, the salient poles on

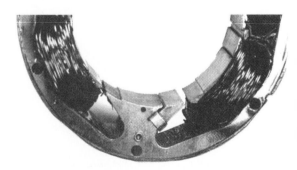

Fig 14-32 A stator that uses reluctance-start and a shading coil to produce starting torque.

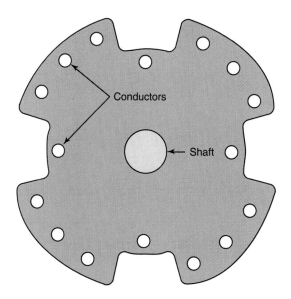

Fig. 14-33 Salient-pole rotor lamination for the rotor of a reluctance motor.

14-6 Other Types of Motors

YOU MAY RECALL . . . that the operating principle of a *dc motor* was discussed in Sec. 7-15. In a dc motor, the stationary winding is called the *field winding* and the rotating winding is called the *armature winding*. The rotating part of the motor, which consists of the shaft, iron-core laminations, winding, and *commutator,* is referred to as the *armature*.

DC motor

Field winding

Armature winding

Commutator

Armature

Shunt DC Motor

Figure 14-35(*a*) shows the diagram for a *shunt dc motor.* In this motor, the field winding can use many turns of small-diameter wire because the large source current needed under full-load conditions flows through the armature wind-

Shunt dc motor

the rotor are almost lined up with magnetic poles in the stator. At this point, the rotor jumps up to the synchronous speed as the flux tries to align the rotor and stator poles to provide the shortest possible flux path which also has the minimum possible reluctance. This idea is illustrated in Fig. 14-34, where it can be seen that the flux path in Fig. 14-34(*b*) is shorter and involves no more air than in Fig. 14-34(*a*). The rotor never quite achieves perfect pole alignment as shown in Fig. 14-34(*b*) because if it did there would be no rotational torque due to the reluctance effect. At a given motor load, the salient rotor poles are always the same distance from perfect alignment with the stator poles when the stator poles are at their maximum strength. As the motor load increases, the distance from perfect alignment increases. Finally, if the load becomes too great, the reluctance effect can no longer hold the rotor in approximate alignment; the motor drops out of synchronization.

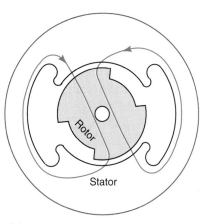

(*a*) Torque caused by misaligned poles

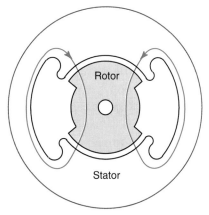

(*b*) Rotor and stator poles aligned

Fig. 14-34 Salient rotor poles try to align themselves with the magnetic poles in the stator.

◼ TEST

Answer the following questions.

73. True or false. Salient rotor poles are needed to start a reluctance motor.
74. The reluctance motor uses a modified _____ rotor.
75. The reluctance motor is classified as a _____ motor.

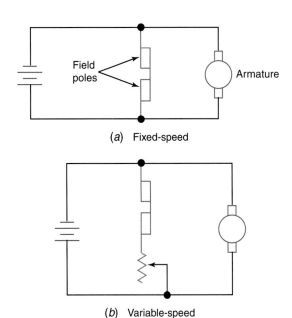

(a) Fixed-speed

Series dc motor

(b) Variable-speed

Fig. 14-35 Shunt dc motor. The rheostat in (b) controls the speed of the motor.

ing. The shunt motor has a fairly constant speed over a wide load range. However, the speed does decrease some as the load increases.

The speed of the armature is always such that the induced voltage (cemf) in the armature winding plus the *IR* voltage (due to the resistance of armature wire) equals the source voltage. Since the armature winding uses large-diameter wire, its resistance is very small. Thus, the *IR* voltage is very small compared with the induced voltage. As the load on the motor increases, so does the armature current. This increases the *IR* voltage and decreases the cemf needed to equal the source voltage. Therefore, if the field flux remains constant, the armature must slow down or the combined *IR* voltage and cemf voltage would exceed the source voltage. Although the field flux changes slightly because of interaction between the field flux and the armature flux, the change is not significant except when the motor is operated with a very weak field flux.

Armature current in any dc motor is a function of the load on the motor. When the load is increased, armature current must increase so that a stronger armature flux will provide more torque as it reacts with the field flux.

The speed of a shunt motor can be controlled over a fairly wide range (about 1:4) by controlling the field current with a series rheostat as shown in Fig. 14-35(*b*). Decreas-

ing the field current (and thus the field flux) forces the armature speed to increase in order to produce the required cemf. For a given load, the armature current must increase to provide the additional flux needed to increase the armature speed. Thus, the amount the speed can be increased depends on the load on the motor shaft. If the motor is operating at rated horsepower, no speed adjustment can be made.

Series DC Motor

Figure 14-36 shows a diagram of a *series dc motor*. This motor is noted for its high starting torque and its load-dependent speed.

The series motor has high starting torque because both the armature and the field are wound with large-diameter wire. When power is first applied, the current is limited only by the resistance of the windings, so a large current can flow. This large current, flowing through both windings, produces strong poles in both the armature and the field, so the resulting torque is very high. Of course, if the load is too large, the magnetic materials saturate before torque sufficient to turn the load can be developed.

With a light load on the series motor, little current is required to develop the necessary torque. However, with little current in the field winding, the field flux is weak, so the armature must rotate at a high speed to produce enough cemf to counter the source voltage. With large loads, the current is high, the field flux is strong, and the armature can, therefore, turn much slower and still produce the required cemf. With no load, the speed of a large series dc motor can become so high that centrifugal forces destroy the armature. Therefore, these motors are only used where there is no chance of the load being disengaged.

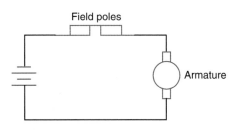

Fig. 14-36 Series dc motor.

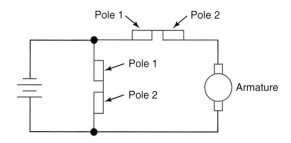

Fig. 14-37 Compound dc motor. Part of each field pole is in series with the armature, and part is in parallel with the supply.

Compound DC Motor

The *compound motor,* shown in Fig. 14-37, has part of the field winding in series with the armature and part in parallel with the source voltage. The series part of the field winding can either aid (called *cumulative compounding*) or oppose (called *differential compounding*) the parallel part of the field winding. Since differential compounding is rarely used, we will only concern ourselves with the characteristics of the cumulative compound motor.

As expected, the cumulative compound motor has speed and starting-torque characteristics somewhere between the series and the shunt motor. Because of the series part of the field, the compound motor has more starting torque than the shunt motor. Because of the parallel part of the field, the speed of this motor is less dependent on the load. Even with no load, the parallel portion of the field provides sufficient flux to limit the speed to a nondestructive value.

Universal Motor

The *universal motor* has the same electrical diagram as a series dc motor. In fact, it has the same characteristics as the series dc motor, and the armature and the field are quite similar in appearance to those of the series dc motor.

The universal motor is designed to operate equally well from either an ac or a dc supply—thus the name "universal." Operating a series motor on alternating current requires a few design changes from the series dc motor. For example, the universal motor uses thin laminations in both the armature core and the field poles (core) to reduce eddy current loss when operating on alternating current.

Three-Phase Motors

Three-phase motor

The stator of a *three-phase* induction *motor* has a separate winding for each phase. Therefore, the stator has a constantly rotating magnetic field and, like the permanent-split capacitor motor, needs no start switch or switch activator.

The technical details and characteristics of the many types and designs of three-phase motors are too complex to cover in this chapter. However, we can, in general terms, compare three-phase motors with single-phase motors.

Because the rotating field in three-phase motors has constant amplitude, three-phase motors produce less noise than their single-phase counterparts. Also, they provide a constant shaft torque throughout the full cycle of the source voltage, whereas the torque of the single-phase motor pulsates during each cycle.

Compound motor

Cumulative compounding

Differential compounding

The efficiency of a three-phase motor is higher than that of a comparable-horsepower single-phase motor. For this and other reasons, the three-phase motor is generally smaller and lighter.

If a single-phase motor is reversible at all, the leads to one of the windings must be switched to achieve reversal. The direction of rotation of any three-phase motor is reversed by merely switching any two of the three supply lines.

Finally, three-phase motors are available in a much wider range of power ratings. Whereas single-phase motors above 10 hp are not common, three-phase motors rated in hundreds of horsepower are commonly used in industry.

■ TEST

Answer the following questions.

76. The windings in dc and universal motors are the _____ and _____ windings.
77. List the three types of dc motors in order of decreasing starting torque.
78. List the three types of dc motors in order of decreasing speed regulation.
79. The universal motor has characteristics like those of the _____ dc motor.
80. How many windings does a three-phase motor have?
81. List as many differences as you can between three-phase motors and single-phase motors.

Universal motor

Summary

1. Motors can be classified by the type of power required or by the intended use.
2. Motors are available in sfhp, fhp, and ihp sizes.
3. Motors are rated for voltage, current, power, temperature, frequency, torque, duty cycle, service factor, and efficiency.
4. The National Electrical Manufacturers Association (NEMA) establishes standards for motor ratings, enclosures, and frame sizes.
5. The voltage-rating tolerance of a motor is typically ± 10 percent, while the tolerance for frequency is usually ± 5 percent.
6. Both voltage-source variation and line voltage drops must be of concern when determining the appropriate voltage rating for a motor.
7. If possible, a motor should be operated on the positive side of its voltage tolerance rather than on the negative side.
8. The locked-rotor current is many times larger than the full-load current.
9. A motor with an SF greater than 1.0 can be operated above its rated horsepower under specified conditions.
10. Ambient temperature for motor-design purposes is usually considered to be 40°C.
11. The maximum operating temperature of a motor is determined by the class of insulation used. The classes are A (105°C), B (130°C), F (155°C), and H (180°C).
12. Three common torque ratings of a motor are locked-rotor (starting), stall (breakdown or pullout), and running (full-load).
13. Torque, speed, and power are related by

$$P = \frac{TS}{5252}$$

	Characteristic				
Type of Motor	Locked-Rotor Torque	Starting Current	Running Current	Efficiency	Power Factor
Split-phase	Moderate	High	Moderate	Moderate	Moderate
Capacitor-starting	Very high	Moderate	Moderate	Moderate	Moderate
Permanent-capacitor	Weak	Low	Low	High	High
Two-value-capacitor	Very high	Moderate	Low	High	High
Shaded-pole	Very weak	Very low	High	Very low	Very low
Reluctance-starting	Very weak	Very low	High	Low	Very low
Reluctance	*	*	High	Low	Low

	Characteristic				
Type of Motor	Auxiliary Winding	Starting Switch	Common Sizes†	Usually Reversible	Variable-Speed	Synchronous
Split-phase	Yes	Yes	1 & 2	Yes	No	No
Capacitor-starting	Yes	Yes	2 & 3	Yes	No	No
Permanent-capacitor	Yes	No	1 & 2	Yes‡	Yes	No
Two-value-capacitor	Yes	Yes	2 & 3	Yes	No	No
Shaded-pole	No	No	1 & 2	No	Yes	No
Reluctance-starting	No	No	1 & 2	No	Yes	No
Reluctance	*	*	1 & 2 & 3	*	No	Yes

* Depends on technique used to start motor.
† 1 = sfhp, 2 = fhp, and 3 = ihp.
‡ Can be reversed while power is applied.

Fig. 14-38 Generalized comparison of single-phase induction motor characteristics.

where T is in pound-feet, S is in r/min, and P is in horsepower.

14. The efficiency and PF of a motor are maximum at rated horsepower.

15. Some common types of motor enclosures are dripproof, splashproof, guarded, TENV, TEFC, explosionproof, waterproof, and dust-ignitionproof.

16. Induction motors have no electric connection between the source and the rotor. Most of these motors use a squirrel-cage rotor.

17. Developing starting torque in an induction motor requires a rotating magnetic field in the stator. This field can be created by an auxiliary (start) winding, shaded poles, or differential reluctance in the poles.

18. Maximum torque results when the two-winding currents are equal in amplitude and 90° out of phase.

19. The characteristics of single-phase induction motors are summarized and compared in Fig. 14-38.

20. Induction motors can be designed for either dual speed or dual voltage.

21. The synchronous-speed formula for a motor is

$$\text{Synchronous speed} = \frac{120\,f}{\text{no. of poles}}$$

where f is in hertz and speed is in r/min.

22. The formula for determining the percent slip is

$$\text{Percent slip} = \frac{\text{synchronous speed} - \text{rated speed}}{\text{synchronous speed}} \times 100$$

23. A start switch may be either mechanically or electrically operated. This switch opens at about 75 percent of the rated speed.

24. A reluctance motor has a squirrel-cage rotor with salient poles.

25. Figure 14-39 compares the characteristics of the series, shunt, and cumulative-compound dc motors.

Type of DC Motor	Characteristics		
	Speed Regulation	Starting Torque	Variable-speed
Series	Very poor	Very high	No
Shunt	Good	Moderate	Yes
Cumulative-compound	Moderate	High	No

Fig. 14-39 Comparison of the characteristics of three types of dc motors.

26. The speed of the shunt dc motor can be varied by varying the field current.

27. The universal motor operates on either alternating or direct current and has characteristics similar to those of the series dc motor.

28. Compared with single-phase motors, three-phase motors are more efficient, make less noise, provide more constant torque, are lighter and smaller, and are easier to reverse.

Chapter Review Questions

For questions 14-1 to 14-14, determine whether each statement is true or false.

14-1. Regardless of the SF rating, a motor should not be continuously operated above its rated horsepower.

14-2. Tolerance for the voltage rating of a motor is typically ± 5 percent.

14-3. The frequency tolerance of a motor rating is of primary concern when a motor is operated from a commercial supply.

14-4. The run-winding current in an induction motor decreases as the motor speeds up.

14-5. The temperature-rise rating of a motor is usually based on a 60°C ambient temperature.

14-6. The efficiency of a motor is usually greatest at its rated power.

14-7. The voltage drop in a line feeding a motor is greatest when the motor is at about 50 percent of its rated speed.

14-8. An explosionproof motor prevents gas and vapors from exploding inside the motor enclosure.

14-9. Since a squirrel-cage rotor is not connected to the power source, it does not need any conducting circuits.

14-10. The start switch in a motor opens at about 75 percent of the rated speed.

14-11. "Reluctance" and "reluctance-start" are two names for the same type of motor.

14-12. The cumulative-compound dc motor has better speed regulation than the shunt dc motor.

14-13. The compound dc motor is often operated as a variable-speed motor.

14-14. All single-phase induction motors have a starting torque which exceeds their running torque.

For questions 14-15 to 14-19, choose the letter that best completes each statement.

14-15. Greater starting torque is provided by a
  a. Shunt dc motor
  b. Series dc motor
  c. Differential compound dc motor
  d. Cumulative compound dc motor

14-16. Which of these motors provides the greater starting torque?
  a. Split-phase
  b. Shaded-pole
  c. Permanent-split capacitor
  d. Capacitor-start

14-17. Which of these motors provides the quieter operation?
  a. Split-phase
  b. Capacitor-start
  c. Two-value capacitor
  d. Universal

14-18. Which of these motors has the greater efficiency?
  a. Reluctance-start
  b. Shaded-pole
  c. Split-phase
  d. Permanent capacitor

14-19. Which of these motors would be available in a 5-hp size?
  a. Split-phase
  b. Two-value capacitor
  c. Permanent capacitor
  d. Shaded-pole

Answer the following questions.

14-20. List three categories of motors which are based on the type of power required.

14-21. List three categories of motors which are based on a range of horsepower.

14-22. What is NEMA the abbreviation for?

14-23. List three torque ratings for motors.

14-24. A 240-V motor is drawing 20 A at a PF of 0.65 from a power source which is providing 225 V. The motor is producing 2.8 hp. It is connected to the power source by 50 ft of 14-gage, two-conductor copper cable. Determine the voltage at the motor terminals and the efficiency of the motor. If this is a typical motor, is it operating within voltage tolerance limits?

14-25. Given a choice, would you operate a 230-V motor from a 220-V or a 240-V supply? Why?

14-26. Determine the running torque of a ¾-hp motor rated at 1725 r/min.

14-27. What are TEFC and TENV the abbreviations for?

14-28. What type of action induces voltage into a rotating rotor?

14-29. List three techniques for producing a rotating field in a stator.

14-30. What relationships should two winding currents have to produce maximum torque?

14-31. Differentiate between a variable-speed and a dual-speed motor.

14-32. Determine the synchronous speed of an eight-pole motor designed to operate from a 240-V, 50-Hz supply.

14-33. A two-pole, 60-Hz motor is operating at 3400 r/min. What is the percent slip for this motor?

14-34. Why does a three-phase motor provide a non-pulsating torque?

14-35. Is a single-phase motor or a three-phase motor of the same horsepower more efficient?

Critical Thinking Questions

14-1. A 120-V, ¾-hp motor has an SF of 1.1, an efficiency of 82 percent, and a PF of 0.77 when loaded to the value allowed by its SF. At 300 cm/A, what gage wire is required to operate this motor when it is located 95 ft from a source that has a minimum voltage of 114 V?

14-2. What features would be required in an apparatus designed to measure the running torque of a motor?

14-3. Suppose the starting capacitor and the running capacitor in a two-value capacitor motor were accidentally interchanged. How would this affect the operation of the motor?

14-4. Why are shaded-pole motors very inefficient?

14-5. A motor is 87 percent efficient, has a PF of 0.82, and produces 2.8 hp at 220 V, which is its minimum operating voltage. What size wire is required if the motor is to be located 100 ft from a source with minimum voltage of 225 V?

14-6. What do you think would be the results of connecting a 240-V, 60-Hz motor to a 220-V, 50-Hz source?

14-7. What are the characteristics of a motor-start capacitor, and the circuit it is used in, that limit its use to intermittent duty?

14-8. How would you expect the speed regulation and the starting torque of a differential compound motor to compare to a cumulative compound motor?

14-9. We know that $P = TS/5252$ when P is in horsepower, T is in lb-ft, and S is in revolutions per minute. We also know that one newton-meter (Nm) = 0.7376 lb-ft and horsepower = 746 W. Derive the formula for calculating P in watts, when T is in Nm, and S is in revolutions per minute.

Answers to Tests

1. DC, ac, and universal
2. F
3. F
4. T
5. National Electrical Manufacturers Association
6. any five of the following: voltage, current, power, speed, temperature, frequency, torque, duty cycle, service factor, or efficiency
7. ± 10 percent
8. 216 V and 264 V
9. 120 V, because it would have more starting torque, require less current while running cooler, and be more efficient at full power output
10. The motor produces more noise at elevated voltages.
11. 13 gage. No; use the next-smaller gage number.
12. running current and starting current
13. starting current
14. T
15. F
16. 40°C
17. the class of insulation used in the motor
18. Allowable temperature rise is the temperature difference between ambient temperature and maximum permissible operating temperature.
19. Service factor specifies the factor (multiplier) by which the rated horsepower can be increased under specified conditions.
20. 1.07
21. 2.875 hp
22. F
23. F
24. 11.7 lb-in. or 0.975 lb-ft
25. 1.5 hp
26. Stall torque is the torque a motor produces just before the motor stalls out or stops rotating.
27. 74.6 percent
28. F
29. T
30. by frame number
31. totally enclosed fan-cooled
32. open and totally enclosed
33. F
34. no electrical connection between the rotor and the power source
35. voltage induced by the magnetic field created by the stator
36. F
37. F
38. T
39. F
40. by making the start-winding Q much lower than the run-winding Q
41. a. start winding
    b. run winding
    c. start winding
    d. run winding
42. a group of coils all wound in the same direction to create a magnetic field
43. opposite
44. 60, 30

45. 30
46. 90
47. F
48. F
49. F
50. 1150 r/min
51. 900 r/min
52. 4.2 percent
53. 70, 80
54. a centrifugally operated switch or a magnetically operated switch
55. by reversing the leads of either winding
56. 90
57. 300 to 450 percent
58. an intermittent-duty, ac electrolytic capacitor
59. greater starting torque and less starting current per unit of starting torque
60. very low starting torque
61. the permanent-split capacitor motor
62. quieter, higher PF, less source current, more efficient, and capability of variable-speed operation
63. It adds a large electrolytic capacitor in parallel with the run capacitor to provide excellent starting torque.
64. continuous duty
65. F
66. salient-
67. low starting torque, very low efficiency, and low PF
68. It is a shorted single turn of large copper wire around part of a salient pole.
69. the differential flux lag between the wide-air-gap and narrow-air-gap portions of the salient pole
70. toward
71. F
72. F
73. F
74. Squirrel-cage
75. synchronous
76. armature, field
77. series, compound, shunt
78. shunt, compound, series
79. series
80. three
81. Three-phase motors are quieter, smaller, lighter, more efficient, and easier to reverse. They also provide a constant torque.

Chapter 15

Instruments and Measurements

Chapter Objectives

This chapter will help you to:

1. *Understand* how a digital multimeter measures current, voltage, and resistance.
2. *Understand* how capacitance and inductance are measured with a digital meter.
3. *Understand* how an analog meter movement can be converted to a VOM.
4. *Calculate* shunt and multiplier resistor values.
5. *Measure* power in three-phase circuits with either balanced or unbalanced loads.
6. *Use* a bridge to measure capacitance, inductance, and resistance.
7. *Calculate* the amount of meter loading when measuring current and voltage.
8. *Understand* why some ohmmeters are reverse reading and nonlinear.
9. *Identify* various types of analog meter movements.

 YOU MAY RECALL . . . that in previous chapters you learned how to use meters to measure electrical quantities. In this chapter you will learn how a meter measures such things as current, voltage, resistance, power, frequency, capacitance, and inductance.

15-1 Digital Multimeter

A digital meter indirectly measures an unknown voltage by measuring the time it takes a capacitor to charge to a voltage equal to the unknown voltage. The capacitor is charged by a constant-current source, so the voltage rise on the capacitor is a linear function of time. Why the voltage rise is linear can be seen from the formulas $C = Q/V$ and $I = Q/t$, which, when rearranged and combined, yield $V = It/C$. Since I and C are constants, V is directly proportional to t.

The readout of the digital meter is connected to a circuit which counts and stores the number of cycles a reference frequency produces in the time it takes the capacitor to charge to the unknown voltage. When the unknown voltage and the capacitor voltage are equal, the stored count is displayed on the digital readout, the capacitor is discharged, and the counter is re-

set to zero. Then the voltage is measured again and the readout is updated.

Let us go through an example to see how the cycle count can represent the amount of unknown voltage. Suppose the capacitor is charging at a rate of 10 V/s, the reference frequency is 10 Hz, and the unknown voltage is 4 V. At 10 V/s, the capacitor voltage will reach 4 V in 0.4 s. At the end of 0.4 s, the reference frequency will have produced four cycles. The stored, and displayed, count will be 4—the value of the unknown voltage.

Although the above explanation is extremely simplified, it is representative of the technique used in many digital meters. Of course, the exact circuits are quite complex, operate at much higher frequencies, provide more than 1 V resolution, and provide multiple ranges and functions.

A *digital multimeter* (DMM) indirectly measures current with the voltmeter circuit ex-

Digital multimeter (DMM)

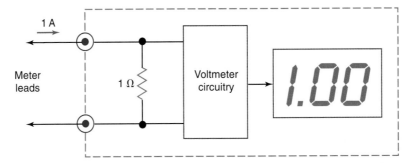

Fig. 15-1 Current function of a digital multimeter.

plained above. Figure 15-1 illustrates how current is measured. As shown, 1 A flowing through a 1-Ω resistor drops 1 V, which is measured and displayed by the voltmeter circuitry. By simultaneously switching precision resistors and voltage ranges of the voltmeter circuitry, one can build various current ranges into the digital multimeter.

Figure 15-2 illustrates how the voltmeter circuitry of the DMM can be used to measure an unknown resistance. If the 1-mA constant current in Fig. 15-2 drops 1 V across the unknown resistance, the unknown resistance must

Analog meters

be 1 kΩ. Thus, the 1-V drop across the unknown resistor, which is measured by the voltmeter circuitry, represents 1 kΩ. Various resistance ranges can be obtained by changing the value of the constant current. For example, a readout of 1.50 would represent 1.5 MΩ if the constant current were 1 μA.

◼ TEST

Answer the following questions.

Full-scale current rating

1. True or false. The voltmeter circuitry of the DMM is used when measuring current with the DMM.
2. True or false. The readout of a DMM dis-

plays the number of cycles produced by a reference frequency.
3. What is the measured resistance when the voltmeter circuitry of the DMM indicates 1.20 and the constant current source across the resistor is 1 μA?
4. What does a digital voltmeter measure to indirectly determine an unknown voltage?
5. What type of current sources are used in a DMM?

15-2 Meter Movements

Meters with moving pointers are called *analog meters*. They measure quantities by moving through an infinite number of points on a scale. The major part of any analog multimeter is the basic meter movement. Basic meter movements utilize the interaction of two magnetic fields. At least one of the fields is created by a current passing through a coil in the meter movement.

Ratings of Meter Movements

All basic meter movements have a *full-scale current rating*. This is the coil current needed to cause the meter to deflect to the maximum

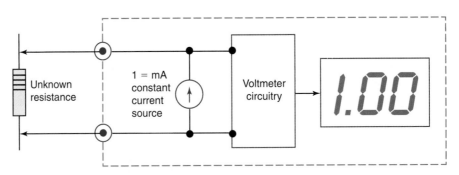

Fig. 15-2 Resistance function of a digital multimeter.

value of the scale. Meter movements with full-scale deflection currents as low as 5 μA are commonly available.

Another important rating of a meter movement is its *internal resistance*. Basic meter movements have appreciable resistance because of the small-diameter wire used in the coil. In general, the lower the full-scale current rating, the higher the internal resistance. A typical 1-mA meter has less than 100 Ω of resistance. A typical 50-μA meter has more than 900 Ω of resistance.

Since a meter movement has both a current and a resistance rating, it must also have a voltage rating. Usually, the manufacturer specifies only two of these three ratings. However, the third rating can easily be determined by using Ohm's law. The voltage across a meter movement (V_m) must be equal to the product of the full-scale current (I_m) and the internal resistance (R_m).

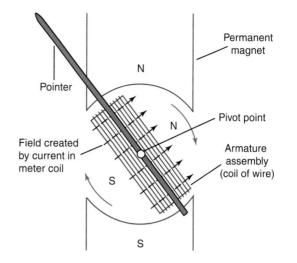

Internal resistance

Fig. 15-3 The d'Arsonval movement principle. The electromagnetic field of the armature tries to line up with the field of the permanent magnet.

D'Arsonval Meter Movement

The most common type of meter movement is the *d'Arsonval movement*, which was invented by the French physicist Jacques d'Arsonval (1851–1940). This meter movement is based on the interaction of the fields of a permanent magnet and an electromagnet.

The armature assembly in Fig. 15-3 contains a coil of very thin wire wound on a lightweight supporting form. The coil, coil form, and attached pointer rotate on pivot points. When current flows through the coil, a magnetic field is created, as seen in Fig. 15-3. The coil's magnetic field reacts with the permanent magnet's field to create a clockwise rotational force. This rotational force reacts against spiral springs. These springs attach to the pivot shaft on which the armature rotates. The points of the shaft rest in jewel bearings to keep friction as low as possible. This arrangement is known as a *jewel-and-pivot* suspension system.

D'Arsonval meter movements are *polarized*. If reverse current flows in the coil, a reverse torque is applied to the armature. This reverse torque can damage the meter movement and the pointer if it is excessive. Even though the d'Arsonval meter movement is polarized, you will see later how it can be used in ac as well as dc meters.

Jewel-and-pivot

Polarized

d'Arsonval movement

Iron-Vane Meter Movement

In an *iron-vane meter* movement, the coil which receives the current to be measured is stationary. The field set up by the coil magnetizes two iron vanes, which then become temporary magnets (Fig. 15-4). Since the same field magnetizes both vanes, both vanes have the same magnetic polarity. Consequently, there is a force of repulsion between the two

Iron-vane meter

Did You Know?

Beware of Drooping Power Lines   Too much current in an overhead power line can make it heat up, enlarge, and droop into tree branches. Devices at crucial points can be on guard for gasses emitted by overheated transformers. Also, thyristors can move current from a failed unit to another source of power.

Nonlinear scale

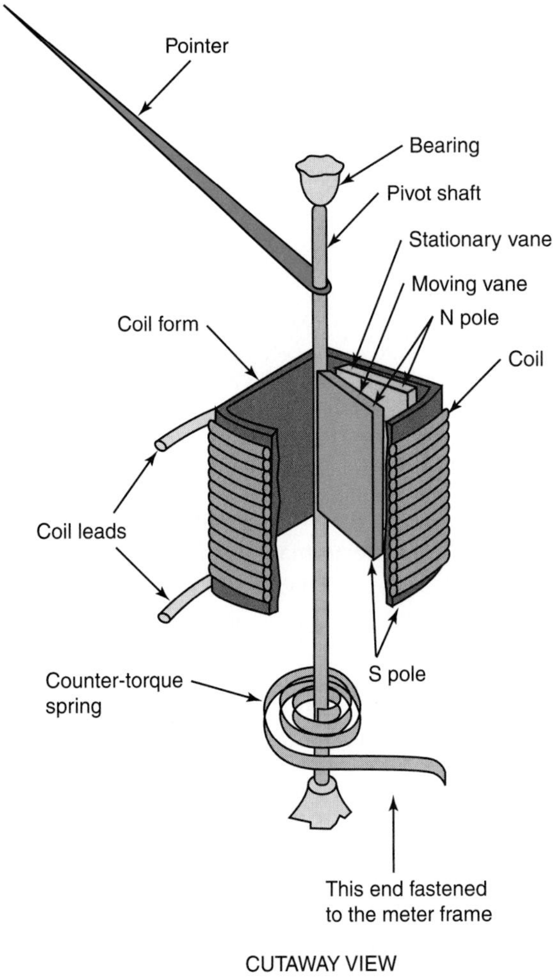

**Electrodynam-
ometer**

**Two
electromagnetic
fields**

Pointer

Bearing

Pivot shaft

Stationary vane

Moving vane

N pole

Coil

Coil form

Coil leads

S pole

Counter-torque
spring

This end fastened
to the meter frame

CUTAWAY VIEW

Fig. 15-4   Iron-vane meter movement. Since the
movement uses no permanent mag-
net, it responds to alternating current
as well as to direct current.

Stationary vane

Moving vane

**Lower
sensitivity**

vanes. One of the vanes (the *stationary vane*)
is attached to the coil form. The other vane
(the *moving vane*) is mounted on the pivot
shaft to which the meter pointer is attached.
Thus, the magnetic force of repulsion pushes
the moving vane away from the stationary
vane. Of course, this force is offset by the
countertorque of the spiral springs attached to
the pivot shaft. The greater the current through
the coil in Fig. 15-4, the stronger the magnetic
repelling force; thus, the farther the moving
vane rotates and the more current the pointer
indicates.

The iron-vane meter movement can operate
on either alternating or direct current. When
the alternating current reverses, the magnetic
polarities of both vanes reverse simultaneously.
Therefore, a force of repulsion is maintained
throughout the cycle.

The iron-vane meter movement has two
shortcomings: it has an extremely *nonlinear
scale* that is very crowded at the low end, and
it requires considerable current for full-scale
deflection (low sensitivity). Its most common
use is an ac ammeter.

The advantages of the iron-vane meter
movement are its ease of construction, rugged-
ness, and dependability.

Electrodynamometer
Meter Movement

The *electrodynamometer* (Fig. 15-5) uses *two
electromagnetic fields* in its operation. One
field is created by current flowing through a
pair of series-connected stationary coils. The
other field is caused by current flowing through
a movable coil that is attached to the pivot
shaft. If the currents in the coils are in the correct directions, the pointer rotates clockwise.
The rotational torque on the movable coil is
caused by the opposing magnetic forces [Fig.
15-5(*b*)] of the three coils.

The directions of the magnetic fields shown
in Fig. 15-5(*b*) were arbitrarily chosen; all the
arrowheads could just as well be reversed. It is
only the polarity of one coil relative to the
other two that is important. (The two stationary coils are permanently connected so that
their fields are always aiding each other.) The
electrodynamometer can operate with alternating current because the alternating current reverses direction simultaneously in all three
coils. The electrodynamometer can also operate on direct current.

Electrodynamometer movements have *lower
sensitivity* than d'Arsonval movements. However, electrodynamometers can be very accurate, and they are quite stable. The accuracy of
the electrodynamometer movement does not
depend on the strength of a permanent magnet
or the permeability of the iron parts.

Although the electrodynamometer is used
for measuring current and voltage, its most
common use is for measuring power.

■ TEST_____

Answer the following questions.

6. Meter movements are rated for
   _____, _____, and _____.
7. The most common type of meter movement is the _____.

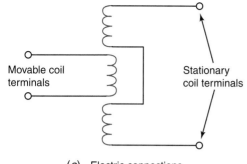

(a) Electric connections

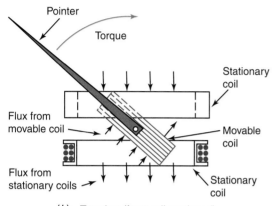

(b) Top view (front coil sectioned)

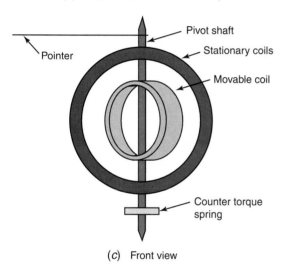

(c) Front view

Fig. 15-5 Electrodynamometer. This meter uses two electromagnetic fields and can respond to either alternating or direct current.

8. The rotational force created by the magnetic field in a meter movement is counteracted by the reverse rotational force of a _____.

9. What is the internal resistance of a meter movement that has a full-scale current of 100 μA and a full-scale voltage of 200 mV?

10. Does a d'Arsonval meter movement respond to alternating current, direct current, or both?
11. Does an iron-vane movement respond to alternating current, direct current, or both.
12. Which of the following meter movements is the most sensitive: iron-vane, d'Arsonval, or electrodynamometer?
13. Which is more linear, a d'Arsonval movement or an iron-vane movement?
14. Does an iron-vane movement use a permanent magnet?
15. Does an electrodynamometer movement respond to alternating current, direct current, or both?
16. What causes the rotary torque in an electrodynamometer movement?
17. How many coils create the stationary magnetic field in an electrodynamometer?

15-3 Analog Ammeters

The d'Arsonval movement can be used to measure alternating current if it is connected in series with a rectifier, as illustrated in Fig. 15-6. A *rectifier* is a device that allows current to flow in only one direction. In other words, it has almost infinite resistance to current trying to flow in one direction and practically no resistance to current flowing in the opposite direction. Thus, the rectifier of Fig. 15-6 allows only one-half of an ac cycle to pass. It converts alternating current to pulsating direct current. The pulsating direct current provides the electromagnetic field of constant polarity needed to operate a d'Arsonval meter movement. The combination of a d'Arsonval movement and a rectifier produces a rectifier-type of ac meter.

Rectifier

> ## JOB TIP
>
> Many jobs require equal doses of human skills and technical skills. Both can be improved with time and effort.

Shunts

When used as an ammeter, a basic meter movement has a range equal to its full-scale deflection current. The range of an ammeter can be made larger by adding a shunt to the meter. A *shunt* is a resistor of very low resistance connected in parallel with the basic meter movement (Fig. 15-7). Shunts are usually made from materials with very low *temperature co-*

Shunt

Temperature coefficients

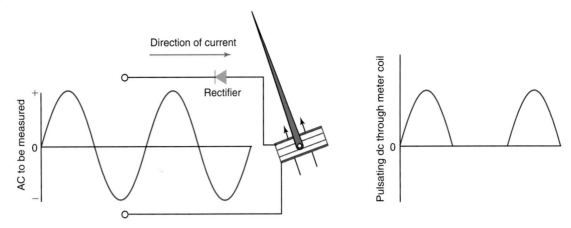

Fig. 15-6   Principle of rectification. The rectified current flows only in one direction.

efficients. They are generally precision, low-tolerance (± 2 percent or less) resistors.

A shunt extends the range of an ammeter by diverting most of the current around the meter movement. For example, a 100-μA movement is converted to a 1-mA ammeter by shunting 900 μA around the movement. As shown in Fig. 15-7, the 1 mA ($I_{circuit}$) splits at the junction of the meter movement and the shunt. Then 100 μA (I_m) goes through the movement and causes full-scale deflection of the pointer. The other 900 μA (I_s) goes through the shunt.

External shunt

Example 15-1

It is desired to make a 1-A ammeter from a 1-mA, 80-Ω movement. Determine the resistance and power dissipation of the required shunt.

Given: $I_m = 1$ mA
$\quad\quad\quad\quad R_m = 80\Omega$

Find: R_s, P_s

Known: $R_s = \dfrac{V_s}{I_s}, V_s = V_m, V_m = I_m R_m$
$\quad\quad\quad\quad I_s = I_{circuit} - I_m, P_s = I_s V_s$

Solution: $I_s = 1$ A $- 0.001$ A $= 0.999$ A
$\quad\quad\quad\quad V_m = 0.001$ A $\times 80\ \Omega = 0.08$ V
$\quad\quad\quad\quad R_s = \dfrac{0.08\ \text{V}}{0.999\ \text{A}} = 0.08\ \Omega$
$\quad\quad\quad\quad P_s = 0.999$ A $\times 0.08$ V
$\quad\quad\quad\quad\quad\ = 0.0799$ W
$\quad\quad\quad\quad\quad\ = 79.9$ mW

Answer: The shunt's resistance must be 0.08 Ω, and its power dissipation must be 79.9 mW.

Meter shunts are often enclosed in the same housing as the basic meter movement. However, for large dc currents (above 10 A) an external shunt is often used. *External shunts,* as shown in Fig. 15-8, are built to handle large currents without heating up or changing resistance. Therefore, they are quite large (5 to 15 cm long).

External shunts have both a current rating and a voltage rating. Current ratings ranging from 10 to several thousand amperes are common. Voltage ratings are usually either 50 or 100 mV. The voltage rating of a shunt specifies how much voltage the shunt drops when it is carrying its rated current. The rated voltage appears between the screw terminals shown on the shunts in Fig. 15-8 (the small screws in the shunt shown on top). When these shunts are used, leads are run from the screw terminals to a basic meter movement. Of course, the basic meter movement must have the same voltage rating as the external shunt. The heavier terminals provided on some shunts are bolted into

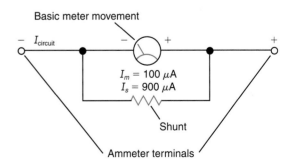

Fig. 15-7   Ammeter with shunt. The shunt extends the range of the basic meter movement.

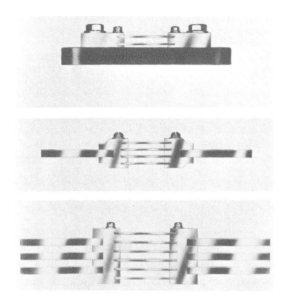

Fig. 15-8 External shunts. *(Courtesy of Triplett Corporation)*

the circuit in which the current is to be measured.

Current Transformer

When measuring large values of alternating current, a current transformer (Fig. 15-9) is used. The *current transformer* is essentially a *toroid-core transformer* that has no primary. When a current-carrying conductor is placed in the center of the current transformer, the conductor becomes the primary. An equivalent electric circuit is shown in Fig. 15-10. The current in the conductor is stepped down by transformer action.

A current transformer may have several secondaries. This allows it to be used with a variety of ammeters having different ratings.

Fig. 15-9 Current transformer. *(Courtesy of F. W. Bell, Inc.)*

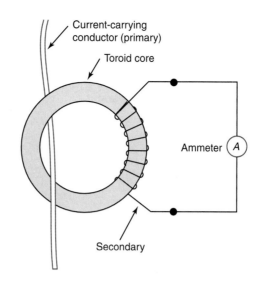

Fig. 15-10 Current transformer principle. The conductor in which current is being measured becomes a single-turn primary.

Multirange Ammeters

A multirange ammeter (or the ammeter section of a multimeter) is illustrated in Fig. 15-11. It consists of a meter movement, a number of shunts, and a rotary switch. The switch is a shorting (make-before-break) type. A *shorting switch* is necessary so that the meter movement is shunted even while it is switching ranges. With a nonshorting switch, the meter movement would carry all the current for the instant it takes to change ranges. This instant would be sufficient time to damage or burn out the meter movement.

Clamp-On Ammeter

A clamp-on ammeter (also called a clamp meter) is pictured in Fig. 15-12. This uses the

Current transformer

Toroid-core transformer

Shorting switch

Clamp-on ammeter

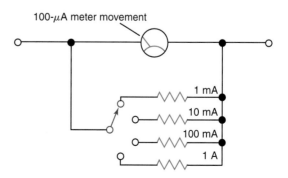

Fig. 15-11 Multirange ammeter. Switching to a smaller shunt increases the range.

RF currents

Thermocouple meter

Fig. 15-12 Clamp-on accessory for a DMM. *(Courtesy of Fluke Corporation).*

same principle as the current transformer-ammeter circuit of Fig. 15-10. However, with the clamp-on meter, the toroid core is made in two halves. The halves are hinged at one end and held together by a spring mechanism. The other end of the two halves can be separated to insert the conductor in which current is to be measured.

A clamp-on ammeter can have either a digital or an analog readout. The clamp-on mechanism of the clamp-on ammeter can be an accessory for use with a multimeter, or it can be a permanent part of a separate unit (clamp meter) that has its own readout, power source, and so on. Many clamp meters also measure other quantities (voltage, resistance, and so on) by using test leads rather than the clamp-on mechanism.

Figure 15-12 shows a clamp-on accessory being used to measure current in a large conductor without interrupting the circuit. The amount of current flow is indicated on the portable DMM to which the accessory is connected.

Thermocouple Meter

Radio-frequency currents are hard to measure with the meters discussed thus far. At these high frequencies, the inductive reactance of the meter coil is high and the capacitive reactance of the capacitance between the turns of the coil is low. The combination of these two reactances make some meters useless at higher frequencies. However, the thermocouple meter avoids the reactance problem by isolating the basic meter movement from the RF currents.

The *thermocouple meter,* shown in Fig. 15-13, uses a d'Arsonval meter movement connected to a thermocouple. When current flows through the meter, the resistive element heats up and increases the temperature of the thermocouple junction.

YOU MAY RECALL . . . that a thermocouple is a device that converts heat energy into electric energy. Therefore, the heated thermocouple produces a current in the meter-movement circuit. The higher the current through the resistive element, the hotter the thermocouple gets. The hotter the thermocouple, the higher the voltage it produces and the greater the current through the d'Arsonval meter movement.

■ TEST

Answer the following questions.

18. What type of meter movement is used in a rectifier-type meter?
19. What is a rectifier?

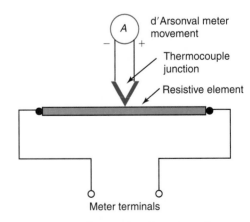

Fig. 15-13 Thermocouple meter. The current being measured does not flow through the meter movement.

20. List two desirable characteristics of a shunt resistor.
21. Refer to Fig. 15-11. How much current is carried by the shunt for the 10-mA range?
22. The meter movement shown in Fig. 15-11 has 1200 Ω of resistance.
    a. What is the resistance of the shunt for the 1-mA range?
    b. What is the power used by the shunt?
23. What type of rotary switch is used in a multirange ammeter?
24. Should the internal resistance of an ammeter be as low or as high as possible?
25. When are external shunts used?
26. How are external shunts rated?
27. What is a current transformer?
28. Does the use of a clamp-on meter require interruption of the circuit in which it is used?
29. What principle is used in the clamp-on ammeter?
30. What are thermocouple meters used for?
31. A thermocouple converts _____ energy into _____ energy.

15-4 Analog Voltmeters

A voltmeter consists of a basic meter movement and a *multiplier* resistor connected in a series circuit (Fig. 15-14). The resistance of the multiplier can be determined by using *series circuit relationships* and Ohm's law because the current through the multiplier (I_{mt}) is the same as the current through the meter movement (I_m), and the multiplier voltage (V_{mt}) plus the meter-movement voltage (V_m) equals the voltmeter voltage (V_t).

Like shunts, multiplier resistors have low tolerances and temperature coefficients. Unlike shunts, multipliers have very high resistances.

Example 15-2

What value multiplier is needed to make a 5-V voltmeter from a 1-mA, 100-Ω meter movement?

Given: $I_m = 1$ mA, $R_m = 100$ Ω
Find: R_{mt}, P_{mt}

Known: $P_{mt} = I_{mt}V_{mt}$, $R_{mt} = \dfrac{V_{mt}}{I_{mt}}$,
   $I_{mt} = I_m$, $V_{mt} = V_t - V_m$,
   $V_m = I_m R_m$

Solution: $V_m = 0.001$ A $\times$ 100 Ω $= 0.1$ V
   $V_{mt} = 5$ V $- 0.1$ V $= 4.9$ V
   $R_{mt} = \dfrac{4.9 \text{ V}}{0.001 \text{ A}} = 4900$ Ω
   $P_{mt} = 0.001$ A $\times 4.9$ V
   $= 0.0049$ W $= 4.9$ mW

Answer: The multiplier's resistance must be 4900 Ω, and its power rating must be greater than 4.9 mW.

A multirange voltmeter, typical of those used in the voltmeter section of a VOM, is shown in Fig. 15-15. The selector switch in Fig. 15-15 is the nonshorting type. When the meter is switching ranges, the voltmeter is momentarily opened. A shorting switch would not be appropriate for a voltmeter. It would momentarily connect two multipliers in parallel. The total resistance of the paralleled multipliers would be less than that of the lowest-value multiplier. This could cause the meter movement to be overloaded.

Multiplier

Series-circuit relationship

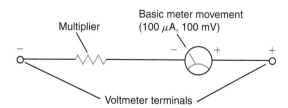

Fig. 15-14 The voltmeter multiplier extends the voltage range of the basic meter movement.

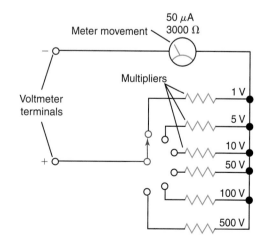

Fig. 15-15 Multirange voltmeter. Switching to a larger multiplier increases the range.

Voltmeter Rating

Input resistance

The total internal resistance of a voltmeter is an important rating of the voltmeter. The internal resistance of a voltmeter is called its *input resistance*. For a multirange meter like the one in Fig. 15-15, the input resistance is different for each range. Rather than an input resistance for each range being specified, one general rating for all ranges is used. This rating is specified in *ohms per volt*. The *ohms-per-volt rating* indicates the input resistance for each volt of that range. For example, a 1000-Ω/V voltmeter would have 1000 Ω of input resistance on the 1-V range. On the 10-V range, it would have 10,000 Ω of input resistance. The input resistance of a voltmeter on a given range is found by multiplying the range by the ohms-per-volt rating. The ohms-per-volt rating of a voltmeter is often referred to as its *sensitivity*.

Ohms per volt

Ohms-per-volt rating

Sensitivity

Example 15-3

What is the input resistance of a 10,000-Ω/V voltmeter on the 20-V range?

Given: 10,000-Ω/V voltmeter
Find: Input resistance on 20-V range
Known: Input resistance = sensitivity $\times$ range
Solution:
Input resistance = $(10,000\Omega/V) \times (20\ V)$
= 200,000 Ω
= 200 kΩ
Answer: The input resistance on the 20-V range is 200 kΩ.

The sensitivity (ohms-per-volt rating) of a voltmeter is determined by the full-scale current of the meter movement. Mathematically, the relationship is

$$\text{Sensitivity} = \frac{1}{\text{full-scale current}}$$

Therefore, the sensitivity of the voltmeter of Fig. 15-15 is

$$\text{Sensitivity} = \frac{1}{0.00005\ A} = 20,000\ \Omega/V$$

High internal resistances

Significant meter loading

Notice that voltmeters have very *high internal resistances* and ammeters have very low internal resistances. A voltmeter should not

significantly change the voltage distribution (or the load) of the circuit in which it is used. Therefore, its internal resistance must be high relative to the resistance of the load.

DMMs have the same input resistance for all voltage ranges. On the lower ranges, they usually have higher input resistance than the VOM. However, on the higher ranges, the opposite is often true. For example, a 100-kΩ/V VOM on the 500-V range has 50 MΩ of input resistance. This exceeds the input resistance of most DMMs.

Often, ac voltmeters have much lower sensitivity than dc voltmeters do. For example, a multimeter may have a 20-kΩ/V rating on the dc ranges and only a 5-kΩ/V rating on the ac ranges.

■ TEST

Answer the following questions.

32. True or false. The resistor used to extend the range of a voltmeter is called a multiplier.
33. True or false. Ammeters have higher internal resistance than voltmeters do.
34. True or false. The internal resistance of the DMM is the same on all dc voltage ranges.
35. True or false. The ac voltage ranges of a multimeter usually have more internal resistance than the comparable dc voltage ranges.
36. What is the sensitivity of a voltmeter which uses a 25-μA meter movement?
37. A 150-μA, 200-Ω meter movement is used in a 50-V voltmeter. What is the power dissipation and the resistance of the multiplier?
38. What is the internal resistance of a 250-V voltmeter which uses a 50-μA meter movement?
39. Multirange voltmeters use a _____ range switch.

15-5 Meter Loading

Whenever a meter is connected to a circuit, the circuit currents (and often the circuit voltages) change. This is what is meant by *meter loading*. *Significant meter loading* occurs when the current or voltage changes exceed those that might be caused by component tolerances.

In general, ac meters tend to be more prone to load a circuit than dc meters are. This is because ac voltmeters generally have less input resistance than dc voltmeters with the same meter-movement rating have.

Meter loading in ac circuits is also caused by the *internal reactances* of the meter movement, rectifier, leads, and so forth. Of course, these reactances vary with frequency. Therefore, ac meters have *frequency ratings* in addition to all the ratings that dc meters have. The frequency rating tells the range of frequency over which the meter is accurate. It is important that ac meters be used only within their frequency ratings.

Ammeter Loading

Ammeter loading in a dc circuit is illustrated in Fig. 15-16. In Fig. 15-16(*a*), the nominal current in the circuit is

$$I = \frac{1.5 \text{ V}}{1500 \text{ } \Omega} = 0.001 \text{ A} = 1 \text{ mA}$$

If the resistor is on either end of its tolerance, the current will be only 1 percent more or less (0.99 to 1.01 mA). In Fig. 15-16(*b*), an ammeter has been inserted into the circuit. The measured current, if both meter and resistor were completely accurate, would be 0.95 mA. The meter has loaded the circuit.

Significant ammeter loading does not occur very often. If the resistor in Fig. 15-16 had 10 percent tolerance, the loading would not be noticed. Also, many 1-mA ammeters have less

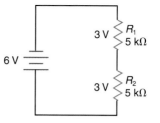

(*a*) Voltages in original circuit

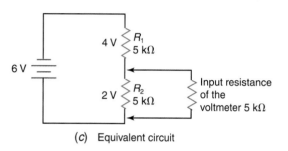

(*b*) Measured voltages

(*c*) Equivalent circuit

Fig. 15-17 Voltmeter loading.

than 85 Ω of internal resistance. When ammeter loading does occur, it is usually in a *low-voltage, low-resistance circuit.*

Voltmeter Loading

Voltmeter loading is much more common than ammeter loading. It can occur in series and series-parallel circuits. It does not occur in parallel circuits because the voltmeter is connected across the power source. That is, it is in parallel with all other parts of the circuit. In general, voltmeter loading occurs in *high-resistance circuits,* such as those found in many electronic devices.

The circuits in Fig. 15-17 illustrate voltmeter loading. In Fig. 15-17(*a*), the source voltage is split into equal parts by the equal resistors. However, the voltmeter in Fig. 15-17(*b*) indicates only 2 V across R_2. This is an obvious case of voltmeter loading. The input resistance of the voltmeter in Fig. 15-17(*b*) (on the 5-V range) can be calculated as follows:

$$\text{Input resistance} = \text{sensitivity} \times \text{range}$$
$$= 1000 \text{ } \Omega/\text{V} \times 5 \text{ V}$$
$$= 5000 \text{ } \Omega = 5 \text{ k}\Omega$$

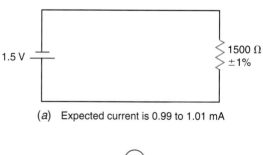

(*a*) Expected current is 0.99 to 1.01 mA

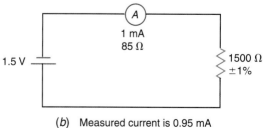

(*b*) Measured current is 0.95 mA

Fig. 15-16 Ammeter loading.

Internal reactances

Frequency ratings

Ammeter loading

Low-voltage, low-resistance circuit

Voltmeter loading

High-resistance circuits

As shown in the equivalent circuit of Fig. 15-17(c), this 5 kΩ is in parallel with the 5 kΩ of R_2. These two 5-kΩ parallel resistances have a combined resistance of 2.5 kΩ. Therefore, R_1 drops twice as much voltage as R_2.

If the voltmeter shown in Fig. 15-17 were a 20-kΩ/V voltmeter, the loading would be minimal. The input resistance would be 100 kΩ, and the parallel resistance of R_2 and the meter would be 4.76 kΩ. Even a 5 percent resistor at R_2 could have a lower resistance than this.

Manufacturers of electric and electronic equipment often specify the voltages at various points in their circuits. Usually, they specify either the sensitivity or the input resistance of the meter used in measuring these voltages.

When using a meter different from the one specified, technicians must be aware of possible meter loading and must be sure that any *observed low voltages* are not due to meter loading rather than component failures. This can be done by analyzing the circuit diagram to determine the value of the resistances in the circuit. The voltmeter's input resistance should be 20 times greater than the resistance across which the voltage is measured. Under these conditions, the loading will change the resistance of the circuit less than 5 percent.

Example 15-4

Assume the voltage across R_2 in Fig. 15-17 was measured with a 2000-Ω/V voltmeter on the 5-V range. How much voltage would the meter indicate?

Given: Sensitivity = 2000 Ω/V
Range = 5 V

Find: Measured voltage across R_2

Known: Input resistance (R_1) = sensitivity × range
Voltmeter loading
Series-parallel circuits

Solution: $R_1 = 2000 \ \Omega/\text{V} \times 5 \ \text{V} = 10 \ \text{k}\Omega$

$$R_{1,2} = \frac{5 \ \text{k}\Omega \times 10 \ \text{k}\Omega}{5 \ \text{k}\Omega + 10 \ \text{k}\Omega} = 3.33 \ \text{k}\Omega$$

$$V_{R1,2} = \frac{6 \ \text{V} \times 3.33 \ \text{k}\Omega}{5 \ \text{k}\Omega + 3.33 \ \text{k}\Omega} = 2.4 \ \text{V}$$

Answer: The measured voltage is 2.4 V.

About ⬤▬ Electronics

Reef Relief
Coral reefs take thousands of years to grow and are home to fish and other aquatic life. They also protect islands from rising sea levels due to global warming. Recognizing these benefits, Dr. Thomas Goreau and the Global Reef Alliance are attempting to grow new reefs to replace damaged ones. To accomplish this, they place steel frames on the ocean floor and connect them by wires to solar panels floating on the surface of the water. A mild electric current from the panels runs inside the frame, causing limestone to form on the frame. Coral grows well on limestone. In fact, in this environment, the coral can grow faster than it can be ruined by sewage. This reef, photographed by architect Wolf Hilbertz, is being grown to help protect the Maldives, the earth's lowest country, from the encroaching ocean.

▪ TEST

Answer the following questions.

40. What is meter loading?
41. Which is more common, ammeter loading or voltmeter loading?
42. Is meter loading as much of a problem with ac meters as it is with dc meters?
43. Does meter loading in ac circuits depend on frequency?
44. A 27-kΩ resistor in a series circuit drops 8 V. Will significant voltmeter loading occur if the voltage is measured with
    a. A 20-kΩ/V voltmeter on the 10-V range
    b. A 50-kΩ/V voltmeter on the 10-V range
    c. A DMM with 10-MΩ input resistance

45. A 4.7-MΩ resistor in a complex electronic circuit has 450 V across it. Which will give a more accurate voltage reading, a 20-kΩ/V voltmeter on the 1000-V range or a DMM with 10-MΩ input resistance?

15-6 Analog Ohmmeters

A simple *ohmmeter* (Fig. 15-18) consists of a cell, meter movement, and two resistors. The rheostat (R_1 in Fig. 15-18) is the ohms-adjust control. Adjustment of this control compensates for changes in the voltage of the cell. When the ohmmeter terminals are short-circuited together as in Fig. 15-19(a), current flows through the meter movement. The ohmmeter is then measuring zero resistance. If R_1 is adjusted until the meter movement indicates full scale, then full scale indicates 0 Ω. This is why the ohmmeter scale of a VOM is *reverse-reading*.

When the ohmmeter of Fig. 15-19(a) is ohms-adjusted, the internal resistance is

$$R_{int} = \frac{V}{I_m} = \frac{1.5 \text{ V}}{0.0001 \text{ A}} = 15{,}000 \text{ } \Omega = 15 \text{ k}\Omega$$

This internal resistance includes the resistance of the meter movement and both resistors (R_1 and R_2). Suppose a 15-kΩ resistor is placed between the ohmmeter terminals, as in Fig. 15-19(b). Now the total resistance of the circuit is 30 kΩ ($R_T = R_{int} + R_x$). The current through the meter movement is cut in half because the resistance is doubled:

$$I_m = \frac{V}{R_T} = \frac{1.5 \text{ V}}{30{,}000 \text{ } \Omega}$$
$$= 0.00005 \text{ A} = 50 \text{ } \mu\text{A}$$

When R_x in Fig. 15-19 is replaced by a 45-kΩ resistor, the total resistance is 60 kΩ

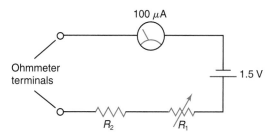

Fig. 15-18 Basic ohmmeter circuit. R_1 is adjusted to provide full-scale current when the terminals are shorted together.

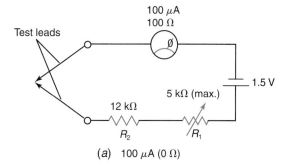

(a) 100 μA (0 Ω)

Ohmmeter

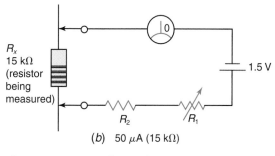

(b) 50 μA (15 kΩ)

Fig. 15-19 Measuring resistance.

Reverse-reading

(R_T = 15 kΩ + 45 kΩ). Now the meter current is

$$I_m = \frac{1.5 \text{ V}}{60{,}000 \text{ } \Omega}$$
$$= 0.000025 \text{ A} = 25 \text{ } \mu\text{A}$$

This is one-fourth of the full-scale current of the meter. Therefore, the quarter-scale point on the ohmmeter is 45 kΩ.

The above calculations illustrate how nonlinear an ohmmeter scale is. The first half of the scale covers 0 to 15 kΩ (Fig. 15-20). The next quarter of the scale covers 15 to 45 kΩ. And the last quarter has to cover the range from 45 kΩ to infinity.

The range of the ohmmeter circuit in Fig. 15-18 can be changed by either of two methods. First, the voltage of the cell *and* the resistance of R_2 can be increased. Second, the full-scale current of the movement can be increased by shunting, *and* the resistance of R_2 can be decreased. The first method increases the range of the ohmmeter, and the second method decreases the range.

Nonlinear ohmmeter scale

JOB TIP

Some field service positions entail both national and international travel.

■ TEST_____

Answer the following questions.

46. What components are used in a simple ohmmeter?

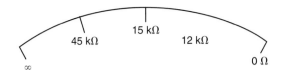

Fig. 15-20 Nonlinear ohmmeter scale.

47. What is the function of the rheostat in an ohmmeter?
48. What is the center-scale resistance of an ohmmeter that has 40 Ω of internal resistance?
49. What is the internal resistance of a properly ohms-adjusted ohmmeter that has a 3-mA meter movement and a 9-V battery?
50. What are two methods of changing the range of an ohmmeter?

15-7 Insulation Testers

Insulating materials, which must withstand high voltages, have extremely high resistances. They must also be able to stand the electric stress created by the high voltage. An ordinary ohmmeter cannot do an adequate job of checking the condition of insulation which is subjected to high voltages. Therefore, the insulation in some equipment, such as transformers and motors, is checked with *insulation testers,* or *Meggers.* These instruments test the insulation's resistance by applying a high voltage across the insulation and checking for a minute leakage current. Very often the test is made between ground (the frame or chassis of the equipment) and the wiring within the equipment. The insulation is tested at a voltage greater than the highest voltage to which the insulation is normally subjected. Engineering test procedures often specify a method for determining the test voltage. For example, certain equipment must be tested at twice the operating voltage plus 5000 V.

Because of the high voltage, insulation testers must be used with great care. Only the test leads supplied (or recommended) by the manufacturer should be used with an insulation tester. The leads should be checked regularly for signs of deterioration (checking and cracking). The manufacturer's directions and safety rules should be followed when using an insulation tester.

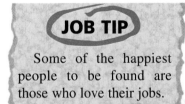

JOB TIP

Some of the happiest people to be found are those who love their jobs.

15-8 Wheatstone Bridges

One common use of the Wheatstone bridge (Fig. 15-21) is to measure resistance with great precision. Let us see how the bridge works for this application.

The meter movement used in the bridge is a *zero-center-scale meter.* When the current or voltage is zero, the meter pointer is at the center of the scale. The type of meter used is called a *galvanometer.* It can measure very low current flow in either direction.

The bridge circuit in Fig. 15-21 is balanced. When the bridge is balanced, no current flows through the meter. Therefore, there is no potential difference (voltage) between points *A* and *B*. The voltage across R_1 must equal the voltage across R_2, and, of course, V_{R_u} must equal V_{R_3}. Also notice that, when the bridge is balanced,

$$I_{R_1} = I_{R_u} \quad \text{and} \quad I_{R_2} = I_{R_3}$$

where I_{R_u} is the current through a resistor of unknown resistance. Because of these voltage and current relationships, a definite relationship exists between the resistances of the circuit. The ratio of R_1/R_u must be equal to the ratio of R_2/R_3. Therefore, for a *balanced bridge,* we can write

$$\frac{R_1}{R_u} = \frac{R_2}{R_3}$$

Rearranging the formula to solve for R_u yields

$$R_u = \frac{R_1}{R_2} \times R_3$$

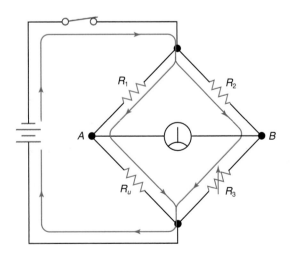

Fig. 15-21 Wheatstone bridge circuit (balanced). There is no voltage between points *A* and *B* when the bridge is balanced.

This formula can be used to measure an unknown resistance R_u. All we need to know is the ratio of R_1/R_2 and the value of R_3.

When a Wheatstone bridge is used, the resistance to be measured is connected into the circuit and power is applied. Then R_3 is adjusted to achieve a balance (zero current through the meter). Resistor R_3 is a precision, calibrated rheostat. Its value can be read from a dial attached to its shaft. Resistors R_1 and R_2 are precision resistors of known value. The value of R_u is then found by multiplying the dial reading of R_3 by the ratio R_1/R_2. The precision with which R_u can be measured is determined by the precision (tolerances) of the other three resistors.

When a bridge is not balanced, none of the voltage, current, and resistance relationships discussed above are valid. Unbalanced bridges can be analyzed using the techniques covered in Chap. 6.

Example 15-5

Refer to Fig. 15-21. Resistor $R_1 = 10,000$ Ω, $R_2 = 1000 \ \Omega$, and $R_3 = 250 \ \Omega$ when the bridge is balanced. What is the value of R_u?

Given: $R_1 = 10,000 \ \Omega$
$R_2 = 1000 \ \Omega$
$R_3 = 250 \ \Omega$

Find: R_u

Known: $R_u = \dfrac{R_1}{R_2} \times R_3$

Solution: $R_u = \dfrac{10,000}{1000} \times 250 = 10 \times 250$
$= 2500 \ \Omega$

Answer: The unknown resistance is 2500 Ω.

■ TEST

Answer the following questions.

51. True or false. The Wheatstone bridge uses a zero-center-scale meter.
52. True or false. No current flows anywhere in a balanced bridge.
53. Refer to Fig. 15-21. Which resistors must drop equal voltages when the bridge is balanced?
54. What determines the accuracy of a measurement made with a bridge?
55. Refer to Fig. 15-21. Assume $R_1 = 300 \ \Omega$, $R_2 = 600 \ \Omega$, and $R_3 = 2000 \ \Omega$ at balance. What is the value of R_u?

15-9 Wattmeters

An electrodynamometer movement (Fig. 15-5) makes a perfect wattmeter. The moving coil (called the *voltage coil*) is used to detect the magnitude of the circuit voltage. The stationary coils are referred to as the *current coils*. The circuit current is detected by the current coils, which are connected in series with the load (Fig. 15-22).

Wattmeter

Voltage coil

Current coils

The stationary (current) coils are wound with large-diameter wire. This keeps the resistance that is in series with the load as low as possible. The movable (voltage) coil is wound with thin wire to keep it as light as possible. Since the movable coil responds to voltage, it has a multiplier in series with it.

Although the wattmeter in Fig. 15-22 is measuring ac power, relative polarity must still be observed. Connecting the plus-minus ($\pm$) terminals of both the voltage coil and the current coils to the power source provides correct polarity. The other end of the current coils

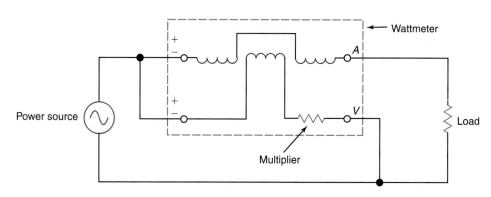

Fig. 15-22 Wattmeter connections. For clockwise torque, the instantaneous currents in the current coils and in the voltage coil must be in the same direction.

(marked *A*) goes to the load. Thus, all the load current flows through the current coils. The other end of the voltage coil (marked *V*) returns to the power source. This puts the full source voltage across the voltage coil-multiplier circuit.

Correct polarity — *Correct polarity* can also be achieved by connecting the plus-minus end of the voltage coil to the *A* end of the current coils. Referring to Fig. 15-22, you can see that either connection gives the same current direction in the voltage coil.

True power — The wattmeter indicates *true power* under all conditions. It automatically compensates for phase differences between the circuit voltage and current. How this happens can be understood by referring to Fig. 15-23. In Fig. 15-23(*a*), the voltage and current are in phase; that is, PF = 1. Therefore, the currents through the voltage coil and current coil change at the same instant. Thus, the magnetic fields of the two coils are opposing and creating a clockwise torque during the complete cycle. However, when the current and the voltage are out of phase [Fig. 15-23(*b*)], the situation changes. Now for a portion of the cycle, the magnetic fields of the two coils are aiding and develop a counterclockwise torque. This reduces the net clockwise torque and thus the power indicated by the meter. When the current and voltage are 90° out of phase [Fig. 15-23(*c*)], the clockwise and counterclockwise torques are exactly equal. The net torque is zero, and the wattmeter indicates zero power.

Wattmeters are rated for maximum current and voltage as well as maximum power. These ratings are necessary to protect the coils in the electrodynamometer movement. If the *voltage rating* is exceeded, the movable coil gets too hot. Exceeding the *current rating* of the meter overloads the stationary coils. When the power factor of the load is low, the current rating is exceeded long before the meter indicates full-scale power.

Voltage rating

Current rating

Measuring Three-Phase Power

Balanced load

Three-phase power

Two-wattmeter method

Balanced Load With a balanced three-phase load, the power in each of the three phases must be equal. Therefore, the total power of the three-phase system is equal to three times the power of any one phase. Consequently, a single wattmeter can be used to measure the power in

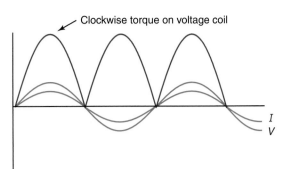

(a) In phase (PF = 1)

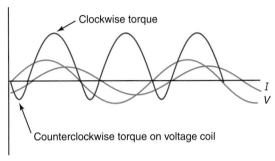

(b) 45° phase shift (PF = 0.707)

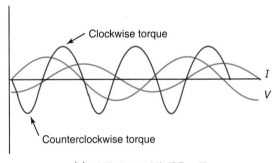

(c) 90° phase shift (PF = 0)

Fig. 15-23 How a wattmeter compensates for power factor.

any one of the phase loads (Fig. 15-24). The measured power is multiplied by 3 to give the total power of the system.

Notice in Fig. 15-24(*a*) that one must be able to connect the voltage coil of the wattmeter to the neutral, or star, point. Likewise, in Fig. 15-24(*b*), one has to break the junction of two of the phase loads to measure the current in a phase leg. In many three-phase loads, these junction points are not readily accessible. In such cases, the power can be measured by using two wattmeters, as described in the next section.

Unbalanced Load The *two-wattmeter* method of measuring *three-phase power* is shown in Fig. 15-25. It works equally well with

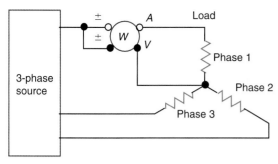

(a) Wye-connected load

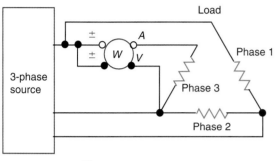

(b) Delta-connected load

Fig. 15-24 Measuring power with balanced loads. Only one wattmeter is needed when the phase loads are equal.

wye or delta loads and with balanced or *unbalanced loads*.

Notice in Fig. 15-25 that the wattmeters are monitoring line currents and line voltages. Yet, the *algebraic sum* of the two meter readings equals the total power of the system. (Proof of this statement requires phasor analysis beyond the scope of this text.) The phrase "algebraic sum of the two meter readings" needs some explanation. Normally, the meters are connected with the polarities shown in Fig. 15-25(a). This connection usually results in clockwise deflection of both meters. If it does, the algebraic sum of the two readings is obtained by simply adding the two readings. For example, if one meter indicates 2800 W and the other indicates 1600 W, the power of the three-phase system is 4400 W (2800 W + 1600 W).

Sometimes, when the loads are extremely unbalanced, the polarities in Fig. 15-25(a) result in a *reverse deflection* of one of the meters. In this case, the polarity of the reverse-deflecting meter must be changed. This can be done by reversing the connections to the voltage coil, as shown in Fig. 15-25(b). Now the algebraic sum of the two meter readings is

found by subtracting the smaller reading from the larger reading. Let us look at an example. Suppose we connect the meters as shown in Fig. 15-25(a) and W_1 deflects counterclockwise. Therefore, we reverse the voltage coil connection on W_1. This yields the connections shown in Fig. 15-25(b). Now suppose W_1 indicates 300 W and W_2 indicates 2600 W. The total power of the system is 2300 W (2600 W − 300 W).

■ TEST

Answer the following questions.

56. Why does a wattmeter use an electrodynamometer movement?
57. Which coil of a wattmeter is wound with the larger-diameter wire?
58. Why does an ac wattmeter require polarity markings?
59. Does a wattmeter indicate true power or apparent power?
60. A wattmeter is connected to a 240-V circuit that draws 10 A. The power factor is 0.5. How much power does the meter indicate?
61. A wattmeter in a balanced three-phase circuit indicates 1265 W. What is the total power of the circuit?

Unbalanced loads

Algebraic sum

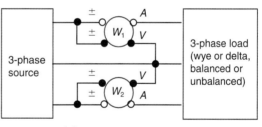

(a) Add the meter readings

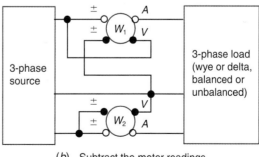

(b) Subtract the meter readings

Fig. 15-25 Power measurement with unbalanced or balanced loads. The connections shown in (b) are needed with some unbalanced loads.

Reverse deflection

62. Two wattmeters are connected to a three-phase system. Both are connected for normal polarity. One meter indicates 2600 W, and the other meter indicates 2200 W. What is the power of the three-phase system?

63. If the 2200-W meter in question 62 required reverse polarity connections, what would the power be?

64. How is it possible to exceed the current rating of a wattmeter without causing the pointer to deflect beyond full scale?

Answer the following questions.

65. True or false. Digital frequency meters are limited to measuring power and audio frequencies.

66. True or false. The digital frequency meter counts the cycles produced by the unknown frequency during a precisely controlled period of time.

67. True or false. A vibrating-reed meter usually covers the entire power and audio-frequency range.

15-10 Frequency Meters

Frequency can be measured with a variety of electric and electronic devices. Electronically, frequency can be measured with such devices as digital frequency counters and heterodyne *frequency meters*. These devices are capable of measuring a wide range of frequencies extending to hundreds of megahertz.

Electric frequency meters can only measure a narrow range of frequencies in the *power frequency range*. A *vibrating-reed meter,* a common electric frequency meter, may only measure frequencies between 58 and 62 Hz.

A *digital frequency meter* measures an unknown frequency by counting the number of cycles the frequency produces in a precisely controlled period of time. The counter circuit is incremented one count for each cycle. At the end of the time period, the final count, which represents the frequency, is displayed by the digital readout. For the next sampling of the unknown frequency, the counter is cleared, the time period is started over, and the final count in the counter is again displayed. If the measured frequency is stable, the readout does not change from sample to sample. Because the range switch selects the time period and places the decimal point in the readout, the indicated frequency is in the units specified by the range switch.

When the time period is 1 ms, the readout is in kilohertz and the range switch indicates kilohertz. For example, if the count at the end of the 1-ms period is 100, the unknown (measured) frequency must be 100 kHz because 100 counts per millisecond is equal to 100,000 counts per second.

15-11 Measuring Impedance

Impedance and reactance must be measured by some *indirect technique*. Of course, both impedance and reactance are functions of frequency, and so they are measured and specified at some stated frequency. In this section we will concern ourselves only with *measuring impedance* (and reactance) at power and audio frequencies.

Current Voltage Method

One of the simplest ways to measure impedance is to measure the current and voltage and then use Ohm's law. This method is illustrated in Fig. 15-26. It works well at any low frequency, but it is especially well suited to measuring impedance at 60 Hz. The current meter must be accurate at the frequency at which the impedance is being measured. Also, the current meter must have a very low current range if high impedances are being measured.

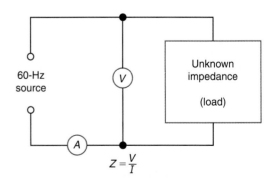

$$Z = \frac{V}{I}$$

Fig. 15-26 Determining an unknown impedance. After the current and voltage are measured, Ohm's law is used to calculate impedance.

Indirect technique

Measuring impedance

Frequency meters

Power frequency range

Vibrating-reed meter

Digital frequency meter

Current-voltage method

Equivalent-resistance method

Equivalent-Resistance Method

A more versatile technique of measuring impedance is shown in Fig. 15-27(*a*). When this technique is used, the source (generator) is first set to the frequency at which the impedance is to be measured. Any convenient output-voltage level (within the rating of the load) is selected. Next, the voltage across the load is noted. Then the switch is moved to its upper position and the voltage across the resistor is measured. If the voltage across *R* is less than the voltage across the load, the value of *R* is increased. If the reverse is true, the value of *R* is decreased. The switch is then returned to its original position to measure the new value of the load voltage. Then the switch is again thrown to the up position and the resistor and load voltages are again compared. If they are not equal, *R* is again adjusted. This process is repeated until the measured voltages across *R* and the load are equal. When these voltages are equal, the impedance and resistance in this series circuit must also be equal. If *R* is a cal-

ibrated rheostat, the value of the impedance of the load can be read from the dial. If not, then the source voltage is removed and the value of *R* (which equals *Z*) can be measured.

When two identical voltmeters are available, the circuit in Fig. 15-27(*b*) is easier and faster to use. Once the generator's output voltage and frequency have been set, *R* is adjusted until the two meters indicate equal voltages. Then the value of *R* is determined.

◼ TEST_____

Answer the following questions.

68. What is the advantage of using two voltmeters when determining impedance by the equivalent-resistance method?
69. A speaker and a 15-Ω resistor are connected in series to a generator. They each develop 1 V at 2000 Hz. What is the impedance of the speaker at 2000 Hz?

15-12 Measuring Inductance and Capacitance

Inductance and capacitance can be measured in a variety of ways. Two common methods are presented in the following discussions.

Digital Method

Digital-voltmeter circuitry (Sec. 15-1) can be used to measure inductance indirectly by measuring the induced cemf under controlled conditions.

 . . . that in Chap. 11, 1 H of inductance was shown to produce 1 V of cemf when the current changed at a rate of 1 A/s. Therefore, in general terms, we can write

$$L = \frac{V}{I/t}$$

Now, suppose the digital inductance meter forces a current which is rising at a rate of 1 *A/s* through an unknown inductor. Further, suppose the digital-voltmeter circuitry measures 2 V across the inductor. By applying the above formula, we can show that this 2 V represents 2 H of inductance. Thus,

$$L = \frac{2\ \text{V}}{(1\ \text{A})/(1\ \text{s})} = \frac{2\ \text{Vs}}{1\ \text{A}} = 2\ \text{H}$$

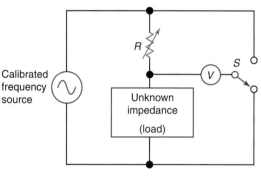

(*a*) With a single voltmeter

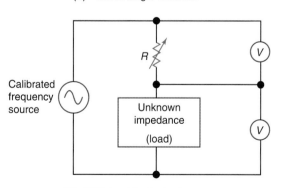

(*b*) With two identical voltmeters

Fig. 15-27 Impedance-measuring technique. When the voltmeter readings in (*b*) are equal, the resistance of *R* will have the same value as the unknown impedance.

In an actual digital inductance meter, the level of current must be kept very small. Then the ohmic voltage drop (due to the resistance of the wire in the inductor) is very small compared with the cemf produced by the inductor.

A digital capacitance meter indirectly measures capacitance by using the time it takes a capacitor to charge to a specified voltage under controlled conditions.

By combining two fundamental formulas, $I = Q/t$ and $C = Q/V$, we can show the relationship between capacitance, voltage, current, and time. Rearranging $I = Q/t$ to $Q = It$ and then substituting It for Q changes $C = Q/V$ into $C = It/V$. This formula shows that C and t are directly proportional when I and V are held constant. The current I is held constant by charging the capacitor from a constant current source. The voltage V is held constant by letting the capacitor charge until the capacitor's voltage is just equal to that of a stable reference voltage. Thus, the time required to charge an unknown capacitor is a direct function of the unknown capacitance. Section 15-1 discussed how to measure time digitally by counting the cycles produced by a stable reference frequency.

Capacitor bridge

When the current, voltage, and reference frequency are selected properly, the digital readout can be in microfarads, or picofarads. When the constant current is 10 mA, the reference voltage is 1 V, and the reference frequency is 10 kHz, the readout is in microfarads. For example, with these constants, the readout would be 10 if it took the capacitor 1 ms to charge (10 counts per millisecond = 10,000 counts per second = 10 kHz). Solving for C with these values of t, V, and I yields

$$C = \frac{It}{V} = \frac{(10 \text{ mA}) \times (1 \text{ ms})}{1 \text{ V}} = 10 \ \mu F$$

Thus, the readout of 10 is in microfarads.

AC-Bridge Method

AC bridge

The value of a capacitor or an inductor can also be determined with an ac bridge. The ac-bridge circuit in Fig. 15-28 is used to measure the value of the unknown capacitor C_u. To use the bridge, R_1 and R_3 are adjusted for the lowest possible indication on the voltmeter. There will be some interaction between R_1 and R_3, and so one and then the other resistor must be

$$C_u = \left(\frac{C_1}{R_2}\right) R_1$$

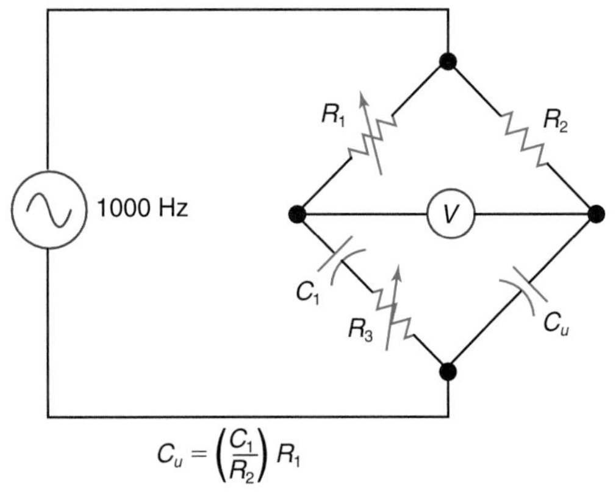

Fig. 15-28 Measuring C with a bridge. The variable resistors are adjusted to obtain the lowest possible voltmeter reading.

adjusted several times to achieve minimum voltmeter reading. When properly adjusted, the meter should indicate essentially 0 V.

The purpose of R_3 is to make the resistive losses in C_1 equal to the resistive losses in C_u. If C_u has as high a quality factor as C_1, then R_3 will be approximately 0 Ω. In fact, R_3 is needed in the bridge only when low-quality capacitors are to be measured. For our analysis of the *capacitor bridge*, we will assume that high-quality capacitors are being measured. Thus, R_3 is 0 Ω.

When R_1 and R_3 are adjusted for minimum voltmeter indication, the bridge is balanced. Under balanced conditions, the ratio of R_1 to X_{C_1} is equal to the ratio of R_2 to X_{C_u}. Mathematically, we can write

$$\frac{R_1}{X_{C_1}} = \frac{R_2}{X_{C_u}}$$

Since capacitive reactance is inversely proportional to capacitance.

$$R_1 C_1 = R_2 C_u$$

or

$$C_u = \frac{C_1}{R_2} \times R_1$$

In the above formula, C_u will have the same unit as $C_1 \cdot R_1$ and R_2 must also have the same unit of resistance.

Example 15-6

In Fig. 15-28, $R_2 = 2000$ Ω, and $C_1 = 0.002$ μF. If the balance occurs when R_1 is adjusted to 1700 Ω, what is the capacitance of C_u?

Given: $R_2 = 2000\ \Omega$
$C_1 = 0.002\ \mu F$
$R_1 = 1700\ \Omega$

Find: C_u

Known: $C_u = \dfrac{C_1}{R_2} \times R_1$

Solution: $C_u = \dfrac{0.002 \times 10^{-6}\ \text{F}}{2000\ \Omega} \times 1700\ \Omega$

$= 0.0017 \times 10^{-6}\ \text{F}$

$= 0.0017\ \mu F$

Answer: The unknown capacitance is

The *inductance bridge* in Fig. 15-29 works on the same principles as the capacitance bridge of Fig. 15-28. The formula $L_u = CR_2R_1$ can be verified by applying the rationale used to derive the capacitance-bridge formula.

Notice in Fig. 15-29 that R_3 is in parallel with C. (In some inductance bridges, it is in series.) Again, R_3 is adjusted to provide the same amount of resistance loss in the capacitor as there is in the inductor. Since R_3 is in parallel with C, the resistive losses increase as R_3 is decreased. If the unknown inductance has a high quality factor, the value of R_3 will be very high.

The resistance of the inductor (R_{ind}) can also be determined from the bridge circuit of Fig. 15-29 by the formula

$$R_{\text{ind}} = \frac{R_2R_1}{R_3}$$

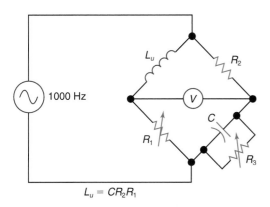

$L_u = CR_2R_1$

Fig. 15-29 Measuring L with a bridge. The bridge is balanced when further adjustments of R_1 and R_2 cannot reduce the voltage indicated by the voltmeter.

Once the resistance and inductance of the inductor are known, it is easy to determine the quality of the inductor. (The generator frequency is known, and $Q = X_L/R = 6.28\,f\,L/R$.) Many inductance bridges have the dial attached to R_3 calibrated directly in Q values.

Example 15-7

Suppose the bridge in Fig. 15-29 balances when $R_1 = 2\ \text{k}\Omega$, $R_2 = 5\ \text{k}\Omega$, $R_3 = 500\ \text{k}\Omega$, and $C = 0.01\ \mu F$. What are the inductance and quality of the inductor?

Given: $R_1 = 2\ \text{k}\Omega$
$R_2 = 5\ \text{k}\Omega$
$R_3 = 500\ \text{k}\Omega$
$C = 0.01\ \mu F$
$f = 1000\ \text{Hz}$

Find: L and Q
Known: $L_u = CR_2R_1$

$$Q = \frac{X_L}{R}$$

$$X_L = 6.28fL$$

$$R_{\text{ind}} = \frac{R_2R_1}{R_3}$$

Inductance bridge

Solution:

$L_u = 0.01 \times 10^{-6} \times 5 \times 10^3 \times 2 \times 10^3$
$= 0.1\ \text{H} = 100\ \text{mH}$

$R_{\text{ind}} = \dfrac{5 \times 10^3 \times 2 \times 10^3}{500 \times 10^3} = 20\ \Omega$

$X_L = 6.28 \times 1000 \times 0.1$
$= 628\ \Omega$

$Q = \dfrac{628}{20} = 31.4$

Answer: The inductance is 100 mH, and the quality is 31.4

◼ TEST_____

Answer the following questions.

70. Name two methods of measuring capacitance and inductance.
71. What quantity is measured by a digital inductance meter to determine the unknown inductor's value?
72. Refer to Fig. 15-28 and find C_u when $R_2 = 2\ \text{k}\Omega$, $R_1 = 4\ \text{k}\Omega$, and $C = 1200\ \text{pF}$.
73. Refer to Fig. 15-29 and find L_u when $R_2 = 2\ \text{k}\Omega$, $R_1 = 4\ \text{k}\Omega$, and $C = 1200\ \text{pF}$.
74. Determine Q for question 73 if $R_3 = 2\ \text{M}\Omega$.

Summary

1. The d'Arsonval meter movement is commonly used in voltmeters, ammeters, and ohmmeters. It responds only to direct current. It is used in rectifier-type instruments to measure alternating current and voltage.

2. Electrodynamometer meter movements use stationary and moving coils to develop interacting magnetic fields. They respond to alternating or direct current and are used in wattmeters. Electrodynamometer meters have low sensitivity and high accuracy.

3. A rectifier allows current to flow in only one direction. It converts alternating current to pulsating direct current.

4. Digital voltmeters measure the time required for a capacitor to charge to the value of the unknown voltage.

5. The iron-vane meter movement has no moving coil or permanent magnet.

6. The iron-vane meter movement responds to both alternating and direct current. It has a nonlinear scale and low sensitivity.

7. Basic meter movements are rated for voltage, current, and resistance.

8. Shunts and multipliers can be used with d'Arsonval, rectifier-type, iron-vane, and electrodynamometer meters.

9. Values for shunts and multipliers can be calculated if the meter-movement ratings and the ranges are known.

10. Ammeters have very low internal resistance.

11. Voltmeters have very high internal resistance.

12. External shunts are rated in amperes and millivolts.

13. Shunts and multipliers are precision resistors having a low temperature coefficient.

14. Shunts are in parallel with the meter movement. They extend the range of ammeters.

15. Multipliers are in series with the meter movement. They extend the range of voltmeters.

16. Current transformers are used to extend the range of ac ammeters.

17. Clamp-on ammeters can measure current in an ac circuit without interrupting the circuit.

18. Thermocouple meters are used to measure high-frequency currents. They use a d'Arsonval meter movement.

19. VOMs have an ohms-per-volt rating. This rating is also called the VOM's sensitivity.

20. Input resistance = sensitivity × range

21. Ohms per volt = $\dfrac{1}{\text{full-scale current}}$

22. DMMs have an input-resistance rating which is independent of the range.

23. Meter loading causes changes in circuit currents and voltages when measurements are made.

24. Ohmmeters have a power source, rheostat, meter movement, and fixed resistor.

25. Ohmmeter scales are nonlinear and often reverse-reading.

26. Ohmmeter ranges are changed by switching voltages or shunts.

27. Insulation testers (Meggers) use high voltages to measure very high resistances (insulation).

28. Wheatstone bridges use a galvanometer which indicates zero when a bridge is balanced.

29. A bridge is balanced when the ratio of rheostat to unknown resistance equals the ratio of the two known resistors.

30. The accuracy of a bridge is determined by the tolerances of the resistors used in it.

31. In a wattmeter the moving coil responds to voltage and the stationary coil responds to current.

32. Wattmeters are polarized instruments. They indicate true power. They have current, voltage, and power ratings.

33. A single wattmeter can measure power in a balanced three-phase load.

34. Two wattmeters are required to measure power in an unbalanced three-phase load. The wattmeter readings are algebraically added.

35. A vibrating-reed meter measures only a narrow range of frequencies.

36. Impedance can be measured by the current-voltage method or by the equivalent-resistance method.
37. Inductance and capacitance can be measured by the digital method or by the ac-bridge method.
38. The digital inductance meter measures the cemf of the inductor.
39. The digital capacitor meter measures the time required to charge the capacitor.

Chapter Review Questions

For questions 15-1 to 15-16, supply the missing word or phrase in each statement.

15-1. The _____ meter movement is polarized.

15-2. The _____ meter movement does not use a moving coil.

15-3. The analog ohmmeter uses a _____ to compensate for small changes in cell voltage.

15-4. A _____ resistor extends the range of a voltmeter.

15-5. A _____ resistor extends the range of an ammeter.

15-6. A _____ switch is used to change ranges in a multirange ammeter.

15-7. The internal resistance of the _____ is different on each voltage range.

15-8. A vibrating-reed meter is used to measure _____.

15-9. A _____ can measure alternating current without interrupting the circuit.

15-10. The moving coil in a wattmeter responds to _____.

15-11. Meter loading in an ac circuit is _____-dependent.

15-12. The DMM measures _____ when measuring an unknown voltage.

15-13. The DMM measures _____ when measuring an unknown resistance.

15-14. The digital inductance meter measures _____ when measuring an unknown inductance.

15-15. The digital capacitance meter measures _____ when measuring an unknown capacitance.

15-16. A _____ converts alternating to direct current.

For questions 15-17 to 15-24, determine whether each statement is true or false.

15-17. A 100-μA meter movement produces a more sensitive voltmeter than a 50-μA meter movement.

15-18. Analog ohmmeters have a nonlinear scale.

15-19. The range of an ohmmeter can be changed only by shunting the meter movement.

15-20. Bridge balance is obtained when the null indicator (voltmeter) indicates its lowest voltage.

15-21. Insulation testers measure resistance under high-voltage conditions.

15-22. When a Wheatstone bridge is in balance, no current flows in any part of the circuit.

15-23. Ammeter loading occurs more often than voltmeter loading.

15-24. Two wattmeters are needed to measure ac power in a balanced three-phase circuit.

For questions 15-25 to 15-29, choose the letter that best completes each statement.

15-25. Which of these meter movements is the most sensitive?
  a. d'Arsonval
  b. Electrodynamometer
  c. Iron-vane

15-26. Which of these meter movements cannot directly measure alternating current?
  a. d'Arsonval
  b. Electrodynamometer
  c. Iron-vane

15-27. Which of these meter movements is used to measure power?
  a. d'Arsonval
  b. Electrodynamometer
  c. Iron-vane

15-28. Which of these meter movements uses a permanent magnet?
  a. d'Arsonval
  b. Electrodynamometer
  c. Iron-vane

15-29. Which of these meter movements uses a moving coil and a fixed coil?
  a. d'Arsonval
  b. Electrodynamometer
  c. Iron-vane

Answer the following questions.

15-30. What is the input resistance of a 20-kΩ/V voltmeter on the 60-V range?

15-31. Would the meter in question 15-30 cause significant loading if used to measure the voltage across a 470-kΩ resistor in a series circuit?

15-32. What is the value of the internal resistance of a meter movement which is rated 50 μA, 200 mV?

15-33. A 100-mA meter indicates 36 mA. Its accuracy is ±2 percent. What are the minimum and maximum values of the current?

15-34. You need to convert a 50-μA, 1000-Ω meter movement to a 1-mA meter. What value of resistor do you need?

15-35. You need to convert the meter movement in question 15-34 to a 5-V meter. What value of resistor do you need?

15-36. What is the ohms-per-volt rating of a voltmeter which uses a 100-μA meter movement?

15-37. List two methods of measuring impedance.

15-38. A digital ohmmeter uses a 10-μA constant current source. How much resistance would be represented by a readout of 2.50?

15-39. A 100-μA constant current source charges a capacitor to 2 V in 4 ms. Determine the capacitance.

15-40. Refer to Fig. 15-21 and determine the value of R_u when $R_1 = 5$ kΩ, $R_2 = 10$ kΩ, and $R_3 = 9.4$ kΩ.

Critical Thinking Questions

15-1. Two 10-kΩ, ±10 percent series resistors are connected to a 10-V source. The voltage across one of the resistors is measured with a 3-kΩ/V voltmeter on the 5-V range. What will the meter indicate under worst-case conditions?

15-2. Why is the iron-vane meter movement so nonlinear?

15-3. Why are shunts and multipliers made from materials with very low temperature coefficients?

15-4. Why might the ac voltage ranges of a VOM have lower sensitivity than the dc voltage ranges?

15-5. An ohmmeter uses a 200-μA meter movement, a 9-V battery, and appropriate values of a series resistor and rheostat. How much resistance is represented by 15 percent deflection of the meter movement?

15-6. How long will it take a 3.3-μF capacitor to charge to 5 V when connected to a 2-mA constant-current source?

15-7. What are the characteristics of a cell or battery that requires an analog ohmmeter to have an ohms-adjust rheostat?

15-8. A precision 4.7-MΩ resistor is series connected with a precision 6.8-MΩ resistor to a 90-V source. Determine the input resistance of a DMM that indicates 46.73 V across the 6.8-MΩ resistor.

15-9. Determine the constant current source needed for a capacitance meter to readout in picofarads if the reference voltage is 10 V and the reference frequency is 500 kHz.

15-10. Two 100-kΩ, ±10 percent resistors are series-connected to a 100-V source. The resistor voltages are measured with a 50-V meter that has a tolerance of ±2 percent of full-scale voltage. What is the minimum voltage reading that could occur without any meter loading?

Answers to Tests

1. T
2. T
3. 1.2 MΩ
4. the time required for a capacitor to charge to the unknown voltage
5. constant current source
6. full-scale current, full-scale voltage, internal resistance
7. d'Arsonval
8. spiral spring
9. 2000 Ω
10. direct current
11. both
12. the d'Arsonval
13. a d'Arsonval
14. no
15. both
16. the interaction of the magnetic fields of the stationary coils and the movable coil
17. two
18. a d'Arsonval movement
19. a device that allows current to flow only in one direction
20. low temperature coefficient and low resistance tolerance
21. 9.9 mA

22. a. 133.3 Ω
    b. 108 μW
23. a shorting type
    (make-before-break)
24. low
25. when measuring cur-
    rents of 10 A or
    higher
26. by current capacity
    and voltage drop
27. a transformer used in
    conjunction with an
    ammeter to measure
    large values of alter-
    nating current
28. no
29. the current-
    transformer principle
30. measuring high-
    frequency currents
31. heat, electric
32. T
33. F
34. T
35. F
36. 40 kΩ/V

37. 7.5 mW, 333.133 kΩ
38. 5 MΩ
39. nonshorting
40. Meter loading means
    that the meter's
    internal resistance
    appreciably changes
    the circuit's current
    or voltage.
41. voltmeter loading
42. yes
43. yes
44. a. yes
    b. no
    c. no
45. a 20-kΩ/V voltmeter
    on the 1000-V range
46. a meter movement, a
    cell, a fixed resistor,
    and a variable resistor
47. The rheostat adjusts
    the ohmmeter for
    zero-resistance read-
    ings.
48. 40 Ω
49. 3000 Ω

50. Change the voltage
    and the series resis-
    tance, or change the
    series resistance and
    add a shunt.
51. T
52. F
53. R_1 and R_2, and R_3
    and R_u
54. the tolerance of the
    resistors in the
    bridge
55. 1000 Ω
56. because the electro-
    dynamometer move-
    ment reads true
    power regardless of
    the value of angle θ
57. the stationary, or
    current, coil
58. so that the current in
    the stationary coils
    will be in the correct
    direction relative to
    the current in the
    movable coil

59. true power
60. 1200 W
61. 3795 W
62. 4.8 kW
63. 400 W
64. The voltage may be
    below rated value, or
    the load may have a
    low power factor.
65. F
66. T
67. F
68. It is much faster. No
    switching of meter
    leads is required.
69. 15 Ω
70. digital method and
    ac-bridge method.
71. cemf
72. 0.0024 μF
73. 9.6 mH
74. 15.1

Term	Definition	Symbol or Abbreviation
Alternation	One-half of a cycle	
Ammeter	Device used to measure current	
Ampere	Base unit of current (coulomb per second)	A
Ampere-hour	Unit used to show energy storage capacity of cell or battery	Ah
Ampere-turn	Base unit of magnetomotive force	A · t
Ampere-turn per meter	Base unit of magnetic field strength	A · t/m
Apparent power	The product of current times voltage in a circuit containing reactance and resistance	
Atom	Building block of all matter	
Bandwidth	The space, expressed in hertz, between the lowest and highest frequencies an *LC* circuit will respond to with at least 70.7 percent of its maximum response	BW
Battery	Two or more cells electrically connected together	
Bemf	Back electromotive force; another name for cemf	
Bimetallic strip	Sandwich of two metals with different coefficients of expansion	
Capacitance	Ability to store energy in the form of electric energy	C
Capacitive reactance	The opposition capacitance offers to alternating current	X_c
Capacitor	An electrical component that possesses capacitance	
Cell	Chemical system that produces dc voltage	
Cemf	Counter electromotive force; the voltage produced by self-inductance	
Charge	Electrical property of electrons and protons	Q
Choke	Another name for an inductor	
Circuit breaker	Device which protects a circuit from excessive currents	
Clamp-on meter	An ammeter used to measure current without physically interrupting the circuit	
Coefficient of coupling	Denotes the portion of flux from one coil that links with another coil	
Coil	Another name for an inductor	
Compound	Matter composed of two or more elements	
Condenser	Outdated name for capacitor	
Conductance	Ability to conduct current	G
Conductors	Materials that have very low resistivity	
Constant current source	A power source which maintains a constant terminal current for all load values	
Constant voltage source	A power source which maintains a constant terminal voltage for all load values	
Continuity	Continuous path for current	
Copper loss	Refers to the conversion of electric energy to heat energy in the windings of magnetic devices	
Core loss	Refers to the conversion of electric energy to heat energy in the core material of magnetic devices	

Term	Definition	Symbol or Abbreviation
Coulomb	Base unit of charge (6.25 $\times$ 10^{18} electrons)	C
Current	Movement of charge in a specified direction	I
Current carrier	Charged particle (electron or ion)	
Current transformer	A transformer used to extend the range of an ac ammeter	
Cycle	That part of a periodic waveform that does not repeat itself	
Delta connection	A method of connecting a three-phase system so that the line and phase voltage are equal	
Dielectric	Insulation used between the plates of a capacitor	
Dielectric constant	A number which compares a material's ability to store energy to the ability of air to store energy	
Dissipation factor	A number obtained by dividing resistance by reactance. The reciprocal of Q. Used to indicate relative energy loss in a capacitor	DF
Eddy current	Current induced into the core of a magnetic device. Causes part of the core losses	
Electric field	Invisible field of force that exists between electric charges	
Electrical degree	One three-hundred-sixtieth ($\frac{1}{360}$) of an ac cycle	
Electron	Negatively charged particle of the atom	
Element	Matter composed entirely of one type of atom	
Energizing current	The primary current in an unloaded transformer	
Energy	Ability to do work	W
Farad	Base unit of capacitance; equal to one coulomb per volt	F
Filter	A circuit designed to separate one frequency, or group of frequencies, from all other frequencies	
Fluctuating direct current	Direct current that varies in amplitude but does not periodically drop to zero	
Flux	Lines of force around a magnet	ϕ
Flux density	Amount of flux per unit cross-sectional area	B
Free electrons	Electrons that are not attached (held) to any atom	
Frequency	Rapidity with which a periodic waveform repeats itself	f
Fuse	Device which protects a circuit from excessive currents	
Henry	Base unit of inductance; one volt of cemf for a current change of one ampere per second	H
Hertz	The base unit of frequency. One cycle per second.	Hz
Horsepower	Unit of power (1 hp = 746 watts)	hp
Hydrometer	Device used to measure specific gravity	
Hysteresis	Magnetic effect caused by residual magnetism in ac-operated magnetic devices; causes part of the core losses; also causes flux to lag behind the magnetomotive force	
Impedance	The total opposition of a circuit consisting of resistance and reactance. The base unit is ohms.	Z
Induced voltage	Voltage created in a conductor when the conductor interacts with a magnetic field	
Inductance	Electrical property which opposes changes in current	L
Inductive kick	The high cemf that is generated when an inductive circuit is opened	
Inductive reactance	The opposition inductance offers to alternating current	X_L
Inductor	An electrical component that possesses inductance	
Insulators	Materials that have very high resistivity	

Term	Definition	Symbol or Abbreviation
Internal resistance	Resistance contained within a power or energy source	
Ion	An atom which has an excess or deficiency of electrons	
Iron-vane meter	A type of meter which uses a moving iron vane and a stationary coil; commonly used to measure alternating current	
Joule	Base unit of energy (newton-meter)	J
Kilowatthour	Unit of energy (1 kWh = 3,600,000 joules)	kWh
Lenz' law	States that the cemf will always oppose the force which created it	
Litz wire	A special wire used to reduce the skin effect	
Load	Device that converts electrical energy to some other form of energy	
Magnetic field strength	Amount of magnetomotive force per unit length. Other terms for magnetic field strength are magnetic field intensity and magnetizing force	H
Magnetomotive force	Force that creates a magnetic field	mmf
Meter loading	Changes in circuit current or voltage caused by putting a meter in a circuit	
Multimeter	Electric instrument designed to measure two or more electrical quantities	
Multiplier	Precision resistor used to extend the voltage range of a meter movement	
Mutual inductance	Inductance caused by the flux from one circuit inducing a voltage into another circuit	
Neutron	Electrically neutral portion of an atom	
Norton's theorem	A method for reducing a circuit to a two-terminal current source	
Nucleus	Center of atom which contains protons and neutrons	
Ohm	Base unit of resistance (volt per ampere)	Ω
Ohmic resistance	The dc resistance of an inductor	
Ohm-meter	Base unit of resistivity	$\Omega \cdot m$
Ohmmeter	Device used to measure resistance	
Period	Time required to complete one cycle	T
Permeability	Ease with which flux is created in a material	μ
Phase	A time relationship between two electrical quantities	ϕ
Phase shift	The result of two waveforms being out of step with each other	
Phasing	Interconnecting transformer, generator, or motor windings so that they have the correct time (phase) relationships between them	
Phasor	A line representing alternating current or voltage at some instant of time	
Photoconductive	Changing conductance, or resistance, by changing the light energy level	
Piezoelectric	Producing voltage by applying pressure to a crystal	
Polarity	Electrical characteristic (negative or positive) of a charge	
Polarization	Accumulation of gas ions around electrode of a cell	
Potentiometer	Three-terminal variable resistor	
Power	Rate of doing work or using energy	P
Power factor	The cosine of theta; the ratio of the true power over the apparent power; also equal to resistance divided by impedance	PF

Term	Definition	Symbol or Abbreviation
Primary cell	Cell that is not meant to be recharged	
Proton	Positively charged particle of the atom	
Pulsating direct current	A direct current which periodically returns to zero	
Quality	A number which indicates the ratio of reactance to resistance	Q
Reactance	Opposition to current which converts no energy (uses no power); a property of both inductance and capacitance	X
Reactor	Another name for an inductor	
Rectification	The process of converting ac to pulsating dc	
Relative permeability	Permeability of a material compared with permeability of air	μ_r
Reluctance	Opposition to creation of magnetic flux	$\mathcal{R}$
Residual magnetism	Magnetic flux left in temporary magnet after magnetizing force has been removed	
Resistance	Opposition to current which converts electric energy into heat energy	R
Resistivity	Characteristic resistance of a material (resistance of a cubic meter of the material)	
Resonance	A circuit condition in which $X_L = X_C$	
Resonant frequency	That frequency at which $X_L = X_C$ for a given value of L and C	f_r
Rheostat	Two-terminal variable resistor	
Secondary cell	Cell that can be recharged	
Selectivity	The ability of a circuit to separate signals (voltages) that are at different frequencies	
Self-inductance	The process of a conductor inducing a cemf in itself	
Semiconductor	An element with four valence electrons	
Service factor	A safety factor rating for electric motors	SF
Shunt	A parallel branch or component; precision resistor used to extend current range of a meter movement	
Siemen	Base unit of conductance (ampere per volt)	S
Sine wave	A symmetrical waveform whose instantaneous value is related to the trigonometric sine function	
Skin effect	Concentration of current near the surface of a conductor that causes the resistance of the conductor to increase with increased frequency	
Slip	Percent of difference between synchronous and operating speed of a motor	
Specific gravity	Weight of a substance compared with weight of equal volume of water	
Superposition theorem	A method for determining unknown currents in complex multiple-source circuits	
Tank circuit	A parallel LC circuit	
Temperature coefficient	Number of units change per degree Celsius change from a specified temperature	
Tesla	Base unit of flux density	T
Theta	The phase angle between phasors	θ
Thevenin's theorem	A method for reducing a circuit to a two-terminal voltage source	
Three-phase	Three voltages or currents displaced from each other by 120 electrical degrees	$3\text{-}\phi$

Term	Definition	Symbol or Abbreviation
Time constant	The time a capacitor requires to charge or discharge (through a resistor) 63.2 percent of the available voltage	T
Torque	Force × distance which produces, or tends to produce, rotation	
Transformer losses	Power losses in a transformer; caused by hysteresis, eddy current, and winding resistance	
Trigonometric functions	The relationships between the angles and sides of a right triangle. The three common functions are sine, cosine, and tangent.	
Turns-per-volt ratio	The number of turns for each volt in a transformer winding	
Valence electrons	Electrons in outermost shell of an atom	
Vector	A line that represents both the direction and the magnitude of a quantity. Vectors of electrical quantities are called phasors	
Vibrating-reed meter	A meter used to measure frequency—especially the low frequencies used in power systems	
Volt	Base unit of voltage (joule per coulomb)	V
Voltage	Potential energy difference (electrical pressure)	V
Voltampere	Base unit of apparent power	VA
Voltmeter	Device used to measure voltage	
Voltmeter sensitivity	Ohms-per-volt rating of voltmeter. Numerically equal to reciprocal of full-scale current of the meter movement used in voltmeter	
Watt	Base unit of power (joule per second)	W
Watthour	Unit of energy (3600 joules)	
Wattmeter	An electrical meter that measures true power	
Wattsecond	Unit of energy (1 joule)	
Weber	Base unit of flux	Wb
Wheatstone bridge	Circuit configuration used to measure electrical qualities such as resistance	
Wye connection	Connecting the phases of a three-phase system at a common point so that the line and phase currents are equal	

Appendix B
Formulas and Conversions

Amplitude Conversions

$$V_{\text{p-p}} = 2 V_{\text{p}} \quad\quad \text{and} \quad\quad V_{\text{p}} = \frac{V_{\text{p-p}}}{2}$$

$$V_{\text{p}} = 1.414 V_{\text{rms}} \quad\quad \text{and} \quad\quad V_{\text{rms}} = 0.707 V_{\text{p}}$$

$$V_{\text{p}} = 1.57 V_{\text{av}} \quad\quad \text{and} \quad\quad V_{\text{av}} = 0.637 V_{\text{p}}$$

$$V_{\text{av}} = 0.9 V_{\text{rms}} \quad\quad \text{and} \quad\quad V_{\text{rms}} = 1.11 V_{\text{av}}$$

V_{rms} = effective value

These formulas can also apply to current by substituting I for V.

Bandwidth

$$\text{BW} \approx \frac{f_r}{Q}$$

Capacitance

$$C = \frac{Q}{V}$$

In Parallel

$$C_T = C_1 + C_2 + C_3 + \text{etc.}$$

In Series

$$C_T = \frac{1}{\dfrac{1}{C_1} + \dfrac{1}{C_2} + \dfrac{1}{C_3} + \text{etc.}}$$

$$C_T = \frac{C_1 \times C_2}{C_1 + C_2}$$

Energy Storage

$$W = 0.5CV^2$$

For Resonance

$$C = \frac{0.02533}{f_r^2 L}$$

Capacitive Reactance

$$X_C = \frac{1}{6.28fC}$$

In Parallel

$$X_{C_T} = \frac{1}{\dfrac{1}{X_{C_1}} + \dfrac{1}{X_{C_2}} + \dfrac{1}{X_{C_3}} + \text{etc.}}$$

$$X_{C_T} = \frac{X_{C_1} \times X_{C_2}}{X_{C_1} + X_{C_2}}$$

In Series

$$X_{C_T} = X_{C_1} + X_{C_2} + X_{C_3} + \text{etc.}$$

Conductance

$$G = \frac{1}{R}$$

Current (Definition)

$$I = \frac{Q}{t}$$

Current (in Impedance Circuits)

For Any Circuit

$$I_T = \frac{V_T}{Z}$$

For Parallel Circuits

$$I_T = \sqrt{I_R^2 + I_C^2}$$
$$I_T = \sqrt{I_R^2 + I_L^2}$$
$$I_T = \sqrt{I_R^2 + (I_L - I_C)^2}$$

Efficiency

$$\% \text{ eff.} = \frac{W_{\text{out}}}{W_{\text{in}}} \times 100$$

$$\% \text{ eff.} = \frac{P_{\text{out}}}{P_{\text{in}}} \times 100$$

Equivalent Circuit Conversions

$$V_{TH} = I_N R_N$$

$$I_N = \frac{V_{TH}}{R_{TH}}$$

Frequency-Period Conversions

$$T = \frac{1}{f}$$

$$f = \frac{1}{T}$$

Generator Frequency

$$f = \frac{r/min}{60} \times \text{pairs of poles}$$

Impedance

For Any Circuit

$$Z = \frac{V_T}{I_T}$$

For Series Circuits

$$Z = \sqrt{R^2 + X_L^2}$$
$$Z = \sqrt{R^2 + X_C^2}$$
$$Z = \sqrt{(X_L - X_C)^2 + R^2}$$

For Parallel Circuits

$$Z = \frac{RX_L}{\sqrt{R^2 + X_L^2}}$$

$$Z = \frac{RX_C}{\sqrt{R^2 + X_C^2}}$$

$$Z = \frac{RX_L X_C}{\sqrt{(RX_L - RX_C)^2 + (X_L^2 X_C^2)}}$$

Inductance

Quality

$$Q = \frac{X_L}{R}$$

In Series

$$L_T = L_1 + L_2 + L_3 + \text{etc.}$$

For Resonance

$$L = \frac{0.02533}{f_r^2 C}$$

In Parallel

$$L_T = \frac{1}{\dfrac{1}{L_1} + \dfrac{1}{L_2} + \dfrac{1}{L_3} + \text{etc.}}$$

$$L_T = \frac{L_1 \times L_2}{L_1 + L_2}$$

Inductive Reactance

$$X_L = 6.28fL$$

In Parallel

$$X_{L_T} = \frac{1}{\dfrac{1}{X_{L_1}} + \dfrac{1}{X_{L_2}} + \dfrac{1}{X_{L_3}} + \text{etc.}}$$

$$X_{L_T} = \frac{X_{L_1} \times X_{L_2}}{X_{L_1} + X_{L_2}}$$

In Series

$$X_{L_T} = X_{L_1} + X_{L_2} + X_{L_3} + \text{etc.}$$

Magnetism

$$\text{mmf} = \text{turns} \times \text{current}$$

$$H = \frac{\text{mmf}}{\text{length}}$$

$$B = \frac{\phi}{\text{area}}$$

$$\mu = \frac{B}{H}$$

Motor

$$\text{Synchronous speed} = \frac{120f}{\text{no. of poles}}$$

$$\% \text{ slip} = \frac{\text{synchronous speed} - \text{rated speed}}{\text{synchronous speed}}$$

$$\text{hp} = \frac{\text{lb-ft} \times r/min}{5252}$$

$$\text{hp} = 746 \text{ W}$$

Ohm's Law

$$V = IR$$

$$I = \frac{V}{R}$$

$$R = \frac{V}{I}$$

Parallel Circuits

$$P_T = P_{R_1} + P_{R_2} + P_{R_3} + \text{etc.}$$

$$R_T = \frac{1}{\dfrac{1}{R_1} + \dfrac{1}{R_2} + \dfrac{1}{R_3} + \text{etc.}}$$

$$R_T = \frac{R_1 \times R_1}{R_2 + R_2} \text{ for two resistances}$$

$$R_T = \frac{R}{n} \text{ for } n \text{ equal resistances}$$

$$V_T = V_{R_1} = V_{R_2} = V_{R_3} = \text{etc.}$$

$$I_T = I_{R_1} + I_{R_2} + I_{R_3} + \text{etc.}$$

$$I_{R_1} = \frac{I_T R_2}{R_1 + R_2}$$

Power

$$P = IV \cos \theta$$

$$P_{\text{app}} = IV$$

$$P = I_R^2 R$$

$$P = V_R I_R$$

$$P = \frac{W}{t}$$

Power Factor

$$\text{PF} = \cos \theta$$

$$\text{PF} = \frac{P}{P_{\text{app}}}$$

Quality

$$Q = \frac{X}{R}$$

Resistance

$$R = \frac{\text{resistivity} \times \text{length}}{\text{area}}$$

$$= \frac{Kl}{A}$$

where K = resistivity

l = length

A = area

Resonant Frequency

$$f_r = \frac{1}{6.28\sqrt{LC}}$$

Series Circuits

$$P_T = P_{R_1} + P_{R_2} + P_{R_3} + \text{etc.}$$

$$R_T = R_1 + R_2 + R_3 + \text{etc.}$$

$$V_T = V_{R_1} + V_{R_2} + V_{R_3} + \text{etc.}$$

$$I_T = I_{R_1} = I_{R_2} = I_{R_3} = \text{etc.}$$

$$V_{R_1} = \frac{V_T R_1}{R_1 + R_2 + \text{etc.}}$$

Time Constant

$$T = RC$$

$$T = \frac{L}{R}$$

Transformers

$$\frac{N_{\text{pri}}}{N_{\text{sec}}} = \frac{V_{\text{pri}}}{V_{\text{sec}}}$$

$$\left(\frac{N_{\text{pri}}}{N_{\text{sec}}}\right)^2 = \frac{Z_{\text{pri}}}{Z_{\text{sec}}}$$

$$\% \text{ eff.} = \frac{P_{\text{sec}}}{P_{\text{pri}}} \times 100$$

Trigonometric Functions

$$\cos\theta = \frac{\text{adjacent}}{\text{hypotenuse}}$$

$$\sin\theta = \frac{\text{opposite}}{\text{hypotenuse}}$$

$$\tan\theta = \frac{\text{opposite}}{\text{adjacent}}$$

For Any Circuit

$$\cos\theta = \frac{P}{P_{\text{app}}} = \text{PF}$$

For Series Circuits

$$\cos\theta = \frac{V_R}{V_T} = \frac{R}{Z}$$

$$\sin\theta = \frac{V_X}{V_T} = \frac{X}{Z}$$

$$\tan\theta = \frac{V_X}{V_R} = \frac{X}{R}$$

For Parallel Circuits

$$\cos\theta = \frac{I_R}{I_T}$$

$$\sin\theta = \frac{I_X}{I_T}$$

$$\tan\theta = \frac{I_X}{I_R}$$

Voltage (Definition)

$$V = \frac{W}{Q}$$

Voltage (in Impedance Circuits)

For Any Circuit

$$V_T = I_T Z$$

For Series Circuits

$$V_T = \sqrt{V_R^2 + V_C^2}$$
$$V_T = \sqrt{V_R^2 + V_L^2}$$
$$V_T = \sqrt{V_R^2 + (V_L - V_C)^2}$$

AWG (B & S) gage	Standard metric size (mm)	Diameter in mils	Cross-sectional area		Ohms per 1000 ft at 20°C [68°F]	lbs per 1000 ft	ft per lb
			Circular mils	Square inches			
0000	11.8	460.0	211,600	0.1662	0.04901	640.5	1.561
000	11.0	409.6	167,800	0.1318	0.06180	507.9	1.968
00	9.0	364.8	133,100	0.1045	0.07793	402.8	2.482
0	8.0	324.9	105,500	0.08289	0.09827	319.5	3.130
1	7.1	289.3	83,690	0.06573	0.1239	253.3	3.947
2	6.3	257.6	66,370	0.05213	0.1563	200.9	4.977
3	5.6	229.4	52,640	0.04134	0.1970	159.3	6.276
4	5.0	204.3	41,740	0.03278	0.2485	126.4	7.914
5	4.5	181.9	33,100	0.02600	0.3133	100.2	9.980
6	4.0	162.0	26,250	0.02062	0.3951	79.46	12.58
7	3.55	144.3	20,820	0.01635	0.4982	63.02	15.87
8	3.15	128.5	16,510	0.01297	0.6282	49.98	20.01
9	2.80	114.4	13,090	0.01028	0.7921	39.63	25.23
10	2.50	101.9	10,380	0.008155	0.9989	31.43	31.82
11	2.24	90.74	8,234	0.006467	1.260	24.92	40.12
12	2.00	80.81	6,530	0.005129	1.588	19.77	50.59
13	1.80	71.96	5,178	0.004067	2.003	15.68	63.80
14	1.60	64.08	4,107	0.003225	2.525	12.43	80.44
15	1.40	57.07	3,257	0.002558	3.184	9.858	101.4
16	1.25	50.82	2,583	0.002028	4.016	7.818	127.9
17	1.12	45.26	2,048	0.001609	5.064	6.200	161.3
18	1.00	40.30	1,624	0.001276	6.385	4.917	203.4
19	0.90	35.89	1,288	0.001012	8.051	3.899	256.5
20	0.80	31.96	1,022	0.0008023	10.15	3.092	323.4
21	0.71	28.46	810.1	0.0006363	12.80	2.452	407.8
22	0.63	25.35	642.4	0.0005046	16.14	1.945	514.2
23	0.56	22.57	509.5	0.0004002	20.36	1.542	648.4
24	0.50	20.10	404.0	0.0003173	25.67	1.223	817.7
25	0.45	17.90	320.4	0.0002517	32.37	0.9699	1,031.0
26	0.40	15.94	254.1	0.0001996	40.81	0.7692	1,300
27	0.355	14.20	201.5	0.0001583	51.47	0.6100	1,639
28	0.315	12.64	159.8	0.0001255	64.90	0.4837	2,067
29	0.280	11.26	126.7	0.00009953	81.83	0.3836	2,607
30	0.250	10.03	100.5	0.00007894	103.2	0.3042	3,287
31	0.224	8.928	79.70	0.00006260	130.1	0.2413	4,145
32	0.200	7.950	63.21	0.00004964	164.1	0.1913	5,227
33	0.180	7.080	50.13	0.00003937	206.9	0.1517	6,591
34	0.160	6.305	39.75	0.00003122	260.9	0.1203	8,310
35	0.140	5.615	31.52	0.00002476	329.0	0.09542	10,480
36	0.125	5.000	25.00	0.00001964	414.8	0.07568	13,210
37	0.112	4.453	19.83	0.00001557	523.1	0.06001	16,660
38	0.100	3.965	15.72	0.00001235	659.6	0.04759	21,010
39	0.090	3.531	12.47	0.000009793	831.8	0.03774	26,500
40	0.080	3.145	9.888	0.000007766	1049.0	0.02993	33,410

Material	*Resistivity at 20°C ($\Omega \cdot cm$)
Aluminum	0.00000262
Brass (66% Cu, 34% Zn)	0.0000039
Carbon (graphite form)	0.0014
Constantan (55% Cu, 45% Ni)	0.0000442
Copper (commercial annealed)	0.0000017241
German silver (18% Ni)	0.000033
Iron	0.00000971
Lead	0.0000219
Mercury	0.0000958
Monel metal (67% Ni, 30% Cu, 1.4% Fe, 1% Mn)	0.000042
Nichrome (65% Ni, 12% Cr, 23% Fe)	0.000100
Nickel	0.0000069
Phosphor bronze (4% Sn, 0.5% P, 95.5% Cu)	0.0000094
Silver	0.00000162
Steel (0.4–0.5% C)	0.000013–0.000022
Tantalum	0.0000131
Tin	0.0000114
Tungsten	0.00000548
Zinc	0.000006

Source: *Reference Data for Radio Engineers*, 6th ed., 1975, Howard W. Sams & Co., Inc.
*To convert to ohm-circular mils per foot multiply the numbers in this column by 6.02×10^6.

Material	*Temperature Coefficient at 20°C (ohm per ohm/°C)
Aluminum	0.0039
Brass (66% Cu, 34% Zn)	0.002
Carbon (graphite form)	−0.0005
Copper (commercial annealed)	0.0039
German silver (18% Ni)	0.0004
Iron	0.0052–0.0062
Lead	0.004
Mercury	0.00089
Monel metal (67% Ni, 30% Cu, 1.4% Fe, 1% Mn)	0.002
Nichrome (65% Ni, 12% Cr, 23% Fe)	0.00017
Nickel	0.0047
Phosphor bronze (4% Sn, 0.5% P, 95.5% Cu)	0.003
Silver	0.0038
Steel (0.4–0.5% C)	0.003
Tantalum	0.003
Tin	0.0042
Tungsten	0.0045
Zinc	0.0037

Source: *Reference Data for Radio Engineers*, 6th ed., 1975, Howard W. Sams & Co., Inc.
*To convert to ppm/°C multiply the numbers in this column by 1×10^6.

Angle (degrees)	Sin	Cos	Tan	Angle (degrees)	Sin	Cos	Tan
0	0.000	1.000	0.000	46	.719	.695	1.036
1	.018	1.000	.018	47	.731	.682	1.072
2	.035	.999	.035	48	.743	.669	1.111
3	.052	.999	.052	49	.755	.656	1.150
4	.070	.998	.070	50	.766	.643	1.192
5	.087	.996	.088	51	.777	.629	1.235
6	.105	.995	.105	52	.788	.616	1.280
7	.122	.993	.123	53	.799	.602	1.327
8	.139	.990	.141	54	.809	.588	1.376
9	.156	.988	.158	55	.819	.574	1.428
10	.174	.985	.176	56	.829	.559	1.483
11	.191	.982	.194	57	.839	.545	1.540
12	.208	.978	.213	58	.848	.530	1.600
13	.225	.974	.231	59	.857	.515	1.664
14	.242	.970	.249	60	.866	.500	1.732
15	.259	.966	.268	61	.875	.485	1.804
16	.276	.961	.287	62	.883	.470	1.881
17	.292	.956	.306	63	.891	.454	1.963
18	.309	.951	.325	64	.899	.438	2.050
19	.326	.946	.344	65	.906	.423	2.145
20	.342	.940	.364	66	.914	.407	2.246
21	.358	.934	.384	67	.921	.391	2.356
22	.375	.927	.404	68	.927	.375	2.475
23	.391	.921	.425	69	.934	.358	2.605
24	.407	.914	.445	70	.940	.342	2.748
25	.423	.906	.466	71	.946	.326	2.904
26	.438	.899	.488	72	.951	.309	3.078
27	.454	.891	.510	73	.956	.292	3.271
28	.470	.883	.532	74	.961	.276	3.487
29	.485	.875	.554	75	.966	.259	3.732
30	.500	.866	.577	76	.970	.242	4.011
31	.515	.857	.601	77	.974	.225	4.332
32	.530	.848	.625	78	.978	.208	4.705
33	.545	.839	.649	79	.982	.191	5.145
34	.559	.829	.675	80	.985	.174	5.671
35	.574	.819	.700	81	.988	.156	6.314
36	.588	.809	.727	82	.990	.139	7.115
37	.602	.799	.754	83	.993	.1229	8.144
38	.616	.788	.781	84	.995	.105	9.514
39	.629	.777	.810	85	.996	.087	11.43
40	.643	.766	.839	86	.998	.070	14.30
41	.656	.755	.869	87	.999	.052	19.08
42	.669	.743	.900	88	.999	.035	28.64
43	.682	.731	.933	89	1.000	.018	57.29
44	.695	.719	.966	90	1.000	.000	—
45	.707	.707	1.000				

Index

ex

Index